移动开发经典丛书

Android 5.0 开发范例
代码大全
（第 4 版）

［美］ Dave Smith 著
       Jeff Friesen

       张永强 译

清华大学出版社
北　京

Dave Smith, Jeff Friesen
Android Recipes: A Problem-Solution Approach, Third Edition
EISBN：978-1-4302-6322-7

Original English language edition published by Apress Media. Copyright © 2014 by Apress Media. Simplified Chinese-Language edition copyright © 2015 by Tsinghua University Press. All rights reserved.

本书中文简体字版由 Apress 出版公司授权清华大学出版社出版。未经出版者书面许可，不得以任何方式复制或抄袭本书内容。

北京市版权局著作权合同登记号　图字：01-2015-0650

本书封面贴有清华大学出版社防伪标签，无标签者不得销售。
版权所有，侵权必究。侵权举报电话：010-62782989　13701121933

图书在版编目(CIP)数据

Android 5.0 开发范例代码大全：第 4 版 / (美) 史密斯(Smith, D.)，(美) 弗里森(Friesen, J.) 著；张永强 译. —北京：清华大学出版社，2015 (2016.5重印)
(移动开发经典丛书)
书名原文：Android Recipes: A Problem-Solution Approach, Third Edition
ISBN 978-7-302-39621-5

Ⅰ.①A…　Ⅱ.①史…②弗…③张…　Ⅲ.①移动终端－应用程序－程序设计　Ⅳ.①TN929.53

中国版本图书馆 CIP 数据核字(2015)第 049230 号

责任编辑：王　军　李维杰
装帧设计：牛静敏
责任校对：成凤进
责任印制：杨　艳

出版发行：清华大学出版社
网　　址：http://www.tup.com.cn，http://www.wqbook.com
地　　址：北京清华大学学研大厦 A 座　　邮　编：100084
社 总 机：010-62770175　　　　　　　　　邮　购：010-62786544
投稿与读者服务：010-62776969，c-service@tup.tsinghua.edu.cn
质 量 反 馈：010-62772015，zhiliang@tup.tsinghua.edu.cn

印 装 者：清华大学印刷厂
经　　销：全国新华书店
开　　本：185mm×260mm　　印　张：44.25　　字　数：1077 千字
版　　次：2015 年 9 月第 1 版　　印　次：2016 年 5 月第 2 次印刷
印　　数：3001~5000
定　　价：98.00 元

产品编号：060955-01

译 者 序

Android 是以 Linux 为基础的开源移动设备操作系统，主要用于智能手机和平板电脑，由 Google 成立的开放手持设备联盟(Open Handset Alliance，OHA)持续领导与开发。经过这么多年的发展，Android 已成为全球市场占有量最大的智能手机操作系统，该系统目前发布的最新版本为 Android 5.1.1。

Android 目前是业界领先的移动操作系统和开发平台，推动着目前移动领域的创新活动和应用生态系统的发展。Android 看起来复杂，但却为带着不同编程语言技能集进入 Android 领域的开发人员提供了丰富的有组织开发工具包。本书旨在帮助开发人员解决实际开发中的问题，通过直观的例子告诉读者如何利用工具编写 Android 平台上的应用程序。

本书在上一版的基础上进行了更新，首先删除了 Android 入门章节，改为直接介绍各种处理实际问题的主题。其次，增加了许多采用 API Level 22 的范例，指导读者在最新的平台上处理实际问题。最后，加入了一些关于 NDK 和 RenderScript 的实际范例，帮助读者更深入地了解它们。

然而，纸上得来终觉浅，绝知此事要躬行。因此，建议读者在学习本书的过程中打开 Android SDK、NDK 或是其他工具，边学习书中给出的范例边动手操作，如此方能成为理论和行动上的"巨人"。有人问大师，如何能技近乎道？大师曰：读书，读好书，然后实践之。万事无他，惟手熟尔！

在这里要感谢清华大学出版社的编辑们，她们为本书的翻译投入了巨大的热情并付出了很多心血。没有你们的帮助和鼓励，本书不可能顺利付梓。

对于这本经典之作，译者本着"诚惶诚恐"的态度，在翻译过程中力求"信、达、雅"，但是鉴于译者水平有限，错误和失误在所难免，如有任何意见和建议，请不吝指正。感激不尽！本书全部章节由张永强翻译，参与翻译活动的还有孔祥亮、陈跃华、杜思明、熊晓磊、曹汉鸣、陶晓云、王通、方峻、李小凤、曹晓松、蒋晓冬、邱培强、洪妍、李亮辉、高娟妮、曹小震、陈笑。

读者可以把本书当作一本可供随时查询的参考书、一本资源丰富的示例手册，随时都可以从中找到有助于高效完成工作的实用建议。

作者简介

Dave Smith 是专业的工程师,一直从事移动和嵌入式平台的软件与硬件开发。目前,Dave 全身心地投入到 Android 开发领域。从 2009 年开始,Dave 就从事 Android 平台各个版本上的开发,包括使用 SDK 编写用户应用程序以及构建和定制 Android 源代码。Dave 会定期通过他的开发博客(http://wiresareobsolete.com)和 Twitter 流(@devunwired)分享自己的想法。

技术评审人员简介

　　Paul Trebilbox-Ruiz 是科罗拉多博尔德的 Android 开发人员，并且是博尔德/丹佛当地技术项目的活跃成员。自从来到科罗拉多州，Paul 已经参与了多次由丹佛谷歌开发人员小组(GDG)主办的黑客马拉松活动并获得胜利，并且从事了多个公民编码项目。他目前在 SportsLabs 从事 Android 平台的相关编程工作，负责为跨美国全境的大学体育项目构建和设计应用程序。

　　《Android 5.0 开发范例大全(第 4 版)》是 Paul 参与评审的第一本书籍，本书较早的一个版本是他在学习 Android 平台时购买的第一本 Android 书籍。在此期间，他正在佛雷斯诺市的加利福尼亚州立大学攻读计算机科学学士学位。

致　　谢

首先，我要感谢我的妻子 Lorie，感谢她在我撰写和构建本书所涉及的各种素材时给予我的支持和耐心。

其次，我还要诚挚地感谢 Apress 给我安排的编辑团队，是他们使本书尽善尽美。没有他们所投入的时间和精力，本书就不可能顺利付梓。

—Dave Smith

前　　言

欢迎阅读《Android 5.0 开发范例代码大全(第 4 版)》!

如果你正在阅读本书,那么移动设备给软件开发人员和用户带来的无限机遇就不用我在此赘述了。近年来,Android 已经成为最主要的移动平台之一。对于开发人员而言,必须了解如何利用 Android,才能确保自己跟得上市场的变化,从而把握各种潜在的机会。但是任何新平台在常见需求的开发和常见问题的解决方案上都会有不确定性。

我们撰写本书旨在帮助开发人员解决实际开发中的问题,通过直观的例子告诉读者如何利用工具编写 Android 平台上的应用程序。本书不会很深入地介绍 Android SDK、NDK 或是其他工具。我们不会让隐藏其中的各种琐碎细节和高深理论打击读者的积极性。但这不意味着这些细节没意思或是不重要。读者应该花时间研究这些细节,以避免在开发中犯错误。但在解决迫在眉睫的问题时,这些东西通常只会让人分心。

本书不会讲解 Java 编程,也不会介绍如何构建 Android 应用程序的代码块。本书略去了很多基础知识(例如,如何使用 TextView 显示文本),因为我们觉得这些知识在学过之后就不会遗忘。相反,本书会帮助熟悉 Android 的开发人员解决很多实际开发中经常要完成的任务,而这些复杂的任务不是寥寥几行代码就能完成的,自然也很难记住。

读者可以把本书当作可供随时查阅的参考书、资源丰富的示例手册,随时都可以从中找到有助于高效完成工作的实用建议。

本书主要内容

本书深入介绍使用 Android SDK 解决实际问题。你将学习高效创建在不同设备上都可良好运行的用户界面的技巧。你将熟练掌握如何合并各种硬件(音频设备、传感器和摄像头),正是这些硬件使得移动设备成为独特的平台。我们甚至会介绍如何整合 Google 和各种服务制造商提供的服务与应用程序,从而使系统真正服务于用户。

如果想开发成功的应用程序,性能问题是不可忽视的。大部分时候,这都不是问题,因为 Android 运行时引擎日渐完善,可将字节码编译成设备的原生代码。然而,你可能需要利用 Android NDK 以进一步提升性能。第 8 章详述了 NDK,并用 Java 原生接口(Java Native Interface,JNI)绑定将原生代码整合到应用程序中。

NDK 是一种比较复杂的技术,它也会降低应用程序的可移植性。此外,虽然能够提升性能,但在应对繁重工作时,NDK 也不能很好地处理多个 CPU 内核。幸运的是,Google

通过引入 RenderScript 已经消除了这种冗长编码并简化了多核执行任务,另外还实现了可移植性。第 8 章介绍 RenderScript 并演示如何使用它的计算引擎(并自动使用 CPU 的多核)来处理图片。

注意目标 API 级别

在本书中,读者会看到绝大部分的解决方案都有相应的最低 API 级别要求。本书中的大部分解决方案都只需要 API Level 1,换言之就是这些代码能在目标版本为 Android 1.0 以上的任何应用程序中运行。但是,有些地方也用到了较新版本中引入的 API。注意各个范例的 API 级别,确保代码与应用程序要支持的 Android 版本相匹配。

本书在线资源

www.apress.com
www.tupwk.com.cn/downpage

目 录

第 1 章 布局和视图 ············ 1
1.1 样式化常见组件 ············ 1
1.1.1 问题 ············ 1
1.1.1 解决方案 ············ 1
1.1.3 实现机制 ············ 2
1.2 切换系统 UI 元素 ············ 10
1.2.1 问题 ············ 10
1.2.2 解决方案 ············ 10
1.2.3 实现机制 ············ 11
1.3 创建并显示视图 ············ 14
1.3.1 问题 ············ 14
1.3.2 解决方案 ············ 14
1.3.3 实现机制 ············ 14
1.4 动画视图 ············ 20
1.4.1 问题 ············ 20
1.4.2 解决方案 ············ 21
1.4.3 实现机制 ············ 21
1.5 布局变化时的动画 ············ 26
1.5.1 问题 ············ 26
1.5.2 解决方案 ············ 26
1.5.3 实现机制 ············ 27
1.6 实现针对具体场景的布局 ············ 30
1.6.1 问题 ············ 30
1.6.2 解决方案 ············ 30
1.6.3 实现机制 ············ 30
1.7 自定义 AdapterView 的空视图 ············ 38
1.7.1 问题 ············ 38
1.7.2 解决方案 ············ 38
1.7.3 实现机制 ············ 38
1.8 自定义 ListView 中的行 ············ 40
1.8.1 问题 ············ 40
1.8.2 解决方案 ············ 40
1.8.3 实现机制 ············ 40
1.9 制作 ListView 的节头部 ············ 44
1.9.1 问题 ············ 44
1.9.2 解决方案 ············ 44
1.9.3 实现机制 ············ 44
1.10 创建组合控件 ············ 52
1.10.1 问题 ············ 52
1.10.2 解决方案 ············ 52
1.10.3 实现机制 ············ 52
1.11 自定义过渡动画 ············ 56
1.11.1 问题 ············ 56
1.11.2 解决方案 ············ 56
1.11.3 实现机制 ············ 56
1.12 创建视图变换 ············ 65
1.12.1 问题 ············ 65
1.12.2 解决方案 ············ 65
1.12.3 实现机制 ············ 65
1.13 建立可扩展的集合视图 ············ 71
1.13.1 问题 ············ 71
1.13.2 解决方案 ············ 72
1.13.3 实现机制 ············ 72
1.14 小结 ············ 82

第 2 章 用户交互 ············ 83
2.1 利用 Action Bar ············ 83
2.1.1 问题 ············ 83
2.1.2 解决方案 ············ 83
2.1.3 实现机制 ············ 84
2.2 锁定 Activity 方向 ············ 91

- 2.2.1 问题 ·················· 91
- 2.2.2 解决方案 ·············· 91
- 2.2.3 实现机制 ·············· 91
- 2.3 动态方向锁定 ················ 92
  - 2.3.1 问题 ·················· 92
  - 2.3.2 解决方案 ·············· 92
  - 2.3.3 实现机制 ·············· 92
- 2.4 手动处理旋转 ················ 94
  - 2.4.1 问题 ·················· 94
  - 2.4.2 解决方案 ·············· 94
  - 2.4.3 实现机制 ·············· 95
- 2.5 创建上下文动作 ·············· 98
  - 2.5.1 问题 ·················· 98
  - 2.5.2 解决方案 ·············· 98
  - 2.5.3 实现机制 ·············· 98
- 2.6 显示一个用户对话框 ········· 103
  - 2.6.1 问题 ················· 103
  - 2.6.2 解决方案 ············· 103
  - 2.6.3 实现机制 ············· 103
- 2.7 自定义菜单和动作 ············ 108
  - 2.7.1 问题 ················· 108
  - 2.7.2 解决方案 ············· 108
  - 2.7.3 实现机制 ············· 109
- 2.8 自定义BACK按键 ············ 114
  - 2.8.1 问题 ················· 114
  - 2.8.2 解决方案 ············· 114
  - 2.8.3 实现机制 ············· 114
- 2.9 模拟HOME按键 ············· 117
  - 2.9.1 问题 ················· 117
  - 2.9.2 解决方案 ············· 117
  - 2.9.3 实现机制 ············· 118
- 2.10 监控TextView的变动 ······· 118
  - 2.10.1 问题 ················ 118
  - 2.10.2 解决方案 ············ 118
  - 2.10.3 实现机制 ············ 119
- 2.11 自定义键盘动作 ············ 121
  - 2.11.1 问题 ················ 121
  - 2.11.2 解决方案 ············ 121
  - 2.11.3 实现机制 ············ 121
- 2.12 消除软键盘 ················ 124
  - 2.12.1 问题 ················ 124
  - 2.12.2 解决方案 ············ 124
  - 2.12.3 实现机制 ············ 124
- 2.13 处理复杂的触摸事件 ········ 125
  - 2.13.1 问题 ················ 125
  - 2.13.2 解决方案 ············ 125
  - 2.13.3 实现机制 ············ 126
- 2.14 转发触摸事件 ·············· 142
  - 2.14.1 问题 ················ 142
  - 2.14.2 解决方案 ············ 142
  - 2.14.3 实现机制 ············ 142
- 2.15 阻止触摸窃贼 ·············· 146
  - 2.15.1 问题 ················ 146
  - 2.15.2 解决方案 ············ 146
  - 2.15.3 实现机制 ············ 146
- 2.16 创建拖放视图 ·············· 149
  - 2.16.1 问题 ················ 149
  - 2.16.2 解决方案 ············ 150
  - 2.16.3 实现机制 ············ 151
- 2.17 构建导航Drawer ············ 157
  - 2.17.1 问题 ················ 157
  - 2.17.2 解决方案 ············ 157
  - 2.17.3 实现机制 ············ 157
- 2.18 在视图之间滑动 ············ 167
  - 2.18.1 问题 ················ 167
  - 2.18.2 解决方案 ············ 167
  - 2.18.3 实现机制 ············ 168
- 2.19 使用选项卡导航 ············ 177
  - 2.19.1 问题 ················ 177
  - 2.19.2 解决方案 ············ 177
  - 2.19.3 实现机制 ············ 178
- 2.20 小结 ······················ 185
- 第3章 通信和联网 ··············· 187
  - 3.1 显示Web信息 ··············· 187
    - 3.1.1 问题 ················· 187
    - 3.1.2 解决方案 ············· 187
    - 3.1.3 实现机制 ············· 187

3.2	拦截 WebView 事件	192	3.11.2 解决方案	241
	3.2.1 问题	192	3.11.3 实现机制	241
	3.2.2 解决方案	192	3.12 查询网络连接状态	250
	3.2.3 实现机制	192	3.12.1 问题	250
3.3	访问带 JavaScript 的 WebView	193	3.12.2 解决方案	250
	3.3.1 问题	193	3.12.3 实现机制	250
	3.3.2 解决方案	194	3.13 使用 NFC 传输数据	253
	3.3.3 实现机制	194	3.13.1 问题	253
3.4	下载图片文件	196	3.13.2 解决方案	253
	3.4.1 问题	196	3.13.3 实现机制	253
	3.4.2 解决方案	197	3.14 USB 连接	260
	3.4.3 实现机制	197	3.14.1 问题	260
3.5	完全在后台下载	200	3.14.2 解决方案	261
	3.5.1 问题	200	3.14.3 实现机制	261
	3.5.2 解决方案	200	3.15 小结	270
	3.5.3 实现机制	200		

第4章 实现设备硬件交互与媒体交互 ... 271

3.6	访问 REST API	203
	3.6.1 问题	203
	3.6.2 解决方案	204
	3.6.3 实现机制	204
3.7	解析 JSON	222
	3.7.1 问题	222
	3.7.2 解决方案	222
	3.7.3 实现机制	222
3.8	解析 XML	225
	3.8.1 问题	225
	3.8.2 解决方案	225
	3.8.3 实现机制	226
3.9	接收短信	235
	3.9.1 问题	235
	3.9.2 解决方案	235
	3.9.3 实现机制	236
3.10	发送短信	238
	3.10.1 问题	238
	3.10.2 解决方案	238
	3.10.3 实现机制	239
3.11	蓝牙通信	241
	3.11.1 问题	241

4.1	整合设备位置	271
	4.1.1 问题	271
	4.1.2 解决方案	271
	4.1.3 实现机制	272
4.2	地图位置	277
	4.2.1 问题	277
	4.2.2 解决方案	277
	4.2.3 实现机制	280
4.3	在地图上标记位置	285
	4.3.1 问题	285
	4.3.2 解决方案	285
	4.3.3 实现机制	286
4.4	监控位置地区	301
	4.4.1 问题	301
	4.4.2 解决方案	301
	4.4.3 实现机制	302
4.5	拍摄照片和视频	311
	4.5.1 问题	311
	4.5.2 解决方案	311
	4.5.3 实现机制	311
4.6	自定义摄像头覆盖层	316

	4.6.1 问题	316
	4.6.2 解决方案	316
	4.6.3 实现机制	317
4.7	录制音频	323
	4.7.1 问题	323
	4.7.2 解决方案	323
	4.7.3 实现机制	323
4.8	自定义视频采集	325
	4.8.1 问题	325
	4.8.2 解决方案	325
	4.8.3 实现机制	326
4.9	添加语音识别	330
	4.9.1 问题	330
	4.9.2 解决方案	330
	4.9.3 实现机制	330
4.10	播放音频/视频	332
	4.10.1 问题	332
	4.10.2 解决方案	332
	4.10.3 实现机制	332
4.11	播放音效	341
	4.11.1 问题	341
	4.11.2 解决方案	341
	4.11.3 实现机制	341
4.12	创建倾斜监控器	344
	4.12.1 问题	344
	4.12.2 解决方案	344
	4.12.3 实现机制	344
4.13	监控罗盘的方向	347
	4.13.1 问题	347
	4.13.2 解决方案	348
	4.13.3 实现机制	348
4.14	从媒体内容中获取元数据	351
	4.14.1 问题	351
	4.14.2 解决方案	351
	4.14.3 实现机制	352
4.15	检测用户移动	355
	4.15.1 问题	355
	4.15.2 解决方案	355
	4.15.3 实现机制	356
4.16	小结	366
第5章	数据持久化	367
5.1	制作首选项界面	367
	5.1.1 问题	367
	5.1.2 解决方案	367
	5.1.3 实现机制	367
5.2	显示自定义首选项	373
	5.2.1 问题	373
	5.2.2 解决方案	373
	5.2.3 实现机制	374
5.3	简单数据存储	378
	5.3.1 问题	378
	5.3.2 解决方案	379
	5.3.3 实现机制	379
5.4	读写文件	383
	5.4.1 问题	383
	5.4.2 解决方案	383
	5.4.3 实现机制	383
5.5	以资源的形式使用文件	390
	5.5.1 问题	390
	5.5.2 解决方案	390
	5.5.3 实现机制	391
5.6	管理数据库	393
	5.6.1 问题	393
	5.6.2 解决方案	393
	5.6.3 实现机制	393
5.7	查询数据库	398
	5.7.1 问题	398
	5.7.2 解决方案	398
	5.7.3 实现机制	399
5.8	备份数据	400
	5.8.1 问题	400
	5.8.2 解决方案	400
	5.8.3 实现机制	400
5.9	分享数据库	405
	5.9.1 问题	405
	5.9.2 解决方案	405
	5.9.3 实现机制	405

5.10 分享SharedPreference……412
  5.10.1 问题……412
  5.10.2 解决方案……412
  5.10.3 实现机制……412
5.11 分享其他数据……421
  5.11.1 问题……421
  5.11.2 解决方案……421
  5.11.3 实现机制……422
5.12 集成系统文档……428
  5.12.1 问题……428
  5.12.2 解决方案……428
  5.12.3 实现机制……429
5.13 小结……442

第6章 与系统交互……443

6.1 后台通知……443
  6.1.1 问题……443
  6.1.2 解决方案……443
  6.1.3 实现机制……443
6.2 创建定时和周期任务……459
  6.2.1 问题……459
  6.2.2 解决方案……460
  6.2.3 实现机制……460
6.3 定时执行周期任务……461
  6.3.1 问题……461
  6.3.2 解决方案……461
  6.3.3 实现机制……462
6.4 创建粘性操作……469
  6.4.1 问题……469
  6.4.2 解决方案……469
  6.4.3 实现机制……470
6.5 长时间运行的后台操作……474
  6.5.1 问题……474
  6.5.2 解决方案……474
  6.5.3 实现机制……475
6.6 启动其他应用程序……480
  6.6.1 问题……480
  6.6.2 解决方案……480
  6.6.3 实现机制……481

6.7 启动系统应用程序……484
  6.7.1 问题……484
  6.7.2 解决方案……484
  6.7.3 实现机制……485
6.8 让其他应用程序启动你的应用程序……489
  6.8.1 问题……489
  6.8.2 解决方案……489
  6.8.3 实现机制……489
6.9 与联系人交互……491
  6.9.1 问题……491
  6.9.2 解决方案……491
  6.9.3 实现机制……492
6.10 读取设备媒体和文档……500
  6.10.1 问题……500
  6.10.2 解决方案……500
  6.10.3 实现机制……500
6.11 保存设备媒体和文档……504
  6.11.1 问题……504
  6.11.2 解决方案……504
  6.11.3 实现机制……504
6.12 读取消息数据……509
  6.12.1 问题……509
  6.12.2 解决方案……509
  6.12.3 实现机制……510
6.13 与日历交互……521
  6.13.1 问题……521
  6.13.2 解决方案……521
  6.13.3 实现机制……521
6.14 执行日志代码……527
  6.14.1 问题……527
  6.14.2 解决方案……528
  6.14.3 实现机制……528
6.15 创建后台工作线程……530
  6.15.1 问题……530
  6.15.2 解决方案……530
  6.15.3 实现机制……530
6.16 自定义任务栈……535
  6.16.1 问题……535

		6.16.2	解决方案	535
		6.16.3	实现机制	535
	6.17	实现 AppWidget		543
		6.17.1	问题	543
		6.17.2	解决方案	543
		6.17.3	实现机制	544
	6.18	支持受限制的配置文件		564
		6.18.1	问题	564
		6.18.2	解决方案	564
		6.18.3	实现机制	565
	6.19	小结		577

第 7 章 图形和绘图 579

	7.1	用 Drawable 做背景		579
		7.1.1	问题	579
		7.1.2	解决方案	579
		7.1.3	实现机制	580
	7.2	创建自定义状态的 Drawable		586
		7.2.1	问题	586
		7.2.2	解决方案	586
		7.2.3	实现机制	586
	7.3	将遮罩应用于图片		591
		7.3.1	问题	591
		7.3.2	解决方案	591
		7.3.3	实现机制	592
	7.4	在视图内容上绘制		601
		7.4.1	问题	601
		7.4.2	解决方案	601
		7.4.3	实现机制	601
	7.5	高性能绘制		617
		7.5.1	问题	617
		7.5.2	解决方案	617
		7.5.3	实现机制	617
	7.6	提取图片调色板		628
		7.6.1	问题	628
		7.6.2	解决方案	628
		7.6.3	实现机制	629

	7.7	平铺 Drawable 元素		633
		7.7.1	问题	633
		7.7.2	解决方案	633
		7.7.3	实现机制	634
	7.8	使用可缩放的向量资源		639
		7.8.1	问题	639
		7.8.2	解决方案	639
		7.8.3	实现机制	639
	7.9	小结		648

第 8 章 使用 Android NDK 和 RenderScript 649

	8.1	Android NDK		649
	8.2	使用 JNI 添加原生位		651
		8.2.1	问题	651
		8.2.2	解决方案	651
		8.2.3	实现机制	652
	8.3	构建纯原生 Activity		660
		8.3.1	问题	660
		8.3.2	解决方案	660
		8.3.3	实现机制	660
	8.4	RenderScript		670
	8.5	使用 RenderScript 过滤图片		671
		8.5.1	问题	671
		8.5.2	解决方案	672
		8.5.3	实现机制	672
	8.6	使用 RenderScript 操作图片		677
		8.6.1	问题	677
		8.6.2	解决方案	677
		8.6.3	实现机制	677
	8.7	使用模糊滤镜仿造透明覆盖层		683
		8.7.1	问题	683
		8.7.2	解决方案	683
		8.7.3	实现机制	683
	8.8	小结		693

第 1 章

布局和视图

Android 平台被设计为能运行在各种类型的设备上,这些设备会有各种各样的屏幕尺寸和分辨率。为了帮助开发人员应对这个挑战,Android 提供了大量的用户界面组件工具集,开发人员可以根据具体的应用程序选择使用组件或自定义组件。Android 还非常依赖于可扩展的 XML 框架和设置资源限定符以实现能够兼容各种环境变化的浮动布局。本章中,我们会学习如何使用这个框架以满足具体的开发需求。

1.1 样式化常见组件

1.1.1 问题

你要让自己的应用程序在所有用户可能运行的 Android 版本上创建一致的外观和体验,同时减少维护这些自定义元素所需的代码量。

1.1.1 解决方案

(API Level 1)

可以将定义应用程序外观的常见属性抽象化到 XML 样式中。样式是视图自定义属性的集合,如文本大小或背景色,这些属性应该应用于应用程序内的多个视图。将这些属性抽象化到样式中,就可以在单个位置定义公共的元素,使得代码更易于更新和维护。

Android 还支持将多个样式共同分组到称为"主题"的全局元素中。主题被应用于整个上下文(如 Activity 或应用程序),并且定义了应适用于该上下文中所有视图的样式。在应用程序中启动的每个 Activity 都应用了一个主题,即使你没有定义任何主题。在此情况下,改为应用默认的系统主题。

1.1.3 实现机制

为研究样式的概念,接下来创建如图 1-1 所示的 Activity 布局。

图 1-1 样式化的小部件

从图中可以看到,此视图中一些元素的外观需要定制,使其不同于通过所应用的默认系统主题样式化的常见外观。一种方法是直接在 Activity 布局中定义适用于全部视图的所有属性。如果这样做的话,则使用的代码如代码清单 1-1 所示。

代码清单 1-1 res/layout/activity_styled.xml

```xml
<?xml version="1.0" encoding="utf-8"?>
<TableLayout xmlns:android="http://schemas.android.com/apk/res/android"
    android:layout_width="match_parent"
    android:layout_height="match_parent"
    android:padding="8dp">
    <TextView
        android:layout_width="wrap_content"
        android:layout_height="wrap_content"
        android:textSize="22sp"
        android:textStyle="bold"
        android:text="Select One"/>
    <RadioGroup
        android:layout_width="match_parent"
        android:layout_height="wrap_content"
        android:orientation="horizontal">
        <RadioButton
            android:layout_width="0dp"
            android:layout_height="wrap_content"
```

```xml
            android:layout_weight="1"
            android:minHeight="@dimen/buttonHeight"
            android:button="@null"
            android:background="@drawable/background_radio"
            android:gravity="center"
            android:text="One"/>
        <RadioButton
            android:layout_width="0dp"
            android:layout_height="wrap_content"
            android:layout_weight="1"
            android:minHeight="@dimen/buttonHeight"
            android:button="@null"
            android:background="@drawable/background_radio"
            android:gravity="center"
            android:text="Two"/>
        <RadioButton
            android:layout_width="0dp"
            android:layout_height="wrap_content"
            android:layout_weight="1"
            android:minHeight="@dimen/buttonHeight"
            android:button="@null"
            android:background="@drawable/background_radio"
            android:gravity="center"
            android:text="Three"/>
</RadioGroup>

<TextView
    android:layout_width="wrap_content"
    android:layout_height="wrap_content"
    android:textSize="22sp"
    android:textStyle="bold"
    android:text="Select All"/>
<TableRow>
    <CheckBox
        android:layout_width="wrap_content"
        android:layout_height="wrap_content"
        android:minHeight="@dimen/buttonHeight"
        android:minWidth="@dimen/checkboxWidth"
        android:button="@null"
        android:gravity="center"
        android:textStyle="italic"
        android:textColor="@color/text_checkbox"
        android:text="One"/>
    <CheckBox
        android:layout_width="wrap_content"
        android:layout_height="wrap_content"
        android:minHeight="@dimen/buttonHeight"
        android:minWidth="@dimen/checkboxWidth"
        android:button="@null"
        android:gravity="center"
```

```xml
            android:textStyle="italic"
            android:textColor="@color/text_checkbox"
            android:text="Two"/>
        <CheckBox
            android:layout_width="wrap_content"
            android:layout_height="wrap_content"
            android:minHeight="@dimen/buttonHeight"
            android:minWidth="@dimen/checkboxWidth"
            android:button="@null"
            android:gravity="center"
            android:textStyle="italic"
            android:textColor="@color/text_checkbox"
            android:text="Three"/>
    </TableRow>

    <TableRow>
        <Button
            android:layout_width="wrap_content"
            android:layout_height="wrap_content"
            android:minWidth="@dimen/buttonWidth"
            android:background="@drawable/background_button"
            android:textColor="@color/accentPink"
            android:text="@android:string/ok"/>
        <Button
            android:layout_width="wrap_content"
            android:layout_height="wrap_content"
            android:minWidth="@dimen/buttonWidth"
            android:background="@drawable/background_button"
            android:textColor="@color/accentPink"
            android:text="@android:string/cancel"/>
    </TableRow>
</TableLayout>
```

在此代码中，我们突出强调了每个视图中与其他相同类型视图共有的属性。这些属性使按钮、文本标题和可选中的元素具有相同的外观。其中有很多重复出现的代码，我们可以通过样式进行简化。

首先，我们需要创建新的资源文件，并且使用<style>标记定义每个属性组。代码清单1-2 显示了完整的抽象化代码。

代码清单 1-2　res/values/styles.xml

```xml
<resources>
    <!--小部件的样式-->
    <style name="LabelText" parent="android:TextAppearance.Large">
        <item name="android:textStyle">bold</item>
    </style>

    <style name="FormButton" parent="android:Widget.Button">
        <item name="android:minWidth">@dimen/buttonWidth</item>
```

```xml
        <item name="android:background">@drawable/background_button</item>
        <item name="android:textColor">@color/accentPink</item>
    </style>

    <style name="FormRadioButton" parent="android:Widget.CompoundButton.
                                    RadioButton">
        <item name="android:minHeight">@dimen/buttonHeight</item>
        <item name="android:button">@null</item>
        <item name="android:background">@drawable/background_radio</item>
        <item name="android:gravity">center</item>
    </style>

    <style name="FormCheckBox" parent="android:Widget.CompoundButton.CheckBox">
        <item name="android:minHeight">@dimen/buttonHeight</item>
        <item name="android:minWidth">@dimen/checkboxWidth</item>
        <item name="android:button">@null</item>
        <item name="android:gravity">center</item>
        <item name="android:textStyle">italic</item>
        <item name="android:textColor">@color/text_checkbox</item>
    </style>

</resources>
```

<style>组将需要应用于每个视图类型的公共属性分组在一起。视图仅可以接受单个样式定义，因此必须在一个组中聚集用于此视图的所有属性。然而，样式支持继承性，这就使我们可以级联每个样式的定义，之后再将它们应用于视图。

请注意每个样式如何声明父样式，父样式是我们应继承的基础框架样式。父样式不是必需的，但因为每个视图上存在的单一样式规则，使用自定义版本覆盖默认样式可替换主题的默认值。如果没有继承基础父样式，则必须定义视图需要的所有属性。通过框架的基础样式扩展小部件的样式，可确保我们只需要添加希望定制的、默认主题外观之外的属性。

显式或隐式的父样式声明

样式继承采用两种形式之一。如前所示，样式可以显式声明其父样式：

```xml
<style name="BaseStyle" />
<style name="NewStyle" parent="BaseStyle" />
```

NewStyle 是 BaseStyle 的扩展，包括在父样式中定义的所有属性。样式还支持隐式父样式声明语法，如下所示：

```xml
<style name="BaseStyle" />
<style name="BaseStyle.Extended" />
```

BaseStyle.Extended 以相同的方式从 BaseStyle 继承其属性。此版本的功能与显式示例相同，只是更加简洁。两种形式不应混用，如果混用，就无法实现在单个样式中采用多个父样式。最终，人们始终优先选择显式父样式声明，而代码的可读性就会降低。

我们可以对原始布局文件应用新的样式，得到的简化版本如代码清单 1-3 所示。

代码清单 1-3　res/layout/activity_styled.xml

```xml
<?xml version="1.0" encoding="utf-8"?>
<TableLayout xmlns:android="http://schemas.android.com/apk/res/android"
    android:layout_width="match_parent"
    android:layout_height="match_parent"
    android:padding="8dp">
    <TextView
        android:layout_width="wrap_content"
        android:layout_height="wrap_content"
        android:textAppearance="@style/LabelText"
        android:text="Select One"/>
    <RadioGroup
        android:layout_width="match_parent"
        android:layout_height="wrap_content"
        android:orientation="horizontal">
        <RadioButton
            style="@style/FormRadioButton"
            android:layout_width="0dp"
            android:layout_height="wrap_content"
            android:layout_weight="1"
            android:text="One"/>
        <RadioButton
            style="@style/FormRadioButton"
            android:layout_width="0dp"
            android:layout_height="wrap_content"
            android:layout_weight="1"
            android:text="Two"/>
        <RadioButton
            style="@style/FormRadioButton"
            android:layout_width="0dp"
            android:layout_height="wrap_content"
            android:layout_weight="1"
            android:text="Three"/>
    </RadioGroup>

    <TextView
        android:layout_width="wrap_content"
        android:layout_height="wrap_content"
        android:textAppearance="@style/LabelText"
        android:text="Select All"/>
    <TableRow>
        <CheckBox
            style="@style/FormCheckBox"
            android:layout_width="wrap_content"
            android:layout_height="wrap_content"
            android:text="One"/>
        <CheckBox
```

```
            style="@style/FormCheckBox"
            android:layout_width="wrap_content"
            android:layout_height="wrap_content"
            android:text="Two"/>
        <CheckBox
            style="@style/FormCheckBox"
            android:layout_width="wrap_content"
            android:layout_height="wrap_content"
            android:text="Three"/>
    </TableRow>

    <TableRow>
        <Button
            style="@style/FormButton"
            android:layout_width="wrap_content"
            android:layout_height="wrap_content"
            android:text="@android:string/ok"/>
        <Button
            style="@style/FormButton"
            android:layout_width="wrap_content"
            android:layout_height="wrap_content"
            android:text="@android:string/cancel"/>
    </TableRow>
</TableLayout>
```

通过对每个视图应用样式属性,就可以避免重复的显式属性引用,而改为对每个元素只引用一次。此行为的一个例外情况是 TextView 头部,它接受特殊的 android:textAppearance 属性。此属性获取一个样式引用,并且仅应用于文本格式化属性(大小、样式、颜色等)。使用 TextView 时,仍然可以同时应用单独的样式属性。这样,TextView 实例在对单个视图使用多种样式的框架中就可以得到支持。

主题

在 Android 中,主题(Theme)就是一种应用到整个应用程序或某个 Activity 的外观风格。使用主题有两个选择,使用系统主题或创建自定义主题。无论采用哪种方法,都要在 AndroidManifest.xml 文件中设置主题,如代码清单 1-4 所示。

代码清单 1-4 AndroidManifest.xml

```
<?xml version="1.0" encoding="utf-8"?>
<manifest xmlns:android="http://schemas.android.com/apk/res/android"
    ...>
    <!--通过 application 标签来设置全局主题-->
    <application android:theme="APPLICATION_THEME_NAME"
        ...>
        <!--通过 activity 标签来设置单个主题-->
        <activity android:name=".Activity"
            android:theme="ACTIVITY_THEME_NAME"
            ...>
```

```
            <intent-filter>
                ...
            </intent-filter>
        </activity>
    </application>
</manifest>
```

(1) 系统主题

Android 框架中打包的 styles.xml 和 themes.xml 文件中包含了一些主题选项，其中是一些有用的自定义属性。完整的清单请查阅 SDK 文档中的 R.style，下面是几个常用示例：

- Theme.Light：标准主题的变体，该主题的背景和用户元素使用相反的颜色主题。它是 Android 3.0 以前版本的应用程序默认推荐使用的基础主题。
- Theme.NoTitleBar.Fullscreen：移除标题栏和状态栏，全屏显示(去掉屏幕上所有的组件)。
- Theme.Dialog：让 Activity 看起来像对话框的有用主题。
- Theme.Holo.Light：(API Level 11)使用逆配色方案的主题并默认拥有一个 Action Bar。这是 Android 3.0 上应用程序默认推荐的基本主题。
- Theme.Holo.Light.DarkActionBar：(API Level 14)使用逆配色方案的主题，但 Action Bar 是黑色实线的。这是 Android 4.0 上应用程序默认推荐的基本主题。
- Theme.Material.Light：(API Level 21)通过小型的原色调色板控制的简化颜色方案主题。此主题还支持使用提供的原色对标准小部件着色。这是 Android 5.0 上应用程序默认推荐的基本主题。

注意：

使用 AppCompat 库时，应改为使用这些主题的每个主题的其他版本(例如，Theme.AppCompat.Light.DarkActionBar)。

代码清单 1-5 中的示例通过设置 AndroidManifest.xml 文件中的 android:theme 属性，将一个系统主题应用到整个应用程序。

代码清单 1-5　设置为应用程序主题的清单文件

```xml
<?xml version="1.0" encoding="utf-8"?>
<manifest xmlns:android="http://schemas.android.com/apk/res/android"
    ...>
    <!--通过application标签来设置全局主题-->
    <application android:theme="Theme.NoTitleBar"
        ...>
        ...
    </application>
</manifest>
```

(2) 自定义主题

有时候系统提供的主题还不能满足需求。毕竟，系统提供的主题并不能自定义窗口中的所有元素。定义自定义主题能方便地解决这个问题。

找到项目目录 res/values 下的 styles.xml 文件,如果没有就创建一个。记住,主题就是应用范围更广的风格样式,所以两者是在同一个地方定义的。与窗口自定义有关的主题元素可以在 SDK 的 R.attr 引用中找到,下面是常用的一些元素:

- android:windowNoTitle:控制是否要移除默认的标题栏;设为 true 以移除标题栏。
- android:windowFullscreen:控制是否移除系统状态栏;设为 true 以移除状态栏并全屏显示。
- android:windowBackground:将某个颜色或 Drawable 资源设为背景。
- android:windowContentOverlay:窗口内容的前景之上放置的 Drawable 资源。默认情况下,就是状态栏下的阴影;可以用任何资源代替默认的状态栏,或者设为 null(XML 中为@null)以将其移除。

此外,Material 主题接受一系列颜色属性,这些属性用于对应用程序界面小部件着色:

- android:colorPrimary:用于对主要的界面元素着色,如 Action Bar 和滚动边界发光特效。同样也影响最近对标题栏颜色的操作。
- android:colorPrimaryDark:对系统控件着色,如状态栏的背景。
- android:colorAccent:应用于拥有焦点或已激活控件的默认颜色。
- android:colorControlNormal:重写没有焦点或未激活控件的颜色。
- android:colorControlActivated:重写拥有焦点或已激活控件的颜色。如果同时定义了强调色,则替换该颜色。
- android:colorControlHighlight:重写正在按下的控件的颜色。

代码清单 1-6 就是一个 styles.xml 文件示例,其中创建了一个自定义主题,以便为应用程序界面提供品牌特有的颜色。

代码清单 1-6 res/values/styles.xml

```xml
<?xml version="1.0" encoding="utf-8"?>
<resources>

    <style name="BaseAppTheme" parent="@style/Theme.AppCompat.Light.
                                DarkActionBar">
        <!-- Action Bar 的背景色 -->
        <item name="colorPrimary">@color/primaryBlue</item>
        <!-- 状态栏的着色 -->
        <item name="colorPrimaryDark">@color/primaryDarkBlue</item>
        <!-- 应用于所有拥有焦点/已激活控件的默认颜色 -->
        <item name="colorAccent">@color/accentPink</item>

        <!-- 未选择控件的颜色 -->
        <item name="colorControlNormal">@color/controlNormalGreen</item>
        <!-- 已激活控件的颜色,重写强调色 -->
        <item name="colorControlActivated">@color/controlActivatedGreen</item>
    </style>

</resources>
```

注意，主题也可以从父主题继承属性，所以并不需要从头创建整个主题。在这个示例中，我们选择了继承 Android 默认的系统主题，只自定义我们要修改的属性。所有平台主题都定义在 Android 包的 res/values/themes.xml 文件中。关于样式和主题的更多细节请查阅 SDK 文档。

代码清单 1-7 展示了如何在 AndroidManufest.xml 中对单个 Activity 实例应用这些主题。

代码清单 1-7　在清单文件中为每个 Activity 设置主题

```xml
<?xml version="1.0" encoding="utf-8"?>
<manifest xmlns:android="http://schemas.android.com/apk/res/android"
    …>
    <!--通过 application 标签来设置全局主题-->
    <application
        …>
        <!--通过 activity 标签来设置单独的主题 -->
        <activity android:name=".ActivityOne"
            android:theme="@style/AppTheme"
            …>
            <intent-filter>
                <action android:name="android.intent.action.MAIN" />
                <category android:name="android.intent.category.LAUNCHER" />
            </intent-filter>
        </activity>
    </application>
</manifest>
```

1.2　切换系统 UI 元素

1.2.1　问题

您的应用程序体验需要对显示进行控制，移除各种系统修饰，例如状态栏和软件导航按钮。

1.2.2　解决方案

(API Level 11)

在应用程序的内容可见时，通过暂时隐藏系统 UI 组件，从而尽可能地提供更大的屏幕空间，可以让很多应用程序(例如阅读器或视频播放器)呈现更好的内容体验。从 Android 3.0 开始，开发人员可以在运行时动态地调整这些属性，而不必再静态地请求窗口特性或声明主题内的值。

1.2.3 实现机制

夜间模式

夜间模式通常也称为"熄灯模式",指的是调暗屏幕导航控件(以及稍后发布版本中的系统状态栏),而不是真正移除它们来减少屏幕上的系统元素(这些元素可能会将用户的注意力从应用程序的当前视图中分散出去)。

要启用这个模式,只需要简单地在视图结构中的任何视图中使用 SYSTEM_UI_FLAG_LOW_PROFILE 标识调用 setSystemUiVisibility()即可。而要想恢复到默认模式,需要以 SYSTEM_UI_FLAG_VISIBLE 调用同样的方法。通过调用 getSystemUiVisibility()并检查标识的当前状态,就可以知道我们现在所处的模式了(参见代码清单1-8 和 1-9)。

注意:

这些标识名称都是在 API Level 14(Android 4.0)中引入的,在之前的版本中,它们名为 STATUS_BAR_HIDDEN 和 STATUS_BAR_VISIBLE。它们的值都是一样的,所以新的标识在 Android 3.x 设备上也可以实现相同的功能。

代码清单 1-8 res/layout/main.xml

```xml
<?xml version="1.0" encoding="utf-8"?>
<RelativeLayout
    xmlns:android="http://schemas.android.com/apk/res/android"
    android:layout_width="match_parent"
    android:layout_height="match_parent" >
    <Button
        android:layout_width="match_parent"
        android:layout_height="wrap_content"
        android:layout_centerVertical="true"
        android:text="Toggle Mode"
        android:onClick="onToggleClick" />
</RelativeLayout>
```

代码清单 1-9 开关夜间模式的 Activity

```java
public class DarkActivity extends Activity {

    @Override
    protected void onCreate(Bundle savedInstanceState) {
        super.onCreate(savedInstanceState);
        setContentView(R.layout.main);
    }

    public void onToggleClick(View v) {
        int currentVis = v.getSystemUiVisibility();
        int newVis;
        if ((currentVis & View.SYSTEM_UI_FLAG_LOW_PROFILE)
                == View.SYSTEM_UI_FLAG_LOW_PROFILE) {
            newVis = View.SYSTEM_UI_FLAG_VISIBLE;
```

```
    } clse {
        newVis = View.SYSTEM_UI_FLAG_LOW_PROFILE;
    }
    v.setSystemUiVisibility(newVis);
}
```

要调节这些参数的窗口中所有可见的视图,可以调用 setSystemUiVisibility()和 getSystemUiVisibility()方法。

1. 隐藏导航控件

(API Level 14)

SYSTEM_UI_FLAG_HIDE_NAVIGATION 标识会移除没有物理按钮的设备屏幕上的 HOME 和 BACK 控件。Android 赋予开发人员这个功能时是非常谨慎的,因为这个功能对于用户来说太重要了。如果导航控件被手动地隐藏,屏幕上的任何点击都会回到上一界面。代码清单 1-10 显示了这个功能的示例。

代码清单 1-10 开关导航控件的 Activity

```
public class HideActivity extends Activity {

    @Override
    protected void onCreate(Bundle savedInstanceState) {
        super.onCreate(savedInstanceState);
        setContentView(R.layout.main);

    }

    public void onToggleClick(View v) {
        //这里我们只需要点击屏幕即可隐藏控件,因为控件隐藏后,
        //只要再点击一下屏幕,Android 系统就会让控件自动重新出现
        v.setSystemUiVisibility(
                View.SYSTEM_UI_FLAG_HIDE_NAVIGATION);
    }
}
```

运行这个示例时还需要注意的是,因为我们是在根布局中进行设置,所以按钮会上下移动来适应内容区域的改变。如果你打算使用这个标识,需要注意在布局改变时所有位于屏幕底部的视图都会移动。

2. 全屏 UI 模式

Android 4.1 之前是没有方法动态地隐藏系统状态栏的,只能通过设置静态主题来实现。在隐藏和显示 Action Bar 时,ActionBar.show()和 ActionBar.hide()会动态地显示和隐藏 Action Bar 元素。如果请求的是 FEATURE_ACTION_BAR_OVERLAY,页面的变化将不会影响到 Activity 的内容;否则,视图的内容会上下移动来适应这种变化。

(API Level 16)

代码清单 1-11 演示了一个如何暂时隐藏所有系统 UI 控件的示例。

代码清单 1-11　开关所有系统 UI 控件的 Activity

```java
public class FullActivity extends Activity {

    @Override
    protected void onCreate(Bundle savedInstanceState) {
        super.onCreate(savedInstanceState);
        //请求这个特性，这样 Action Bar 就会隐藏起来
        requestWindowFeature(Window.FEATURE_ACTION_BAR_OVERLAY);
        setContentView(R.layout.main);
    }

    public void onToggleClick(View v) {
        //这里我们只需要点击屏幕即可隐藏 UI，因为控件隐藏后，
        //只要再点击一下屏幕，Android 系统就会让控件自动重新出现
        v.setSystemUiVisibility(
                    /*这个标识会告诉 Android 在改变窗口大小来
                     *隐藏/显示系统元素时不要移动我们的布局
                     */
                    View.SYSTEM_UI_FLAG_LAYOUT_STABLE
                    /* 这个标识会隐藏系统状态栏。如果请求 ACTION_BAR_OVERLAY，
                     * 同时会隐藏 Action Bar
                     */
                    | View.SYSTEM_UI_FLAG_FULLSCREEN
                    /* 这个标识会隐藏屏幕上的所有控件
                     */
                    | View.SYSTEM_UI_FLAG_HIDE_NAVIGATION);
    }
}
```

与只隐藏导航控件的示例类似，我们不需要再次显示控件，因为任何屏幕上的点击都会让它们再次显示出来。作为 Android 4.1 的一个便捷之处，在系统通过这种方式清除 SYSTEM_UI_FLAG_HIDE_NAVIGATION 后，同时会清除 SYSTEM_UI_FLAG_FULLSCREEN 标识，所以顶部和底部的元素会一起可见。如果我们请求了 FEATURE_ACTION_BAR_OVERLAY，Android 将会隐藏作为全屏标识一部分的 Action Bar；否则，只会影响到状态栏。

我们在这个示例中添加另一个有趣的标识：SYSTEM_UI_LAYOUT_STABLE。这个标识会通知 Android 在添加和移除系统 UI 时不要移动我们的内容视图。正因为如此，我们的按钮在开关 UI 时会一直位于中间。

1.3 创建并显示视图

1.3.1 问题

应用程序需要视图元素来显示信息并与用户交互。

1.3.2 解决方案

(API Level 1)

无论是使用 Android SDK 中的各种视图和小部件,还是创建自定义显示,所有的应用程序都需要使用视图来与用户进行交互。在 Android 中构建用户界面的首选方法是,在 XML 中将其定义好,然后在运行时调用。

Android 中的视图结构是树状的,根部通常是 Activity 或窗口的内容视图。ViewGroup 是一种特殊的视图,用于管理一个或多个子视图的显示方式。子视图可以是另一个 ViewGroup,整棵视图树就这样继续生长。所有的标准布局类都源自 ViewGroup,经常作为 XML 布局文件的根节点。

1.3.3 实现机制

下面定义一个有两个 Button 实例和一个 EditText 的布局来接收用户输入。我们可以在 res/layout 中定义一个名为 main.xml 的文件,参见代码清单 1-12。

代码清单 1-12　res/layout/main.xml

```xml
<LinearLayout
    xmlns:android="http://schemas.android.com/apk/res/android"
    android:layout_width="fill_parent"
    android:layout_height="fill_parent"
    android:orientation="vertical">
    <EditText
        android:id="@+id/editText"
        android:layout_width="fill_parent"
        android:layout_height="wrap_content"/>
    <LinearLayout
        android:layout_width="fill_parent"
        android:layout_height="wrap_content"
        android:orientation="horizontal">
        <Button
            android:id="@+id/save"
            android:layout_width="wrap_content"
            android:layout_height="wrap_content"
            android:text="Save"/>
        <Button
            android:id="@+id/cancel"
            android:layout_width="wrap_content"
```

```
            android:layout_height="wrap_content"
            android:text="Cancel"/>
    </LinearLayout>
</LinearLayout>
```

LinearLayout 是一个 ViewGroup，它将元素横向或纵向排列。在 main.xml 中，EditText 和其中的 LinearLayout 是按序纵向排列的。内部的 LinearLayout(里面是按钮)的内容是横向排列的。带有 android:id 值的视图元素可以在 Java 代码中引用，以备进一步自定义或显示之用。

为了用这个布局显示 Activity 的内容，必须在运行时将其填充。经过重载的 Activity.setContentView()方法可以很方便地完成这个工作，只需要提供布局的 ID 值即可。在 Activity 中设置布局就是这样简单：

```
public void onCreate(Bundle savedInstanceState) {
    super.onCreate(savedInstanceState);
    setContentView(R.layout.main);
    //继续初始化 Activity
}
```

除了提供 ID 值(main.xml 有一个自动生成的 ID——R.layout.main)，不需要其他内容。如果在将布局附加到窗口之前还需要进一步自定义，可以手动将其填充，在完成所需的自定义后再将其作为内容视图添加。代码清单 1-13 填充了同一个布局，但在显示之前加上了第三个按钮。

代码清单 1-13　在显示之前修改布局

```
public void onCreate(Bundle savedInstanceState) {
    super.onCreate(savedInstanceState);
    //填充布局文件
    LinearLayout layout = (LinearLayout)getLayoutInflater()
            .inflate(R.layout.main,null);
    //添加一个按钮
    Button reset = new Button(this);
    reset.setText("Reset Form");
    layout.addView(reset,
        new LinearLayout.LayoutParams(LayoutParams.FILL_PARENT,
            LayoutParams.WRAP_CONTENT));

    //将视图关联到窗口
    setContentView(layout);
}
```

在这个示例中，这个 XML 布局是在 Activity 的代码中用 LayoutInflater 填充的，它的 inflate()方法会返回一个指向填充后的视图的句柄。因为 LayoutInflater.inflate()返回的是视图，所以我们必须将其转换成 XML 中的某个子类，这样才能在将其关联到窗口之前进行修改。

注意：
XML 布局文件中的根元素是 LayoutInflater.inflate()返回的 View 元素。

inflate()的第二个参数代表父 ViewGroup，这个参数非常重要，因为它定义了如何解释被填充布局中的 LayoutParams。可能的话，只要你知道被填充视图的父视图，就应该把它传进来；否则，XML 中根视图的 LayoutParams 会被忽略。当传入一个父视图后，还要注意 inflate()的第三个参数，该参数决定了被填充的布局是否会自动关联到父视图上。在后面的范例中会看到这种机制对于自定义视图是非常有用的。但在本例中，我们填充的是 Activity 最顶层的视图，因此这里传递了 null。

完全自定义视图

有时，SDK 中的可用小部件不足以提供你所需的输出。或许是要将多个显示元素结合到单个视图中，减少层次结构中视图的数量以提升性能。对于这些情况，就要创建自己的 View 子类。创建 View 子类之后，类和框架之间就有两个主要的交互方面需要关注：测量和绘制。

测量

自定义视图必须满足的第一个要求是向框架提供其内容的测量。在显示视图的层次结构之前，Android 会为每个元素(布局和视图节点)调用 onMeasure()并向该方法传递两个约束，视图应该使用这两个约束来管理如何报告其应该具备的大小。每个约束是一个称为 MeasureSpec 的封包整数，它包含模式标记和大小值。其中，模式采用如下值之一：

- AT_MOST：如果视图的布局参数是 match_parent 或存在其他的大小上限，则通常使用此模式。该模式告诉视图，其应该报告所需的大小，前提是不超出规范中规定的值。
- EXACTLY：如果视图的布局参数是固定值，则通常使用此模式。框架期望视图自动设置其大小以匹配规范——不多不少。
- UNSPECIFIED：该值通常用于指出视图在无约束时所需的大小。它可能是另一个具有不同约束的测量的前置模式，或者可能只是因为布局参数被设置为 wrap_content 且父节点中没有其他约束。在此模式中，视图可能报告其在任何情况下所需的大小。此规范中的大小通常为 0。

完成所报告大小的计算之后，必须在 onMeasure()返回之前将这些值传入 setMeasuredDimension()调用。如果没有这样做，框架将报告严重的错误。

通过测量还可以基于可用空间配置视图的输出。测量约束基本上表明在布局内分配了多少空间，因此如果要创建的视图在方向上与其所包含的内容不同，例如垂直空间或多或少，onMeasure()将提供决策所需的信息。

注意：
在测量期间，视图实际上还没有确定的大小；它只有已测量的尺寸。如果在分配大小后需要对视图做一些自定义工作，则应该重写 onSizeChanged()并添加适当的代码。

绘制

自定义视图的第二个步骤就是绘制内容,这可能是最重要的步骤。对视图进行测量并将其放置在布局层次结构中之后,框架将为该视图构造一个 Canvas 实例,调整其大小并放置在适当位置,然后通过 onDraw()传递该实例以供视图使用。Canvas 对象驻留单独的绘制调用,因此它包括的 drawLine()、drawBitmap()和 drawText()等方法用于独立地布局视图内容。如同其名称所暗示的那样,Canvas 使用 Painter 的算法,因此最后绘制的项将放在第一个绘制项的顶部。

绘制的内容会依附到通过测量和布局提供的视图的边界上,因此虽然可以对 Canvas 元素进行平移、缩放、旋转等操作,但不能在放置视图的矩形外部绘制内容。

最后,在 onDraw()中提供的内容不包括视图的背景,可以使用 setBackgroundColor()或 setBackgroundResource()等方法设置该背景。如果在视图上设置背景,则背景会自动绘制,不需要在 onDraw()中进行处理。

代码清单 1-14 显示了应用程序可以遵循的非常简单的定制视图模板。至于其中的内容,我们绘制了一系列同心圆来表示靶心目标。

代码清单 1-14 自定义视图的示例

```java
public class BullsEyeView extends View {

    private Paint mPaint;

    private Point mCenter;
    private float mRadius;

    /*
     * Java 构造函数
     */
    public BullsEyeView(Context context) {
        this(context, null);
    }

    /*
     * XML 构造函数
     */
    public BullsEyeView(Context context, AttributeSet attrs) {
        this(context, attrs, 0);
    }

    /*
     * 带有样式的 XML 构造函数
     */
    public BullsEyeView(Context context, AttributeSet attrs,
            int defStyle) {
        super(context, attrs, defStyle);
```

```java
        // 在此构造函数中进行视图的初始化工作

        // 创建用于绘制的画刷
        mPaint = new Paint(Paint.ANTI_ALIAS_FLAG);
        // 我们要绘制填充的圆
        mPaint.setStyle(Style.FILL);
        // 创建圆的中心点
        mCenter = new Point();
    }

    @Override
    protected void onMeasure(int widthMeasureSpec,
            int heightMeasureSpec) {
        int width, height;
        // 确定内容的理想大小, 无约束
        int contentWidth = 100;
        int contentHeight = 100;

        width = getMeasurement(widthMeasureSpec, contentWidth);
        height = getMeasurement(heightMeasureSpec, contentHeight);
        // 必须使用测量值调用此方法!
        setMeasuredDimension(width, height);
    }

    /*
     * 用于测量宽度和高度的辅助方法
     */
    private int getMeasurement(int measureSpec, int contentSize) {
        int specSize = MeasureSpec.getSize(measureSpec);
        switch (MeasureSpec.getMode(measureSpec)) {
            case MeasureSpec.AT_MOST:
                return Math.min(specSize, contentSize);
            case MeasureSpec.UNSPECIFIED:
                return contentSize;
            case MeasureSpec.EXACTLY:
                return specSize;
            default:
                return 0;
        }
    }

    @Override
    protected void onSizeChanged(int w, int h,
            int oldw, int oldh) {
        if (w != oldw || h != oldh) {
            // 如果有变化, 则复位参数
```

```java
            mCenter.x = w / 2;
            mCenter.y = h / 2;
            mRadius = Math.min(mCenter.x, mCenter.y);
        }
    }

    @Override
    protected void onDraw(Canvas canvas) {
        // 绘制一系列从小到大且颜色交替变换的同心圆
        mPaint.setColor(Color.RED);
        canvas.drawCircle(mCenter.x, mCenter.y, mRadius, mPaint);

        mPaint.setColor(Color.WHITE);
        canvas.drawCircle(mCenter.x, mCenter.y, mRadius * 0.8f,
            mPaint);

        mPaint.setColor(Color.BLUE);
        canvas.drawCircle(mCenter.x, mCenter.y, mRadius * 0.6f,
            mPaint);

        mPaint.setColor(Color.WHITE);
        canvas.drawCircle(mCenter.x, mCenter.y, mRadius * 0.4f,
            mPaint);

        mPaint.setColor(Color.RED);
        canvas.drawCircle(mCenter.x, mCenter.y, mRadius * 0.1f,
            mPaint);
    }
}
```

首先可以注意到该视图有如下 3 个构造函数：
- View(Context context)：通过 Java 代码构造视图时使用该版本。
- View(Context, AttributeSet)：从 XML 填充视图时使用该版本。AttributeSet 包括附加到视图的 XML 元素的所有属性。
- View(Context,AttributeSet,int)：该版本类似于上一个版本，但在将样式属性添加到 XML 元素时被调用。

常见的方案是将所有 3 个构造函数链接在一起，并且仅在最后一个构造函数中实现定制，这就是我们在视图示例中完成的工作。

在 onMeasure()中，我们使用一种简单的实用方法，基于测量约束返回正确的尺寸。我们基本上可以在所需的内容大小(在此任意选择大小，但应该表示真实应用程序中的视图内容)和所提供的大小之间选择。对于 AT_MOST，我们选择两者中较小的值；即视图的大小应该适合我们的内容，前提是不超出规范的大小。完成测量后，我们调用 onSizeChanged() 收集一些所需的基本数据来绘制目标圆。我们等到此处才调用该方法，这是为了确保使用确切符合视图布局的值。

在 onDraw()内部,我们构造显示内容。在 Canvas 上绘制 5 个逐步递减半径且颜色交替变换的同心圆。Paint 元素控制所绘制内容样式的相关信息,例如笔画宽度、文本大小和颜色。在为此视图声明 Paint 时,设置样式为 FILL,这就确保使用每种颜色填充圆。根据 Painter 的算法,在较大圆的顶部绘制较小的圆,这就提供了我们所需的目标效果。

将此视图添加到 XML 布局非常简单,但因为视图没有驻留在 android.view 或 android.widget 包中,我们需要使用类的完全限定包名命名元素。例如,如果应用程序包是 com.androidrecipes. customwidgets,则 XML 代码如下所示:

```
<com.androidrecipes.customwidgets.BullsEyeView
    android:layout_width="match_parent"
    android:layout_height="match_parent" />
```

图 1-2 显示了将此视图添加到 Activity 的结果。

图 1-2　靶心定制视图

1.4　动画视图

1.4.1　问题

应用程序要让视图对象运动起来,实现变化或其他特效。

1.4.2 解决方案

(API Level 12)

ObjectAnimator 实例,例如 ViewPropertyAnimator,可以用来操作 View 对象的属性,例如视图的位置或旋转。ViewPropertyAnimator 是通过 View.animate()获得的,然后根据动画的特征进行修改。通过这个 API 进行的修改会影响到 View 对象本身的真实属性。

1.4.3 实现机制

ViewPropertyAnimator 是对视图内容制作动画的最便利方法。此 API 的工作方式就像生成器一样,所有对不同属性修改的调用都可以连接起来组成一个动画。在当前线程的 Looper 的相同迭代中,对 ViewPropertyAnimator 的所有调用都会汇集到一个动画中。代码清单 1-15 和 1-16 演示了一个简单的视图过渡 Activity。

代码清单 1-15　res/layout/main.xml

```xml
<?xml version="1.0" encoding="utf-8"?>
<LinearLayout xmlns:android="http://schemas.android.com/apk/res/android"
    android:orientation="vertical"
    android:layout_width="fill_parent"
    android:layout_height="fill_parent">
    <Button
        android:id="@+id/toggleButton"
        android:layout_width="fill_parent"
        android:layout_height="wrap_content"
        android:text="Click to Toggle" />
    <View
        android:id="@+id/theView"
        android:layout_width="fill_parent"
        android:layout_height="wrap_content"
        android:background="#AAA" />
</LinearLayout>
```

代码清单 1-16　使用了 ViewPropertyAnimator 的 Activity

```java
public class AnimateActivity extends Activity implements View.OnClickListener {

    View viewToAnimate;

    @Override
    public void onCreate(Bundle savedInstanceState) {
        super.onCreate(savedInstanceState);
        setContentView(R.layout.main);

        Button button = (Button)findViewById(R.id.toggleButton);
        button.setOnClickListener(this);

        viewToAnimate = findViewById(R.id.theView);
```

```
        }
        @Override
        public void onClick(View v) {
            if(viewToAnimate.getAlpha() > 0f) {
                //如果视图已经可见,将其从右侧滑出
                viewToAnimate.animate().alpha(0f).translationX(1000f);
            } else {
                //如果视图是隐藏的,原地做渐显动画
                //Property Animation 会实际修改视图,因此必须首先恢复视图的位置
                viewToAnimate.setTranslationX(0f);
                viewToAnimate.animate().alpha(1f);
            }
        }
    }
}
```

在这个示例中,滑动动画和渐显动画是通过修改 alpha 和 translationX(这个过渡值需要足够大才能够让视图移出屏幕)属性一起实现的。我们不需要将这些方法调用链接到一起,从而组成一个动画。即使我们在不同的地方调用,它们还是会一起执行,因为它们都是在主线程的 Looper 的相同迭代中设置的。

注意,这里我们首先恢复了 View 的过渡属性,然后运行了没有滑动效果的渐显动画。这是因为属性动画会修改视图本身,而不只是暂时地绘制(之前的动画 API 的机制)。如果不恢复这个属性,依然会有渐显动画,但会在屏幕外右侧 1000 像素处进行。

1. ObjectAnimator

虽然 ViewPropertyAnimator 可以很方便且快速地实现简单的属性动画,但对于一些更加复杂的工作,例如将多个动画链接到一起,这种方式会受到一定的限制。这时我们可以使用它的父类 ObjectAnimator。通过 ObjectAnimator,我们可以设置监听器,从而在动画开始和结束时得到相应的通知;另外在动画做增量更新时也可以得到通知。

代码清单 1-17 和 1-18 显示了如何使用 ObjectAnimator 来修改我们的 Flipper 动画代码。

代码清单 1-17 res/layout/main.xml

```xml
<?xml version="1.0" encoding="utf-8"?>
<FrameLayout xmlns:android="http://schemas.android.com/apk/res/android"
    android:layout_width="fill_parent"
    android:layout_height="fill_parent">

    <ImageView
        android:id="@+id/flip_image"
        android:layout_width="wrap_content"
        android:layout_height="wrap_content"
        android:layout_gravity="center"/>
</FrameLayout>
```

代码清单 1-18　使用 ObjectAnimator 实现掷硬币动画

```java
public class FlipperActivity extends Activity {

    private boolean mIsHeads;
    private ObjectAnimator mFlipper;
    private Bitmap mHeadsImage, mTailsImage;
    private ImageView mFlipImage;

    @Override
    public void onCreate(Bundle savedInstanceState) {
        super.onCreate(savedInstanceState);
        setContentView(R.layout.main);

        mHeadsImage = BitmapFactory.decodeResource(getResources(),
        R.drawable.heads);
        mTailsImage = BitmapFactory.decodeResource(getResources(),
        R.drawable.tails);

        mFlipImage = (ImageView)findViewById(R.id.flip_image);
        mFlipImage.setImageBitmap(mHeadsImage);
        mIsHeads = true;

        mFlipper = ObjectAnimator.ofFloat(mFlipImage, "rotationY", 0f, 360f);
        mFlipper.setDuration(500);
        mFlipper.addUpdateListener(new AnimatorUpdateListener() {
            @Override
            public void onAnimationUpdate(ValueAnimator animation) {
                if (animation.getAnimatedFraction() >= 0.25f && mIsHeads) {
                    mFlipImage.setImageBitmap(mTailsImage);
                    mIsHeads = false;
                }
                if (animation.getAnimatedFraction() >= 0.75f && !mIsHeads) {
                    mFlipImage.setImageBitmap(mHeadsImage);
                    mIsHeads = true;
                }
            }
        });
    }

    @Override
    public boolean onTouchEvent(MotionEvent event) {
        if(event.getAction() == MotionEvent.ACTION_DOWN) {
            mFlipper.start();
            return true;
        }
        return super.onTouchEvent(event);
    }
}
```

属性动画提供了一些之前旧动画系统没有的变换功能，例如 x 轴和 y 轴的旋转效果，从而实现三维变换效果。本例中，我们不需要计算缩放比例来实现旋转，只需要告诉视图沿着 y 轴旋转即可。正因为如此，我们不再需要使用两个动画来操作硬币，整个旋转过程只需要操作视图的 rotationY 属性即可。

另一个强大之处就是 AnimationUpdateListener，它提供了动画运行过程中的常规回调方法。getAnimatedFraction()方法会返回当前动画完成的百分比。还可以通过 getAnimatedValue()得到当前动画中某个属性的准确值。

本例中，我们使用第一个方法在动画运行到两个时刻(硬币可以换面，即 90°和 270°或者动画时长的 25%和 75%)时，更换正反面的图片。因为并不能保证从每个角度我们都可以得到通知，所以当达到阈值后，我们会立即更换图片。我们还设置了一个布尔标识来避免在旋转过程中对同一个值进行重复的图片设置(这会产生不必要的性能损耗)。

如果应用程序需要链接多个动画，ObjectAnimator 还可以支持更加传统的 AnimationListener 来响应动画的主要事件，例如开始、结束和重复。

提示：
在 Android 4.4 上，Animator 还支持 pause()和 resume()方法以挂起运行中的动画而不用完全取消它。

2. AnimationSet

如果需要执行多个动画，则可以在 AnimationSet 中聚集这些动画。可以同时播放 AnimationSet 中的所有动画，或者依次播放每个动画。定义动画集合的 Java 代码可能稍微有点冗长，因此我们将在此例中改为使用 XML 动画格式。代码清单 1-19 定义了一组将应用于硬币翻转的动画。

代码清单 1-19　res/animator/flip.xml

```xml
<?xml version="1.0" encoding="utf-8"?>
<set xmlns:android="http://schemas.android.com/apk/res/android"
    android:ordering="together">
    <!-- 建立硬币旋转的线性重复 -->
    <objectAnimator
    android:propertyName="rotationX"
    android:duration="400"
    android:valueFrom="0"
    android:valueTo="360"
    android:valueType="floatType"
    android:repeatMode="restart"
    android:repeatCount="3"
    android:interpolator="@android:interpolator/linear"/>
    <!-- 添加一个提升动画以显示硬币在空中上升 -->
    <objectAnimator
        android:propertyName="translationY"
        android:duration="800"
        android:valueTo="-200"
```

```
            android:valueType="floatType"
            android:repeatMode="reverse"
            android:repeatCount="1" />
</set>
```

此处我们在<set>中定义了两个同时播放的动画(通过 android:ordering="together")。第一个动画我们前面已经看过,用于旋转硬币图片一次。此动画设置为重复 3 次,提供 3 个完整的旋转。图片的默认插值器是加速/减速时间曲线,这看起来会结束硬币翻转动画。为提供全程一致的速度,我们改为对动画应用系统的线性插值器。

第二个动画使硬币在旋转期间沿着视图向上滑动,看起来就像是硬币被扔到空中的效果。因为硬币还必须落回来,将此动画设置为在完成后反向运行一次。

代码清单 1-20 显示了附加到翻转器 Activity 的新动画。

代码清单 1-20　带有 XML 版本 AnimatorSet 的翻转器动画

```java
public class FlipperActivity extends Activity {

    private boolean mIsHeads;
    private AnimatorSet mFlipper;
    private Bitmap mHeadsImage, mTailsImage;
    private ImageView mFlipImage;

    @Override
    public void onCreate(Bundle savedInstanceState) {
        super.onCreate(savedInstanceState);
        setContentView(R.layout.main);

        mHeadsImage = BitmapFactory.decodeResource(getResources(),
R.drawable.heads);
        mTailsImage = BitmapFactory.decodeResource(getResources(),
R.drawable.tails);

        mFlipImage = (ImageView)findViewById(R.id.flip_image);
        mFlipImage.setImageResource(R.drawable.heads);
        mIsHeads = true;

        mFlipper = (AnimatorSet) AnimatorInflater.loadAnimator(this,
R.animator.flip);
        mFlipper.setTarget(mFlipImage);

        ObjectAnimator flipAnimator = (ObjectAnimator) mFlipper.
getChildAnimations().get(0);
        flipAnimator.addUpdateListener(new ValueAnimator.
AnimatorUpdateListener() {
            @Override
            public void onAnimationUpdate(ValueAnimator animation) {
                if (animation.getAnimatedFraction() >= 0.25f && mIsHeads) {
                    mFlipImage.setImageBitmap(mTailsImage);
                    mIsHeads = false;
```

```
                }
                if (animation.getAnimatedFraction() >= 0.75f && !mIsHeads) {
                    mFlipImage.setImageBitmap(mHeadsImage);
                    mIsHeads = true;
                }
            }
        });
    }

    @Override
    public boolean onTouchEvent(MotionEvent event) {
        if(event.getAction() == MotionEvent.ACTION_DOWN) {
            mFlipper.start();
            return true;
        }
        return super.onTouchEvent(event);
    }
}
```

在此例中，我们在 XML 中使用 AnimatorInflater 构造 AnimatorSet 对象。产生的动画必须通过 setTarget()附加到适当的目标视图，这是 ObjectAnimator.ofFloat()隐式完成的工作。我们仍然需要 AnimatorUpdateListener 确定何时从头部切换到尾部，但该侦听器不能直接应用于集合对象。因此，我们必须使用 getChildAnimations()查找集合内的旋转动画，以便在适当的位置附加此侦听器。

运行此新的示例将产生更加真实的硬币翻转动画。

1.5 布局变化时的动画

1.5.1 问题

应用程序动态地添加或移除布局中的视图，你希望这种变化能够以动画的形式展示出来。

1.5.2 解决方案

(API Level 11)

使用 LayoutTransition 对象自定义在给定布局中对视图结构修改后的动画效果。在 Android 3.0 及以后版本中，只需要简单地在 XML 中设置 android:animateLayoutChanges 标识或者在 Java 代码中添加一个 LayoutTransition 对象，即可实现任何 ViewGroup 改变布局时的动画效果。

布局中的每个 View 对象在布局变换时有 5 种状态。应用程序可以为下面任何一种状态设置自定义动画：

- APPEARING：容器中出现一个视图。
- DISAPPEARING：容器中消失一个视图。

- CHANGING:布局改变导致某个视图随之改变,例如调整大小,但不包括添加或移除视图。
- CHANGE_APPEARING:其他视图的出现导致某个视图改变。
- CHANGE_DISAPPEARING:其他视图的消失导致某个视图改变。

1.5.3 实现机制

代码清单 1-21 和 1-22 演示了一个应用程序,该应用程序在基本的 LinearLayout 改变时创建动画。

代码清单 1-21 res/layout/main.xml

```xml
<?xml version="1.0" encoding="utf-8"?>
<LinearLayout
    xmlns:android="http://schemas.android.com/apk/res/android"
    android:layout_width="match_parent"
    android:layout_height="match_parent"
    android:gravity="center_horizontal"
    android:orientation="vertical" >

    <Button
        android:id="@+id/button_add"
        android:layout_width="wrap_content"
        android:layout_height="wrap_content"
        android:onClick="onAddClick"
        android:text="Click To Add Item" />

    <LinearLayout
        android:id="@+id/verticalContainer"
        android:layout_width="match_parent"
        android:layout_height="match_parent"
        android:animateLayoutChanges="true"
        android:orientation="vertical" />

</LinearLayout>
```

代码清单 1-22 添加和移除视图的 Activity

```java
public class MainActivity extends Activity {

    LinearLayout mContainer;

    @Override
    public void onCreate(Bundle savedInstanceState) {
        super.onCreate(savedInstanceState);
        setContentView(R.layout.main);
        mContainer =
```

```
            (LinearLayout) findViewById(R.id.verticalContainer);
    }

    //添加可以移除自身的按钮
    public void onAddClick(View v) {
        Button button = new Button(this);
        button.setText("Click To Remove");
        button.setOnClickListener(new View.OnClickListener() {
            @Override
            public void onClick(View v) {
                mContainer.removeView(v);
            }
        });

        mContainer.addView(button, new LinearLayout.LayoutParams(
                LayoutParams.MATCH_PARENT,
                LayoutParams.WRAP_CONTENT));
    }
}
```

这个简单的示例在单击 Add Item 按钮时会在 LinearLayout 中添加 Button 实例。每个新的按钮在单击时都具有将自己从布局中移除的功能。要使此过程动起来，我们只需要在 LinearLayout 上设置 android:animateLayoutChanges="true"，之后框架会进行接下来的工作。默认情况下，新按钮会渐入到新的位置，而不会干扰其他视图；移除时按钮会出现渐出动画，而周围的视图则会平滑地填充移除时产生的空隙。

我们可以为每个过程自定义过渡动画来实现自定义的动画效果。参见代码清单 1-23，会向之前的 Activity 中添加一些自定义过渡动画。

代码清单 1-23　使用了自定义 LayoutTransition 的 Activity

```
public class MainActivity extends Activity {

    LinearLayout mContainer;

    @Override
    public void onCreate(Bundle savedInstanceState) {
        super.onCreate(savedInstanceState);
        setContentView(R.layout.main);

        //布局改变时的动画
        mContainer = (LinearLayout) findViewById(R.id.verticalContainer);
        LayoutTransition transition = new LayoutTransition();
        mContainer.setLayoutTransition(transition);

        //通过翻转进入的动画代替默认的出现动画
        Animator appearAnim = ObjectAnimator.ofFloat(null,
            "rotationY", 90f, 0f).setDuration(
            transition.getDuration(LayoutTransition.APPEARING));
```

```java
        transition.setAnimator(LayoutTransition.APPEARING, appearAnim);

        //通过翻转消失的动画代替默认的消失动画
        Animator disappearAnim = ObjectAnimator.ofFloat(null,
            "rotationX", 0f, 90f).setDuration(
            transition.getDuration(LayoutTransition.DISAPPEARING));
        transition.setAnimator(LayoutTransition.DISAPPEARING, disappearAnim);

        //通过滑动动画代替默认的布局改变时的动画
        //我们需要立即设置一些动画属性，所以创建了多个
        //PropertyValueHolder 对象的动画
        //这个动画会让视图滑动进入并短暂地缩小一半长度
        PropertyValuesHolder pvhSlide =
            PropertyValuesHolder.ofFloat("y", 0, 1);
        PropertyValuesHolder pvhScaleY =
            PropertyValuesHolder.ofFloat("scaleY", 1f, 0.5f, 1f);
        PropertyValuesHolder pvhScaleX =
            PropertyValuesHolder.ofFloat("scaleX", 1f, 0.5f, 1f);
        Animator changingAppearingAnim =
            ObjectAnimator.ofPropertyValuesHolder(
                this, pvhSlide, pvhScaleY, pvhScaleX);
        changingAppearingAnim.setDuration(
            transition.getDuration(LayoutTransition.CHANGE_DISAPPEARING)
        );
        transition.setAnimator(LayoutTransition.CHANGE_DISAPPEARING,
            changingAppearingAnim);
    }

    public void onAddClick(View v) {
        Button button = new Button(this);
        button.setText("Click To Remove");
        button.setOnClickListener(new View.OnClickListener() {
            @Override
            public void onClick(View v) {
                mContainer.removeView(v);
            }
        });

        mContainer.addView(button, new LinearLayout.LayoutParams(
            LayoutParams.MATCH_PARENT, LayoutParams.WRAP_CONTENT));
    }
}
```

本例中，我们在 Button 布局中修改了 APPEARING、DISAPPEARING 和 CHANGE_DISAPPEARING 时的过渡动画。前两个过渡动画会影响到视图添加或移除时的效果。当单击 Add Item 按钮后，新增的按钮会水平旋转进入现有视图中。当单击 Remove 按钮时，该

按钮会在视图中垂直旋转地消失。这两个动画都是通过创建新的 ObjectAnimator 对象并设置自定义旋转属性来实现的，而动画持续的时间则是每种过渡类型的默认时间(通过一个特定过渡类型的键值设置到我们的 LayoutTransition 实例上)。最后一种过渡动画稍微有点复杂，需要创建一个动画，让周围的视图可以平滑地运动到新位置，滑动的同时会产生缩放效果。

注意：
在自定义视图改变的过渡动画时，添加移动到视图新位置的动画非常重要，否则在创建视图或填充视图消失区域时可能会出现闪烁的现象。

为了实现这种效果，需要通过 PropertyValuesHolder 实例创建一个 ObjectAnimator 来设置一些属性。动画的每个属性都是单独的 PropertyValuesHolder，并且通过 ofPropertyValues-Holder()工厂方法添加到 animator 对象中。最后这个过渡动画使 Remove 按钮下面的所有按钮向上滑动到刚刚空出的位置，同时稍微收缩一下。

1.6 实现针对具体场景的布局

1.6.1 问题

应用必须是通用的，能在各种尺寸和方向的屏幕上运行。你需要为各种不同的情况准备好相应的布局资源。

1.6.2 解决方案

(API Level 4)
构建多个布局文件，然后通过资源限定符让 Android 选择合适的布局。下面看看如何使用资源构建特定于不同大小和方向屏幕的布局，你还将学习在多个配置使用同一个布局时如何使用布局别名来减少重复代码。

1.6.3 实现机制

1. 针对不同方向

用下面的限定符可以为 Activity 的横屏和竖屏方向创建不同的资源：
- resource-land
- resource-port

这适用于所有类型的资源，但通常都是用于布局。因此，在项目中不能只有 res/layout/一个目录，而是应该有 res/layout-port/和 res/layout-land/两个目录。

注意：
在实际开发中最好有一个不带标识符的默认资源目录。这样在 Android 设备不能匹配任何给定的设置时也能有可用的资源。

2. 针对不同尺寸

屏幕尺寸标识符(物理尺寸，不要与像素密度混淆)可以用于针对平板电脑等大屏幕设备的资源。大部分情况下，一个布局就能满足所有手机设备的屏幕尺寸，但如果在平板电脑上，就可能需要专门的平板电脑布局来填充用户所面对的大屏幕。

在 Android 3.2 (API Level 13)之前，下面的资源标识符用于标识屏幕的物理尺寸：

- resource-small：屏幕尺寸最小为 426dp×320dp
- resource-medium：屏幕尺寸最小为 470dp×320dp
- resource-large：屏幕尺寸最小为 640dp×480dp
- resource-xlarge：屏幕尺寸最小为 960dp×720dp

随着手持设备和平板电脑中大屏幕设备越来越普及，显然只有 4 种广义的类型不能完全避免资源定义上的重叠。因此，在 Android 3.2 中引入了一个新的基于屏幕真实尺寸(单位为 dp(density-independent pixels))的系统。通过这个新系统，可以对物理屏幕使用以下资源标识符：

- 最小宽度(resource-sw_dp)：屏幕在最短方向上(即方向无关)至少具有指定的 dp。
  - 640dp×480dp 的屏幕通常最小宽度为 480dp。
- 宽度(resource-w_dp)：屏幕在当前水平方向上(即方向无关)至少具有指定的 dp。
  - 640dp×480dp 的屏幕横屏时宽度为 640dp，而竖屏时宽度为 480dp。
- 高度(resource-h_dp)：屏幕在当前垂直方向上(即方向无关)至少具有指定的 dp。
  - 640dp×480dp 的屏幕竖屏时高度为 640dp，而横屏时高度为 480dp。

所以，要想在整个应用程序中包含平板电脑专用的布局，只需要为旧版本的平板电脑添加/res/layout-large/目录，为新版本的平板电脑添加 res/layout-sw720dp/目录即可。

3. 布局别名

在创建通用的应用程序 UI 时，还有最后一个概念需要讨论，那就是布局的别名。通常同一个布局要用于多个不同的设备配置，但在同一个资源目录中使用多个资源限定符(如最小宽度限定符和传统的尺寸限定符)会产生问题。这种情况常常会导致开发人员要在不同的目录中创建同一个布局的多个副本，这维护起来非常困难。

我们可以通过别名解决这个问题。首先在默认资源目录中创建一个布局文件，然后就可以在以"资源-限定符"命名的目录下为这个文件创建多个别名。下面的代码片段演示了 res/layout/main_tablet.xml 文件的别名：

```
<resources>
    <item name="main" type="layout">@layout/main_tablet</item>
</resources>
```

name 属性表示别名的名称，表示这个别名代表的资源会用在特定的配置上。当在代码中使用 R.layout.main 时，这个别名就会链接到 main_tablet.xml 文件。这段代码可以放到 res/values-xlarge/layout.xml 和 res/values-sw720dp/layout.xml 中，这样这两个配置都会链接到同一个布局。

4. 综合示例

接下来看一个综合使用这些技术的简单示例。定义一个 Activity，在其中用代码加载一个布局资源。不过，在该资源中这个布局被定义了三次，分别针对竖屏、横屏和平板电脑产生不同的结果。先看看代码清单 1-24 中的 Activity。

代码清单 1-24　加载一个布局的简单 Activity

```java
public class UniversalActivity extends Activity {

    @Override
    public void onCreate(Bundle savedInstanceState) {
        super.onCreate(savedInstanceState);
        setContentView(R.layout.main);
    }
}
```

现在我们将为该 Activity 的不同配置定义三个单独的布局。代码清单 1-25 到 1-27 显示了默认的布局、横屏时的布局和平板电脑配置的 UI 布局。

代码清单 1-25　res/layout/main.xml

```xml
<?xml version="1.0" encoding="utf-8"?>
<!--默认布局-->
<LinearLayout xmlns:android="http://schemas.android.com/apk/res/android"
    android:layout_width="match_parent"
    android:layout_height="match_parent"
    android:orientation="vertical" >
    <TextView
        android:layout_width="match_parent"
        android:layout_height="wrap_content"
        android:text="This is the default layout" />
    <Button
        android:layout_width="match_parent"
        android:layout_height="wrap_content"
        android:text="Button One" />
    <Button
        android:layout_width="match_parent"
        android:layout_height="wrap_content"
        android:text="Button Two" />
    <Button
        android:layout_width="match_parent"
        android:layout_height="wrap_content"
        android:text="Button Three" />
</LinearLayout>
```

代码清单 1-26　res/layout-land/main.xml

```xml
<?xml version="1.0" encoding="utf-8"?>
<!--横屏时的布局 -->
```

```xml
<LinearLayout xmlns:android="http://schemas.android.com/apk/res/android"
    android:layout_width="fill_parent"
    android:layout_height="fill_parent"
    android:orientation="vertical" >
    <TextView
        android:layout_width="wrap_content"
        android:layout_height="wrap_content"
        android:text="This is a horizontal layout for LANDSCAPE" />
    <!--三个按钮会平均大小地填满整个屏幕 -->
    <LinearLayout
        android:layout_width="match_parent"
        android:layout_height="wrap_content"
        android:orientation="horizontal" >
        <Button
            android:layout_width="0dp"
            android:layout_height="wrap_content"
            android:layout_weight="1"
            android:text="Button One" />
        <Button
            android:layout_width="0dp"
            android:layout_height="wrap_content"
            android:layout_weight="1"
            android:text="Button Two" />
        <Button
            android:layout_width="0dp"
            android:layout_height="wrap_content"
            android:layout_weight="1"
            android:text="Button Three" />
    </LinearLayout>
</LinearLayout>
```

代码清单 1-27 res/layout/main_tablet.xml

```xml
<?xml version="1.0" encoding="utf-8"?>
<!--平板电脑的布局 -->
<LinearLayout xmlns:android="http://schemas.android.com/apk/res/android"
    android:layout_width="match_parent"
    android:layout_height="match_parent"
    android:orientation="horizontal" >
    <!--所有的用户按钮占用25%的屏幕宽度-->
    <LinearLayout
        android:layout_width="0dp"
        android:layout_height="match_parent"
        android:layout_weight="1"
        android:orientation="vertical">
        <TextView
            android:layout_width="match_parent"
            android:layout_height="wrap_content"
            android:text="This is the layout for TABLETS" />
        <Button
            android:layout_width="match_parent"
```

```xml
        android:layout_height="wrap_content"
        android:text="Button One" />
    <Button
        android:layout_width="match_parent"
        android:layout_height="wrap_content"
        android:text="Button Two" />
    <Button
        android:layout_width="match_parent"
        android:layout_height="wrap_content"
        android:text="Button Three" />
    <Button
        android:layout_width="match_parent"
        android:layout_height="wrap_content"
        android:text="Button Four" />
</LinearLayout>

<!--显示详细内容的额外视图 -->
<TextView
    android:layout_width="0dp"
    android:layout_height="match_parent"
    android:layout_weight="3"
    android:text="Detail View"
    android:background="#CCC" />
</LinearLayout>
```

一种方式是创建具有相同文件名的三个文件并把它们放到各自的配置目录中，例如用于横屏的 res/layout-land 目录和用于平板电脑的 res/layout-large 目录。如果每个布局文件只使用一次，这种方案是非常好的，但我们想要在多个配置下重用每个布局，因此在这个示例中采用了为这三个布局创建限定别名的方式。代码清单 1-28 到 1-31 显示了如何将每个布局链接到正确的配置。

代码清单 1-28 res/values-large-land/layout.xml

```xml
<?xml version="1.0" encoding="utf-8"?>
<resources
    xmlns:android="http://schemas.android.com/apk/res/android">
    <item name="main" type="layout">@layout/main_tablet</item>
</resources>
```

代码清单 1-29 res/value-sw600dp-land/layout.xml

```xml
<?xml version="1.0" encoding="utf-8"?>
<resources
    xmlns:android="http://schemas.android.com/apk/res/android">
    <item name="main" type="layout">@layout/main_tablet</item>
</resources>
```

代码清单 1-30 res/values-xlarge/layout.xml

```xml
<?xml version="1.0" encoding="utf-8"?>
<resources
```

```
    xmlns:android="http://schemas.android.com/apk/res/android">
    <item name="main" type="layout">@layout/main_tablet</item>
</resources>
```

代码清单1-31   res/values-sw720dp/layout.xml

```
<?xml version="1.0" encoding="utf-8"?>
<resources
    xmlns:android="http://schemas.android.com/apk/res/android">
    <item name="main" type="layout">@layout/main_tablet</item>
</resources>
```

我们定义了一组配置来适应三种类型的设备：手机、7英寸平板设备、10英寸平板设备。手机设备会在竖屏模式时加载默认的布局，在屏幕旋转时加载横屏布局。因为只有一种配置使用这些文件，所以文件会被直接分别放置到res/layout和res/layout-land目录下。

7英寸的平板设备在之前的尺寸方案中通常会被定义为大屏幕设备，而在新的方案中它们是最小宽度为大约600dp的设备。竖屏时，我们决定让应用程序都使用默认的布局，而在横屏时会得到更大的空间，因此我们改为加载平板电脑的布局。要想实现该功能，我们为横屏模式创建了多个配置目录来匹配设备的类型。同时使用了最小宽度限定符和广义的尺寸限定符，这样就可以兼容旧平板电脑和新平板电脑。

10英寸的平板设备在之前的尺寸方案中通常会被定义为超大屏幕设备，而在新的方案中它们是最小宽度为大约720dp的设备。对于这些设备，屏幕大到可以在两个方向都使用平板电脑布局，因此只通过屏幕尺寸定义了配置目录。和小平板电脑一样，同时使用最小宽度和广义尺寸限定符来保证可以兼容所有的平板电脑版本。

在所有引用平板电脑布局的场景中，我们只需要创建一个布局文件即可管理，这多亏使用了别名。现在，运行应用程序后，会看到Android是如何选择合适的布局来匹配我们的配置的。图1-3显示了手机上默认的布局和横屏时的布局。

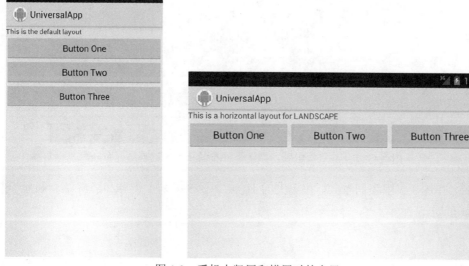

图1-3   手机上竖屏和横屏时的布局

同样的应用程序运行在 7 英寸的平板设备上会在竖屏时显示默认的布局，但在横屏上会显示全屏的平板电脑布局(参见图 1-4)。

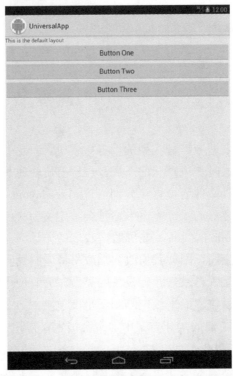

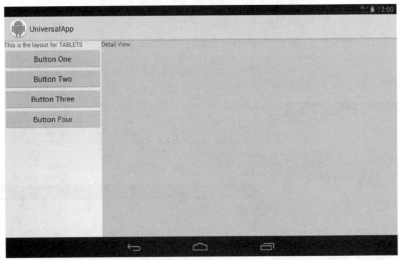

图 1-4　7 英寸平板电脑：默认竖屏布局和平板电脑横屏布局

最后，在图 1-5 中我们看到，在更大的 10 英寸平板电脑上横屏和竖屏都会显示全屏的平板电脑布局。

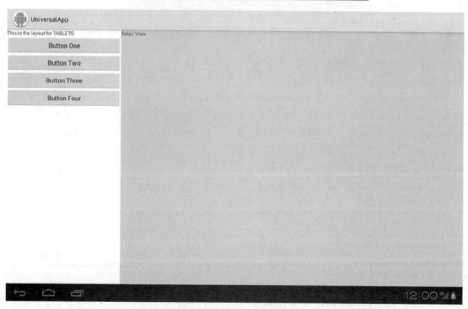

图 1-5　10 英寸平板电脑：两个方向都显示全屏平板电脑布局

通过 Android 资源选择系统的强大能力，大大减少了为不同设备类型提供不同的优化 UI 布局的难度。

1.7 自定义 AdapterView 的空视图

1.7.1 问题

要在 AdapterView(ListView、GridView 等诸如此类的视图)没有数据时显示自定义的视图。

1.7.2 解决方案

(API Level 1)

把要显示的视图跟 AdapterView 放在同一个布局树中,然后调用 AdapterView.setEmptyView()自行处理。AdapterView 会根据其中 ListAdapter 的 isEmpty()方法的返回值选择显示其自身还是显示空视图。

重点:

一定要将 AdapterView 和空视图放入布局中,AdapterView 仅仅只是变换这两个对象是否可见的参数,而绝对不会在布局树中插入或删除某个视图。

1.7.3 实现机制

下面将一个简单的 TextView 用作空视图。首先,在布局中放入这两个视图,参见代码清单 1-32。

代码清单 1-32　res/layout/empty.xml

```xml
<?xml version="1.0" encoding="utf-8"?>
<FrameLayout
    xmlns:android="http://schemas.android.com/apk/res/android"
    android:layout_width="fill_parent"
    android:layout_height="fill_parent">
    <TextView
        android:id="@+id/myempty"
        android:layout_width="fill_parent"
        android:layout_height="wrap_content"
        android:text="No Items to Display"
    />
    <ListView
        android:id="@+id/mylist"
        android:layout_width="fill_parent"
        android:layout_height="fill_parent"
    />
</FrameLayout>
```

然后在 Activity 中将空视图的引用提供给 ListView，让其进行管理(参见代码清单 1-33)。

代码清单 1-33　将空视图链接到列表的 Activity

```java
public void onCreate(Bundle savedInstanceState) {
    super.onCreate(savedInstanceState);
    ListView list = (ListView)findViewById(R.id.mylist);
    TextView empty = (TextView)findViewById(R.id.myempty);
    /*
     * 附加空视图。
     * 框架将在ListView的Adapter没有元素时显示此视图。
     */
    list.setEmptyView(empty);

    //继续给列表添加 Adapter 和数据

}
```

让空视图更有趣

空视图不一定非得是简单无趣的 TextView。接下来做点儿对用户更有用的工作，并且在列表为空时添加 Refresh 按钮(参加代码清单 1-34)。

代码清单 1-34　交互式空布局

```xml
<?xml version="1.0" encoding="utf-8"?>
<FrameLayout
    xmlns:android="http://schemas.android.com/apk/res/android"
    android:layout_width="match_parent"
    android:layout_height="match_parent">
    <LinearLayout
        android:id="@+id/myempty"
        android:layout_width="match_parent"
        android:layout_height="wrap_content"
        android:orientation="vertical">
        <TextView
            android:layout_width="match_parent"
            android:layout_height="wrap_content"
            android:text="No Items to Display"
        />
        <Button
            android:layout_width="match_parent"
            android:layout_height="wrap_content"
            android:text="Tap Here to Refresh"
        />
    </LinearLayout>
    <ListView
        android:id="@+id/mylist"
        android:layout_width="match_parent"
        android:layout_height="match_parent"
```

```
    />
</FrameLayout>
```

现在，一样是用前面的 Activity 代码，将一个完整的布局设为空视图，让用户可以对空数据进行一些操作。

1.8 自定义 ListView 中的行

1.8.1 问题

应用程序需要自定义 ListView 中各行的外观。

1.8.2 解决方案

(API Level 1)
创建一个自定义的 XML 布局，将其传递给某个常见的适配器，或是扩展你自己的适配器，然后用自定义的状态 Drawable 覆盖背景和选中状态下的行。

1.8.3 实现机制

1. 简单的自定义

如果需要简单，那么创建一个布局，连接到已有的 ListAdapter 进行填充；我们以 ArrayAdapter 为例加以介绍。ArrayAdapter 接受的参数是自定义的布局资源和用于使用数据填充该布局的 TextView 控件的 ID。让我们创建一些用作背景的自定义 Drawable 和一个满足这些要求的布局(参见代码清单 1-35 到代码清单 1-37)。

代码清单 1-35　res/drawable/row_background_default.xml

```xml
<?xml version="1.0" encoding="utf-8"?>
<shape xmlns:android="http://schemas.android.com/apk/res/android"
    android:shape="rectangle">
    <gradient
        android:startColor="#EFEFEF"
        android:endColor="#989898"
        android:type="linear"
        android:angle="270"
    />
</shape>
```

代码清单 1-36　res/drawable/row_background_pressed.xml

```xml
<?xml version="1.0" encoding="utf-8"?>
<shape xmlns:android="http://schemas.android.com/apk/res/android"
    android:shape="rectangle">
    <gradient
```

```xml
        android:startColor="#0B8CF2"
        android:endColor="#0661E5"
        android:type="linear"
        android:angle="270"
    />
</shape>
```

代码清单 1-37　res/drawable/row_background.xml

```xml
<?xml version="1.0" encoding="utf-8"?>
<selector
    xmlns:android="http://schemas.android.com/apk/res/android">
    <item android:state_pressed="true"
        android:drawable="@drawable/row_background_pressed"/>
    <item android:drawable="@drawable/row_background_default"/>
</selector>
```

代码清单 1-38 是一个自定义布局，其中的文本是居中排列的，而不是左对齐的。

代码清单 1-38　res/layout/custom_row.xml

```xml
<?xml version="1.0" encoding="utf-8"?>
<LinearLayout
    xmlns:android="http://schemas.android.com/apk/res/android"
    android:layout_width="match_parent"
    android:layout_height="wrap_content"
    android:padding="10dip"
    android:background="@drawable/row_background">
    <TextView
        android:id="@+id/line1"
        android:layout_width="wrap_content"
        android:layout_height="wrap_content"
        android:layout_gravity="center"
    />
</LinearLayout>
```

这个布局的背景是自定义的渐变状态列表，在这个列表中为每一行设置了默认状态和按下状态的外观。我们已经定义好一个能与 **ArrayAdapter** 所期望的行为契合的布局，现在可以创建一个 Activity，调用这个列表，不必再进行其他任何自定义(参见代码清单 1-39)。

代码清单 1-39　使用了自定义行布局的 Activity

```java
public void onCreate(Bundle savedInstanceState) {
    super.onCreate(savedInstanceState);
    ListView list = new ListView(this);
    ArrayAdapter<String> adapter = new ArrayAdapter<String>(this,
            R.layout.custom_row,
            R.id.line1,
            new String[] {"Bill","Tom","Sally","Jenny"});
    list.setAdapter(adapter);
```

```
        setContentView(list);
}
```

2. 适应更复杂的选项

有时自定义列表中的行意味着要扩展 ListAdapter。常见的情况是要在一行中放入多项数据或是有些数据并不是文本。在这个示例中，我们再次用自定义的 Drawable 作为背景，不过这里的布局会变得更有意思(参见代码清单 1-40)。

代码清单 1-40　修改后的 res/layout/custom_row.xml

```xml
<?xml version="1.0" encoding="utf-8"?>
<RelativeLayout
    xmlns:android="http://schemas.android.com/apk/res/android"
    android:layout_width="match_parent"
    android:layout_height="wrap_content"
    android:orientation="horizontal"
    android:padding="10dip">
    <ImageView
        android:id="@+id/leftimage"
        android:layout_width="32dip"
        android:layout_height="32dip"
    />
    <ImageView
        android:id="@+id/rightimage"
        android:layout_width="32dip"
        android:layout_height="32dip"
        android:layout_alignParentRight="true"
    />

    <TextView
        android:id="@+id/line1"
        android:layout_width="match_parent"
        android:layout_height="wrap_content"
        android:layout_toLeftOf="@id/rightimage"
        android:layout_toRightOf="@id/leftimage"
        android:layout_centerVertical="true"
        android:gravity="center_horizontal"
    />
</RelativeLayout>
```

这个布局中同样有一个居中对齐的 TextView，但它的每条边都是一个 ImageView。要将这个布局应用到 ListView，我们需要扩展 SDK 中的某个 ListAdapter。具体扩展哪个 ListAdapter 取决于要在列表中呈现的数据源。如果数据只是简单的字符串数组，扩展 ArrayAdapter 就足够了。如果数据更复杂些，就有可能需要扩展抽象的 BaseAdapter。需要扩展的方法只有 getView()，这个方法负责控制列表中每一行的显示方式。

在这个示例中，数据只是简单的字符串数组，所以只需要简单扩展 ArrayAdapter 即可（参见代码清单 1-41）。

代码清单 1-41　显示新布局的 Activity 和自定义的 ListAdapter

```java
public class MyActivity extends Activity {

    public void onCreate(Bundle savedInstanceState) {
        super.onCreate(savedInstanceState);
        ListView list = new ListView(this);
        setContentView(list);

        CustomAdapter adapter = new CustomAdapter(this,
                R.layout.custom_row,
                R.id.line1,
                new String[] {"Bill","Tom","Sally","Jenny"});
        list.setAdapter(adapter);

    }

    private static class CustomAdapter extends ArrayAdapter<String> {

        public CustomAdapter(Context context, int layout, int resId,
                String[] items) {
            //调用 ArrayAdapter 的实现
            super(context, layout, resId, items);
        }

        @Override
        public View getView(int position, View convertView,
                ViewGroup parent) {
            View row = convertView;
            //如果某行没有被回收的话，填充该行
            if(row == null) {
                row = LayoutInflater.from(getContext())
                        .inflate(R.layout.custom_row, parent, false);
            }
            String item = getItem(position);
            ImageView left =
                    (ImageView)row.findViewById(R.id.leftimage);
            ImageView right =
                    (ImageView)row.findViewById(R.id.rightimage);
            TextView text = (TextView)row.findViewById(R.id.line1);

            left.setImageResource(R.drawable.icon);
            right.setImageResource(R.drawable.icon);
            text.setText(item);
```

```
            return row;
        }
    }
}
```

注意，这里用来创建适配器实例的构造函数与前面使用的构造函数相同，因为它们都继承自 ArrayAdapter。因为重载了适配器的视图显示机制，所以在这里将 R.layout.custom_row 和 R.id.line1 传递给构造函数只是因为这是构造函数的必要参数；在这个示例中，这两个参数是不起作用的。

现在，当 ListView 要显示一行内容时，就会调用其适配器的 getView()方法。我们已经对这个方法进行了自定义，以便能控制每一行内容的返回方式。getView()方法有一个名为 convertView 的参数，这个参数对性能有重要影响。将 XML 文件转换为布局是一个代价很高的操作，为了尽量减少其对系统的影响，在滚动时 ListView 会回收视图。如果回收的视图还能重用，就作为 convertView 传递给 getView()方法。所以，为了确保列表滚动的响应速度，要尽可能重用视图，而不是生成新的视图。

在这个示例中，调用 getItem()得到列表(字符串数组)中位置的当前值，然后将 TextView 设为这个值并应用于该行。我们还可以给每行设置图片来凸显数据，不过这里只使用了应用程序的图标。

1.9 制作 ListView 的节头部

1.9.1 问题

需要创建一个有若干节内容的列表，其中每一节的顶部都有各自的头部。

1.9.2 解决方案

(API Level 1)

我们可以通过构建自定义列表适配器来实现此效果，此适配器利用了对多种视图类型的支持。适配器依赖 getViewTypeCount()和 getItemViewType()来确定将多少种视图用作列表中的行。在大多数情况下，如果所有的行类型相同，就忽略上述方法。然而，在此可以使用这些回调为头部行和内容行定义独特的类型。

1.9.3 实现机制

图 1-6 显示了带有节头部的示例列表的预览效果。

图 1-6 分节的列表

我们首先在代码清单 1-42 中定义 SectionItem 数据结构，用于表示列表中的每一节。此项将保存节标题以及此标题下列出的数据数组子集。

代码清单 1-42　包含每一节数据的结构

```java
public class SectionItem<T> {
    private String mTitle;
    private T[] mItems;

    public SectionItem(String title, T[] items) {
        if (title == null) title = "";

        mTitle = title;
        mItems = items;
    }

    public String getTitle() {
        return mTitle;
    }

    public T getItem(int position) {
        return mItems[position];
    }

    public int getCount() {
        //为节标题包含额外的项
        return (mItems == null ? 1 : 1 + mItems.length);
```

```java
        }

        @Override
        public boolean equals(Object object) {
            //如果两个节有相同的标题,则它们相等
            if (object != null && object instanceof SectionItem) {
                return ((SectionItem) object).getTitle().equals(mTitle);
            }

            return false;
        }
    }
```

此结构将使列表适配器中的逻辑更易于管理。在代码清单 1-43 中,我们可以看到提供分节列表视图的适配器。此适配器的任务是将每个节项(包括它们的头部)的位置映射到适配器视图(如 ListView)所了解的可见列表的全局位置。

代码清单 1-43 用于显示多个节的 ListAdapter

```java
public abstract class SimpleSectionsAdapter<T> extends BaseAdapter implements
        AdapterView.OnItemClickListener {

    /*为每个视图类型定义常量*/
    private static final int TYPE_HEADER = 0;
    private static final int TYPE_ITEM = 1;

    private LayoutInflater mLayoutInflater;
    private int mHeaderResource;
    private int mItemResource;

    /*所有节的唯一集合*/
    private List<SectionItem<T>> mSections;
    /*节的分组,按其初始位置设置键*/
    private SparseArray<SectionItem<T>> mKeyedSections;

    public SimpleSectionsAdapter(ListView parent, int headerResId, int itemResId) {
        mLayoutInflater = LayoutInflater.from(parent.getContext());
        mHeaderResource = headerResId;
        mItemResource = itemResId;

        //创建包含自动排序键的集合
        mSections = new ArrayList<SectionItem<T>>();
        mKeyedSections = new SparseArray<SectionItem<T>>();

        //将自身附加为列表的单击处理程序
        parent.setOnItemClickListener(this);
    }

    /*
```

```java
 * 向列表添加新的带标题的节，
 * 或者更新现有的节
 */
public void addSection(String title, T[] items) {
    SectionItem<T> sectionItem = new SectionItem<T>(title, items);
    //添加节，替换具有相同标题的现有节
    int currentIndex = mSections.indexOf(sectionItem);
    if (currentIndex >= 0) {
        mSections.remove(sectionItem);
        mSections.add(currentIndex, sectionItem);
    } else {
        mSections.add(sectionItem);
    }

    //对最新的集合排序
    reorderSections();
    //表明视图数据已改变
    notifyDataSetChanged();
}

/*
 *将带有初始全局位置的节标记为可引用的键
 */
private void reorderSections() {
    mKeyedSections.clear();
    int startPosition = 0;
    for (SectionItem<T> item : mSections) {
        mKeyedSections.put(startPosition, item);
        //此计数包括头部视图
        startPosition += item.getCount();
    }
}

@Override
public int getCount() {
    int count = 0;
    for (SectionItem<T> item : mSections) {
        //添加项的计数
        count += item.getCount();
    }

    return count;
}

@Override
public int getViewTypeCount() {
    //两种视图类型：头部和项
    return 2;
}
```

```java
    @Override
    public int getItemViewType(int position) {
        if (isHeaderAtPosition(position)) {
            return TYPE_HEADER;
        } else {
            return TYPE_ITEM;
        }
    }

    @Override
    public T getItem(int position) {
        return findSectionItemAtPosition(position);
    }

    @Override
    public long getItemId(int position) {
        return position;
    }

    /*
     *重写并返回false，告诉ListView有一些项(头部)不可点击
     */
    @Override
    public boolean areAllItemsEnabled() {
        return false;
    }

    /*
     *重写以告诉ListView哪些项(头部)是不可点击的
     */
    @Override
    public boolean isEnabled(int position) {
        return !isHeaderAtPosition(position);
    }

    @Override
    public View getView(int position, View convertView, ViewGroup parent) {
        switch (getItemViewType(position)) {
            case TYPE_HEADER:
                return getHeaderView(position, convertView, parent);
            case TYPE_ITEM:
                return getItemView(position, convertView, parent);
            default:
                return convertView;
        }
    }

    private View getHeaderView(int position, View convertView, ViewGroup parent) {
        if (convertView == null) {
```

```java
            convertView = mLayoutInflater.inflate(mHeaderResource, parent,
               false);
        }

        SectionItem<T> item = mKeyedSections.get(position);
        TextView textView = (TextView) convertView.
           findViewById(android.R.id.text1);

        textView.setText(item.getTitle());

        return convertView;
    }

    private View getItemView(int position, View convertView, ViewGroup
parent) {
        if (convertView == null) {
            convertView = mLayoutInflater.inflate(mItemResource, parent,
               false);
        }

        T item = findSectionItemAtPosition(position);
        TextView textView = (TextView) convertView.
            findViewById(android.R.id.text1);

        textView.setText(item.toString());

        return convertView;
    }

    /** OnItemClickListener 方法 */

    @Override
    public void onItemClick(AdapterView<?> parent, View view, int position,
long id) {
        T item = findSectionItemAtPosition(position);
        if (item != null) {
            onSectionItemClick(item);
        }
    }

    /**
     * 重写方法以处理特定元素的单击事件，即用户单击的@param 项列表项
     */
    public abstract void onSectionItemClick(T item);

    /* 用于将项映射到节的辅助方法 */

    /*
     * 检查是否是代表节标题的全局位置值
     */
```

```java
        private boolean isHeaderAtPosition(int position) {
            for (int i=0; i < mKeyedSections.size(); i++) {
                //如果此位置是键值,则它就是头部位置
                if (position == mKeyedSections.keyAt(i)) {
                    return true;
                }
            }

            return false;
        }

        /*
         * 返回给定全局位置的显式列表项
         */
        private T findSectionItemAtPosition(int position) {
            int firstIndex, lastIndex;
            for (int i=0; i < mKeyedSections.size(); i++) {
                firstIndex = mKeyedSections.keyAt(i);
                lastIndex = firstIndex + mKeyedSections.valueAt(i).getCount();
                if (position >= firstIndex && position < lastIndex) {
                    int sectionPosition = position - firstIndex - 1;
                    return mKeyedSections.valueAt(i).getItem(sectionPosition);
                }
            }

            return null;
        }
    }
```

SimpleSectionsAdapter 通过 getViewTypeCount()返回 2,这样我们就可以分别支持头部视图和内容视图。我们需要为每个视图提供独特的类型标识符,内部列表将这些值用作索引,因此它们应始终从 0 开始并递增,从 TYPE_HEADER 和 TYPE_ITEM 常量中可以看出这一点。

我们通过 addSection()方法将数据按节提供给此适配器,该方法接受的参数是节标题和此部分的项的数组。适配器使每个节标题仅出现一次,因此每次尝试添加新的节都会删除具有相同标题的现有节。

添加新的节时,列举所有的节以确定它们的全局开始位置是否在列表内。为了在后面进行更快速的访问,在 SparseArray 中将这些值存储为键。此集合将用于搜索给定位置的节。最后,每个节改动需要我们调用 notifyDataSetChanged(),该方法告诉视图需要再次查询适配器并刷新显示。

通过对每个节的计数求和来确定所有项的计数。这包括头部视图,列表将其与其他行同等对待。这在传统上意味着这些项也是可交互的,会将单击事件传送给附加的侦听器。我们不需要头部的此行为,因此还必须重写 areAllItemsEnabled()和 isEnabled()方法,向视图表明这些头部应是不可交互的。

在 getView()方法内部,我们传递一个视图类型,该视图类型用于确定应返回何种类型的视图。我们知道,对于 TYPE_HEADER 需要返回设置了标题文本的头部行。对于

TYPE_ITEM，我们返回显示正确项的内容行。Android 框架通过实现 getItemViewType()了解在每个位置应存在何种视图类型，该方法将位置与类型标识符进行关联。我们创建了简单的 isHeaderAtPosition()辅助方法来做出决定。通过检查给定位置是否与某个节键匹配(这使其成为节中的第一个位置，即头部)，我们可以快速确定类型。

获得每个项的视图之后，我们必须确定给定位置的节。另一个辅助方法 findSectionItemAtPosition()通过根据前面计算出来的节键验证位置来执行此搜索。在用户点击项时，我们也利用此方法将项自身返回给侦听器，而不是仅返回位置值。在简单列表中，可能只需要位置就能找到正确的数据，但采用位置映射之后，侦听器就可以更轻松地直接获得所需的项。

我们已将此适配器定义为抽象类型，这只需要应用程序实现为列表项单击事件提供处理程序。代码清单 1-44 和 1-45 显示了如何将此适配器与数据绑定并显示在 Activity 中。

代码清单 1-44   res/layout/list_header.xml

```xml
<?xml version="1.0" encoding="utf-8"?>
<LinearLayout xmlns:android="http://schemas.android.com/apk/res/android"
    android:layout_width="match_parent"
    android:layout_height="wrap_content"
    android:padding="8dp"
    android:background="#CCF">
    <TextView
        android:id="@android:id/text1"
        android:layout_width="wrap_content"
        android:layout_height="wrap_content"/>
</LinearLayout>
```

代码清单 1-45   显示分节列表的 Activity

```java
public class SectionsActivity extends ActionBarActivity {

    @Override
    public void onCreate(Bundle savedInstanceState) {
        super.onCreate(savedInstanceState);
        ListView list = new ListView(this);

        SimpleSectionsAdapter<String> adapter = new
          SimpleSectionsAdapter<String>(
                list, /* 资源扩充的上下文 */
                R.layout.list_header, /* 头部视图的布局 */
                android.R.layout.simple_list_item_1 /* Layout for item views */
        ) {
            //适用于项点击的单击处理程序
            @Override
            public void onSectionItemClick(String item) {
                Toast.makeText(SectionsActivity.this, item,
                    Toast.LENGTH_SHORT).show();
            }
        };
```

```
        adapter.addSection("Fruits", new String[]{"Apples", "Oranges",
"Bananas", "Mangos"});
        adapter.addSection("Vegetables", new String[]{"Carrots", "Peas",
"Broccoli"});
        adapter.addSection("Meats", new String[]{"Pork", "Chicken", "Beef",
"Lamb"});

        list.setAdapter(adapter);
        setContentView(list);
    }
}
```

在 Activity 内部，建立 ListView 以显示 3 个节：水果、蔬菜和肉类。SimpleSectionsAdapter 获取布局的两个资源 ID，适配器应该为头部和内容扩充这些布局。此例中的头部布局是居中显示单个 TextView 的自定义布局，而内容布局是框架中的标准布局(android.R.id.simple_list_item_1)。因为适配器是抽象的，我们需要提供 onSectionClick()的定义，在本例中，该定义仅在 Toast 中显示所选项的名称。

1.10 创建组合控件

1.10.1 问题

需要通过组合现有的元素来创建自定义的小部件。

1.10.2 解决方案

(API Level 1)

通过扩展通用的 ViewGroup 并添加所需的功能就能创建自定义的小部件。创建自定义小部件或可重用用户界面元素的最简单、最实用的方法就是利用 Android SDK 提供的现有小部件来创建组合控件。

1.10.3 实现机制

ViewGroup 及其子类 LinearLayout、RelativeLayout 等能帮助你摆放控件的位置，这样你就可以专注于添加所需的功能。

TextImageButton

下面将创建一个 Android SDK 中没有原生提供的小部件：含有图片或文字的按钮。为了实现这个小部件，我们创建 TextImageButton 类，这是对 FrameLayout 的扩展。其中包含一个用于放置文本内容的 TextView，以及一个用于放置图片内容的 ImageView(参见代码清单 1-46)。

代码清单 1-46　自定义 TextImageButton 小部件

```java
public class TextImageButton extends FrameLayout {

    private ImageView imageView;
    private TextView textView;
    /* 构造函数 */
    public TextImageButton(Context context) {
        this(context, null);
    }

    public TextImageButton(Context context, AttributeSet attrs) {
        this(context, attrs, 0);
    }

    public TextImageButton(Context context, AttributeSet attrs, int defaultStyle) {
        //通过系统的按钮样式初始化父布局
        //这样会设置clickable属性和按钮背景来匹配当前的主题
        super(context, attrs, android.R.attr.buttonStyle);
        //创建子视图
        imageView = new ImageView(context, attrs, defaultStyle);
        textView = new TextView(context, attrs, defaultStyle);
        //创建子视图的LayoutParams 为 WRAP_CONTENT 并居中显示
        FrameLayout.LayoutParams params = new FrameLayout.LayoutParams(
                LayoutParams.WRAP_CONTENT,
                LayoutParams.WRAP_CONTENT,
                Gravity.CENTER);
        //添加视图
        this.addView(imageView, params);
        this.addView(textView, params);

        //如果有图片，切换到图片模式
        if(imageView.getDrawable() != null) {
            textView.setVisibility(View.GONE);
            imageView.setVisibility(View.VISIBLE);
        } else {
            textView.setVisibility(View.VISIBLE);
            imageView.setVisibility(View.GONE);
        }
    }

    /*存取器*/
    public void setText(CharSequence text) {
        //切换到文字模式
        textView.setVisibility(View.VISIBLE);
        imageView.setVisibility(View.GONE);
        //设置文字
        textView.setText(text);
    }
```

```java
public void setImageResource(int resId) {
    //切换到图片模式
    textView.setVisibility(View.GONE);
    imageView.setVisibility(View.VISIBLE);
    //设置图片
    imageView.setImageResource(resId);
}

public void setImageDrawable(Drawable drawable) {
    //切换到图片模式
    textView.setVisibility(View.GONE);
    imageView.setVisibility(View.VISIBLE);
    //设置图片
    imageView.setImageDrawable(drawable);
}
}
```

SDK 中所有的小部件都有至少两个(通常为 3 个)构造函数。第一个构造函数的参数只有一个 Context，通常用于在代码中新建视图。另外两个构造函数用于将 XML 转换为视图，XML 文件中定义的属性会传递给 AttributeSet 参数。这里我们用 Java 的 this()符号调用实际完成所有工作的函数，从而实现了这两个构造函数。以这种方式构建自定义控件就确保了我们能在 XML 布局中定义此视图。如果不实现这两个属性化的构造函数(attributed constructor)，就不能在 XML 布局中使用自定义的控件。

为了让 FrameLayout 看起来更像一个标准的按钮，我们在构造函数中传入了 android.R.attr.buttonStyle 属性。这里定义了和当前主题匹配的样式并设置到了视图上。这不仅设置了背景来匹配其他的按钮实例，还让视图变得可单击和可获得焦点(因为这些样式就是系统样式的一部分)。只要有可能，都应该从当前主题中加载自定义小部件的外观，从而可轻松定制并与应用程序的其他部分保持统一。

构造函数还创建了一个 TextView 和一个 ImageView，将它们放入布局中。向每个子构造函数传递相同的属性集，从而可以正确读取特定于每个构造函数设置的 XML 属性(例如文字和图片状态)。其他的代码根据属性中传递的数据设置默认的显示模式(文字或图片)。

加入存取器函数是为了方便以后改变按钮的内容。这些函数还负责在内容变化时在文字和图片模式间进行切换。

因为这个自定义的控件并没有在 android.view 或 android.widget 包中，所以在 XML 布局中使用该控件必须使用全名。代码清单 1-47 和 1-48 演示了含有该自定义小部件的 Activity。

代码清单 1-47　res/layout/main.xml

```xml
<?xml version="1.0" encoding="utf-8"?>
<LinearLayout
    xmlns:android="http://schemas.android.com/apk/res/android"
    android:layout_width="match_parent"
    android:layout_height="match_parent"
```

```
    android:orientation="vertical" >
<com.examples.customwidgets.TextImageButton
    android:layout_width="match_parent"
    android:layout_height="wrap_content"
    android:text="Click Me!"
    android:textColor="#000" />
<com.examples.customwidgets.TextImageButton
    android:layout_width="match_parent"
    android:layout_height="wrap_content"
    android:src="@drawable/ic_launcher" />
</LinearLayout>
```

代码清单 1-48　使用了新的自定义小部件的 Activity

```
public class MyActivity extends Activity {

    @Override
    public void onCreate(Bundle savedInstanceState) {
        super.onCreate(savedInstanceState);
        setContentView(R.layout.main);
    }
}
```

注意，我们还是可以使用传统的属性来设置要显示的文字或图片。这是因为我们用属性化构造函数来构造各个元素(FrameLayout、TextView 和 ImageView)，所以每个视图都会根据自己的需求设置参数，忽略其他参数。

如果我们定义使用该布局的 Activity，效果如图 1-7 所示。

图 1-7　同时显示文字和图片的 TextImageButton

1.11 自定义过渡动画

1.11.1 问题

应用程序需要自定义 Activity 切换或 Fragment 切换时产生的过渡动画。

1.11.2 解决方案

(API Level 5)

要修改 Activity 间的过渡动画，可以使用 overridePendingTransition() API 进行某次切换时的动画，或者在应用程序的主题中声明自定义动画值来进行更多全局设置。要修改 Fragment 间的过渡动画，可以使用 onCreateAnimation()或 onCreateAnimator() API 方法。

1.11.3 实现机制

1. Activity

要自定义 Activity 切换时的过渡动画，可以考虑 4 种动画：打开一个新 Activity 时的进入动画和退出动画，以及当前 Activity 关闭时的进入动画和退出动画。每种动画都会应用到过渡动画中所涉及的两个 Activity 之一。例如，当打开一个新的 Activity 时，当前 Activity 将会运行"打开退出"动画，而新 Activity 会运行"打开进入"动画。由于这些动画都是同时运行的，因此动画间应该是互补的，否则看起来会不太协调。代码清单 1-49 到 1-52 演示了这 4 种动画。

代码清单 1-49　res/anim/activity_open_enter.xml

```xml
<?xml version="1.0" encoding="utf-8"?>
<set xmlns:android="http://schemas.android.com/apk/res/android">
    <rotate
        android:fromDegrees="90" android:toDegrees="0"
        android:pivotX="0%" android:pivotY="0%"
        android:fillEnabled="true"
        android:fillBefore="true" android:fillAfter="true"
        android:duration="500" />
    <alpha
        android:fromAlpha="0.0" android:toAlpha="1.0"
        android:fillEnabled="true"
        android:fillBefore="true" android:fillAfter="true"
        android:duration="500" />
</set>
```

代码清单 1-50　res/anim/activity_open_exit.xml

```xml
<?xml version="1.0" encoding="utf-8"?>
<set xmlns:android="http://schemas.android.com/apk/res/android">
```

```xml
    <rotate
        android:fromDegrees="0" android:toDegrees="-90"
        android:pivotX="0%" android:pivotY="0%"
        android:fillEnabled="true"
        android:fillBefore="true" android:fillAfter="true"
        android:duration="500" />
    <alpha
        android:fromAlpha="1.0" android:toAlpha="0.0"
        android:fillEnabled="true"
        android:fillBefore="true" android:fillAfter="true"
        android:duration="500" />
</set>
```

代码清单 1-51　res/anim/activity_close_enter.xml

```xml
<?xml version="1.0" encoding="utf-8"?>
<set xmlns:android="http://schemas.android.com/apk/res/android">
    <rotate
        android:fromDegrees="-90" android:toDegrees="0"
        android:pivotX="0%p" android:pivotY="0%p"
        android:fillEnabled="true"
        android:fillBefore="true" android:fillAfter="true"
        android:duration="500" />
    <alpha
        android:fromAlpha="0.0" android:toAlpha="1.0"
        android:fillEnabled="true"
        android:fillBefore="true" android:fillAfter="true"
        android:duration="500" />
</set>
```

代码清单 1-52　res/anim/activity_close_exit.xml

```xml
<?xml version="1.0" encoding="utf-8"?>
<set xmlns:android="http://schemas.android.com/apk/res/android" >
    <rotate
        android:fromDegrees="0" android:toDegrees="90"
        android:pivotX="0%p" android:pivotY="0%p"
        android:fillEnabled="true"
        android:fillBefore="true" android:fillAfter="true"
        android:duration="500" />
    <alpha
        android:fromAlpha="1.0" android:toAlpha="0.0"
        android:fillEnabled="true"
        android:fillBefore="true" android:fillAfter="true"
        android:duration="500" />
</set>
```

我们创建了两个"打开"动画，即旧的 Activity 顺时针旋转消失、新的 Activity 顺时针旋转进入。补足的"关闭"动画会将当前 Activity 逆时针旋转退出、之前的 Activity 逆时针旋转进入。每个动画还有渐出或渐入的效果，这样过渡动画看起来会更加流畅。

要在特定时刻应用这些自定义动画,可以在 startActivity() 或 finish() 后立刻调用 overridePendingTransition()方法,如下所示:

```
//使用自定义过渡动画启动一个新的Activity
Intent intent = new Intent(...);
startActivity(intent);
overridePendingTransition(R.anim.activity_open_enter,
    R.anim.activity_open_exit);

//使用自定义过渡动画关闭当前Activity
finish();
overridePendingTransition(R.anim.activity_close_enter,
    R.anim.activity_close_exit);
```

这种方式在只希望在某些场合使用自定义过渡动画的情况下非常有用。但如果希望在应用程序中自定义每个 Activity 的过渡动画,到处调用这个方法可能会有点麻烦。反之,最好在应用程序的主题中使用自定义的动画。代码清单 1-53 演示了一个自定义的主题,该主题可以全局使用这些过渡动画。

代码清单 1-53　res/values/styles.xml

```xml
<resources>
    <style name="AppTheme" parent="android:Theme.Holo.Light">
        <item name="android:windowAnimationStyle">
            @style/ActivityAnimation</item>
    </style>

    <style name="ActivityAnimation"
        parent="@android:style/Animation.Activity">
        <item name="android:activityOpenEnterAnimation">
            @anim/activity_open_enter</item>
        <item name="android:activityOpenExitAnimation">
            @anim/activity_open_exit</item>
        <item name="android:activityCloseEnterAnimation">
            @anim/activity_close_enter</item>
        <item name="android:activityCloseExitAnimation">
            @anim/activity_close_exit</item>
    </style>

</resources>
```

通过提供一个主题的自定义属性 android:windowAnimationStyle 值,我们可以自定义这些过渡动画。引用框架的父样式也很重要,因为这 4 种动画并不是该样式中唯一定义的内容,否则可能会无意中去除现有的一些窗口动画。

2. 支持 Fragment

自定义 Fragment 间的过渡动画会有些不同,这取决于你是否使用了支持库。这是因为原生的 Fragment 使用了新的 Animator 对象,该对象在支持库的 Fragment 版本中是不支持的。

使用支持库时，可以通过调用 setCustomAnimations()覆写单个 FragmentTransaction 的过渡动画。该方法的接受两个参数的版本可以设置添加/替换/移除动作时的动画效果，但在界面栈回退时不能设置相应的动画。该方法的接受 4 个参数的版本则可以为界面栈的回退添加自定义的动画。还是使用之前示例中一样的 Animation 对象，下面的代码显示了如何将这些动画添加到 FragmentTransaction 中：

```
FragmentTransaction ft = getSupportFragmentManager().beginTransaction();
    //首先必须调用该方法！
    ft.setCustomAnimations(R.anim.activity_open_enter,
            R.anim.activity_open_exit,
            R.anim.activity_close_enter,
            R.anim.activity_close_exit);
    ft.replace(R.id.container_fragment, fragment);
    ft.addToBackStack(null);
ft.commit();
```

重点：

setCustomAnimations()必须在 add()、replace()和其他动作方法之前调用，否则动画将不会运行。最好是在每个事务代码块开始时就调用该方法。

如果希望对某个 Fragment 一直使用同样的动画，可能需要覆写 Fragment 中的 onCreateAnimation()方法。代码清单 1-54 显示了使用这种方式定义的 Fragment 动画。

代码清单 1-54　使用自定义动画的 Fragment

```java
public class SupportFragment extends Fragment {

    @Override
    public View onCreateView(LayoutInflater inflater,
            ViewGroup container, Bundle savedInstanceState) {
        TextView tv = new TextView(getActivity());
        tv.setText("Fragment");
        tv.setBackgroundColor(Color.RED);
        return tv;
    }

    @Override
    public Animation onCreateAnimation(int transit, boolean enter,
            int nextAnim) {
        switch (transit) {
        case FragmentTransaction.TRANSIT_FRAGMENT_FADE:
            if(enter) {
                return AnimationUtils.loadAnimation(getActivity(),
                        android.R.anim.fade_in);
            } else {
                return AnimationUtils.loadAnimation(getActivity(),
                        android.R.anim.fade_out);
```

```
            }
        case FragmentTransaction.TRANSIT_FRAGMENT_CLOSE:
            if(enter) {
                return AnimationUtils.loadAnimation(getActivity(),
                    R.anim.activity_close_enter);
            } else {
                return AnimationUtils.loadAnimation(getActivity(),
                    R.anim.activity_close_exit);
            }
        case FragmentTransaction.TRANSIT_FRAGMENT_OPEN:
        default:
            if(enter) {
                return AnimationUtils.loadAnimation(getActivity(),
                    R.anim.activity_open_enter);
            } else {
                return AnimationUtils.loadAnimation(getActivity(),
                    R.anim.activity_open_exit);
            }
        }
    }
}
```

Fragment 的动画行为和 FragmentTransaction 的设置有很大的关系。有很多的过渡值可以通过 setTransition()方法关联到事务上。如果没有调用 setTransition()，Fragment 就无法知道打开动画集和关闭动画集的区别，因此我们唯一知道的就是运行进入动画还是退出动画。

要获得之前通过 setCustomAnimations()实现的相同效果，需要将事务的过渡值设为 TRANSIT_FRAGMENT_OPEN。这时会使用这个过渡值调用初始的事务，但同时会通过 TRANSIT_FRAGMENT_CLOSE 调用界面栈回退动作，这样就允许 Fragment 提供不同的动画。下面的代码片段演示了如何用这种方式构造一个事务：

```
FragmentTransaction ft = getSupportFragmentManager().beginTransaction();
    //设置过渡值来触发相应的动画
    ft.setTransition(FragmentTransaction.TRANSIT_FRAGMENT_OPEN);
    ft.replace(R.id.container_fragment, fragment);
    ft.addToBackStack(null);
ft.commit();
```

Fragment 还有第三种状态，这在 Activity 中是没有的，它是通过 TRANSIT_FRAGMENT_FADE 过渡值定义的。这个动画是在过渡行为不再是变化的一部分时出现的，例如添加或替换，但在 Fragment 隐藏或显示时不会出现。在这个示例中，我们使用标准的系统渐变动画来诠释这种情形。

3. 本地 Fragment

如果应用程序的目标版本是 API Level 11 或之后版本，则不必使用支持库中的 Fragment，而且这种情况下的自定义动画代码会稍微有些不同。本地 Fragment 实现使用相对较新的 Animator 对象(而非旧的 Animation 对象)来创建过渡动画。

这需要对代码做一些修改；首先，需要使用 Animator 来定义所有的 XML 动画。参见代码清单 1-55 到代码清单 1-58。

代码清单 1-55　res/animator/fragment_exit.xml

```xml
<?xml version="1.0" encoding="utf-8"?>
<set xmlns:android="http://schemas.android.com/apk/res/android" >
    <objectAnimator
        android:valueFrom="0" android:valueTo="-90"
        android:valueType="floatType"
        android:propertyName="rotation"
        android:duration="500"/>
    <objectAnimator
        android:valueFrom="1.0" android:valueTo="0.0"
        android:valueType="floatType"
        android:propertyName="alpha"
        android:duration="500"/>
</set>
```

代码清单 1-56　res/animator/fragment_enter.xml

```xml
<?xml version="1.0" encoding="utf-8"?>
<set xmlns:android="http://schemas.android.com/apk/res/android" >
    <objectAnimator
        android:valueFrom="90" android:valueTo="0"
        android:valueType="floatType"
        android:propertyName="rotation"
        android:duration="500"/>
    <objectAnimator
        android:valueFrom="0.0" android:valueTo="1.0"
        android:valueType="floatType"
        android:propertyName="alpha"
        android:duration="500"/>
</set>
```

代码清单 1-57　res/animator/fragment_pop_exit.xml

```xml
<?xml version="1.0" encoding="utf-8"?>
<set xmlns:android="http://schemas.android.com/apk/res/android" >
    <objectAnimator
        android:valueFrom="0" android:valueTo="90"
        android:valueType="floatType"
        android:propertyName="rotation"
        android:duration="500"/>
    <objectAnimator
        android:valueFrom="1.0" android:valueTo="0.0"
        android:valueType="floatType"
        android:propertyName="alpha"
        android:duration="500"/>
</set>
```

代码清单 1-58   res/animator/fragment_pop_enter.xml

```xml
<?xml version="1.0" encoding="utf-8"?>
<set xmlns:android="http://schemas.android.com/apk/res/android" >
    <objectAnimator
        android:valueFrom="-90" android:valueTo="0"
        android:valueType="floatType"
        android:propertyName="rotation"
        android:duration="500"/>
    <objectAnimator
        android:valueFrom="0.0" android:valueTo="1.0"
        android:valueType="floatType"
        android:propertyName="alpha"
        android:duration="500"/>
</set>
```

除了语法上的细微区别，这些动画几乎和之前创建的动画一模一样。其他仅有的差别是，这些动画被设置为围绕在视图的中心(默认行为)而不是左上角。

和以前一样，可以通过 setCustomAnimations()直接设置某个 FragmentTransaction 单独的过渡动画；但是，新的版本使用了我们的 Animator 实例。下面的代码片段使用新 API 实现了这一过程：

```java
FragmentTransaction ft = getFragmentManager().beginTransaction();
    //首先必须调用该方法!
    ft.setCustomAnimations(R.animator.fragment_enter,
        R.animator.fragment_exit,
        R.animator.fragment_pop_enter,
        R.animator.fragment_pop_exit);
    ft.replace(R.id.container_fragment, fragment);
    ft.addToBackStack(null);
ft.commit();
```

如果想要对某一特定子类总是使用相同的过渡动画,可以像以前一样自定义 Fragment。但本地 Fragment 没有 onCreateAnimation()方法，而是使用了 onCreateAnimator()。查看代码清单 1-59，它使用新的 API 重新定义了我们创建的 Fragment。

代码清单 1-59   自定义过渡动画的本地 Fragment

```java
public class NativeFragment extends Fragment {

@Override
    public View onCreateView(LayoutInflater inflater,
        ViewGroup container, Bundle savedInstanceState) {
        TextView tv = new TextView(getActivity());
        tv.setText("Fragment");
        tv.setBackgroundColor(Color.BLUE);
        return tv;
    }
```

```java
@Override
public Animator onCreateAnimator(int transit, boolean enter,
        int nextAnim) {
    switch(transit) {
    case FragmentTransaction.TRANSIT_FRAGMENT_FADE:
        if(enter) {
            return AnimatorInflater.loadAnimator(
                    getActivity(),
                    android.R.animator.fade_in);
        } else {
            return AnimatorInflater.loadAnimator(
                    getActivity(),
                    android.R.animator.fade_out);
        }
    case FragmentTransaction.TRANSIT_FRAGMENT_CLOSE:
        if(enter) {
            return AnimatorInflater.loadAnimator(
                    getActivity(),
                    R.animator.fragment_pop_enter);
        } else {
            return AnimatorInflater.loadAnimator(
                    getActivity(),
                    R.animator.fragment_pop_exit);
        }
    case FragmentTransaction.TRANSIT_FRAGMENT_OPEN:
    default:
        if(enter) {
            return AnimatorInflater.loadAnimator(
                    getActivity(),
                    R.animator.fragment_enter);
        } else {
            return AnimatorInflater.loadAnimator(
                    getActivity(),
                    R.animator.fragment_exit);
        }
    }
}
```

同样，会像支持库示例一样检查同样的过渡值，我们只是返回 Animator 实例。下面同样的代码片段可以使用过渡值集开始一个事务：

```java
FragmentTransaction ft = getFragmentManager().beginTransaction();
    //设置过渡值来触发相应的动画
    ft.setTransition(FragmentTransaction.TRANSIT_FRAGMENT_OPEN);
    ft.replace(R.id.container_fragment, fragment);
    ft.addToBackStack(null);
ft.commit();
```

将这些自定义动画全局地应用到整个应用程序的最终方式就是将这些动画关联到应

用程序的主题上。代码清单 1-60 显示了使用我们的 Fragment 动画的自定义主题。

代码清单 1-60　res/values/styles.xml

```xml
<resources>
    <style name="AppTheme" parent="android:Theme.Holo.Light">
        <item name="android:windowAnimationStyle">
            @style/FragmentAnimation</item>
    </style>

    <style name="FragmentAnimation"
        parent="@android:style/Animation.Activity">
        <item name="android:fragmentOpenEnterAnimation">
            @animator/fragment_enter</item>
        <item name="android:fragmentOpenExitAnimation">
            @animator/fragment_exit</item>
        <item name="android:fragmentCloseEnterAnimation">
            @animator/fragment_pop_enter</item>
        <item name="android:fragmentCloseExitAnimation">
            @animator/fragment_pop_exit</item>
        <item name="android:fragmentFadeEnterAnimation">
            @android:animator/fade_in</item>
        <item name="android:fragmentFadeExitAnimation">
            @android:animator/fade_out</item>
    </style>
</resources>
```

正如你所看到的，一个主题默认的 Fragment 动画属性就是相同的 windowAnimationStyle 属性的一部分。因此，我们在自定义时要保证继承自同样的父样式，否则会移除其他的一些系统默认效果(例如 Activity 过渡动画)。必须依然要在 FragmentTransaction 中请求相应的过渡类型，从而触发相应的动画。

如果想要在主题中同时自定义 Activity 和 Fragment 的过渡动画，可以将它们一起放到一个相同的自定义样式中(参见代码清单 1-61)。

代码清单 1-61　res/values/styles.xml

```xml
<resources>
    <style name="AppTheme" parent="android:Theme.Holo.Light">
        <item name="android:windowAnimationStyle">
            @style/TransitionAnimation</item>
    </style>

    <style name="TransitionAnimation"
        parent="@android:style/Animation.Activity">
        <item name="android:activityOpenEnterAnimation">
            @anim/activity_open_enter</item>
        <item name="android:activityOpenExitAnimation">
            @anim/activity_open_exit</item>
        <item name="android:activityCloseEnterAnimation">
```

```xml
            @anim/activity_close_enter</item>
        <item name="android:activityCloseExitAnimation">
            @anim/activity_close_exit</item>
        <item name="android:fragmentOpenEnterAnimation">
            @animator/fragment_enter</item>
        <item name="android:fragmentOpenExitAnimation">
            @animator/fragment_exit</item>
        <item name="android:fragmentCloseEnterAnimation">
            @animator/fragment_pop_enter</item>
        <item name="android:fragmentCloseExitAnimation">
            @animator/fragment_pop_exit</item>
        <item name="android:fragmentFadeEnterAnimation">
            @android:animator/fade_in</item>
        <item name="android:fragmentFadeExitAnimation">
            @android:animator/fade_out</item>
    </style>
</resources>
```

警告：

对主题添加的 Fragment 过渡动画只会作用于本地 Fragment 实现。支持库中的 Fragment 因为早期的平台版本并没有这些属性，所以也找不到这些属性。

1.12 创建视图变换

1.12.1 问题

应用程序需要动态变换视图的外观，从而为视图添加一些视觉效果，例如视角变换效果。

1.12.2 解决方案

(API Level 1)

ViewGroup 中的静态变换 API 提供了应用视觉效果的简单方法，例如旋转、缩放、透明度变化，而且不必依靠动画。使用它也很容易使用父视图的上下文来应用变换，例如根据位置的变化而缩放。

在初始化过程中调用 setStaticTranformationsEnabled(true)，可以启用任何 ViewGroup 的静态变换。启用此功能后，框架会定期调用每个子视图的 getChildStaticTransformation()，从而允许应用程序设置变换。

1.12.3 实现机制

首先看一个示例，在该例中变换被应用一次而且不会改变(参见代码清单 1-62)。

代码清单 1-62　使用静态变换自定义布局

```java
public class PerspectiveLayout extends LinearLayout {

    public PerspectiveLayout(Context context) {
        super(context);
        init();
    }

    public PerspectiveLayout(Context context, AttributeSet attrs) {
        super(context, attrs);
        init();
    }

    public PerspectiveLayout(Context context, AttributeSet attrs, int defStyle) {
        super(context, attrs, defStyle);
        init();
    }

    private void init() {
        //启用静态变换，这样对于每个子视图都会调用getChildStaticTransformation()
        setStaticTransformationsEnabled(true);
    }

    @Override
    protected boolean getChildStaticTransformation(View child, Transformation t) {
        // 清除所有现有的变换
        t.clear();

        if (getOrientation() == HORIZONTAL) {
            //根据到左边缘的距离对子视图进行缩放
            float delta = 1.0f - ((float) child.getLeft() / getWidth());

            t.getMatrix().setScale(delta, delta, child.getWidth() / 2,
                    child.getHeight() / 2);
        } else {
            //根据到顶端边缘的距离对子视图进行缩放
            float delta = 1.0f - ((float) child.getTop() / getHeight());

            t.getMatrix().setScale(delta, delta, child.getWidth() / 2,
                    child.getHeight() / 2);
            //同样也根据它的位置应用颜色淡出效果
            t.setAlpha(delta);
        }
        return true;
    }
}
```

这个示例介绍了一个自定义的 LinearLayout，它根据子视图到父视图起始边缘的距离，对每个子视图做了缩放变换。getChildStaticTransformation()中的代码通过子视图到父视图左边缘或顶端边缘的距离与父视图完整尺寸的比值计算得出应该使用的缩放比例。设定变换之后,这个方法的返回值会通知 Android 框架。任何情况下,只要应用程序中设置了一个自定义变换,这个方法就必须返回 true,以确保它被关联到视图上。

大多数的视觉效果(如旋转或缩放)都实际地应用于 Transformation 的 Matrix 上。在我们的示例中,通过调用 getMatrix().setScale()来调整每个子视图的缩放,同时传入缩放比例和轴心点。轴心点就是缩放发生的位置,我们将这个位置设置在视图的中心点,这样缩放的结果就会居中显示。

如果布局是垂直方向的,我们同样会根据相同的距离值为子视图应用透明渐变效果,只需要直接使用 Transformation 的 setAlpha()方法即可。代码清单 1-63 就是使用这个视图的示例布局。

代码清单 1-63   res/layout/main.xml

```xml
<?xml version="1.0" encoding="utf-8"?>
<LinearLayout xmlns:android="http://schemas.android.com/apk/res/android"
    android:layout_width="match_parent"
    android:layout_height="match_parent"
    android:orientation="vertical">
    <!-- 水平方向自定义布局 -->
    <com.examples.statictransforms.PerspectiveLayout
        android:layout_width="match_parent"
        android:layout_height="wrap_content"
        android:orientation="horizontal" >
        <ImageView
            android:layout_width="wrap_content"
            android:layout_height="wrap_content"
            android:src="@drawable/ic_launcher" />
        <ImageView
            android:layout_width="wrap_content"
            android:layout_height="wrap_content"
            android:src="@drawable/ic_launcher" />
        <ImageView
            android:layout_width="wrap_content"
            android:layout_height="wrap_content"
            android:src="@drawable/ic_launcher" />
        <ImageView
            android:layout_width="wrap_content"
            android:layout_height="wrap_content"
            android:src="@drawable/ic_launcher" />
    </com.examples.statictransforms.PerspectiveLayout>
    <!--垂直方向自定义布局 -->
    <com.examples.statictransforms.PerspectiveLayout
        android:layout_width="wrap_content"
        android:layout_height="match_parent"
        android:orientation="vertical" >
```

```xml
    <ImageView
        android:layout_width="wrap_content"
        android:layout_height="wrap_content"
        android:src="@drawable/ic_launcher" />
    <ImageView
        android:layout_width="wrap_content"
        android:layout_height="wrap_content"
        android:src="@drawable/ic_launcher" />
    <ImageView
        android:layout_width="wrap_content"
        android:layout_height="wrap_content"
        android:src="@drawable/ic_launcher" />
    <ImageView
        android:layout_width="wrap_content"
        android:layout_height="wrap_content"
        android:src="@drawable/ic_launcher" />
    </com.examples.statictransforms.PerspectiveLayout>
</LinearLayout>
```

图 1-8 显示了示例变换的结果。

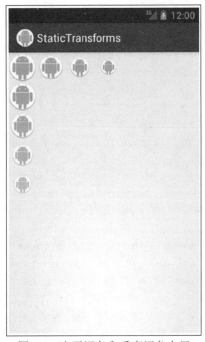

图 1-8　水平视角和垂直视角布局

在水平布局中，越往右的视图使用的缩放比例越小。同样，垂直方向视图的缩放比例也是越往下越小。另外，垂直方向的视图由于添加了透明度变化会出现逐渐淡出的效果。

现在让我们看一个提供了更为动态变化效果的示例。代码清单 1-64 展示了一个封装在 HorizontalScrollView 中的自定义布局。当子视图滚动时，这个布局使用静态变换来缩放子视图。在屏幕中心的视图大小总是正常的，越靠近边缘视图就越小。这样就会出现下面的效果:在滚动的过程中,视图首先会逐渐靠近,然后会逐渐远离。

代码清单 1-64　自定义视角滚动内容

```java
public class PerspectiveScrollContentView extends LinearLayout {

    /* 每个子视图的缩放比例都是可调节的*/
    private static final float SCALE_FACTOR = 0.7f;
    /*
     * 变换的轴心点。(0,0)是左上,(1,1) 是右下。当前设置的是底部中间(0.5, 1)
     */
    private static final float ANCHOR_X = 0.5f;
    private static final float ANCHOR_Y = 1.0f;

    public PerspectiveScrollContentView(Context context) {
        super(context);
        init();
    }

    public PerspectiveScrollContentView(Context context,
        AttributeSet attrs) {
        super(context, attrs);
        init();
    }

    public PerspectiveScrollContentView(Context context,
        AttributeSet attrs,int defStyle) {
        super(context, attrs, defStyle);
        init();
    }

    private void init() {
        //启动静态变换,这样对于每个子视图,getChildStaticTransformation()
        //都会被调用
        setStaticTransformationsEnabled(true);
    }

    /*
     * 工具方法,用于计算屏幕坐标系内所有视图的当前位置
     */
    private int getViewCenter(View view) {
        int[] childCoords = new int[2];
        view.getLocationOnScreen(childCoords);
        int childCenter = childCoords[0] + (view.getWidth() / 2);

        return childCenter;
    }

    @Override
```

```java
        protected boolean getChildStaticTransformation(View child,
    Transformation t) {
        HorizontalScrollView scrollView = null;
        if(getParent() instanceof HorizontalScrollView) {
            scrollView = (HorizontalScrollView) getParent();
        }
        if(scrollView == null) {
            return false;
        }

        int childCenter = getViewCenter(child);
        int viewCenter = getViewCenter(scrollView);

        //计算子视图和父容器中心之间的距离,这会决定应用的缩放比例
        float delta = Math.min(1.0f, Math.abs(childCenter - viewCenter)
                / (float) viewCenter);
        //设置最小缩放比例为0.4
        float scale = Math.max(0.4f, 1.0f - (SCALE_FACTOR * delta));
        float xTrans = child.getWidth() * ANCHOR_X;
        float yTrans = child.getHeight() * ANCHOR_Y;

        //清除现有的所有变换
        t.clear();
        //为子视图设置变换
        t.getMatrix().setScale(scale, scale, xTrans, yTrans);

        return true;
    }
}
```

在这个示例中,自定义布局会根据每个子视图相对于父视图 HorizontalScrollView 中心位置的距离,为每个子视图计算变换。当用户滚动时,每个子视图的变换需要重新计算,从而实现视图移动时子视图的动态放大和缩小。这个示例将变换轴心点设置在每个子视图的底部中间位置,这会创造出如下效果:每个子视图会垂直放大,而且保持水平居中。代码清单 1-65 的 Activity 示例将这个自定义布局付诸实践。

代码清单 1-65　使用了 PerspectiveScrollContentView 的 Activity

```java
public class ScrollActivity extends Activity {

    @Override
    protected void onCreate(Bundle savedInstanceState) {
        super.onCreate(savedInstanceState);
        HorizontalScrollView parentView = new HorizontalScrollView(this);
        PerspectiveScrollContentView contentView =
                new PerspectiveScrollContentView(this);
```

```
            //对此视图禁用硬件加速,因为动态调整每个子视图的变换当前无法通过硬件实现
            //也可以通过在清单文件中添加 android:hardwareAccelerated="false"
            //禁用整个 Activity 或应用程序的硬件加速。但出于性能的考虑,最好尽可能少地
            //禁用硬件加速
            if(Build.VERSION.SDK_INT >= Build.VERSION_CODES.HONEYCOMB) {
                contentView.setLayerType(View.LAYER_TYPE_SOFTWARE, null);
            }

            //向滚动条中添加几张图片
            for(int i = 0; i < 20; i++) {
                ImageView iv = new ImageView(this);
                iv.setImageResource(R.drawable.ic_launcher);
                contentView.addView(iv);
            }
            //添加要显示的视图
            parentView.addView(contentView);
            setContentView(parentView);
        }
    }
```

在这个示例中创建了一个滚动视图,并且关联了一个自定义的包含若干张滚动图片的 PerspectiveScrollContentView。这里的代码不需要太多关注,但有个非常重要的地方需要提一下。虽然一般情况下静态变换都会被支持,但视图刷新过程中动态更新变换效果在当前的 SDK 版本中是不能使用硬件加速的。因此,如果应有程序的目标 SDK 为 11 或以上版本,或者已经在某种程度上启用了硬件加速,这时需要对这个视图禁用硬件加速。

可以在清单文件的<activity>或整个<application>标签中添加 android:hardwareAccelerated="false"属性,对硬件加速进行全局设置;但是我们也可以通过调用 setLayerType()方法并设置 LAYER_TYPE_SOFTWARE,在 Java 代码中对这个自定义视图进行单独设置。如果应用程序的目标 SDK 版本低于此版本,即使是较新的设备,默认情况下硬件加速也是关闭的,出于兼容性考虑,这些代码也许是不必要的。

1.13 建立可扩展的集合视图

1.13.1 问题

你希望以独特的方式展示大型数据集合,而不是以垂直滚动列表显示;或者,你要以 AdapterView 小部件无法轻松支持的方式样式化此集合。

1.13.2 解决方案

(API Level 1)

以 Android 支持库中的 RecyclerView 为基础构建解决方案。RecyclerView 小部件利用与 AdapterView 组件相同的视图回收功能来提供大型数据集的高效内存使用显示方式。然而，与核心框架中的 AdapterView 不同的是，RecyclerView 以更加灵活的模型为基础，在此模型中，子视图组件的放置委托给 LayoutManager 实例完成。Android 库支持两个内置的布局管理器：

- LineLayoutManager：将子视图垂直(自上而下)或水平(自左而右)放置在列表中。垂直布局行为类似于 ListView 框架。
- GridLayoutManager：将子视图垂直(自上而下)或水平(自左而右)放置在网格中。该管理器支持添加行/列跨度值以使网格中的子视图交错显示。对于单一跨度项的垂直布局行为类似于 GridView 框架。

RecyclerView.ItemDecoration 实例使应用程序支持在子视图的上方和下方进行自定义的绘制操作，此外还直接支持页边距以在子视图之间添加间距。该实例可用于绘制像网格线和连接线这样的简单对象，还可以用于在内容区域中绘制更加复杂的图案或图片。

RecyclerView.Adapter 实例还包括用于通知数据集视图变化的新方法，使得该小部件可以更好地处理各种变化的动画，如添加或删除元素，而使用 AdapterView 较难处理这些动画：

- notifyItemInserted()、notifyItemRemoved()、notifyItemChanged()：表明已添加、删除关联数据集中的单个项，或者该项已改变位置。
- notifyItemRangeInserted()、notifyItemRangeRemoved()、notifyItemRangeChanged()：表明关联数据集中已修改的一定位置范围的项。

这些方法接受的参数是特定项的位置，因此 RecycleView 可以智能地判断如何制作变化的动画。标准的 notifyDataSetChanged()方法仍然得到支持，但它不会制作变化的动画。

要点：

RecyclerView 仅在 Android 支持库中提供；它不是任意平台级别中的原生 SDK 的一部分。然而，目标平台为 API Level 7 或之后版本的任意应用程序都可以通过包含支持库使用此小部件。有关在项目中包含支持库的更多信息，请参考如下网址：http://developer.android.com/tools/support-library/index.html。

1.13.3 实现机制

下面的示例使用 4 个不同的 LayoutManager 实例，通过 RecyclerView 显示相同的项数据。图 1-9 显示了在垂直和水平列表中显示的数据。

第 1 章 布局和视图

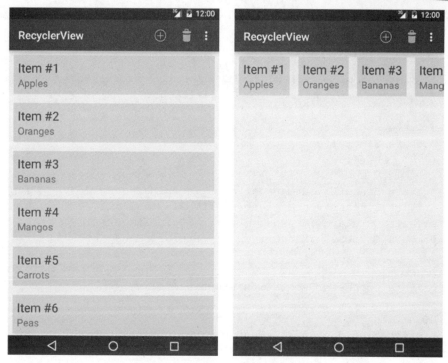

图 1-9 垂直和水平列表集合

图 1-10 分别在交错的垂直网格和均匀的水平网格中显示相同的数据。

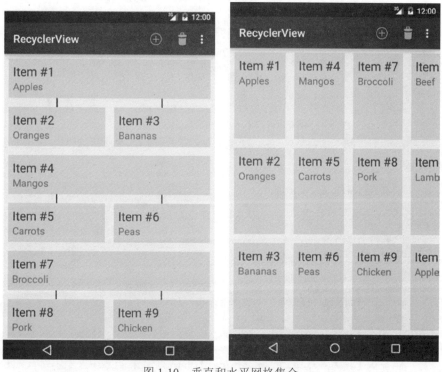

图 1-10 垂直和水平网格集合

73

首先，代码清单 1-66 和 1-67 显示了 Activity 和用于选择布局的选项菜单。

代码清单 1-66　使用了 RecyclerView 的 Activity

```java
public class SimpleRecyclerActivity extends ActionBarActivity implements
        SimpleItemAdapter.OnItemClickListener {

    private RecyclerView mRecyclerView;
    private SimpleItemAdapter mAdapter;

    /* 布局管理器 */
    private LinearLayoutManager mHorizontalManager;
    private LinearLayoutManager mVerticalManager;
    private GridLayoutManager mVerticalGridManager;
    private GridLayoutManager mHorizontalGridManager;

    /* 修饰 */
    private ConnectorDecoration mConnectors;

    @Override
    protected void onCreate(Bundle savedInstanceState) {
        super.onCreate(savedInstanceState);

        mRecyclerView = new RecyclerView(this);
        mHorizontalManager = new LinearLayoutManager(this,
                LinearLayoutManager.HORIZONTAL, false);
        mVerticalManager = new LinearLayoutManager(this,
                LinearLayoutManager.VERTICAL, false);
        mVerticalGridManager = new GridLayoutManager(this,
                2, /* 网格的列数 */
                LinearLayoutManager.VERTICAL, /* 垂直定位网格 */
                false);
        mHorizontalGridManager = new GridLayoutManager(this,
                3, /* 网格的行数 */
                LinearLayoutManager.HORIZONTAL, /* 水平定位网格 */
                false);

        //垂直网格的连接线修饰
        mConnectors = new ConnectorDecoration(this);

        //交错垂直网格
        mVerticalGridManager.setSpanSizeLookup(new GridStaggerLookup());

        mAdapter = new SimpleItemAdapter(this);
        mAdapter.setOnItemClickListener(this);
        mRecyclerView.setAdapter(mAdapter);

        //对所有连接应用边缘修饰
        mRecyclerView.addItemDecoration(new InsetDecoration(this));
```

```java
        //默认为垂直布局
        selectLayoutManager(R.id.action_vertical);
        setContentView(mRecyclerView);
    }

    @Override
    public boolean onCreateOptionsMenu(Menu menu) {
        getMenuInflater().inflate(R.menu.layout_options, menu);
        return true;
    }

    @Override
    public boolean onOptionsItemSelected(MenuItem item) {
        return selectLayoutManager(item.getItemId());
    }

    private boolean selectLayoutManager(int id) {
        switch (id) {
            case R.id.action_vertical:
                mRecyclerView.setLayoutManager(mVerticalManager);
                mRecyclerView.removeItemDecoration(mConnectors);
                return true;
            case R.id.action_horizontal:
                mRecyclerView.setLayoutManager(mHorizontalManager);
                mRecyclerView.removeItemDecoration(mConnectors);
                return true;
            case R.id.action_grid_vertical:
                mRecyclerView.setLayoutManager(mVerticalGridManager);
                mRecyclerView.addItemDecoration(mConnectors);
                return true;
            case R.id.action_grid_horizontal:
                mRecyclerView.setLayoutManager(mHorizontalGridManager);
                mRecyclerView.removeItemDecoration(mConnectors);
                return true;
            case R.id.action_add_item:
                //插入新的项
                mAdapter.insertItemAtIndex("Android Recipes", 1);
                return true;
            case R.id.action_remove_item:
                //删除第一项
                mAdapter.removeItemAtIndex(1);
                return true;
            default:
                return false;
        }
    }

    /** OnItemClickListener 方法 */

    @Override
```

```
        public void onItemClick(SimpleItemAdapter.ItemHolder item, int
position) {
            Toast.makeText(this, item.getSummary(), Toast.LENGTH_SHORT).show();
        }
    }
```

代码清单 1-67   res/menu/layout_options.xml

```xml
<menu xmlns:android="http://schemas.android.com/apk/res/android"
    xmlns:app="http://schemas.android.com/apk/res-auto">
    <item
        android:id="@+id/action_add_item"
        android:title="Add Item"
        android:icon="@android:drawable/ic_menu_add"
        app:showAsAction="ifRoom" />
    <item
        android:id="@+id/action_remove_item"
        android:title="Remove Item"
        android:icon="@android:drawable/ic_menu_delete"
        app:showAsAction="ifRoom" />
    <item
        android:id="@+id/action_vertical"
        android:title="Vertical List"
        app:showAsAction="never"/>
    <item
        android:id="@+id/action_horizontal"
        android:title="Horizontal List"
        app:showAsAction="never"/>
    <item
        android:id="@+id/action_grid_vertical"
        android:title="Vertical Grid"
        app:showAsAction="never"/>
    <item
        android:id="@+id/action_grid_horizontal"
        android:title="Horizontal Grid"
        app:showAsAction="never"/>
</menu>
```

此例使用选项菜单选择应该应用于 RecyclerView 的布局管理器。任何改动都会触发 selectLayoutManager()辅助方法，该方法将请求的管理器传递给 setLayoutManager()。这会从现有的适配器中重新加载目前的数据，因此我们不需要维护多个 RecyclerView 实例即可动态更改布局。

可以看到，利用内置的布局管理器不需要太多的代码。此例构造两个 LinearLayoutManager 实例，它们的构造函数以方向常量为参数(VERTICAL 或 HORIZONTAL)。该管理器还支持(通过最后的布尔参数)翻转布局，以便按照最后项最先显示的顺序布置适配器数据。

同样，我们构造两个 GridLayoutManger 实例，分别用于水平和垂直布局。此对象获取一个附加参数，即 spanCount，表示布局应使用的行数(水平网格)或列数(垂直网格)。此参数与支持交错网格没有任何关系；稍后将对此进行讨论。

与所有集合视图一样，我们需要让适配器将数据项与视图关联。你可能已注意到，在 Activity 的代码清单中创建了 SimpleItemAdapter 类。该适配器的实现如代码清单 1-68 和 1-69 所示。

代码清单 1-68   res/layout/collection_item.xml

```xml
<?xml version="1.0" encoding="utf-8"?>
<LinearLayout xmlns:android="http://schemas.android.com/apk/res/android"
    android:layout_width="match_parent"
    android:layout_height="wrap_content"
    android:orientation="vertical"
    android:padding="8dp"
    android:background="#CCF">
    <TextView
        android:id="@+id/text_title"
        android:layout_width="wrap_content"
        android:layout_height="wrap_content"
        android:textAppearance="?android:textAppearanceLarge"/>
    <TextView
        android:id="@+id/text_summary"
        android:layout_width="wrap_content"
        android:layout_height="wrap_content"
        android:textAppearance="?android:textAppearanceMedium"/>
</LinearLayout>
```

代码清单 1-69   RecyclerView 的适配器实现

```java
public class SimpleItemAdapter extends RecyclerView.Adapter<
SimpleItemAdapter.ItemHolder> {

    /*
     * 单击处理程序接口。与 AdapterView 不同,
     * RecyclerView 没有自己的内置接口
     */
    public interface OnItemClickListener {
        public void onItemClick(ItemHolder item, int position);
    }

    private static final String[] ITEMS = {
            "Apples", "Oranges", "Bananas", "Mangos",
            "Carrots", "Peas", "Broccoli",
            "Pork", "Chicken", "Beef", "Lamb"
    };
    private List<String> mItems;

    private OnItemClickListener mOnItemClickListener;
    private LayoutInflater mLayoutInflater;

    public SimpleItemAdapter(Context context) {
```

```java
        mLayoutInflater = LayoutInflater.from(context);
        //创建虚拟项的静态列表
        mItems = new ArrayList<String>();
        mItems.addAll(Arrays.asList(ITEMS));
        mItems.addAll(Arrays.asList(ITEMS));
    }

    @Override
    public ItemHolder onCreateViewHolder(ViewGroup parent, int viewType) {
        View itemView = mLayoutInflater.inflate(R.layout.collection_item, parent, false);

        return new ItemHolder(itemView, this);
    }

    @Override
    public void onBindViewHolder(ItemHolder holder, int position) {
        holder.setTitle("Item #"+(position+1));
        holder.setSummary(mItems.get(position));
    }

    @Override
    public int getItemCount() {
        return mItems.size();
    }

    public OnItemClickListener getOnItemClickListener() {
        return mOnItemClickListener;
    }

    public void setOnItemClickListener(OnItemClickListener listener) {
        mOnItemClickListener = listener;
    }

    /* 管理数据集修改的方法 */
    public void insertItemAtIndex(String item, int position) {
        mItems.add(position, item);
        //通知视图触发变化动画
        notifyItemInserted(position);
    }

    public void removeItemAtIndex(int position) {
        if (position >= mItems.size()) return;

        mItems.remove(position);
        //通知视图触发变化动画
        notifyItemRemoved(position);
    }

    /* 封装项视图所需的 ViewHolder 实现 */
```

```java
    public static class ItemHolder extends RecyclerView.ViewHolder
implements View.OnClickListener {
        private SimpleItemAdapter mParent;
        private TextView mTitleView, mSummaryView;

        public ItemHolder(View itemView, SimpleItemAdapter parent) {
            super(itemView);
            itemView.setOnClickListener(this);
            mParent = parent;

            mTitleView = (TextView) itemView.findViewById(R.id.text_title);
            mSummaryView = (TextView) itemView.findViewById(R.id.text_summary);
        }

        public void setTitle(CharSequence title) {
            mTitleView.setText(title);
        }

        public void setSummary(CharSequence summary) {
            mSummaryView.setText(summary);
        }

        public CharSequence getSummary() {
            return mSummaryView.getText();
        }

        @Override
        public void onClick(View v) {
            final OnItemClickListener listener = mParent.getOnItemClickListener();
            if (listener != null) {
                listener.onItemClick(this, getPosition());
            }
        }
    }
}
```

RecyclerView.Adapter 重点实施 ViewHolder 设计模式,在此模式中,它要求实现返回 RecyclerView.ViewHolder 类型的子集。该类在内部用作与子项关联的元数据(例如其当前位置和稳定的 ID)的存储位置。具体的实现通常还提供对视图内部字段的直接访问,从而尽量减少对 findViewById()的重复调用,该方法占用大量系统资源,因为它会遍历整个视图层次结构以查找请求的子项。

RecyclerView.Adapter 实现与 CursorAdapter 类似的模式,其中通过 onCreateViewHolder()和 onBindViewHolder()分开执行创建和绑定步骤。如果必须重头创建新的视图,则调用前一个方法,因此我们在此构造一个新的 ItemHolder 返回。如果特定位置的数据(在此是简单的字符串)需要附加到新的视图,则调用后一个方法;这可能是新创建的或回收的视图。这种模式与 ArrayAdapter 相反,后者将两个方法的功能结合到一个 getView()方法中。

在我们的示例中,还利用此适配器提供来自 AdapterView 的一个附加功能,RecyclerView 本质上不支持此功能:项单击侦听器。为了使用最少量的引用交换处理子视图上的单击事件,我们将每个 ViewHolder 设置为根项视图的 OnClickListener。然后,ViewHolder 处理这些事件,并将其发送回在适配器上定义的公共侦听器接口。完成上述操作后,ViewHolder 就可以在最后的侦听器回调中添加位置元数据,我们过去期望在诸如 AdapterView.OnItemClickListener 的类似侦听器中看到这样的结果。

1. 交错网格

在 Activity 示例中,垂直网格布局管理器还配备有 SpanSizeLookup 辅助类,该类用于生成如图 1-10 所示的交错效果。代码清单 1-70 显示了具体的实现。

代码清单 1-70　交错网格 SpanSizeLookup

```java
public class GridStaggerLookup extends GridLayoutManager.SpanSizeLookup {

    @Override
    public int getSpanSize(int position) {
        return (position % 3 == 0 ? 2 : 1);
    }
}
```

getSpanSize()方法用于提供查询,告诉布局管理器给定位置应占据多少跨度(行数或列数,具体取决于布局方向)。该例表明每隔三个位置应占据两列,而所有其他位置应仅占据一列。

2. 修饰项

你可能已注意到,我们在 Activity 示例中还添加了两个 ItemDecoration 实例。第一个修饰实例 InsetDecoration 应用于所有示例布局管理器,为每个子项提供页边距。第二个修饰实例 ConnectorDecoration 仅应用于垂直交错网格,用于在主要和次要网格项之间绘制连接线。代码清单 1-71 到 1-73 定义了这些修饰实例。

代码清单 1-71　提供嵌入式页边距的 ItemDecoration

```java
public class InsetDecoration extends RecyclerView.ItemDecoration {

    private int mInsetMargin;

    public InsetDecoration(Context context) {
        super();
        mInsetMargin = context.getResources()
                .getDimensionPixelOffset(R.dimen.inset_margin);
    }

    @Override
    public void getItemOffsets(Rect outRect, View view, RecyclerView parent,
            RecyclerView.State state) {
```

```
            //对子视图的所有4个边界应用计算得出的页边距
            outRect.set(mInsetMargin, mInsetMargin, mInsetMargin, mInsetMargin);
        }
    }
```

代码清单 1-72 提供连接线的 ItemDecoration

```
public class ConnectorDecoration extends RecyclerView.ItemDecoration {

    private Paint mLinePaint;
    private int mLineLength;

    public ConnectorDecoration(Context context) {
        super();
        mLineLength = context.getResources()
                .getDimensionPixelOffset(R.dimen.inset_margin);
        int connectorStroke = context.getResources()
                .getDimensionPixelSize(R.dimen.connector_stroke);

        mLinePaint = new Paint(Paint.ANTI_ALIAS_FLAG);
        mLinePaint.setColor(Color.BLACK);
        mLinePaint.setStyle(Paint.Style.STROKE);
        mLinePaint.setStrokeWidth(connectorStroke);
    }

    @Override
    public void onDraw(Canvas c, RecyclerView parent, RecyclerView.State state) {
        final RecyclerView.LayoutManager manager = parent.getLayoutManager();

        for (int i=0; i < parent.getChildCount(); i++) {
            final View child = parent.getChildAt(i);
            boolean isLarge =
            parent.getChildViewHolder(child).getPosition() % 3 == 0;

            if (!isLarge) {
                final int childLeft = manager.getDecoratedLeft(child);
                final int childRight = manager.getDecoratedRight(child);
                final int childTop = manager.getDecoratedTop(child);
                final int x = childLeft + ((childRight - childLeft) / 2);

                c.drawLine(x, childTop - mLineLength,
                        x, childTop + mLineLength,
                        mLinePaint);
            }
        }
    }
}
```

代码清单 1-73 res/values/dimens.xml

```
<?xml version="1.0" encoding="utf-8"?>
```

```xml
<resources>
    <dimen name="inset_margin">8dp</dimen>
    <dimen name="connector_stroke">2dp</dimen>
</resources>
```

ItemDecoration 可以实现 3 个主要的回调。第一个回调是 getItemOffsets()，它提供修饰器可用于对给定子视图应用页边距的 Rect 实例。在此例中，我们希望所有子视图具有相同的页边距，因此在每次调用中设置相同的值。

提示：
即使没有在 getItemOffsets()中采用位置参数，也仍然可以在需要此位置确定如何应用页边距时，通过 getChildViewHolder(view).getPostion()从 RecyclerView 获得位置。

另外两个调用—— onDraw()和 onDrawOver()提供修饰器可用于绘制其他内容的 Canvas。这些方法分别在子视图的下方和上方绘制内容。ConnectorDecoration()使用 onDraw()渲染任何可见子项之间的连接线。为此，我们遍历子视图，并且在未占据两列的每个子项(根据前面描述的交错查询)上绘制中心线。

这些绘制回调方法将在 RecyclerView 需要重绘时被调用，例如当内容滚动时，因此我们必须经常了解视图当前所在位置，以便知道在何处绘制线。相比于直接询问子视图左/右/上/下坐标，我们更喜欢通过 getDecoratedXxx()从布局管理器请求此信息。这是因为其他修饰实例(例如 InsetDecoration)可能在事后修改视图的边界，我们的绘制方法需要考虑这些因素。

3. 项动画

支持适配器数据集的变化动画的逻辑内置在每个布局管理器中。为了使管理器适当地确定如何在数据集改变时制作动画，我们必须使用特定于 RecyclerView 的适配器更新方法，而不是传统的 notifyDataSetChanged()。

修改适配器数据含两个步骤：必须首先添加或删除数据项，然后适配器必须通知视图变化发生的确切位置。在本例中，数据项添加选项触发适配器上的 notifyItemInserted()，而数据项删除选项触发 notifyItemRemoved()。

1.14 小结

本章探讨了 Android 框架提供用来向用户显示内容的许多工具，研究了创建、定制视图以及制作视图动画的技术。接下来，介绍了应用程序窗口中的许多定制项，包括从一个屏幕到另一个屏幕的自定义过渡动画。最后，我们了解了如何利用资源限定符系统为不同的屏幕配置创建优化的视图布局。

在第 2 章，将研究 UI 工具箱中的更多元素，这些元素关注与用户的交互以及实现常见的应用导航模式。

第 2 章

用 户 交 互

如果用户发现应用程序使用不便、功能不易找到，那么应用程序具有良好的外观设计就没有任何意义。大多数 Android 应用程序中的用户交互模式旨在设计一种在不同应用程序之间一致的用户体验。通过维持该平台的一致性，即使用户以前从未使用过你的应用程序，他们也会熟悉其功能。本章将探讨一些常见的实现模式，用于向用户展示信息和检索他们的输入。

2.1 利用 Action Bar

2.1.1 问题

你要在应用程序中使用最新的 Action Bar 模式，同时保持向后兼容旧设备；此外，还要定制外观以符合应用程序的主题。

2.1.2 解决方案

(API Level 7)

Action Bar 在 Android 3.0 (API Level 11) SDK 中才被引入，但可以通过 Android 支持库的 AppCompat 组件内的 ActionBarActivity 向后移植到较早的版本。使用 ActionBarActivity 以及 AppCompat 中包含的样式和资源，就可以将操作栏放入目标平台为 Android 2.1 和之后版本的应用程序中。

要点：

ActionBarActivity 仅在 AppCompat 库中提供，AppCompat 库是 Android 支持库的一部分；它不是任意平台级别中的原生 SDK 的一部分。然而，目标平台为 API Level 7 或之后版本的任意应用程序都可以通过包含支持库使用此小部件。有关在项目中包含支持库的更多信息，请参考如下网址：http://developer.android.com/tools/support-library/index.html。

Action Bar 是顶层窗口装饰的一部分,这意味着应用程序内容始终在其下方显示。因此,很难在 Action Bar 上绘制内容或对 Action Bar 的位置制作动画。如果必须执行上述操作,则使用 Toolbar 代替传统的 Action Bar。工具栏位于布局内部,因此可控制它的所在位置。给定工具栏引用,Activity 会将此视图视为装饰用的 Action Bar。

注意:
Toolbar 在 API Level 21 中引入,但它也在 AppCompat 中提供。

2.1.3 实现机制

图 2-1 显示了适当放置标准 Action Bar 的 Activity。

图 2-1　带有 Up 按钮(右图)的标准 Action Bar(左图)

为了将 Action Bar 添加到窗口装饰,只需要对启用 Action Bar 的 Activity 应用某个主题。所有默认的 Holo 和 Material 原生主题都包括 Action Bar,AppCompat 中也是如此。请注意,Action Bar 的左侧是标题和副标题,右侧则是一系列操作按钮。还有一个可选的 Up 导航按钮,可以启用该按钮以提供与 Back 按钮类似的导航。我们将在后面的章节中更详细地讨论这一点。

代码清单 2-1 到 2-3 介绍了如何将 AppCompat 主题附加到 Activity 以实现图 2-1 中的效果。

代码清单 2-1 Action Bar 主题清单

```xml
<?xml version="1.0" encoding="utf-8"?>
<manifest xmlns:android="http://schemas.android.com/apk/res/android"
    package="com.androidrecipes.actionbar">

    <application
        android:allowBackup="true"
        android:icon="@drawable/ic_launcher"
        android:label="@string/app_name">
        <activity
            android:name=".SupportActionActivity"
            android:label="@string/label_actionbar"
            android:theme="@style/AppTheme">
            <intent-filter>
                <action android:name="android.intent.action.MAIN" />
                <category android:name="android.intent.category.LAUNCHER" />
            </intent-filter>
        </activity>
    </application>
</manifest>
```

代码清单 2-2 res/values/styles.xml

```xml
<?xml version="1.0" encoding="utf-8"?>
<resources>
    <!-- 定义将应用于整个应用程序或至少应用于许多 Activity 的"主题" -->
    <style name="AppTheme" parent="@style/Theme.AppCompat.Light.DarkActionBar">
        <!-- 提供装饰主题颜色 -->
        <item name="colorPrimary">@color/primaryGreen</item>
        <item name="colorPrimaryDark">@color/darkGreen</item>
        <item name="colorAccent">@color/accentGreen</item>

    </style>
</resources>
```

代码清单 2-3 res/values/colors.xml

```xml
<?xml version="1.0" encoding="utf-8"?>
<resources>
    <color name="primaryGreen">#259b24</color>
    <color name="darkGreen">#0a7e07</color>
    <color name="accentGreen">#d0f8ce</color>
</resources>
```

通过向 Activity 应用 Theme.AppCompat.Light.DarkActionBar，就可以使 Action Bar 显示出来。我们还使用标准颜色主题属性对 Activity 应用绿色阴影。

右侧的图标由 Activity 的选项菜单生成，该选项菜单的定义如代码清单 2-4 所示。本章后面将讨论选项菜单的使用；在此添加此菜单仅是出于完整性考虑。在代码清单 2-5 中，我们可以看到如何通过代码设置 Action Bar 的基本属性。

代码清单 2-4　res/menu/support.xml

```xml
<?xml version="1.0" encoding="utf-8"?>
<menu xmlns:android="http://schemas.android.com/apk/res/android"
    xmlns:app="http://schemas.android.com/apk/res-auto">
    <item android:id="@+id/action_send"
        android:title="@string/action_send"
        android:icon="@android:drawable/ic_menu_send"
        app:showAsAction="ifRoom" />
    <item android:id="@+id/action_settings"
        android:title="@string/action_settings"
        android:orderInCategory="100"
        app:showAsAction="never"/>
</menu>
```

代码清单 2-5　Action Bar 的设置

```java
public class SupportActionActivity extends ActionBarActivity {

    @Override
    protected void onCreate(Bundle savedInstanceState) {
        super.onCreate(savedInstanceState);

        ActionBar actionBar = getSupportActionBar();

        //通过"Up"箭头指示符显示主页
        actionBar.setDisplayHomeAsUpEnabled(true);
        //设置标题文本
        actionBar.setTitle("Android Recipes");
        //设置副标题文本
        actionBar.setSubtitle("ActionBar Recipes");
    }

    @Override
    public boolean onCreateOptionsMenu(Menu menu) {
        getMenuInflater().inflate(R.menu.support, menu);
        return true;
    }

    @Override
    public boolean onOptionsItemSelected(MenuItem item) {
        switch (item.getItemId()) {
            case android.R.id.home:
                Toast.makeText(this, "Home", Toast.LENGTH_SHORT).show();
            default:
                return super.onOptionsItemSelected(item);
        }
    }
}
```

我们在 onCreate()中通过 getSupportActionBar()获得 Action Bar 的引用，并且开始设置其标题属性。副标题是可选项，但如果未提供标题，Action Bar 将显示 Activity 清单中提供的 android:label 字符串。

使用 setDisplayHomeAsUpEnabled()方法，可以根据需要激活可选的 Up 箭头。该箭头用于向回导航至父 Activity，通常在顶层 Activity 不启用。Up 按钮的行为由应用程序定义。用户单击该按钮将触发 Activity 的 onOptionsItemSelected()方法并包含 android.R.id.home 值。

1. 自定义视图

Action Bar 还支持自定义视图。设置自定义视图时，该视图显示在标题和操作按钮之间。如果视图布局填满屏幕宽度，它将隐藏标题文本。自定义视图不能重叠已启用的 Up 按钮。代码清单 2-6 显示了在 Action Bar 中启用自定义视图的 Activity。

代码清单 2-6　Action Bar 的自定义视图

```
public class SupportActionActivity extends ActionBarActivity {

    @Override
    protected void onCreate(Bundle savedInstanceState) {
        super.onCreate(savedInstanceState);

        ActionBar actionBar = getSupportActionBar();

        actionBar.setDisplayShowCustomEnabled(true);

        //显示图片取代标题
        ImageView imageView = new ImageView(this);
        imageView.setImageResource(R.drawable.ic_launcher);
        imageView.setScaleType(ImageView.ScaleType.CENTER);

        ActionBar.LayoutParams lp = new ActionBar.LayoutParams(
            ActionBar.LayoutParams.MATCH_PARENT,
            ActionBar.LayoutParams.MATCH_PARENT);

        actionBar.setCustomView(imageView, lp);
    }
}
```

这会将应用程序启动图标的图片放入 Action Bar，该位置之前显示的是标题文本，产生的结果如图 2-2 所示。

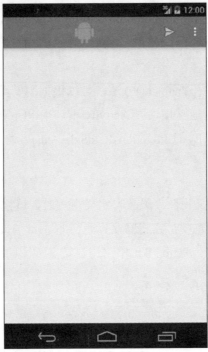

图 2-2　Action Bar 的自定义视图

2. Toolbar

当应用程序的设计要求更多地控制 Action Bar 的放置和顺序时,可以使用 Toolbar 代替。Toolbar 能代替标准的 Action Bar,因此必须对从窗口装饰中移除 Action Bar 的 Activity 应用主题。代码清单 2-7 和 2-8 是应用修改后的主题。

代码清单 2-7　Toolbar 的 Activity 清单

```xml
<?xml version="1.0" encoding="utf-8"?>
<manifest xmlns:android="http://schemas.android.com/apk/res/android"
    package="com.androidrecipes.actionbar">

    <application
        android:allowBackup="true"
        android:icon="@drawable/ic_launcher"
        android:label="@string/app_name">
        <activity
            android:name=".SupportToolbarActivity"
            android:label="@string/label_toolbar"
            android:theme="@style/AppToolbarTheme">
            <intent-filter>
                <action android:name="android.intent.action.MAIN" />
                <category android:name="android.intent.category.LAUNCHER" />
            </intent-filter>
        </activity>

    </application>
```

代码清单 2-8　res/values/styles.xml

```xml
<?xml version="1.0" encoding="utf-8"?>
<resources>
    <!-- Toolbar 代替标准的 Action Bar,
         因此必须对从窗口装饰中移除 Action Bar 的 Activity 应用主题-->
    <style name="AppToolbarTheme" parent="@style/Theme.AppCompat.Light.
                                         NoActionBar">
        <!-- 提供装饰主题颜色 -->
        <item name="colorPrimary">@color/primaryGreen</item>
        <item name="colorPrimaryDark">@color/darkGreen</item>
        <item name="colorAccent">@color/accentGreen</item>

    </style>
</resources>
```

在本例中,我们的样式主题继承自 Theme.AppCompat.Light.NoActionBar,后者禁用窗口中的默认 Action Bar;我们将使用自己的 Toolbar 代替它。代码清单 2-9 显示了 Toolbar Activity 的布局。

代码清单 2-9　res/layout/activity_toolbar.xml

```xml
<?xml version="1.0" encoding="utf-8"?>
<LinearLayout xmlns:android="http://schemas.android.com/apk/res/android"
    xmlns:app="http://schemas.android.com/apk/res-auto"
    android:orientation="vertical"
    android:layout_width="match_parent"
    android:layout_height="match_parent">
    <!-- Toolbar 小部件 -->
    <android.support.v7.widget.Toolbar
        android:id="@+id/toolbar"
        android:layout_height="wrap_content"
        android:layout_width="match_parent"
        android:minHeight="?attr/actionBarSize"
        android:background="?attr/colorPrimary"
        app:theme="@style/ThemeOverlay.AppCompat.Dark.ActionBar"/>

    <!-- 此处加入其他的应用程序视图内容 -->

</LinearLayout>
```

Toolbar 小部件必须位于视图层次结构中的某个位置,通常是在视图的顶部。Toolbar 不会自动接收主题样式,因此需要将背景色设置为当前激活主题中的 colorPrimary 属性。我们还必须向 Toolbar 传递主题属性,Toolbar 使用此属性样式化它创建的扩展资源,如操作按钮和弹出列表菜单。ThemeOverlay.AppCompat.Dark.ActionBar 是特殊的"重叠"主题类的一部分,AppCompat 提供该类以仅应用 Action Bar 或 Toolbar 所需的元素样式化其内部组件。代码清单 2-10 显示了将上述内容结合在一起的 Activity 代码。

代码清单 2-10　Toolbar Activity

```java
public class SupportToolbarActivity extends ActionBarActivity {

    private Toolbar mToolbar;

    @Override
    protected void onCreate(Bundle savedInstanceState) {
        super.onCreate(savedInstanceState);
        setContentView(R.layout.activity_toolbar);

        //我们必须将Toolbar的所在位置告诉Activity
        mToolbar = (Toolbar) findViewById(R.id.toolbar);
        setSupportActionBar(mToolbar);
    }

    @Override
    protected void onPostCreate(Bundle savedInstanceState) {
        super.onPostCreate(savedInstanceState);

        /*
         * 使用Toolbar,我们必须在onCreate()之后设置标题文本,
         * 否则默认标签将覆盖我们的设置
         */

        //设置标题文本
        mToolbar.setTitle("Android Recipes");
        //设置副标题文本
        mToolbar.setSubtitle("Toolbar Recipes");
    }

    @Override
    public boolean onCreateOptionsMenu(Menu menu) {
        getMenuInflater().inflate(R.menu.support, menu);
        return true;
    }
}
```

在 onCreate() 内部,我们负责通过 setSupportActionBar() 向 Activity 传递布局内 Toolbar 的引用。尽管该方法的命名很含糊,但它确实需要 Toolbar 作为参数。这就可以使 Activity 向视图应用选项菜单和其他特定项。

在使用 Toolbar 时,我们必须对前一个示例进行一处修改。Activity 实现在 onCreate() 完成后设置 Toolbar 实例的标题值。这意味着在 onCreate() 中设置的任何标题将重置为清单的 android:label 值。为抵消这种行为,我们必须在 onPostCreate() 中进行修改以使改动得以保持。

如果运行此 Activity,其外观应如图 2-1 所示。本章后面将给出不同 UI 范例如何倾向于使用 Toolbar 而非顶层 Action Bar 的示例。

2.2 锁定 Activity 方向

2.2.1 问题

应用程序中的某个 Activity 不能旋转,或是旋转需要应用程序代码更直接的干预。

2.2.2 解决方案

(API Level 1)

在 AndroidManifest.xml 文件中可以用静态声明将每个 Activity 的方向锁定为横向或纵向。这个声明只能用于<activity>标签,所以不能一次性解决整个应用程序。

只需要在<activity>元素中加上 android:screenOrientation 属性,无论设备处于什么位置,Activity 都会按指定的方向显示。下面是最常用的属性:

- portrait:屏幕向上朝向设备的顶部。
- landscape:屏幕向上朝向设备的右侧。
- sensorPortrait:屏幕的朝向采用纵向模式(最小的水平宽度),设备的短边向上。
- sensorLandscape:屏幕的朝向采用横向模式(最小的垂直宽度),设备的长边向上。
- behind:屏幕朝向与 Activity 栈中前一个 Activity 的方向保持一致。

如需可用选项的完整列表,可查阅 SDK 文档中关于<activity>清单元素的部分。

2.2.3 实现机制

在代码清单 2-11 中,示例 AndroidManifest.xml 文件中有 3 个 Activity,每个 Activity 在不同方向上被锁定。

代码清单 2-11 有一些 Activity 被锁定为纵向的清单文件

```xml
<?xml version="1.0" encoding="utf-8"?>
<manifest
    xmlns:android="http://schemas.android.com/apk/res/android"
    package="com.examples.rotation"
    android:versionCode="1"
    android:versionName="1.0">
    <application android:icon="@drawable/icon"
      android:label="@string/app_name">
      <activity android:name=".MainActivity"
        android:label="@string/app_name"
        android:screenOrientation="portrait">
        <intent-filter>
          <action
            android:name="android.intent.action.MAIN" />
          <category
            android:name="android.intent.category.LAUNCHER" />
```

```xml
            </intent-filter>
        </activity>

        <activity android:name=".ResultActivity"
            android:screenOrientation="landscape" />

        <activity android:name=".UserEntryActivity" />
            android:screenOrientation="sensorLandscape" />
    </application>
</manifest>
```

2.3 动态方向锁定

2.3.1 问题

在某些特定的条件下,不能让屏幕旋转,但这个条件是临时的或是根据用户的意愿决定的。

2.3.2 解决方案

(API Level 1)

借助 Android 的请求方向机制(requested orientation mechanism),应用程序可以调整显示 Activity 的屏幕方向,将其固定为某个方向或是交由设备决定。这是通过调用 Activity.SetRequestedOrientation()方法实现的,该方法的参数是 ActivityInfo.screenOrientation 属性组中的整数常量。

默认情况下,请求的屏幕方向设为 SCREEN_ORIENTATION_UNSPECIFIED,也就是由设备决定屏幕的方向。这通常是根据设备的物理方向来确定的。当前请求的方向可以随时通过调用 Activity.getRequestedOrientation()方法获得。

2.3.3 实现机制

举个示例,让我们创建一个 ToggleButton 实例来控制是否锁定当前屏幕方向,这样就能让用户随时控制 Activity 的方向。代码清单 2-12 描绘了在其中定义 ToggleButton 实例的简单布局。

代码清单 2-12 res/layout/activity_lock.xml

```xml
<?xml version="1.0" encoding="utf-8"?>
<FrameLayout xmlns:android="http://schemas.android.com/apk/res/android"
    android:layout_width="match_parent"
    android:layout_height="match_parent">

    <ToggleButton
```

```xml
        android:id="@+id/toggleButton"
        android:layout_width="match_parent"
        android:layout_height="wrap_content"
        android:layout_gravity="center"
        android:textOff="Lock"
        android:textOn="LOCKED"
        />
</FrameLayout>
```

在 Activity 代码中,我们给这个按钮的状态创建一个监听器,根据按钮的值决定锁定或解锁屏幕的方向(参见代码清单 2-13)。

代码清单 2-13　动态锁定/解锁屏幕方向的 Activity

```java
public class LockActivity extends Activity {

    protected void onCreate(Bundle savedInstanceState) {
        super.onCreate(savedInstanceState);
        setContentView(R.layout.main);
        //获得按钮资源的句柄
        ToggleButton toggle = (ToggleButton)findViewById(R.id.toggleButton);
        //在添加监听器之前设置默认状态
        if(getRequestedOrientation() !=
               ActivityInfo.SCREEN_ORIENTATION_UNSPECIFIED) {
            toggle.setChecked(true);
        } else {
            toggle.setChecked(false);
        }
        //将监听器关联到按钮
        toggle.setOnCheckedChangeListener(listener);
    }

    OnCheckedChangeListener listener = new OnCheckedChangeListener() {
        public void onCheckedChanged(CompoundButton buttonView,
                boolean isChecked) {
            int current = getResources().getConfiguration().orientation;
            if(isChecked) {
                switch(current) {
                case Configuration.ORIENTATION_LANDSCAPE:
                    setRequestedOrientation(
                            ActivityInfo.SCREEN_ORIENTATION_LANDSCAPE);
                    break;
                case Configuration.ORIENTATION_PORTRAIT:
                    setRequestedOrientation(
                            ActivityInfo.SCREEN_ORIENTATION_PORTRAIT);
                    break;
                default:
                    setRequestedOrientation(
                            ActivityInfo.SCREEN_ORIENTATION_UNSPECIFIED);
```

```
                }
            } else {
                setRequestedOrientation(
                    ActivityInfo.SCREEN_ORIENTATION_UNSPECIFIED);
            }
        }
    }
}
```

监听器中的这段代码是本攻略的关键。如果用户按下这个按钮，将其置为 ON 状态，应用程序就通过保存 Resources.getConfiguration() 的 orientation 参数读取当前的屏幕方向。Configuration 对象用来表示屏幕方向的常数与所请求的方向使用的常数不同，所以我们根据当前的屏幕方向进行切换，然后再用合适的常数调用 setRequestedOrientation() 方法。

注意：
如果所请求的方向跟当前状态不一样，且 Activity 在前台，Activity 就会立即切换方向以满足请求。

如果用户按下了这个按钮，将其设为 OFF 状态，我们就不再锁定方向，因此用 SCREEN_ORIENTATION_UNSPECIFIED 常数再调用一次 setRequestedOrientation()，将控制权交还给设备。如果设备当前的物理方向跟移除锁定时的 Activity 方向不一致，也会导致屏幕切换方向。

注意：
设置请求的方向不会保持默认的 Activity 生命周期。如果设备配置发生变化(例如物理键盘弹出，或是设备方向改变)，Activity 依然会被销毁并重新创建，因此所有保持 Activity 状态的规则都适用。

2.4 手动处理旋转

2.4.1 问题

在旋转过程中，默认会将 Activity 销毁，然后再重新创建，这会严重影响应用程序的性能。

如果没有自行修改的话，在配置变化时，Android 会结束当前的 Activity 实例，然后重新创建一个适用于新配置的 Activity 实例，这会带来性能上的损失，因为这需要先保存 UI 状态，然后再完全重新构建 UI。

2.4.2 解决方案

(API Level 1)
利用清单文件中的 android:configChanges 参数，告诉 Android 某个 Activity 在处理旋转

事件时不需要运行时进行干预,这不仅能降低 Android 的工作量,即销毁和重建 Activity 实例,也会降低应用程序的工作量。保持 Activity 实例不变,应用程序就不必为了针对用户维持一致性而花费时间保存和还原应用程序的当前状态。

注册了一个或多个配置变动的 Activity 可以通过 Activity.onConfigurationChanged()回调方法收到通知,在该方法中可以执行各种跟配置变动有关的必要手动操作。

要完全以手动方式处理旋转,Activity 至少要注册三个配置变动参数:orientation、keyboardHidden 和 screenSize。

- orientation:为设备方向变动的事件注册 Activity。
- screenSize:为设备屏幕长宽比变动的事件注册 Activity。在每次方向变动时才会发生此事件。
- keyboardHidden:为用户滑开或关闭物理键盘的事件注册 Activity。

尽管后者看上去跟屏幕旋转没有直接关系,但如果不注册这些事件的话,Android 就会在这些事件发生时重建 Activity,这会使前面手动处理旋转的努力付之东流。

2.4.3 实现机制

将这些参数添加到 AndroidManifest.xml 的任意一个<Activity>元素中,如下所示:

```
<activity android:name=".MyActivity"
    android:configChanges="orientation|keyboardHidden|screenSize" />
```

在一条赋值语句中可以注册多种变动,用 "|" 符号将它们分开即可。因为这些参数都不能应用于<application>元素,所以每个 Activity 都必须在 AndroidManifest.xml 中注册。

注册 Activity 后,配置发生改变时就会调用 Activity 的 onConfigurationChanged()方法。代码清单 2-14 到 2-16 是一个简洁的 Activity 定义,用于处理变动发生时产生的回调。

代码清单 2-14 res/layout/activity_manual.xml

```xml
<?xml version="1.0" encoding="utf-8"?>
<LinearLayout xmlns:android="http://schemas.android.com/apk/res/android"
    android:orientation="vertical"
    android:layout_width="match_parent"
    android:layout_height="match_parent">
    <TextView
        android:layout_width="match_parent"
        android:layout_height="wrap_content"
        android:gravity="center_horizontal"
        android:text="Rotate the device, Activity will remain"/>
    <CheckBox
        android:id="@+id/override"
        android:layout_width="wrap_content"
        android:layout_height="wrap_content"
        android:text="Check to Force Reload View"/>
    <EditText
        android:id="@+id/text"
        android:layout_width="match_parent"
```

```xml
        android:layout_height="wrap_content"/>
</LinearLayout>
```

代码清单 2-15　res/layout-land/activity_manual.xml

```xml
<?xml version="1.0" encoding="utf-8"?>
<LinearLayout xmlns:android="http://schemas.android.com/apk/res/android"
    android:orientation="horizontal"
    android:layout_width="match_parent"
    android:layout_height="match_parent">
    <CheckBox
        android:id="@+id/override"
        android:layout_width="wrap_content"
        android:layout_height="wrap_content"
        android:text="Check to Force Reload View"/>
    <EditText
        android:id="@+id/text"
        android:layout_width="match_parent"
        android:layout_height="wrap_content"/>
</LinearLayout>
```

代码清单 2-16　手动管理旋转的 Activity

```java
public class MyActivity extends Activity {

    //引用视图元素
    private EditText mEditText;
    private CheckBox mCheckBox;

    @Override
    protected void onCreate(Bundle savedInstanceState) {
        //必须调用超类
        super.onCreate(savedInstanceState);
        //加载视图资源
        loadView();
    }

    @Override
    public void onConfigurationChanged(Configuration newConfig) {
        //必须调用超类
        super.onConfigurationChanged(newConfig);

        //如果选中此复选框，则只重新加载新配置下的元素
        if (mCheckBox.isChecked()) {
            final Bundle uiState = new Bundle();
            //保存重要的 UI 状态
            saveState();
            //重新加载视图资源
            loadView();
```

```java
        //恢复UI状态
        restoreState(uiState);
}

    //实现持久化UI状态的代码
    private void saveState() {
        state.putBoolean("checkbox", mCheckBox.isChecked());
        state.putString("text", mEditText.getText().toString());
    }

    //恢复重新加载之前保存的任何元素
    private void restoreState(Bundle state) {
        mCheckBox.setChecked(state.getBoolean("checkbox"));
        mEditText.setText(state.getString("text"));
    }
    //设置内容视图并获得视图引用
    private void loadView() {
        setContentView(R.layout.activity_manual);
        //我们必须在设置新的布局时重置视图引用
        mCheckBox = (CheckBox) findViewById(R.id.override);
        mEditText = (EditText) findViewById(R.id.text);
    }
}
```

注意：

Google 并不推荐这样处理旋转，除非应用程序的性能确实有此要求。在响应各个变动事件时，所有跟配置有关的资源都必须手动加载。

Google 建议允许默认的 Activity 旋转行为重建，除非应用程序的性能对此确实有要求。这主要是因为，阻止默认的 Activity 重建会导致 Android 无法加载在资源条件目录中存放的备选资源(例如 res/layout-land/目录中的横屏布局资源)。

在示例的 Activity 中，所有用于处理视图布局的代码都被抽象成在 onCreate()和 onConfigurationChanged()中调用的私有方法 loadView()中。在该方法中，类似 setContentView()这样的代码可以确保加载与配置相匹配的布局。

调用 setContentView()会重新加载视图，所以一定要在没有诸如 onSaveInstanceState()和 onRestoreInstanceState()等生命周期回调函数的协助下，保存所有的 UI 状态。为了达到该目的，示例中实现了 saveState()和 restoreState()方法。

为了在不添加视图重载代码的情况下演示此 Activity 的行为，我们在布局中关联了一个复选框，用于确定在配置改变时视图是否再次加载。选中此复选框时，Activity 仍然会旋转并在新的方向重绘其内容。然而，相反的配置布局(横向或纵向)将不会重新加载。

2.5 创建上下文动作

2.5.1 问题

需要为用户在用户界面中的选择提供多个可供执行的动作。

2.5.2 解决方案

(API Level 11)
对于与单个项相关的上下文动作，使用 PopupMenu 显示固定到相关视图的这些动作。在应该影响多个项的示例中，可启用 ActionMode 来响应用户的动作。

注意：
该例使用 AppCompat 支持库来实现最佳的版本兼容性。如果应用程序单纯地支持较早的平台版本，可以使用原生的 API 实现相同的结果。关于在项目中包括支持库的更多信息，请参阅 http://developer.android.com/tools/support-library/index.html。

2.5.3 实现机制

对于本例，我们将构造如图 2-3 所示的 Activity。

图 2-3　具有上下文动作(左图)和 ActionMode(右图)的列表视图

在点击项视图右侧的溢出按钮时，列表中的每个项通过弹出列表提供上下文动作。

长按任何项时，Activity 将激活 ActionMode，从而可以将同一个动作应用于多个已选择的项。

1. 上下文弹出菜单

PopupMenu 用于获得一个选项菜单资源并将其作为小型弹出窗口轻松附加到任何视图。首先，我们需要在 res/menu 中创建一个 XML 文件以定义菜单自身；将此文件称为 contextmenu.xml(参见代码清单 2-17)。

代码清单 2-17　res/menu/contextmenu.xml

```xml
<?xml version="1.0" encoding="utf-8"?>
<menu xmlns:android="http://schemas.android.com/apk/res/android">
    <item
        android:id="@+id/menu_delete"
        android:icon="@android:drawable/ic_menu_delete"
        android:title="Delete Item"
    />
    <item
        android:id="@+id/menu_edit"
        android:icon="@android:drawable/ic_menu_edit"
        android:title="Edit Item"
    />
</menu>
```

我们为驻留在此列表中的每个项将此资源扩展到 PopupMenu 实例中。为更好地封装此逻辑，代码清单 2-18 和 2-19 为列表行布局自定义了 ContextListItem 类。

代码清单 2-18　自定义行项列表

```java
public class ContextListItem extends LinearLayout implements
        PopupMenu.OnMenuItemClickListener,
        View.OnClickListener {

    private PopupMenu mPopupMenu;
    private TextView mTextView;

    public ContextListItem(Context context) {
        super(context);
    }

    public ContextListItem(Context context, AttributeSet attrs) {
        super(context, attrs);
    }

    public ContextListItem(Context context, AttributeSet attrs, int defStyleAttr) {
        super(context, attrs, defStyleAttr);
    }
```

```java
    @Override
    protected void onFinishInflate() {
        super.onFinishInflate();
        mTextView = (TextView) findViewById(R.id.text);

        //附加单击处理程序
        View contextButton = findViewById(R.id.context);
        contextButton.setOnClickListener(this);
        //创建上下文菜单
        mPopupMenu = new PopupMenu(getContext(), contextButton);
        mPopupMenu.setOnMenuItemClickListener(this);
        mPopupMenu.inflate(R.menu.contextmenu);
    }

    @Override
    public void onClick(View v) {
        //处理上下文按钮单击以显示菜单
        mPopupMenu.show();
    }

    @Override
    public boolean onMenuItemClick(MenuItem item) {
        String itemText = mTextView.getText().toString();

        switch (item.getItemId()) {
            case R.id.menu_edit:
                Toast.makeText(getContext(), "Edit "+itemText,
                    Toast.LENGTH_SHORT).show();
                break;
            case R.id.menu_delete:
                Toast.makeText(getContext(), "Delete "+itemText,
                    Toast.LENGTH_SHORT).show();
                break;
        }
        return true;
    }
}
```

代码清单 2-19 res/layout/list_item.xml

```xml
<com.examples.popupmenus.ContextListItem
    xmlns:android="http://schemas.android.com/apk/res/android"
    android:orientation="horizontal"
    android:layout_width="match_parent"
    android:layout_height="?android:attr/listPreferredItemHeightSmall"
    android:paddingLeft="?android:attr/listPreferredItemPaddingLeft"
    android:paddingRight="?android:attr/listPreferredItemPaddingRight"
    android:background="?android:attr/activatedBackgroundIndicator" >
    <TextView
        android:id="@+id/text"
        android:layout_width="0dp"
```

```xml
            android:layout_height="wrap_content"
            android:layout_weight="1"
            android:textAppearance="?android:attr/textAppearanceListItemSmall"
            android:layout_gravity="center_vertical" />

    <ImageView
            android:id="@+id/context"
            style="@style/Widget.AppCompat.Light.ActionButton.Overflow"
            android:layout_width="?android:attr/listPreferredItemHeightSmall"
            android:layout_height="match_parent"
            android:clickable="true"
            android:focusable="false"/>
</com.examples.popupmenus.ContextListItem>
```

ContextListItem 是包含上下文菜单的文本项和图片按钮的 LinearLayout。为实现平台一致性，我们对按钮应用 Widget.AppCompat.Light.ActionButton.Overflow 样式，使其外观和行为类似于标准的溢出菜单按钮。创建视图时，我们构建一个 PopupMenu 来显示 R.menu.contextmenu，并且关联该菜单，使其在按钮溢出按钮时始终会显示。行视图也设置为 OnMenuItemClickListener，以便处理从弹出菜单中选择适当的选项。

注意：
ListView 内的可单击项需要将 android:focusable 设置为 false，否则就不能够同时单击顶层列表项。

为了将此视图与列表中的数据绑定，我们需要创建引用列表项布局的基本适配器：

```
ArrayAdapter<String> adapter = new ArrayAdapter<String>(this,
        R.layout.list_item, R.id.text, ITEMS);
```

我们将很快看到将此功能与其他功能联系在一起的完整 Activity 代码，但首先让我们查看一下如何实现多选逻辑。

2. ActionMode

ActionMode API 解决了类似的问题，即允许用户对 UI 中的某个条目执行某种动作，但它的实现方式稍微有点不同。激活 ActionMode 后会在窗口的 Action Bar 上出现一个包含所提供的菜单选项的叠层，并出现一个可以退出 ActionMode 的额外选项。它还允许同时选择多个条目来执行同一个动作。代码清单 2-20 演示了这一功能。

代码清单 2-20　使用了上下文 ActionMode 的 Activity

```java
public class ActionActivity extends Activity implements
        AbsListView.MultiChoiceModeListener {

    private static final String[] ITEMS =
            "Mom", "Dad", "Brother", "Sister", "Uncle", "Aunt",
            "Cousin", "Grandfather", "Grandmother"};
    private ListView mList;
```

```java
protected void onCreate(Bundle savedInstanceState) {
    super.onCreate(savedInstanceState);
    //为上下文事件注册一个按钮
    mList = new ListView(this);
    ArrayAdapter<String> adapter = new ArrayAdapter<String>(this,
            R.layout.list_item, R.id.text, ITEMS);
    mList.setAdapter(adapter);
    mList.setChoiceMode(ListView.CHOICE_MODE_MULTIPLE_MODAL);
    mList.setMultiChoiceModeListener(this);

    setContentView(mList, new ViewGroup.LayoutParams(
            ViewGroup.LayoutParams.MATCH_PARENT,
            ViewGroup.LayoutParams.MATCH_PARENT));
}

@Override
public boolean onPrepareActionMode(ActionMode mode, Menu menu) {
    //如果 ActionMode 一直是无效的，可以在这里做些工作来更新菜单
    return true;
}

@Override
public void onDestroyActionMode(ActionMode mode) {
    //退出 ActionMode 时会调用这个方法
}

@Override
public boolean onCreateActionMode(ActionMode mode, Menu menu) {
    MenuInflater inflater = mode.getMenuInflater();
    inflater.inflate(R.menu.contextmenu, menu);
    return true;
}

@Override
public boolean onActionItemClicked(ActionMode mode, MenuItem item) {
    SparseBooleanArray items = mList.getCheckedItemPositions();
    //通过条目的 ID 得到用户选择的动作
    switch(item.getItemId()) {
    case R.id.menu_delete:
        //执行删除动作
        break;
    case R.id.menu_edit:
        //执行编辑动作
        break;
    default:
        return false;
    }
    return true;
}
```

```
@Override
public void onItemCheckedStateChanged(ActionMode mode, int position,
        long id, boolean checked) {
    int count = mList.getCheckedItemCount();
    mode.setTitle(String.format("%d Selected", count));
}
}
```

要使用我们的 ListView 来激活一个多选择项的 ActionMode，需要将 choiceMode 属性设置为 CHOICE_MODE_MULTIPLE_MODAL。它和传统的 CHOICE_MODE_MULTIPLE 不同，CHOICE_MODE_MULTIPLE 是在每个列表条目上都提供一个可选择的小部件。在 ActionMode 处于激活状态时，它的模式标识只应用这个选择模式。

对于 ActionMode，有很多的回调方法需要实现它，因为它不像 ContextMenu 一样是直接内置在 Activity 中的。我们需要实现 ActionMode.Callback 接口来响应创建菜单和选择选项的事件。ListView 有一个名为 MultiChoiceModeListener 的特殊接口，它是 ActionMode.Callback 的子接口，本例中就实现了这个特殊的接口。

我们在 onCreateActionMode()中的响应和 onCreateContextMenu()类似，只是构建了叠层中要显示的菜单选项。这个菜单不需要包含图标；这时，ActionMode 会显示条目名。选中每个条目后，我们会在 onItemCheckedStateChanged()方法中得到通知。这里，我们会更新 ActionMode 的标题来显示当前选中条目的个数。

当用户结束选择并单击一个菜单选项时会调用 onActionItemClicked()方法。因为会同时选择多个条目，所以我们通过 getCheckedItemPositions()得到所有选择的条目，这样就可以对所有选择的条目进行操作。

2.6 显示一个用户对话框

2.6.1 问题

需要向用户显示一个简单的弹出式对话框来进行事件通知或展示一个选项列表。

2.6.2 解决方案

(API Level 1)

在向用户快速展示重要模态信息的场景中，AlertDialog 是最高效的解决方案。它展示的内容可以很轻松地进行自定义，同时框架还提供了一个方便的 AlertDialog.Builder 类来快速构建弹出式对话框。

2.6.3 实现机制

通过使用 AlertDialog.Builder，可以构建类似的报警对话框，但包含不同的额外选项。AlertDialog 在创建简单的弹出式对话框来获得用户反馈时是一个非常通用的类。通过

AlertDialog.Builder,可以很容易地在一个简洁的小部件中添加单选或多选列表、按钮和消息字符串。

为了说明这一点,让我们用 AlertDialog 创建一个和以前一样的弹出式选择框。这一次,我们将在选项列表的底部增加 Cancel 按钮(参见代码清单 2-21)。

代码清单 2-21　使用了 AlertDialog 的动作菜单

```java
public class DialogActivity extends Activity
        implements DialogInterface.OnClickListener, View.OnClickListener {

    private static final String[] ZONES = {"Pacific Time", "Mountain Time",
            "Central Time", "Eastern Time", "Atlantic Time"};

    Button mButton;
    AlertDialog mActions;

    @Override
    protected void onCreate(Bundle savedInstanceState) {
        super.onCreate(savedInstanceState);

        mButton = new Button(this);
        mButton.setText("Click for Time Zones");
        mButton.setOnClickListener(this);

        AlertDialog.Builder builder = new AlertDialog.Builder(this);
        builder.setTitle("Select Time Zone");
        builder.setItems(ZONES, this);
        //这里的取消动作只会让对话框消失,但在用户单击 Cancel 按钮时,也可以添加另一个
        //监听器来处理一些其他的操作
        builder.setNegativeButton("Cancel", null);
        mActions = builder.create();

        setContentView(mButton);
    }

        //这里处理列表的选择事件
        @Override
        public void onClick(DialogInterface dialog, int which) {
            String selected = ZONES[which];
            mButton.setText(selected);
        }

        //这里处理按钮的单击事件(弹出对话框)
        @Override
        public void onClick(View v) {
            mActions.show();
        }
}
```

本例中,我们创建了一个新的 AlertDialog.Builder 实例并使用它的便捷方法添加了如下

条目：
- 标题，通过 setTitle()设置。
- 选项列表，通过 setItems()和一个字符串数组(也可以是数组资源)设置。
- Cancel 按钮，通过 setNegativeButton()设置。

当选择列表中的一个选项时，我们关联到列表项的监听器会返回所选择的选项(索引是之前提供的数组的索引，从 0 开始)，然后我们就可以根据用户的选择来更新按钮上的文本信息。对于 Cancel 按钮，因为我们只想在按下 Cancel 按钮时关闭对话框，所以我们为 Cancel 按钮传入的监听器为 null。如果在按下 Cancel 按钮时还想处理一些重要的工作，可以向 setNegativeButton()方法传入其他的监听器。

另外，还有其他的一些选项可以设置除了选项列表之外的对话框中的内容：
- setMessage()可以为对话框的内容设置一条简单的文本信息。
- setSingleChoiceItems()和 setMultiChoiceItems()会创建一个与本例类似的列表，但每个选项是以复选框的方式显示的。
- setView()可以为对话框的内容使用任意自定义视图。

现在，按钮按下后得到的应用程序的外观如图 2-4 所示。

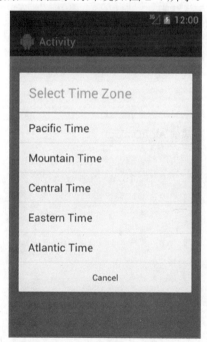

图 2-4　带有选项列表的 AlertDialog

自定义列表条目

AlertDialog.Builder 允许传入一个自定义的 ListAdapter 作为对话框中显示列表条目的数据源，这也就意味着我们可以自定义列表中每一行的布局来向用户显示更加详细的信息。在代码清单 2-22 和 2-23 中我们改进了之前的示例，为每一行使用自定义的行布局来显示每个条目的其他信息。

代码清单 2-22   res/layout/list_item.xml

```xml
<?xml version="1.0" encoding="utf-8"?>
<RelativeLayout
    xmlns:android="http://schemas.android.com/apk/res/android"
    android:layout_width="match_parent"
    android:layout_height="wrap_content"
    android:paddingLeft="10dp"
    android:paddingRight="10dp"
    android:minHeight="?android:attr/listPreferredItemHeight">
    <TextView
        android:id="@+id/text_name"
        android:layout_width="wrap_content"
        android:layout_height="wrap_content"
        android:layout_centerVertical="true"
        android:textAppearance="?android:attr/textAppearanceMedium"
    />
    <TextView
        android:id="@+id/text_detail"
        android:layout_width="wrap_content"
        android:layout_height="wrap_content"
        android:layout_alignParentRight="true"
        android:layout_centerVertical="true"
        android:textAppearance="?android:attr/textAppearanceSmall"
    />
</RelativeLayout>
```

代码清单 2-23   带有自定义布局的 AlertDialog

```java
public class CustomItemActivity extends Activity implements
        DialogInterface.OnClickListener,
        View.OnClickListener {

    private static final String[] ZONES = {
            "Pacific Time", "Mountain Time",
            "Central Time", "Eastern Time",
            "Atlantic Time"};

    private static final String[] OFFSETS = {
            "GMT-08:00", "GMT-07:00", "GMT-06:00",
            "GMT-05:00", "GMT-04:00"};

    Button mButton;
    AlertDialog mActions;

    @Override
    protected void onCreate(Bundle savedInstanceState) {
        super.onCreate(savedInstanceState);

        mButton = new Button(this);
        mButton.setText("Click for Time Zones");
```

```java
        mButton.setOnClickListener(this);

        ArrayAdapter<String> adapter = new ArrayAdapter<String>(
                this,
                R.layout.list_item) {
            @Override
            public View getView(int position, View convertView,
                    ViewGroup parent) {
                View row = convertView;
                if(row == null) {
                    row = getLayoutInflater().inflate(R.layout.list_item,
                            parent, false);
                }

                TextView name =
                        (TextView) row.findViewById(R.id.text_name);
                TextView detail =
                        (TextView) row.findViewById(R.id.text_detail);
                name.setText(ZONES[position]);
                detail.setText(OFFSETS[position]);

                return row;
            }

            @Override
            public int getCount() {
                return ZONES.length;
            }
        };

        AlertDialog.Builder builder = new AlertDialog.Builder(this);
        builder.setTitle("Select Time Zone");
        builder.setAdapter(adapter, this);
        //这里的取消动作只会让对话框消失,但在用户单击Cancel按钮时,也可以添加一个
        //监听器来处理一些其他的操作
        builder.setNegativeButton("Cancel", null);
        mActions = builder.create();

        setContentView(mButton);
    }

    //这里处理列表的选择事件
    @Override
    public void onClick(DialogInterface dialog, int which) {
        String selected = ZONES[which];
        mButton.setText(selected);
    }

    //这里处理按钮的单击事件(弹出对话框)
    @Override
```

```
public void onClick(View v) {
    mActions.show();
}
```
}

这里我们为生成器提供了一个 ArrayAdapter，而不是简单地传入选项数组。ArrayAdapter 有 getView()的自定义实现，会返回一个 XML 中自定义的布局，其中显示了两个文本标签：一个靠左对齐，另一个靠右对齐。通过这个自定义布局，我们现在可以显示 GMT 偏移值和时区名称。本章的后面还会进一步讨论自定义适配器的特性。图 2-5 显示了新的、更加实用的弹出式对话框。

图 2-5　带有自定义条目的 AlertDialog

2.7　自定义菜单和动作

2.7.1　问题

应用程序需要为用户提供一个动作集，但又不想占用视图结构的屏幕空间。

2.7.2　解决方案

(API Level 7)

使用框架中的选项菜单功能在 Action Bar 内提供常用动作，以及在溢出的弹出式菜单中提供额外的选项。此外，通过使用 PopupMenu，可以将菜单附加到现有的视图并显示为浮动下拉菜单。此功能用于在应用程序中除了 Action Bar 之外的任意位置放置菜单，但在用户需要这些菜单之前使它们保持脱离视图。

根据设备平台版本的不同，Android 的菜单功能也有所不同。在早期的版本中，所有的 Android 都有一个物理的 MENU 键可以触发这个功能。从 Android 3.0 开始，出现了没有物理按钮的设备，菜单功能也变成了 Action Bar 的一部分。

驻留在 Action Bar 中的动作项还可以展开以显示称为动作视图的自定义小部件。这可以用于提供搜索字段等功能，该功能需要额外的用户输入，但要将其隐藏在某个动作项的后面，直到用户点击时才会显示。

注意：

该例使用来自 Android 支持库的一些兼容类来向后兼容运行 Android 2.1(API Level 7) 的设备。关于在项目中包括支持库的更多信息，请参阅 http://developer.android.com/tools/support-library/index.html。

2.7.3 实现机制

代码清单 2-24 定义了将在 XML 中使用的选项菜单。

代码清单 2-24　res/menu/options.xml

```xml
<menu xmlns:android="http://schemas.android.com/apk/res/android">
    <item android:id="@+id/menu_add"
        android:title="Add Item"
        android:icon="@android:drawable/ic_menu_add"
        android:showAsAction="always|collapseActionView"
        android:actionLayout="@layout/view_action" />
    <item android:id="@+id/menu_remove"
        android:title="Remove Item"
        android:icon="@android:drawable/ic_menu_delete"
        android:showAsAction="ifRoom" />
    <item android:id="@+id/menu_edit"
        android:title="Edit Item"
        android:icon="@android:drawable/ic_menu_edit"
        android:showAsAction="ifRoom" />
    <item android:id="@+id/menu_settings"
        android:title="Settings"
        android:icon="@android:drawable/ic_menu_preferences"
        android:showAsAction="never" />
</menu>
```

title 和 icon 属性定义了每个条目该如何显示，旧版本平台会显示这两个值，而新版本中会显示一个或根据位置显示另一个。只有 Android 3.0 及以后的设备可以识别 showAsAction 属性，该属性定义了条目是否应该变成 Action Bar 中的一个动作或者放到溢出菜单中。这个属性最常用的值如下：

- always：总是作为一个动作显示其图标。
- never：总是显示在溢出菜单中并显示其名称。
- ifRoom：如果 Action Bar 上有空间，就作为动作显示，否则显示在溢出菜单中。

菜单中的第一个条目还定义了 android:actionLayout 资源，该资源指向在点击此条目时要展开到其中的小部件；此外还定义了额外的显示标志 collapseActionView，该标志告诉框架，此条目有可折叠的动作视图以供显示。代码清单 2-25 显示了动作视图布局，这是带有两个 CheckBox 实例的简单布局。

代码清单 2-25　res/layout/view_action.xml

```xml
<?xml version="1.0" encoding="utf-8"?>
<LinearLayout
    xmlns:android="http://schemas.android.com/apk/res/android"
    android:layout_width="match_parent"
    android:layout_height="wrap_content"
    android:orientation="horizontal">
    <CheckBox
        android:id="@+id/option_first"
        android:layout_width="0dp"
        android:layout_height="wrap_content"
        android:layout_weight="1"
        android:text="First"/>
    <CheckBox
        android:id="@+id/option_second"
        android:layout_width="0dp"
        android:layout_height="wrap_content"
        android:layout_weight="1"
        android:text="Second"/>
</LinearLayout>
```

代码清单 2-26 显示了完整的 Activity，其中将选项菜单填充到 Action Bar 中，同时在某个动作项中放置可展开的动作视图。

代码清单 2-26　覆写菜单动作的 Activity

```java
public class OptionsActivity extends Activity implements
        PopupMenu.OnMenuItemClickListener,
        CompoundButton.OnCheckedChangeListener {

    private MenuItem mOptionsItem;
    private CheckBox mFirstOption, mSecondOption;

    @Override
    public void onCreate(Bundle savedInstanceState) {
        super.onCreate(savedInstanceState);

        // 此例中无附加工作
    }

    @Override
    public boolean onCreateOptionsMenu(Menu menu) {
        // 使用此回调创建菜单并执行任何必要的初始化设置
```

```java
        getMenuInflater().inflate(R.menu.options, menu);

        // 查找并初始化动作项
        mOptionsItem = menu.findItem(R.id.menu_add);
        MenuItemCompat.setOnActionExpandListener(
                new MenuItem.OnActionExpandListener() {

                    @Override
                    public boolean onMenuItemActionExpand(
                            MenuItem item) {
                        // 必须返回true以使项展开
                        return true;
                    }

                    @Override
                    public boolean onMenuItemActionCollapse(
                            MenuItem item) {
                        mFirstOption.setChecked(false);
                        mSecondOption.setChecked(false);
                        // 必须返回true以使项折叠
                        return true;
                    }
                });

        mFirstOption = (CheckBox) MenuItemCompat
                .getActionView(mOptionsItem)
                .findViewById(R.id.option_first);
        mFirstOption
                .setOnCheckedChangeListener(this);
        mSecondOption = (CheckBox) MenuItemCompat
                .getActionView(mOptionsItem)
                .findViewById(R.id.option_second);
        mSecondOption
                .setOnCheckedChangeListener(this);

        return true;
    }

    /* 复选框回调方法 */

    @Override
    public void onCheckedChanged(CompoundButton buttonView,
            boolean isChecked) {
        if(mFirstOption.isChecked()
                && mSecondOption.isChecked()) {
            // 隐藏动作视图
            MenuItemCompat.collapseActionView(mOptionsItem);
        }
    }
}
```

```java
    @Override
    public boolean onPrepareOptionsMenu(Menu menu) {
        // 使用此回调执行每次菜单打开时需要调用的设置
        return super.onPrepareOptionsMenu(menu);
    }

    // 通过PopupMenu点击的回调
    public boolean onMenuItemClick(MenuItem item) {
        menuItemSelected(item);
        return true;
    }

    // 通过标准选项菜单点击的回调
    @Override
    public boolean onOptionsItemSelected(MenuItem item) {
        menuItemSelected(item);
        return true;
    }

    // 私有辅助方法，让每个回调可以触发相同的动作
    private void menuItemSelected(MenuItem item) {
        // 按id获得选择的选项
        switch(item.getItemId()) {
        case R.id.menu_add:
            // 执行添加动作
            break;
        case R.id.menu_remove:
            // 执行删除动作
            break;
        case R.id.menu_edit:
            // 执行编辑动作
            break;
        case R.id.menu_settings:
            // 执行设置动作
            break;
        default:
            break;
        }
    }
}
```

在用户按下设备上的 MENU 键后(或者 Activity 在加载时显示了 Action Bar)，就会调用 onCreateOptionsMenu()方法来构建菜单。有一个名为 MenuInflater 的特殊 LayoutInflater 对象，可以用来根据 XML 创建菜单。这里我们使用 Activity 中已有的 getMenuInflater()方法获得 MenuInflater 实例来创建 XML 菜单。

如果需要在用户每次打开菜单时传递一些动作，可以在 onPrepareOptionsMenu()中完成。这里有一个建议，就是所有变成位于 Action Bar 中的动作在用户选择它们时将不会触发这个回调方法，但在溢出菜单中的动作还会触发该方法。

用户做了选择后，onOptionsItemSelected()回调方法将会触发同时传入选择的菜单条目。因为我们在 XML 中为每个条目都定义了唯一的 ID，所以可以通过 switch 语句判断用户的选择项并进行相应的操作。

最后，在 onCreateOptionsMenu()中有一些用于可展开动作视图的额外设置。在此获得指向包含动作视图布局的菜单项的引用，并且附加一个 OnActionExpandListener 回调。此处使用回调只是为了在条目折叠时清除动作视图中的已选元素。

要点：

如果提供 OnActionExpandListener，就需要在 onMenuItemActionExpand()内返回 true，否则永远不会展开！

可以使用 MenuItem 中的 getActionView()方法获得一个引用，该引用指向在菜单 XML 中设置的已填充动作布局。在我们的示例中，使用此方法对布局内的每个 CheckBox 设置选择的侦听器。在动作视图内同时选择这两个条目时，调用 collapseActionView()将视图转变回单个动作项的图标。

图 2-6 显示了不同版本和配置的设备上这个菜单的样子。依然带有物理按键的设备会在 Action Bar 上显示提示的动作，但溢出菜单还是通过 MENU 键触发的。带有软键的设备会挨着 Action Bar 动作显示一个溢出菜单按钮。

图 2-6 带有物理按键的 Android 设备(左图)和带有软键的 Android 设备(右图)

图 2-7 显示了展开的动作视图，单击 Action Bar 中的 Add 动作时就会显示该视图。

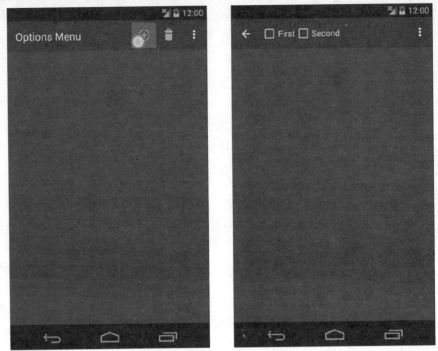

图 2-7　自定义动作视图

2.8　自定义 BACK 按键

2.8.1　问题

应用程序要以自己的方式来处理用户按下物理 BACK 按键后的行为。

2.8.2　解决方案

(API Level 5)
可以在 Activity 中使用 onBackPressed()回调方法或者在 Fragment 中操作回退栈。

2.8.3　实现机制

如果想要用户在你的 Activity 上按下 BACK 按键时可以得到相应的通知,可以覆写 onBackPressed()方法,如下所示:

```
@Override
public void onBackPressed() {
    //实现自定义返回功能

    //调用super以进行常规处理(例如销毁Activity)
    super.onBackPressed();
}
```

这个方法的默认实现会将当前回退栈中的 Fragment 弹出并且销毁 Activity。如果不打算改变这个流程，只需要确保调用父类的实现来保持这种常规的处理方式。

警告：
覆写物理按键事件时应保持谨慎。在 Android 系统中，所有的物理按键都有一致的功能，如果这些按键的功能变化太大，会让用户感到困惑和不满。

BACK 操作和 Fragment

当 UI 中包含 Fragment 时，可以进一步自定义设备的 BACK 按键的行为。默认情况下，在 UI 中添加或替换 Fragment 的操作并不会在任务的回退栈中添加相应的 Fragment，因此当用户按下 BACK 按键后，并不能够回退这些动作。但是，所有的 FragmentTransaction 都可以作为条目通过简单地调用 addToBackStack()(在事务提交前)添加到回退栈中。

默认情况下，当用户按下 BACK 按键后，Activity 会调用 FragmentManager.popBackStack-Immediate()，这样每个通过这种方式添加的 FragmentTransaction 都会在每次点击时弹出，直到栈中一个不剩，然后 Activity 会被销毁。另外，这个方法还有一些变体，允许直接跳到栈中的某个位置。让我们看一下代码清单 2-27 和 2-28。

代码清单 2-27   res/layout/main.xml

```xml
<?xml version="1.0" encoding="utf-8"?>
<LinearLayout xmlns:android="http://schemas.android.com/apk/res/android"
    android:layout_width="match_parent"
    android:layout_height="match_parent"
    android:orientation="vertical">
    <Button
        android:layout_width="match_parent"
        android:layout_height="wrap_content"
        android:text="Go Home"
        android:onClick="onHomeClick" />
    <FrameLayout
        android:id="@+id/container_fragment"
        android:layout_width="match_parent"
        android:layout_height="match_parent"/>
</LinearLayout>
```

代码清单 2-28   自定义 Fragment 回退栈的 Activity

```java
public class MyActivity extends FragmentActivity {

    @Override
    protected void onCreate(Bundle savedInstanceState) {
        super.onCreate(savedInstanceState);
        setContentView(R.layout.main);
        //构建 Fragment 的回退栈
        FragmentTransaction ft = getSupportFragmentManager().beginTransaction();
        ft.add(R.id.container_fragment, MyFragment.newInstance("First Fragment"));
        ft.commit();
```

```java
        ft = getSupportFragmentManager().beginTransaction();
        ft.add(R.id.container_fragment, MyFragment.newInstance("Second Fragment"));
        ft.addToBackStack("second");
        ft.commit();

        ft = getSupportFragmentManager().beginTransaction();
        ft.add(R.id.container_fragment, MyFragment.newInstance("Third Fragment"));
        ft.addToBackStack("third");
        ft.commit();

        ft = getSupportFragmentManager().beginTransaction();
        ft.add(R.id.container_fragment, MyFragment.newInstance("Fourth Fragment"));
        ft.addToBackStack("fourth");
        ft.commit();
    }

    public void onHomeClick(View v) {
        getSupportFragmentManager().popBackStack("second",
            FragmentManager.POP_BACK_STACK_INCLUSIVE);
    }

    public static class MyFragment extends Fragment {
        private CharSequence mTitle;

        public static MyFragment newInstance(String title) {
            MyFragment fragment = new MyFragment();
            fragment.setTitle(title);

            return fragment;
        }

        @Override
        public View onCreateView(LayoutInflater inflater, ViewGroup container,
                Bundle savedInstanceState) {
            TextView text = new TextView(getActivity());
            text.setText(mTitle);
            text.setBackgroundColor(Color.WHITE);

            return text;
        }

        public void setTitle(CharSequence title) {
            mTitle = title;
        }
    }
}
```

注意：

这里我们使用了支持库，允许 Android 3.0 之前的版本使用 Fragment。如果应用程序的目标 API Level 为 11 或更高版本，可以将 FragmentActivity 用 Activity 替换，将 getSupport-FragmentManager() 用 getFragmentManager() 替换。

这个示例会向栈中放入 4 个自定义的 Fragment 实例，所以在应用程序运行时会显示最后添加的那个 Fragment 实例。对于每个事务，我们调用了 addToBackStack()，同时传入一个标记名称来标识这个事务。如果不想直接跳转到栈中某个位置，那么不需要执行以上操作，直接传入 null 即可。每次按下 BACK 按键后，会移除一个 Fragment 实例，直到只剩下第一个 Fragment 实例，这时再按下 BACK 按键，Activity 就会被正常销毁。

注意，第一个事务并没有添加到栈中，这是因为我们希望第一个 Fragment 作为根视图。如果将它也加入到回退栈中，会导致它在 Activity 销毁前被弹出栈，这就使 UI 会有空白状态出现。

这个应用程序还有 "Go Home" 按钮，它可以让用户无论在哪个界面都可以立即回到根 Fragment。这是通过调用 FragmentManager 的 popBackStack() 方法，同时传入希望跳转的事务的标记名称实现的。我们还传入了 POP_BACK_STACK_INCLUSIVE 标识来告诉管理器在栈中移除我们标识的事务。如果没有这个标识，这个示例就会跳转到第二个 Fragment，而不是根视图。

注意：

Android 会跳转到第一个匹配给定标记的事务。如果同一个标记被使用多次，则会跳到第一个添加此标记的事务，而不是最新的事务。

我们不能使用这个方法直接跳到根视图，因为我们无法引用那个事务在回退栈中的标记。这个方法还有一个使用唯一事务 ID(ID 为 FragmentTransaction 的 commit() 方法的返回值) 的版本。使用这个方法，无须包括标记也可以直接跳到根视图。

2.9 模拟 HOME 按键

2.9.1 问题

应用程序需要实现与按下物理 HOME 按键一样的功能。

2.9.2 解决方案

(API Level 5)

用户按下 HOME 按键的行为会发送一个 Intent 给系统，要求系统加载 Home Activity。这与在应用程序中启动其他的 Activity 并没有什么区别，你要做的就是构建合适的 Intent 以实现该效果。

2.9.3 实现机制

把下面这几行代码添加到 Activity 中要实现该功能的地方:

```
Intent intent = new Intent(Intent.ACTION_MAIN);
intent.addCategory(Intent.CATEGORY_HOME);
startActivity(intent);
```

该功能的一个常见用途就是重载 BACK 按键,让用户按下此按键时直接返回主屏幕而不是回到前一个 Activity。当要保护前台 Activity 之前的 Activity(例如登录界面)时,这是很有用的方法。如果执行 BACK 按键的默认行为,就有可能会让用户在未授权的情况下访问系统。

要点:
在修改系统按键的行为时,务必确保不会扰乱用户对此按键所完成动作的预期。

下面这个示例利用前两个范例中介绍的技术,实现了在 Activity 中按下 BACK 按键时返回到主屏幕的行为:

```
@Override
public void onBackPressed() {
    Intent intent = new Intent(Intent.ACTION_MAIN);
    intent.addCategory(Intent.CATEGORY_HOME);
    startActivity(intent);
}
```

2.10 监控 TextView 的变动

2.10.1 问题

应用程序需要持续监控 TextView 小部件(例如 EditText)中文本内容的变动情况。

2.10.2 解决方案

(API Level 1)

实现 android.text.TextWatcher 接口。TextWacher 提供了 3 个文本更新过程中的回调方法:

```
public void beforeTextChanged(CharSequence s, int start, int count, int after);
public void onTextChanged(CharSequence s, int start, int before, int count);
public void afterTextChanged(Editable s);
```

beforeTextChanged()和 onTextChanged()方法主要用于提供提示功能,因为这两个方法实际上都无法修改 CharSequence。如果要截获视图中输入的文本,当 afterTextChanged()方法被调用时文本就有可能发生了变化。

2.10.3 实现机制

调用 TextView.addTextChangedListener()方法将 TextWatcher 注册到 TextView。注意，根据语法，可以将多个 TextWatcher 注册到一个 TextView。

1. 字符计数器示例

TextWatcher 的一个简单应用是创建 EditText 的字符计数器，能随着用户的输入和删除操作实时显示其中的字数。代码清单2-29中的示例Activity就实现了这样一个TextWatcher，注册了一个 EditTex 小部件并在 Activity 的标题上显示字符数。

代码清单 2-29　字符计数器 Activity

```java
public class MyActivity extends Activity implements TextWatcher {

    EditText text;
    int textCount;

    @Override
    protected void onCreate(Bundle savedInstanceState) {
        super.onCreate(savedInstanceState);
        //创建EditText 小部件并添加监控器
        text = new EditText(this);
        text.addTextChangedListener(this);

        setContentView(text);
    }

    /*实现 TextWatcher 的方法*/
    public void beforeTextChanged(CharSequence s, int start, int count,
            int after) { }

    public void onTextChanged(CharSequence s, int start, int before, int count) {
        textCount = text.getText().length();
        setTitle(String.valueOf(textCount));
    }

    @Override
    public void afterTextChanged(Editable s) { }

}
```

因为不需要修改用户正在输入的文本，所以可以调用 onTextChanged()方法以获得字数，只要用户输入的文本发生变化就会回调该方法。其他的方法暂时没用，留空即可。

2. 货币符号格式化示例

SDK 中有一些很方便的用于格式化文本输入的预定义 TextWatcher 实例；PhoneNumberFormattingTextWatcher 就是其中之一。这些实例的任务就是在用户输入时将其按标准格式化，减少输入规范化数据的击键次数。

在代码清单 2-30 中，我们创建了一个 CurrencyTextWatcher，在 TextView 中插入货币符号和分隔点。

代码清单 2-30　货币符号格式器

```java
public class CurrencyTextWatcher implements TextWatcher {

    boolean mEditing;

    public CurrencyTextWatcher() {
        mEditing = false;
    }

    public synchronized void afterTextChanged(Editable s) {
        if(!mEditing) {
            mEditing = true;

            //strip 符号
            String digits = s.toString().replaceAll("\\D", "");
            NumberFormat nf = NumberFormat.getCurrencyInstance();
            try{
                String formatted = nf.format(Double.parseDouble(digits)/100);
                s.replace(0, s.length(), formatted);
            } catch (NumberFormatException nfe) {
                s.clear();
            }

            mEditing = false;
        }
    }

    @Override
    public void beforeTextChanged(CharSequence s, int start, int count,
            int after) { }

    @Override
    public void onTextChanged(CharSequence s, int start, int before, int count) { }

}
```

注意：

在 afterTextChanged()方法中修改 Editable 的值会导致 TextWatcher 方法被再次调用(因为刚刚修改了文本)。考虑到这一点，在实现自定义 TextWatcher 时，应该通过某个布尔变量或是其他跟踪机制来判断文本修改的来源，避免产生无限循环。

我们可以将这个自定义的文本格式化器应用于 Activity 中的 EditText(参见代码清单 2-31)。

代码清单 2-31　使用货币符号格式化器的 Activity

```
public class MyActivity extends Activity {

    EditText text;

    @Override
    protected void onCreate(Bundle savedInstanceState) {
        super.onCreate(savedInstanceState);
        text = new EditText(this);
        text.addTextChangedListener(new CurrencyTextWatcher());

        setContentView(text);
    }
}
```

使用这个格式化器格式化用户输入,就可以很方便地在 XML 中定义 EditText 的 android:inputType 和 android:digits 约束来格式化用户的输入,从而保护字段不受输入错误的影响。特别是,为 EditText 加上 android:digits="0122456789."(注意末尾的小数点),不仅能保护此格式化器,对用户也有好处。

2.11 自定义键盘动作

2.11.1 问题

要自定义软键盘上 Enter 键的外观,或是改变用户按这个键所触发的动作,或是两者都要实现。

2.11.2 解决方案

(API Level 3)
自定义键盘输入数据小部件的输入方法(Input Method,IME)选项。

2.11.3 实现机制

1. 自定义 Enter 键

软键盘出现在屏幕上时,Enter 键上的文字通常显示的是根据当前聚焦的控件在视图中的顺序所执行的动作。在没有特别指定时,如果视图中还有其他可聚焦的控件,这个按键会显示 next 动作;如果当前聚焦的对象已经是最后一个可聚焦对象,则会显示 done 动作。

对于多行字段，该动作为换行。对于每个输入视图，通过视图的 XML 文件中的 android:imeOptions 可以自定义这个值。可用于自定义 Enter 键的值如下所示：

- actionUnspecified：默认值，根据设备的情况显示动作。
    - 动作事件是 IME_NULL。
- actionGo：在 Enter 键上显示 Go。
    - 动作事件是 IME_ACTION_GO。
- ActionSearch：在 Enter 键上显示搜索图标。
    - 动作事件是 IME_ACTION_SEARCH。
- actionSend：在 Enter 键上显示 Send。
    - 动作事件是 IME_ACTION_SEND。
- antionNest：在 Enter 键上显示 Next。
    - 动作事件是 IME_ACTION_NEXT。
- actionDone：在 Enter 键上显示 Done。
    - 动作事件是 IME_ACTION_DONE。

下面看一个有两个可编辑文本框的布局，如代码清单 2-32 所示。第一个文本框在 Enter 键上显示搜索放大镜图标，第二个则显示 Go。

代码清单 2-32　res/layout/main.xml

```
<LinearLayout
    xmlns:android="http://schemas.android.com/apk/res/android"
    android:layout_width="match_parent"
    android:layout_height="match_parent"
    android:orientation="vertical">
    <EditText
        android:id="@+id/text1"
        android:layout_width="match_parent"
        android:layout_height="wrap_content"
        android:singleLine="true"
        android:imeOptions="actionSearch" />
    <EditText
        android:id="@+id/text2"
        android:layout_width="match_parent"
        android:layout_height="wrap_content"
        android:singleLine="true"
        android:imeOptions="actionGo" />
</LinearLayout>
```

最终显示的键盘可能会因为生产商自定义的用户界面工具包而有些许差异，Google 原生用户界面的效果如图 2-8 所示。

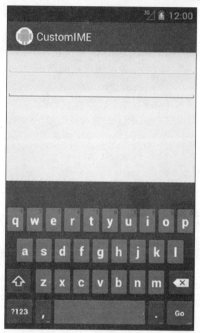

图 2-8　Enter 键上的自定义输入选项

注意：

自定义编辑器选项只会影响软键盘输入。改变这些选项的值不会影响到用户在物理键盘上按 Enter 键时触发的事件。

2. 自定义动作

与自定义 Enter 键上显示的文字一样重要的是自定义用户按下此按键时所触发的动作。重载动作的默认行为只需要给相应的视图加上 TextView.OnEditorActionListener。下面继续修改上面的布局示例，这次给两个视图都加上一个自定义的动作(参见代码清单 2-33)。

代码清单 2-33　实现了自定义按键动作的 Activity

```java
public class MyActivity extends Activity implements OnEditorActionListener {

    @Override
    public void onCreate(Bundle savedInstanceState) {
        super.onCreate(savedInstanceState);
        setContentView(R.layout.main);

        //给视图添加监听器
        EditText text1 = (EditText)findViewById(R.id.text1);
        text1.setOnEditorActionListener(this);
        EditText text2 = (EditText)findViewById(R.id.text2);
        text2.setOnEditorActionListener(this);
    }
```

```java
    @Override
    public boolean onEditorAction(TextView v, int actionId, KeyEvent event) {
        if(actionId == IME_ACTION_SEARCH) {
            //处理搜索按键单击
            return true;
        }
        if(actionId == IME_ACTION_GO) {
            //处理 Go 按键单击
            return true;
        }
        return false;
    }
}
```

onEditorAction()返回的布尔值会告诉系统应用是否处理了这个事件，或者是否应该将其传递给下一个可能的响应者(如果有的话)。所以在实现该方法时一定要返回 true，这样系统就不会再对其他实现进行处理。当然，如果没有处理这个事件，就可以返回 false，这样系统的其他部分就能对其进行处理。

注意：
如果应用程序自定义为某个键盘返回的 actionId 值，则要注意只会在软键盘 IME 上进行这种自定义。如果设备附带物理键盘，Enter 键就会始终返回值为 0 的 actionId 或 IME_NULL。

2.12 消除软键盘

2.12.1 问题

需要通过用户界面上的某个事件隐藏或消除屏幕上的软键盘。

2.12.2 解决方案

(API Level 2)
用 InputMethodManager.hideSoftInputFromWindow()方法可以让输入法管理器显式地隐藏可见的输入法。

2.12.3 实现机制

下面这个示例演示了如何在 View.OnClickListener 中调用该方法：

```java
public void onClick(View view) {
    InputMethodManager imm = (InputMethodManager)getSystemService(
            Context.INPUT_METHOD_SERVICE);
    imm.hideSoftInputFromWindow(view.getWindowToken(), 0);
}
```

hideSofeInputFromWindow()方法的参数是一个 IBinder 窗口令牌。可以用 View.getWindow-Token()从附加到窗口的 View 对象获得该令牌。大部分情况下，每个事件的回调方法都会有一个引用指向正在编辑的 TextView，或是点击以生成事件的 View(如某个按键)。通过调用这些 View 对象获得窗口令牌，再将其传递给 InputMethodManager，这是最方便的做法。

2.13 处理复杂的触摸事件

2.13.1 问题

应用程序需要实现自定义的单点触摸或多点触摸来与 UI 进行交互。

2.13.2 解决方案

(API Level 3)

可以使用框架中的 GestureDetector 和 ScaleGestureDetector，或者干脆通过覆写 onTouchEvent()和 onInterceptTouchEvent()方法来手动处理传递给视图的所有触摸事件。前者可以很容易地在应用程序中添加复杂的手势控制。后者则非常强大，但也有一些需要注意的地方。

Android 通过自上而下的分发系统来处理 UI 上的触摸事件，这是框架在多层结构中发送消息的通用模式。触摸事件源于顶层窗口并首先发送给 Activity。然后，这些事件被分发到已加载视图层次结构中的根视图，并从父视图依次传递到相应的子视图，直到事件被处理或者整个视图链都已经传递。

每个父视图的工作就是确认一个触摸事件应该发送给哪个子视图(通常通过检查视图的边界)以及以正确的顺序将事件分发出去。如果可以分发给多个子视图(例如子视图是重叠的)，父视图会按照子视图的添加顺序反向地将事件分发出去，这样就可以保证叠置顺序中最高级别的视图(顶层视图)可以优先获得触摸事件。如果没有子视图处理事件，则父视图在该事件传回到视图层次结构之前会获得处理该事件的机会。

任何视图都可以通过在其 onTouchEvent()方法中返回 true 来表明已经处理了某个特定的触摸事件，这样该事件就不会再向其他地方分发了。所有 ViewGroup 的额外功能都可以通过 onInterceptTouchEvent()回调方法拦截或窃取传递给其子视图的触摸事件。这在父视图需要控制某个特定用例的场景下非常有用，例如 ScrollView 会在其检测到用户拖动手指之后控制触摸事件。

在手势进行的过程中会有几种不同的触摸事件动作标识符：

- ACTION_DOWN：当第一根手指点击屏幕时的第一个事件。这个事件通常是新手势的开始。
- ACTION_MOVE：当一根手指在屏幕上改变位置时的事件。
- ACTION_UP：最后一根手指离开屏幕时的结束事件。这个事件通常是一个手势的结束。

- ACTION_CANCEL：这个事件被子视图收到，即在子视图接收事件时父视图拦截了手势事件。和 ACTION_UP 一样，这标志着视图上的手势操作已经结束。
- ACTION_POINTER_DOWN：当另一根手指点击屏幕时的事件。在切换为多点触摸时很有用。
- ACTION_POINTER_UP：当另一根手指离开屏幕时的事件。在切换出多点触摸时很有用。

为了提高效率，在一个视图没有处理 ACTION_DOWN 事件的情况下，Android 将不会向该视图传递后续的事件。因此，如果你正在自定义处理触摸事件并希望处理后续的事件，那么必须在 ACTION_DOWN 事件中返回 true。

如果在一个父 ViewGroup 的内部实现自定义触摸事件处理器，你可能还需要在 onInterceptTouchEvent() 方法中编写一些代码。这个方法的工作方式和 onTouchEvent() 类似，如果返回 true，自定义视图就会接管手势后续所有的触摸事件(即 ACTION_UP 和 ACTION_UP 之前的所有事件)。这个操作是不可取消的，在确定要接管所有事件之前不要轻易拦截这些事件。

最后，Android 提供了大量有用的阈值常量，这些值可以根据设备屏幕的分辨率进行缩放，可以用于构建自定义触摸交互。这些常数都保存在 ViewConfiguration 类中。本例中会用到最小和最大急滑(fling)速率值以及触摸倾斜常量，表示 ACTION_MOVE 事件变化到什么程度才表示是用户手指的真实移动动作。

2.13.3 实现机制

代码清单 2-34 演示了一个自定义的 ViewGroup，该 ViewGroup 实现了平面滚动，即在内容足够大的情况下，允许用户在水平方向和垂直方向上进行滚动。该实现使用 GestureDetector 来处理触摸事件。

代码清单 2-34　通过 GestureDetector 自定义 ViewGroup

```java
public class PanGestureScrollView extends FrameLayout {

    private GestureDetector mDetector;
    private Scroller mScroller;

    /*最后位移事件的位置*/
    private float mInitialX, mInitialY;
    /*拖曳阈值*/
    private int mTouchSlop;

    public PanGestureScrollView(Context context) {
        super(context);
        init(context);
    }

    public PanGestureScrollView(Context context, AttributeSet attrs) {
        super(context, attrs);
```

```java
        init(context);
    }

    public PanGestureScrollView(Context context, AttributeSet attrs,
            int defStyle) {
        super(context, attrs, defStyle);
        init(context);
    }

    private void init(Context context) {
        mDetector = new GestureDetector(context, mListener);
        mScroller = new Scroller(context);
        //获得触摸阈值的系统常量
        mTouchSlop = ViewConfiguration.get(context).getScaledTouchSlop();
    }

    /*
     * 覆写measureChild...的实现来保证生成的子视图尽可能大
     * 默认实现会强制一些子视图和该视图一样大
     */
    @Override
    protected void measureChild(View child, int parentWidthMeasureSpec,
            int parentHeightMeasureSpec) {
        int childWidthMeasureSpec;
        int childHeightMeasureSpec;

        childWidthMeasureSpec = MeasureSpec.makeMeasureSpec(0,
                MeasureSpec.UNSPECIFIED);
        childHeightMeasureSpec = MeasureSpec.makeMeasureSpec(0,
                MeasureSpec.UNSPECIFIED);

        child.measure(childWidthMeasureSpec, childHeightMeasureSpec);
    }

    @Override
    protected void measureChildWithMargins(View child,
            int parentWidthMeasureSpec, int widthUsed,
            int parentHeightMeasureSpec, int heightUsed) {
        final MarginLayoutParams lp = (MarginLayoutParams) child.getLayoutParams();

        final int childWidthMeasureSpec = MeasureSpec.makeMeasureSpec(
                lp.leftMargin + lp.rightMargin, MeasureSpec.UNSPECIFIED);
        final int childHeightMeasureSpec = MeasureSpec.makeMeasureSpec(
                lp.topMargin + lp.bottomMargin, MeasureSpec.UNSPECIFIED);

        child.measure(childWidthMeasureSpec, childHeightMeasureSpec);
    }

    //处理所有触摸事件的监听器
    private SimpleOnGestureListener mListener = new SimpleOnGestureListener() {
```

```java
    public boolean onDown(MotionEvent e) {
        //取消当前的急滑动画
        if(!mScroller.isFinished()) {
            mScroller.abortAnimation();
        }
        return true;
    }

    public boolean onFling(MotionEvent e1, MotionEvent e2, float velocityX,
            float velocityY) {
        //调用一个辅助方法来启动滚动动画
        fling((int) -velocityX / 3, (int) -velocityY / 3);
        return true;
    }

    public boolean onScroll(MotionEvent e1, MotionEvent e2,
            float distanceX, float distanceY) {
        //任意视图都可以调用它的scrollBy()进行滚动
        scrollBy((int) distanceX, (int) distanceY);
        return true;
    }
};

@Override
public void computeScroll() {
    if(mScroller.computeScrollOffset()) {
        //会在ViewGroup绘制时调用
        //我们使用这个方法保证急滑动画的顺利完成
        int oldX = getScrollX();
        int oldY = getScrollY();
        int x = mScroller.getCurrX();
        int y = mScroller.getCurrY();

        if(getChildCount() > 0) {
            View child = getChildAt(0);
            x = clamp(x, getWidth() - getPaddingRight() - getPaddingLeft(),
                    child.getWidth());
            y = clamp(y,
                    getHeight() - getPaddingBottom() - getPaddingTop(),
                    child.getHeight());
            if(x != oldX || y != oldY) {
                scrollTo(x, y);
            }
        }
        //在动画完成前会一直绘制
        postInvalidate();
    }
}

//覆写scrollTo方法进行每个滚动请求的边界检查
```

```java
@OverridescrollTo
public void scrollTo(int x, int y) {
    // 我们依赖View.scrollBy调用scrollTo
    if(getChildCount() > 0) {
        View child = getChildAt(0);
        x = clamp(x, getWidth() - getPaddingRight() - getPaddingLeft(),
                child.getWidth());
        y = clamp(y, getHeight() - getPaddingBottom() - getPaddingTop(),
                child.getHeight());
        if(x != getScrollX() || y != getScrollY()) {
            super.scrollTo(x, y);
        }
    }
}

/*
 * 监控传递给子视图的触摸事件,并且一旦确定拖曳就进行拦截
 */
@Override
public boolean onInterceptTouchEvent(MotionEvent event) {
    switch(event.getAction()) {
    case MotionEvent.ACTION_DOWN:
        mInitialX = event.getX();
        mInitialY = event.getY();
        //将按下事件传给手势检测器,这样当/如果拖曳开始就有了上下文
        mDetector.onTouchEvent(event);
        break;
    case MotionEvent.ACTION_MOVE:
        final float x = event.getX();
        final float y = event.getY();
        final int yDiff = (int) Math.abs(y - mInitialY);
        final int xDiff = (int) Math.abs(x - mInitialX);
        //检查x或y上的距离是否适合拖曳
        if(yDiff > mTouchSlop || xDiff > mTouchSlop) {
            //开始捕捉事件
            return true;
        }
        break;
    }
    return super.onInterceptTouchEvent(event);
}

/*
 * 将我们接受的所有触摸事件传给检测器处理
 */
@Override
public boolean onTouchEvent(MotionEvent event) {
    return mDetector.onTouchEvent(event);
}
```

```java
/*
 * 初始化Scroller和开始重新绘制的实用方法
 */
public void fling(int velocityX, int velocityY) {
    if(getChildCount() > 0) {
        int height = getHeight() - getPaddingBottom() - getPaddingTop();
        int width = getWidth() - getPaddingLeft() - getPaddingRight();
        int bottom = getChildAt(0).getHeight();
        int right = getChildAt(0).getWidth();

        mScroller.fling(getScrollX(), getScrollY(), velocityX, velocityY,
                0, Math.max(0, right - width), 0,
                Math.max(0, bottom - height));
        invalidate();
    }
}


/*
 * 用来进行边界检查的辅助实用方法
 */
private int clamp(int n, int my, int child) {
    if(my >= child || n < 0) {
        //子视图超过了父视图的边界或者小于父视图,不能滚动
        return 0;
    }
    if((my + n) > child) {
        //请求的滚动超出了子视图的右边界
        return child - my;
    }
    return n;
}
```

与ScrollView或HorizontalScrollView类似,这个示例有一个子视图并可以根据用户输入滚动它的内容。这个示例的多数代码与触摸事件的处理并没有直接关系,而是处理滚动并让滚动位置不要超过子视图的边界。

作为一个ViewGroup,第一个可以看到所有触摸事件的地方就是onInterceptTouchEvent()。在这个方法中我们必须分析用户的触摸行为,从而确定是否是真正的拖动。这个方法中ACTION_DOWN和ACTION_MOVE的处理一起决定了用户的手指移动了多远,只有该值大于系统的触摸阈值常量,我们才认为是拖动事件并拦截后续触摸事件。这种做法允许子视图接收简单的触摸事件,所以按钮和其他小部件可以放心地作为这个视图的子视图,并且依然会得到触摸事件。如果该视图没有可交互的子视图小部件,事件将会被直接传递到我们的onTouchEvent()方法中,但因为我们允许这种情况发生,所以这里做了初始检查。

这里的onTouchEvent()方法很简单,因为所有的事件都被转发到了GestureDetector中,它会追踪和计算用户正在做的特定动作。然后我们会通过SimpleOnGestureListener对那些事件进行响应,特别是onScroll()和onFling()事件。为了保证GestureDetector能够准确地设置手势的初始触点,我们还在onInterceptTouchEvent()中向它转发了ACTION_DOWN事件。

onScroll()在用户的手指移动一段距离时会被重复调用。所以,在手指拖动时,我们可以很方便地将这些值直接传递给视图的 scrollBy() 来移动视图的内容。

onFling()中需要做稍微多一点的工作。说明一下,急滑(fling)操作就是用户在屏幕上快速移动手指并抬起的动作。这个动作期望的结果就是惯性的滚动动画。同样,当用户手指抬起时会计算手指的速度,但必须依然保持滚动动画。这就是引入 Scroller 的原因。Scroller 是框架的一个组件,用来通过用户的输入值和时间插值动画设置来让视图滚动起来。本例中的动画是通过调用 Scroller 的 fing() 方法并刷新视图实现的。

注意:

如果目标版本为 API Level 9 或更高,可以使用 OverScroller 代替 Scroller,它会为较新的设备提供更好的性能。它还允许包含拉到底发光的动画(overscroll glow)。可以通过传入自定义的 Interpolator 加工急滑动画。

这会启动一个循环进程,在这个进程中框架会定期调用 computeScroll() 来绘制视图,我们刚好通过这个时机来检查 Scroller 当前的状态,并且将视图向前滚动(如果动画未完成的话)。这也是开发人员对 Scroller 感到困惑的地方。该控件是用来让视图动起来的,但实际上却没有制作任何动画。它只是简单地提供了每个绘制帧移动的时机和距离计算。应用程序必须同时调用 computeScrollOffset() 来获得新位置,然后再实际地调用一个方法(本例中为 scrollTo() 方法)渐进地改变视图。

GestureDetector 中使用的最后一个回调方法是 onDown(),它会在侦测器收到 ACTION_DOWN 事件时得到调用。如果用户手指单击屏幕,我们会通过这个回调方法终止所有当前的急滑动画。代码清单 2-35 显示了我们该如何在 Activity 中使用这个自定义视图。

代码清单 2-35 使用了 PanGestureScrollView 的 Activity

```
public class PanScrollActivity extends Activity {

    @Override
    protected void onCreate(Bundle savedInstanceState) {
        super.onCreate(savedInstanceState);

        PanGestureScrollView scrollView = new PanGestureScrollView(this);

        LinearLayout layout = new LinearLayout(this);
        layout.setOrientation(LinearLayout.VERTICAL);
        for(int i=0; i < 5; i++) {
            ImageView iv = new ImageButton(this);
            iv.setImageResource(R.drawable.ic_launcher);
            //让每个视图可以足够大来请求滚动
            layout.addView(iv, new LinearLayout.LayoutParams(1000, 500));
        }

        scrollView.addView(layout);
        setContentView(scrollView);
    }
}
```

我们使用大量的 ImageButton 实例来填充这个自定义的 PanGestureScrollView，这是为了演示这些按钮都是可以单击的，并且可以接收单击事件，但是只要你拖动或急滑手指，视图就会开始滚动。要想了解 GestureDetector 为我们做了多少工作，可查看代码清单 2-36，它实现了相同的功能，但需要在 onTouchEvent()中手动处理所有的触摸事件。

代码清单 2-36　使用了自定义触摸处理的 PanScrollView

```java
public class PanScrollView extends FrameLayout {

    //急滑控件
    private Scroller mScroller;
    private VelocityTracker mVelocityTracker;

    /* 上一个移动事件的位置 */
    private float mLastTouchX, mLastTouchY;
    /* 拖动阈值 */
    private int mTouchSlop;
    /* 急滑的速度 */
    private int mMaximumVelocity, mMinimumVelocity;
    /* 拖动锁 */
    private boolean mDragging = false;

    public PanScrollView(Context context) {
        super(context);
        init(context);
    }

    public PanScrollView(Context context, AttributeSet attrs) {
        super(context, attrs);
        init(context);
    }

    public PanScrollView(Context context, AttributeSet attrs, int defStyle) {
        super(context, attrs, defStyle);
        init(context);
    }

    private void init(Context context) {
        mScroller = new Scroller(context);
        mVelocityTracker = VelocityTracker.obtain();
        //获得触摸阈值的系统常量
        mTouchSlop = ViewConfiguration.get(context).getScaledTouchSlop();
        mMaximumVelocity = ViewConfiguration.get(context)
                .getScaledMaximumFlingVelocity();
        mMinimumVelocity = ViewConfiguration.get(context)
                .getScaledMinimumFlingVelocity();
    }
```

```java
/*
 * 覆写measureChild...的实现来保证子视图可以尽可能的大
 * 默认实现会强制一些子视图和该视图一样大
 */
@Override
protected void measureChild(View child, int parentWidthMeasureSpec,
        int parentHeightMeasureSpec) {
    int childWidthMeasureSpec;
    int childHeightMeasureSpec;

    childWidthMeasureSpec = MeasureSpec.makeMeasureSpec(0,
            MeasureSpec.UNSPECIFIED);
    childHeightMeasureSpec = MeasureSpec.makeMeasureSpec(0,
            MeasureSpec.UNSPECIFIED);

    child.measure(childWidthMeasureSpec, childHeightMeasureSpec);
}

@Override
protected void measureChildWithMargins(View child,
        int parentWidthMeasureSpec, int widthUsed,
        int parentHeightMeasureSpec, int heightUsed) {
    final MarginLayoutParams lp = (MarginLayoutParams)
            child.getLayoutParams();

    final int childWidthMeasureSpec = MeasureSpec.makeMeasureSpec(
            lp.leftMargin + lp.rightMargin, MeasureSpec.UNSPECIFIED);
    final int childHeightMeasureSpec = MeasureSpec.makeMeasureSpec(
            lp.topMargin + lp.bottomMargin, MeasureSpec.UNSPECIFIED);

    child.measure(childWidthMeasureSpec, childHeightMeasureSpec);
}

@Override
public void computeScroll() {
    if(mScroller.computeScrollOffset()) {
        //这个方法会在ViewGroup绘制时调用
        //我们使用这个方法保证急滑动画的完成
        int oldX = getScrollX();
        int oldY = getScrollY();
        int x = mScroller.getCurrX();
        int y = mScroller.getCurrY();

        if(getChildCount() > 0) {
            View child = getChildAt(0);
            x = clamp(x, getWidth() - getPaddingRight() - getPaddingLeft(),
                    child.getWidth());
            y = clamp(y,
                    getHeight() - getPaddingBottom() - getPaddingTop(),
                    child.getHeight());
```

```java
            if(x != oldX || y != oldY) {
                scrollTo(x, y);
            }
        }
        //在动画完成前会一直绘制
        postInvalidate();
    }
}

//覆写scrollTo方法以进行每个滚动请求的边界检查
@Override
public void scrollTo(int x, int y) {
    //我们依赖View.scrollBy调用scrollTo
    if(getChildCount() > 0) {
        View child = getChildAt(0);
        x = clamp(x, getWidth() - getPaddingRight() - getPaddingLeft(),
            child.getWidth());
        y = clamp(y, getHeight() - getPaddingBottom() - getPaddingTop(),
            child.getHeight());
        if(x != getScrollX() || y != getScrollY()) {
            super.scrollTo(x, y);
        }
    }
}

/*
 * 监控传递给子视图的触摸事件，并且一旦确定拖曳就进行拦截
 * 如果子视图是可交互的(如按钮)，那么依然允许子视图接收触摸事件
 */
@Override
public boolean onInterceptTouchEvent(MotionEvent event) {
    switch (event.getAction()) {
    case MotionEvent.ACTION_DOWN:
        //终止所有正在进行的急滑动画
        if(!mScroller.isFinished()) {
            mScroller.abortAnimation();
        }
        //还原速度跟踪器
        mVelocityTracker.clear();
        mVelocityTracker.addMovement(event);
        //保存初始触点
        mLastTouchX = event.getX();
        mLastTouchY = event.getY();
        break;
    case MotionEvent.ACTION_MOVE:
        final float x = event.getX();
        final float y = event.getY();
        final int yDiff = (int) Math.abs(y - mLastTouchY);
        final int xDiff = (int) Math.abs(x - mLastTouchX);
        //检查x或y上的距离是否适合拖曳
```

```
        if(yDiff > mTouchSlop || xDiff > mTouchSlop) {
            mDragging = true;
            mVelocityTracker.addMovement(event);
            //我们自己开始捕捉事件
            return true;
        }
        break;
    case MotionEvent.ACTION_CANCEL:
    case MotionEvent.ACTION_UP:
        mDragging = false;
        mVelocityTracker.clear();
        break;
    }

    return super.onInterceptTouchEvent(event);
}

/*
*/将我们接收到的所有触摸事件传给检测器处理
*/
@Override
public boolean onTouchEvent(MotionEvent event) {
    mVelocityTracker.addMovement(event);

    switch (event.getAction()) {
    case MotionEvent.ACTION_DOWN:
        //我们已经保存了初始触点,但如果这里发现有子视图没有捕捉事件,
        //还是需要返回true的
        return true;
    case MotionEvent.ACTION_MOVE:
        final float x = event.getX();
        final float y = event.getY();
        float deltaY = mLastTouchY - y;
        float deltaX = mLastTouchX - x;
        //检查各个方向事件上的阈值
        if((Math.abs(deltaY) > mTouchSlop || Math.abs(deltaX) > mTouchSlop)
                && !mDragging) {
            mDragging = true;
        }
        if(mDragging) {
            //滚动视图
            scrollBy((int) deltaX, (int) deltaY);
            //更新最后一个触摸事件
            mLastTouchX = x;
            mLastTouchY = y;
        }
        break;
    case MotionEvent.ACTION_CANCEL:
        mDragging = false;
        //终止所有进行的急滑动画
```

```java
                if(!mScroller.isFinished()) {
                    mScroller.abortAnimation();
                }
                break;
            case MotionEvent.ACTION_UP:
                mDragging = false;
                //计算当前的速度,如果高于最小阈值,则启动一个急滑
                mVelocityTracker.computeCurrentVelocity(1000, mMaximumVelocity);
                int velocityX = (int) mVelocityTracker.getXVelocity();
                int velocityY = (int) mVelocityTracker.getYVelocity();
                if(Math.abs(velocityX) > mMinimumVelocity
                        || Math.abs(velocityY) > mMinimumVelocity) {
                    fling(-velocityX, -velocityY);
                }
                break;
        }
        return super.onTouchEvent(event);
    }

    /*
     * 初始化 Scroller 和开始重新绘制的实用方法
     */
    public void fling(int velocityX, int velocityY) {
        if(getChildCount() > 0) {
            int height = getHeight() - getPaddingBottom() - getPaddingTop();
            int width = getWidth() - getPaddingLeft() - getPaddingRight();
            int bottom = getChildAt(0).getHeight();
            int right = getChildAt(0).getWidth();

            mScroller.fling(getScrollX(), getScrollY(), velocityX, velocityY,
                    0, Math.max(0, right - width), 0,
                    Math.max(0, bottom - height));

            invalidate();
        }
    }

    /*
     * 用来进行边界检查的辅助实用方法
     */
    private int clamp(int n, int my, int child) {
        if(my >= child || n < 0) {
            //子视图超过了父视图的边界或者小于父视图,不能滚动
            return 0;
        }
        if((my + n) > child) {
            //请求的滚动超出了子视图的右边界
            return child - my;
        }
```

```
        return n;
    }
}
```

本例中，onInterceptTouchEvent()和 onTouchEvent()中的工作会多一点。如果当前存在子视图处理初始的触摸事件，那么在我们接管事件之前，ACTION_DOWN 和开始的一些移动事件都会通过 onInterceptTouchEvent()进行传递；但是，如果并不存在可交互的子视图，所有这些初始触摸事件都会直接传递到 onTouchEvent()中。在这两个方法中，我们必须都要对初始拖动进行阈值检查，如果确实开始了拖动事件，会设置一个标识。一旦标识用户正在拖动，滚动视图的代码就和之前的一样了，即调用 scrollBy()。

提示:
只要某个 ViewGroup 通过 onTouchEvent()返回了 "true"，即使没有显式地请求拦截，也不会再有事件被传递到 onInterceptTouchEvent()。

要想实现急滑效果，我们必须手动使用 VelocityTracker 对象手动跟踪用户的滚动速度。该对象会将发生的事件通过 addMovement()方法收集起来，然后通过 computeCurrentVelocity()计算相应的平均速度。我们的自定义视图会根据 ViewConfiguration 最小速度在每次用户抬起手指时计算这个速度值，从而决定是否要开始一段急滑动画。

提示:
在不需要显式返回 true 来处理事件的情形下，最好返回父类的实现而不是返回 false。通常父类会有很多关于 View 和 ViewGroup 的隐藏处理(通常不要覆写它们)。

代码清单 2-37 再次显示了示例 Activity，这一次使用了新的自定义视图。

代码清单 2-37　使用了 PanScrollView 的 Activity

```java
public class PanScrollActivity extends Activity {

    @Override
    protected void onCreate(Bundle savedInstanceState) {
        super.onCreate(savedInstanceState);

        PanScrollView scrollView = new PanScrollView(this);

        LinearLayout layout = new LinearLayout(this);
        layout.setOrientation(LinearLayout.VERTICAL);
        for(int i=0; i < 5; i++) {
            ImageView iv = new ImageView(this);
            iv.setImageResource(R.drawable.ic_launcher);
            layout.addView(iv,
                    new LinearLayout.LayoutParams(1000, 500));
        }

        scrollView.addView(layout);
        setContentView(scrollView);
```

            }
        }

我们将视图的内容设定为 ImageView 而非 ImageButton，从而演示子视图不能交互时的对比效果。

多点触摸处理

(API Level 8)

现在，让我们看一个处理多点触摸事件的示例。代码清单 2-38 是一个自定义的添加了多点触摸交互的 ImageView。

代码清单 2-38　带有处理多点触摸的 ImageView

```
public class RotateZoomImageView extends ImageView {

    private ScaleGestureDetector mScaleDetector;
    private Matrix mImageMatrix;
    /*上次的旋转角度*/
    private int mLastAngle = 0;
    /* 变换时的轴点 */
    private int mPivotX, mPivotY;

    public RotateZoomImageView(Context context) {
        super(context);
        init(context);
    }

    public RotateZoomImageView(Context context, AttributeSet attrs) {
        super(context, attrs);
        init(context);
    }

    public RotateZoomImageView(Context context, AttributeSet attrs, int defStyle) {
        super(context, attrs, defStyle);
        init(context);
    }

    private void init(Context context) {
        mScaleDetector = new ScaleGestureDetector(context, mScaleListener);

        setScaleType(ScaleType.MATRIX);
        mImageMatrix = new Matrix();
    }

    /*
    *在 onSizeChanged()中根据视图的尺寸计算一些值
    *这个视图在 init()期间并没有尺寸，因此必须等到这个回调方法才能得到尺寸
    */
```

```java
@Override
protected void onSizeChanged(int w, int h, int oldw, int oldh) {
    if(w != oldw || h != oldh) {
        //将图片移到视图的中央
        int translateX = Math.abs(w - getDrawable().getIntrinsicWidth()) / 2;
        int translateY = Math.abs(h - getDrawable().getIntrinsicHeight()) / 2;
        mImageMatrix.setTranslate(translateX, translateY);
        setImageMatrix(mImageMatrix);
        //得到未来缩放和旋转变换时的中轴点
        mPivotX = w / 2;
        mPivotY = h / 2;
    }
}

private SimpleOnScaleGestureListener mScaleListener =
        new SimpleOnScaleGestureListener() {
    @Override
    public boolean onScale(ScaleGestureDetector detector) {
        //ScaleGestureDetector会根据手指的分开和合拢计算出缩放因子
        float scaleFactor = detector.getScaleFactor();
        //将缩放因子传给图片进行缩放
        mImageMatrix.postScale(scaleFactor, scaleFactor, mPivotX,
            mPivotY);
        setImageMatrix(mImageMatrix);

        return true;
    }
};

/*
 * 处理两根手指的事件来旋转图片
 * 这个方法根据触点间的角度变化对图片进行相应的旋转
 * 当用户旋转手指时，图片也会跟着旋转
 */
private boolean doRotationEvent(MotionEvent event) {
    //计算两根手指间的角度
    float deltaX = event.getX(0) - event.getX(1);
    float deltaY = event.getY(0) - event.getY(1);
    double radians = Math.atan(deltaY / deltaX);
    //转换为角度
    int degrees = (int)(radians * 180 / Math.PI);

    switch(event.getAction()) {
    case MotionEvent.ACTION_DOWN:
        //记住初始角度
        mLastAngle = degrees;
        break;
    case MotionEvent.ACTION_MOVE:
        //返回一个转换后的介于-90°和+90°的值，
        //这样在两根手指垂直触摸时可以得到翻转信号和相应的角度
```

```
        //这种情况下会将图片在我们侦测到的方向上旋转一个很小的角度(5°)
        if((degrees - mLastAngle) > 45) {
            //逆时针旋转(可以超出边界)
            mImageMatrix.postRotate(-5, mPivotX, mPivotY);
        } else if ((degrees - mLastAngle) < -45) {
            //顺时针旋转(可以超出边界)
            mImageMatrix.postRotate(5, mPivotX, mPivotY);
        } else {
            //正常旋转,旋转角度即为手指的旋转角度
            mImageMatrix.postRotate(degrees - mLastAngle, mPivotX,
                mPivotY);
        }
        //将旋转矩阵发送给图片
        setImageMatrix(mImageMatrix);
        //保存当前的角度
        mLastAngle = degrees;
        break;
    }
    
    return true;
}

@Override
public boolean onTouchEvent(MotionEvent event) {
    if(event.getAction() == MotionEvent.ACTION_DOWN) {
        //我们并不直接关心这个事件,但会声明要处理后续的多点触摸
        return true;
    }
    
    switch (event.getPointerCount()) {
    case 2:
        //按下三根手指时,使用ScaleGestureDetector缩放图片
        return mScaleDetector.onTouchEvent(event);
    case 2:
        //按下两根手指时,根据手指操作旋转图片
        return doRotationEvent(event);
    default:
        //忽略这个事件
        return super.onTouchEvent(event);
    }
}
```

这个示例创建了一个自定义的 ImageView 来监听多点触摸事件并及时变换图像的内容。这个视图可以侦测到的两种事件就是两根手指的旋转操作和三根手指的缩放操作。旋

转事件是通过每个 MotionEvent 来处理的，缩放事件则是通过 ScaleGestureDetector 来处理的。这个视图的 ScaleType 被设置为 MATRIX，这样就可以让我们通过应用不同的 Matrix 变换来调整图片的外观。

在该视图构建并布局完成后，就会触发 onSizeChanged()回调方法。这个方法可以被多次调用，所以我们只会在上次值和本次值不同时计算相应的值。这里，我们会根据视图的尺寸设置一些值，以便将图片放置到视图的中央并稍后进行正确的变换。同时我们执行了第一次变换，即将图片移到视图的中央。

我们会分析 onTouchEvent()接收的触摸事件来决定处理哪个事件。通过检查每个 MotionEvent 的 getPointerCount()方法，可以判断按下了几根手指并将事件传递给相应的处理程序。正如之前所说的一样，这里也必须处理第一个 ACTION_DOWN 事件；否则用户其他手指的后续触摸事件将不会传递到这个视图。虽然我们不想对这个事件做任何操作，但仍然需要显式地返回 true。

ScaleGestureDetector 会分析应用程序反馈的每个触摸事件，当出现缩放事件时，就调用一系列的 OnScaleGestureListener 回调方法。最重要的回调方法就是 onScale()，它在用户手指移动时就会被经常调用，但开发人员还可以使用 onScaleBegin()和 onScaleEnd()在手势开始和结束时进行一些操作。

ScaleGestureDetector 提供了很多有用的计算值，应用程序可以使用这些值来修改 UI：
- getCurrentSpan()：获得该手势中两个触点间的距离。
- getFocusX()/getFocusY()：获得当前手势的焦点坐标。它是触点收缩时的平均位置。
- getScaleFactor()：得到当前事件和之前事件之间的变化比例。多根手指分开时，这个值会稍微大于 1，收拢时会稍微小于 1。

这个示例从侦测器中得到缩放因子并使用它通过 postScale()设置图像的 Matrix，从而缩放视图中的图片内容。

我们两根手指的旋转事件是手动处理的。对于每个传入的事件，会通过 getX()和 getY()计算两根手指间 x 和 y 方向的距离。getX()和 getY()方法使用的参数为触点的索引，0 表示第一个触点，1 表示第二个触点。

使用这些距离，我们可以使用一点儿三角函数的知识计算出两根手指之间形成的无形直线的角度。我们将使用这个角度来进行变换。在 ACTION_DOWN 期间，不管角度值是什么，都会把它作为初始值并简单地保存起来。在接下来的 ACTION_MOVE 事件中，会根据每次触摸事件之间的角度变化发送给图片相应的旋转矩阵。

这个示例必须处理一种边界情况，并且必须使用 Math.atan()三角函数。这个函数会返回一个介于-90°和+90°的角度值，而这种翻转发生在一根手指垂直地位于另一根手指上方的情况。这个问题会导致触摸角度不再是逐渐改变的：在手指旋转时，角度值会从+90°立即变为-90°，从而导致图片跳动。为了解决这个问题，我们会检查之前角度和当前角度超过这种边界值的情况，然后在相同的行进方向做 5°的小旋转，从而保证动画的流畅性。

注意，变换图片的所有操作都是使用 postScale()和 postRotate()完成的，而不是之前的这些方法的 setXXX 版本(如 setTranslation())。这是因为每个变换都只是一种新增的变换，这意味着只能适当地改变当前的状态而不是替换。调用 setScale()或 setRotate()将会清除当前的状态，从而导致只剩下 Matrix 中的变换。

这些变换都是围绕我们在 onSizeChanged()中计算出的轴点(视图的中点)进行的。这么做是因为默认情况下变换发生在目标点(0,0)，即视图的左上角。因为我们已经将图片移到视图中央，所以需要保证所有的变换也发生在同样的中央轴点。

2.14 转发触摸事件

2.14.1 问题

应用程序中的一些视图或触摸目标非常小，导致手指很难准确地触摸到。

2.14.2 解决方案

(API Level 1)

使用 TouchDelegate 指定任意的矩形区域来向小视图转发触摸事件。TouchDelegate 的设计宗旨就是为父 ViewGroup 关联特定的区域，该区域侦测到触摸事件后会将该事件转发给它的某个子视图。TouchDelegate 会发送每个事件到目标视图，就像触摸目标视图自己一样。

2.14.3 实现机制

代码清单 2-39 和 2-40 演示了如何在自定义的父 ViewGroup 中使用 TouchDelegate。

代码清单 2-39　自定义父视图实现了 TouchDelegate

```java
public class TouchDelegateLayout extends FrameLayout {

    public TouchDelegateLayout(Context context) {
        super(context);
        init(context);
    }

    public TouchDelegateLayout(Context context, AttributeSet attrs) {
        super(context, attrs);
        init(context);
    }

    public TouchDelegateLayout(Context context, AttributeSet attrs, int defStyle) {
        super(context, attrs, defStyle);
        init(context);
    }

    private CheckBox mButton;
    private void init(Context context) {
        //创建一个很小的子视图，我们要将触摸事件转发给它
```

```java
        mButton = new CheckBox(context);
        mButton.setText("Tap Anywhere");

        LayoutParams lp = new FrameLayout.LayoutParams(LayoutParams.WRAP_CONTENT,
                LayoutParams.WRAP_CONTENT, Gravity.CENTER);
        addView(mButton, lp);
    }

    /*
    *TouchDelegate 会将该视图(父视图)的某个特定矩形区域作为代理区域,将
    *所有触摸事件转发给 CheckBox(子视图)。这里,矩形区域即为父视图的全部大小
    *
    *这个过程必须在视图确定了大小以后进行,这样才能知道矩形应该有多大,
    *所以我们选择在 onSizeChanged()中添加代理区域
    */
    @Override
    protected void onSizeChanged(int w, int h, int oldw, int oldh) {
        if(w != oldw || h != oldh) {
            //将该视图的整个区域作为代理区域
            Rect bounds = new Rect(0, 0, w, h);
            TouchDelegate delegate = new TouchDelegate(bounds, mButton);
            setTouchDelegate(delegate);
        }
    }
}
```

代码清单 2-40　示例 Activity

```java
public class DelegateActivity extends Activity {

    @Override
    protected void onCreate(Bundle savedInstanceState) {
        super.onCreate(savedInstanceState);
        TouchDelegateLayout layout =
                new TouchDelegateLayout(this);
        setContentView(layout);
    }
}
```

在这个示例中,我们创建了一个父视图,其中包含了一个居中显示的复选框。这个视图还包含一个 TouchDelegate,它会将父视图区域内收到的触摸事件转发给复选框。因为我们想让父布局的整个区域转发触摸事件,所以会等到在视图上调用 onSizeChanged()后再构建和关联 TouchDelegate 实例。如果在构造函数中,构建将不会生效,因为在执行构造函数时,视图还没有被测量,并且没有可以读取的尺寸大小。

Android 框架会将没有处理的触摸事件自动从 TouchDelegate 分发到它的代理视图,因此无需额外代码即可转发这些事件。在图 2-9 中可以看到,应用程序在距离复选框很远的地方收到触摸事件后,复选框会做相应的响应,如同它自己直接被触摸了一样。

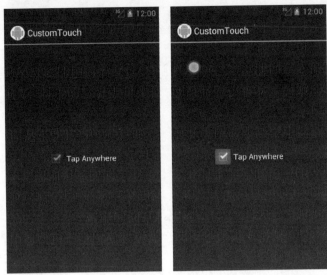

图2-9 带有复选框的示例应用程序(左图),复选框会接收转发的触摸事件(右图)

自定义触摸转发(远程滚动条)

TouchDelegate 非常适合于转发触摸事件,但它有一个缺点,就是每个被转发的事件转发到代理视图后都会定位到代理视图的中间位置。这也就意味着,如果想要通过 TouchDelegate 转发一系列 ACTION_MOVE 事件的话,结果将不会如你所愿,因为这时代理视图会显示手指并没有移动过(每次都定位到同一个点上)。

如果想要以一种更加精确的方式重新路由触摸事件,可以通过手动地调用目标视图的 dispatchTouchEvent()方法来实现。参见代码清单 2-41 和 2-42 以了解相应的实现机制。

代码清单 2-41　res/layout/main.xml

```xml
<LinearLayout
    xmlns:android="http://schemas.android.com/apk/res/android"
    android:layout_width="match_parent"
    android:layout_height="match_parent"
    android:orientation="vertical" >

    <TextView
        android:id="@+id/text_touch"
        android:layout_width="match_parent"
        android:layout_height="0dp"
        android:layout_weight="1"
        android:gravity="center"
        android:text="Scroll Anywhere Here" />

    <HorizontalScrollView
        android:id="@+id/scroll_view"
        android:layout_width="match_parent"
        android:layout_height="0dp"
        android:layout_weight="1"
```

```xml
            android:background="#CCC">
        <LinearLayout
            android:layout_width="wrap_content"
            android:layout_height="match_parent"
            android:orientation="horizontal" >
            <ImageView
                android:layout_width="250dp"
                android:layout_height="match_parent"
                android:scaleType="fitXY"
                android:src="@drawable/ic_launcher" />
            <ImageView
                android:layout_width="250dp"
                android:layout_height="match_parent"
                android:scaleType="fitXY"
                android:src="@drawable/ic_launcher" />
            <ImageView
                android:layout_width="250dp"
                android:layout_height="match_parent"
                android:scaleType="fitXY"
                android:src="@drawable/ic_launcher" />
            <ImageView
                android:layout_width="250dp"
                android:layout_height="match_parent"
                android:scaleType="fitXY"
                android:src="@drawable/ic_launcher" />
        </LinearLayout>
    </HorizontalScrollView>
</LinearLayout>
```

代码清单 2-42 转发触摸事件的 Activity

```java
public class RemoteScrollActivity extends Activity implements
        View.OnTouchListener {

    private TextView mTouchText;
    private HorizontalScrollView mScrollView;

    @Override
    protected void onCreate(Bundle savedInstanceState) {
        super.onCreate(savedInstanceState);
        setContentView(R.layout.main);

        mTouchText = (TextView) findViewById(R.id.text_touch);
        mScrollView = (HorizontalScrollView) findViewById(R.id.scroll_view);
        //为顶层视图关联触摸事件的监听器
        mTouchText.setOnTouchListener(this);
    }

    @Override
    public boolean onTouch(View v, MotionEvent event) {
        //如果需要的话，可以修改事件位置
```

```
            //这里我们将每个事件的垂直方向的位置设置为 HorizontalScrollView 的中间
            //视图需要的事件位置都是相对于自己的坐标
            event.setLocation(event.getX(), mScrollView.getHeight() / 2);

            //将 TextView 上的每个事件转发到 HorizontalScrollView
            mScrollView.dispatchTouchEvent(event);
            return true;
        }
    }
```

这个示例将一个 Activity 一分为二。上半部分是一个 TextView，它会提示你触摸并滑动它；而下半部分是一个内部包含若干张图片的 HorizontalScrollView。Activity 为 TextView 设置了一个 OnTouchListener，这样就可以将它接收的所有触摸事件转发给 HorizontalScrollView。

我们希望触摸事件就像发生在(从 HorizontalScrollView 的角度)HorizontalScrollView 自己的视图内部一样。所以在转发事件之前，我们会调用 setLocation()来修改 x/y 坐标。在本例中，x 坐标就是原来的坐标，y 坐标则调整到了 HorizontalScrollView 的中间。这样，当用户手指向前或向后滚动时，就好像在 HorizontalScrollView 的中间滚动一样。然后，调用 dispatchTouchEvent()将修改后的事件交予 HorizontalScrollView 处理。

注意：

避免直接调用 onTouchEvent()方法转发触摸事件。调用 dispatchTouchEvent()可以使其像常规触摸事件一样处理目标视图的触摸事件，包括必要时的事件拦截。

2.15 阻止触摸窃贼

2.15.1 问题

你在应用程序视图中设计了嵌套的触摸交互，这些交互不能很好地作用于触摸层次结构的标准流程，在此层次结构中，较高层的容器视图通过子视图进行窃取来直接处理触摸事件。

2.15.2 解决方案

(API Level 1)

ViewGroup 是框架中所有布局和容器的基类，它为此提供了描述性命名方法 requestDisallowTouchIntercept()。在任意容器视图上设置此标志会指示框架，在当前手势持续期间，我们更希望它们不会拦截进入其子视图的事件。

2.15.3 实现机制

为展示此方法的实际使用，我们创建了一个示例，其中两个互相竞争的可触摸视图位于同一个位置。外部包含视图是 ListView，它通过滚动内容响应指示垂直拖动的触摸事件。

在 ListView 内部是作为头部添加的 ViewPager，它响应水平拖动触摸事件以在页面之间轻扫。就其本质来说，该例带来了一个问题，水平轻扫在垂直方向上远距离变化的 ViewPager 的尝试会为了支持 ListView 滚动而被取消，因为 ListView 会监控和拦截这些事件。人们无法在垂直或水平运动过程中进行拖动，因此这就产生了可用性问题。

为建立此例，首先需要声明一个维度资源(参见代码清单 2-43)，代码清单 2-44 给出了完整的 Activity。

代码清单 2-43　res/values/dimens.xml

```
<?xml version="1.0" encoding="utf-8"?>
<resources>
    <dimen name="header_height">150dp</dimen>
</resources>
```

代码清单 2-44　管理触摸拦截的 Activity

```java
public class DisallowActivity extends Activity implements
        ViewPager.OnPageChangeListener {
    private static final String[] ITEMS = {
            "Row One", "Row Two", "Row Three", "Row Four",
            "Row Five", "Row Six", "Row Seven", "Row Eight",
            "Row Nine", "Row Ten"
    };

    private ViewPager mViewPager;

    private ListView mListView;

    @Override
    protected void onCreate(Bundle savedInstanceState) {
        super.onCreate(savedInstanceState);

        // 创建水平轻扫条目的头部视图
        mViewPager = new ViewPager(this);
        // 作为 ListView 的头部，ViewPager 必须有固定的高度
        mViewPager.setLayoutParams(new ListView.LayoutParams(
                ListView.LayoutParams.MATCH_PARENT,
                getResources().getDimensionPixelSize(
                        R.dimen.header_height)) );
        // 侦听分页状态变化以禁止父视图触摸
        mViewPager.setOnPageChangeListener(this);
        mViewPager.setAdapter(new HeaderAdapter(this));

        // 创建垂直滚动列表
        mListView = new ListView(this);
        // 添加页面调度程序作为列表头部
        mListView.addHeaderView(mViewPager);
        // 添加列表条目
        mListView.setAdapter(new ArrayAdapter<String>(this,
```

```java
                    android.R.layout.simple_list_item_1, ITEMS));

    setContentView(mListView);
}

/* OnPageChangeListener 方法 */

@Override
public void onPageScrolled(int position,
        float positionOffset, int positionOffsetPixels) { }

@Override
public void onPageSelected(int position) { }

@Override
public void onPageScrollStateChanged(int state) {
    // 当 ViewPager 滚动时，禁止 ScrollView 触摸拦截
    // 从而使它不能接管并尝试垂直滚动
    // 必须为每个要重写的手势设置此标志
    boolean isScrolling =
            state != ViewPager.SCROLL_STATE_IDLE;
    mListView.requestDisallowInterceptTouchEvent(isScrolling);
}

private static class HeaderAdapter extends PagerAdapter {
    private Context mContext;

    public HeaderAdapter(Context context) {
        mContext = context;
    }

    @Override
    public int getCount() {
        return 5;
    }

    @Override
    public Object instantiateItem(ViewGroup container,
            int position) {
        // 创建新的页面视图
        TextView tv = new TextView(mContext);
        tv.setText(String.format("Page %d", position + 1));
        tv.setBackgroundColor((position % 2 == 0) ? Color.RED
                : Color.GREEN);
        tv.setGravity(Gravity.CENTER);
        tv.setTextColor(Color.BLACK);

        // 添加此位置的视图，返回此位置的对象
        container.addView(tv);
        return tv;
```

```
        }

        @Override
        public void destroyItem(ViewGroup container,
                int position, Object object) {
            View page = (View) object;
            container.removeView(page);
        }

        @Override
        public boolean isViewFromObject(View view,
                Object object) {
            return (view == object);
        }
    }
}
```

在此 Activity 中，作为根视图的 ListView 包含一个基本适配器，用于显示字符串条目的静态列表。同样在 onCreate()方法中，创建 ViewPager 实例并作为头部视图添加到列表中。我们将在本章后面更详细地讨论 ViewPager 的工作方式，此处只需要知道我们正在创建一个带有自定义 PagerAdapter 的简单 ViewPager，它显示一些彩色视图作为其页面，以供用户在这些页面之间轻扫。

创建 ViewPager 之后，构造并应用一组 ListView.LayoutParams 来控制 ViewPager 如何作为头部显示。必须执行该操作，因为 ViewPager 自身没有内在的内容大小，并且列表头部不能很好地作用于没有明确高度的视图。通过维度资源应用固定的高度，从而可以轻松获得适当缩放的 dp 值，该值与设备无关。这比完全通过 Java 代码全面构造 dp 值要简单很多。

此例的关键之处在于 Activity 实现的 OnPageChangeListener(该回调在后面会应用于 ViewPager)。当用户与 ViewPager 交互并左右轻扫时，就会触发此回调。在 OnPageScrollStateChanged()方法内部，我们传递一个指示 ViewPager 是否空闲、Activity 正在滚动或在滚动后停到某个页面的值。这是控制父 ListView 的触摸拦截行为的最佳位置。当 ViewPager 的滚动状态不是空闲时，我们不希望 ListView 窃取 ViewPager 正在使用的触摸事件，因此在 requestDisallowTouchIntercept()中设置相应的标志。

连续触发该值还有另一个原因。在原始解决方案中提及，该标志对当前手势有效。这意味着每次新的 ACTION_DOWN 事件发生时，我们需要再次设置该标志。没有添加触摸侦听器来查找特定的事件，我们基于子视图的滚动行为连续设置该标志，这就获得了相同的效果。

2.16 创建拖放视图

2.16.1 问题

应用程序的 UI 需要允许用户将一些视图在屏幕上进行拖动，而且可以将它们放置到其他视图的上面。

2.16.2 解决方案

(API Level 11)

使用 Android 3.0 框架中可用的拖放 API。View 类包含了对管理屏幕上的所有拖动事件的改进，而 OnDragListener 接口则可以关联到任何拖动事件发生时需要得到通知的 View。要想开始拖动事件，只需要在希望用户开始拖动的视图上简单地调用 startDrag() 方法即可。这个方法需要一个 DragShadowBuilder 实例，它用来构建视图中拖动部分的外观，另外还有两个参数将会传递给放置时的目标和监听器。

在所有传递的参数中首先是一个 ClipData 对象，用来传递文本或 Uri 实例。它在传递文件路径或查询 ContentProvider 的场景下非常有用。第二个参数是一个对象，表示拖动事件的"本地状态"。这个参数可以为任意对象，它是一个轻量级的实例，用来对拖动进行一些应用程序相关的描述。ClipData 只会用于拖动视图放下事件的监听器，而本地状态对于所有的监听器都是可访问的(任何时刻调用 DragEvent 的 getLocalState()方法即可)。

在拖放过程中发生的每个特定事件都会调用 OnDragListener.onDrag()方法，同时传回一个描述每个事件特征的 DragEvent。每个 DragEvent 都具有以下动作中的一个：

- ACTION_DRAG_STARTED：当调用 startDrag()以开始一个新的拖动事件时会向所有视图发送该动作。
  - ➢ 位置信息可以通过 getX()和 getY()获得。
- ACTION_DRAG_ENTERED：当拖动事件进入某个视图的边界框时会向该视图发送该动作。
- ACTION_DRAG_EXITED：当拖动事件离开某个视图的边界框时会向该视图发送该动作。
- ACTION_DRAG_LOCATION：当拖动事件介于 ACTION_DRAG_ENTERED 和 ACTION_DRAG_EXITED 之间时会向某个视图发送该动作，同时会传递该视图中拖动的当前位置。
  - ➢ 位置信息可以通过 getX()和 getY()获得。
- ACTION_DROP：当拖动终止并依然位于某个视图的边界中时会向该视图发送该动作。
  - ➢ 位置信息可以通过 getX()和 getY()获得。
  - ➢ 随同事件传入的 ClipData 只能在执行该动作时通过 getClipData()方法获得。
- ACTION_DRAG_ENDED：当前拖动事件完成时会向所有视图发送该动作。
  - ➢ 拖动操作的结果可以在这里通过 getResult()获得。
  - ➢ 该方法的返回值取决于放置时的目标视图是否拥有一个可用的 OnDragListener，其可以针对 ACTION_DROP 事件返回 true。

这个方法和自定义触摸事件的工作方式类似，即监听器中返回的值决定了后续的事件传递。如果某个特殊的 OnDragListener 并没有对 ACTION_DRAG_STARTED 动作返回 true，那么除了 ACTION_DRAG_ENDED 以外，它将不会收到拖动过程中任何后续的事件。

2.16.3 实现机制

让我们看一个拖放功能的示例,首先是代码清单 2-45。这里我们创建了一个自定义的 ImageView,它实现了 OnDragListener 接口。

代码清单 2-45 自定义视图实现了 OnDragListener

```java
public class DropTargetView extends ImageView implements OnDragListener {

    private boolean mDropped;

    public DropTargetView(Context context) {
        super(context);
        init();
    }

    public DropTargetView(Context context, AttributeSet attrs) {
        super(context, attrs);
        init();
    }

    public DropTargetView(Context context, AttributeSet attrs, int defaultStyle) {
        super(context, attrs, defaultStyle);
        init();
    }

    private void init() {
        //我们必须设置一个有效的监听器来接收 DragEvent
        setOnDragListener(this);
    }

    @Override
    public boolean onDrag(android.view.View v, DragEvent event) {
        PropertyValuesHolder pvhX, pvhY;
        switch(event.getAction()) {
        case DragEvent.ACTION_DRAG_STARTED:
            //通过收缩视图响应新的拖动动作
            pvhX = PropertyValuesHolder.ofFloat("scaleX", 0.5f);
            pvhY = PropertyValuesHolder.ofFloat("scaleY", 0.5f);
            ObjectAnimator.ofPropertyValuesHolder(this, pvhX, pvhY).start();
            //新的拖动行为会清空当前的放置图片
            setImageDrawable(null);
            mDropped = false;
            break;
        case DragEvent.ACTION_DRAG_ENDED:
            //拖动结束时,如果没有找到放置目标,视图则恢复到原来的大小
            if(!mDropped) {
                pvhX = PropertyValuesHolder.ofFloat("scaleX", 1f);
                pvhY = PropertyValuesHolder.ofFloat("scaleY", 1f);
```

```java
            ObjectAnimator.ofPropertyValuesHolder(this, pvhX, pvhY).start();
            mDropped = false;
        }
        break;
    case DragEvent.ACTION_DRAG_ENTERED:
        //拖动进入到边界区域时,视图会稍微放大一下
        pvhX = PropertyValuesHolder.ofFloat("scaleX", 0.75f);
        pvhY = PropertyValuesHolder.ofFloat("scaleY", 0.75f);
        ObjectAnimator.ofPropertyValuesHolder(this, pvhX, pvhY).start();
        break;
    case DragEvent.ACTION_DRAG_EXITED:
        //拖动离开视图时,视图会恢复到之前的大小
        pvhX = PropertyValuesHolder.ofFloat("scaleX", 0.5f);
        pvhY = PropertyValuesHolder.ofFloat("scaleY", 0.5f);
        ObjectAnimator.ofPropertyValuesHolder(this, pvhX, pvhY).start();
        break;
    case DragEvent.ACTION_DROP:
        //拖动后放置时会有一段简短的关键帧动画并将视图的图片
        //设置为拖动事件传入的drawable图片

        //这个动画会收缩一下视图,然后再还原
        Keyframe frame0 = Keyframe.ofFloat(0f, 0.75f);
        Keyframe frame1 = Keyframe.ofFloat(0.5f, 0f);
        Keyframe frame2 = Keyframe.ofFloat(1f, 0.75f);
        pvhX = PropertyValuesHolder.ofKeyframe("scaleX", frame0, frame1,
                frame2);
        pvhY = PropertyValuesHolder.ofKeyframe("scaleY", frame0, frame1,
                frame2);
        ObjectAnimator.ofPropertyValuesHolder(this, pvhX, pvhY).start();
        //DragEvent中传递的Object设置我们的图片
        setImageDrawable((Drawable) event.getLocalState());
        //我们设置放置标识让ENDED动画不再运行
        mDropped = true;
        break;
    default:
        //忽略我们不感兴趣的事件
        return false;
    }
    //处理所有感兴趣的事件
    return true;
}
```

这个ImageView用来监控新产生的拖动事件并自己运行相应的动画。每次新的拖动行为出现时,ACTION_DRAG_STARTED事件就会被发送到这里,这个ImageView就会自己缩小50%。这对用户是一个非常好的引导,可以告诉用户他们刚刚选择的视图可以拖动到哪里。这里我们还确保这个监听器会对此事件返回true,这样就可以接收拖动过程中的其他事件了。

如果用户将他们的视图拖动到这个ImageView上,就会触发ACTION_DRAG_ENTERED,

这时 ImageView 会稍微放大一下，表示该 ImageView 可以接收的视图放置行为。当视图被拖离时会触发 ACTION_DRAG_EXITED 事件，这时 ImageView 会恢复到刚进入"拖放模式"时的大小。如果用户在该 ImageView 的上方松手，会触发 ACTION_DROP 事件，同时会运行一段特殊的动画表示放置动作已经收到。这时我们会读取事件中的本地状态变量，如果是一个 Drawable，就把它设置为 ImageView 的图片内容。

ACTION_DRAG_ENDED 会通知该 ImageView 恢复到之前的大小，因为此时已经不再处于"拖动模式"了。但是，如果这个 ImageView 就是放置的目标，我们希望保持它的大小，因此这种情况下会忽略掉这个事件。

代码清单 2-46 和 2-47 显示了一个示例 Activity，该 Activity 允许用户长按一张图片，然后可以拖动该图片到我们自定义的放置目标上。

代码清单 2-46　res/layout/main.xml

```xml
<?xml version="1.0" encoding="utf-8"?>
<RelativeLayout xmlns:android="http://schemas.android.com/apk/res/android"
    android:layout_width="match_parent"
    android:layout_height="match_parent" >

    <!--顶部一行是可拖放的条目-->
    <LinearLayout
        android:layout_width="match_parent"
        android:layout_height="wrap_content"
        android:orientation="horizontal" >
        <ImageView
            android:id="@+id/image1"
            android:layout_width="0dp"
            android:layout_height="wrap_content"
            android:layout_weight="1"
            android:src="@drawable/ic_send" />
        <ImageView
            android:id="@+id/image2"
            android:layout_width="0dp"
            android:layout_height="wrap_content"
            android:layout_weight="1"
            android:src="@drawable/ic_share" />
        <ImageView
            android:id="@+id/image2"
            android:layout_width="0dp"
            android:layout_height="wrap_content"
            android:layout_weight="1"
            android:src="@drawable/ic_favorite" />
    </LinearLayout>

    <!--底部一行是可放置的目标 -->
    <LinearLayout
        android:layout_width="match_parent"
        android:layout_height="wrap_content"
        android:layout_alignParentBottom="true"
```

```xml
            android:orientation="horizontal" >
        <com.examples.dragtouch.DropTargetView
            android:id="@+id/drag_target1"
            android:layout_width="0dp"
            android:layout_height="100dp"
            android:layout_weight="1"
            android:background="#A00" />
        <com.examples.dragtouch.DropTargetView
            android:id="@+id/drag_target2"
            android:layout_width="0dp"
            android:layout_height="100dp"
            android:layout_weight="1"
            android:background="#0A0" />
        <com.examples.dragtouch.DropTargetView
            android:id="@+id/drag_target2"
            android:layout_width="0dp"
            android:layout_height="100dp"
            android:layout_weight="1"
            android:background="#00A" />
    </LinearLayout>

</RelativeLayout>
```

代码清单 2-47　转发触摸事件的 Activity

```java
public class DragTouchActivity extends Activity implements OnLongClickListener {

    @Override
    public void onCreate(Bundle savedInstanceState) {
        super.onCreate(savedInstanceState);
        setContentView(R.layout.main);
        //为每个 ImageView 关联长按监听器
        findViewById(R.id.image1).setOnLongClickListener(this);
        findViewById(R.id.image2).setOnLongClickListener(this);
        findViewById(R.id.image3).setOnLongClickListener(this);
    }
    @Override
    public boolean onLongClick(View v) {
        DragShadowBuilder shadowBuilder = new DragShadowBuilder(v);
        //开始拖动，将 View 的图片作为本地状态传递出去
        v.startDrag(null, shadowBuilder, ((ImageView) v).getDrawable(), 0);

        return true;
    }

}
```

这个示例会在屏幕的顶部显示一行三张图片，同时在屏幕的底部显示三个我们自定义的目标视图。为每张图片都设置了一个长按事件监听器，长按动作会通过 startDrag()触发一个新的拖动事件。拖动事件初始化时传入的 DragShadowBuilder 是框架提供的默认实现。下一节会查看如何自定义 DragShadowBuilder，但本节中视图在拖动时会创建一个透明的副

本并显示在触摸点的正下方。

我们还通过 getDrawable() 得到了用户选择的视图的图片内容并把它作为拖动的本地状态传递出去，自定义的放置目标会使用它来设置图片。这样就会产生视图已经放置到目标上的效果。参见图 2-10，查看加载时的效果、拖动操作过程中的效果以及图片被放置到某个放置目标后的效果。

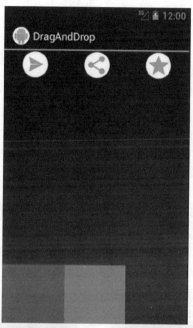

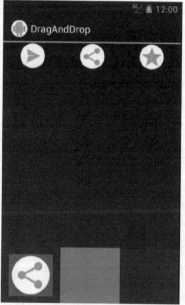

图 2-10　Drag 示例拖动前的效果(上图)，用户拖动并游走于放置目标上方时的效果(左图)以及视图放置后的效果(右图)

自定义 DragShadowBuilder

DragShadowBuilder 的默认实现非常方便，但有可能并不是应用程序需要的。让我们看一下代码清单 2-48，它实现了一个自定义的 DragShadowBuilder。

代码清单 2-48　自定义 DragShadowBuilder

```
public class DrawableDragShadowBuilder extends DragShadowBuilder {
    private Drawable mDrawable;

    public DrawableDragShadowBuilder(View view, Drawable drawable) {
        super(view);
        //设置 Drawable 并使用一个绿色的过滤器
        mDrawable = drawable;
        mDrawable.setColorFilter(
                new PorterDuffColorFilter(Color.GREEN, PorterDuff.Mode.MULTIPLY));
    }

    @Override
    public void onProvideShadowMetrics(Point shadowSize, Point touchPoint) {
        //填充大小
        shadowSize.x = mDrawable.getIntrinsicWidth();
        shadowSize.y = mDrawable.getIntrinsicHeight();
        //设置阴影相对于触摸点的位置
        //这里阴影位于手指下方的中心
        touchPoint.x = mDrawable.getIntrinsicWidth() / 2;
        touchPoint.y = mDrawable.getIntrinsicHeight() / 2;

        mDrawable.setBounds(new Rect(0, 0, shadowSize.x, shadowSize.y));
    }

    @Override
    public void onDrawShadow(Canvas canvas) {
        //在提供的 Canvas 上绘制阴影视图
        mDrawable.draw(canvas);
    }
}
```

此自定义实现会使用一个单独的 Drawable 参数，阴影的显示会使用该图片而不是使用源视图的可见副本。另外，我们还对该图片使用了绿色的 ColorFilter 来增加一些效果。DragShadowBuilder 是一个非常容易扩展的类，只需要有效地覆写两个主要的方法。

第一个方法是 onProvideShadowMetrics()，它会在 DragShadowBuilder 初始化时调用一次并使用两个 Point 对象填充 DragShadowBuilder 的内容。首先会填充阴影使用的图片的大小，即会将期望的宽度设置为 x 值，将期望的高度设置为 y 值。本例中会将该大小设置为图片本来的宽度和高度。另外需要填充的是阴影期望触摸的位置。这里会定义阴影图片相对于用户手指的位置，例如将 x 和 y 都设置为 0 时，手指会位于图片的左上角。在我们的示例中，我们设置到了图片的中心点，因此手指会位于图片中心上方。

第二个方法是 onDrawShadow()，它会被重复调用以渲染阴影图片。这个方法中传入的 Canvas 是由框架根据 onProvideShadowMetrics()中包含的信息创建的。这里你可以像其他自定义视图一样进行各种自定义绘制。我们的示例只是简单地告诉 Drawable 在 Canvas 上绘制它自己。

2.17 构建导航 Drawer

2.17.1 问题

应用程序需要顶层导航菜单，而为了符合最新的 Google 设计指南，你要实现一个这样的菜单，该菜单以动画方式从屏幕的一侧滑进和滑出。

2.17.2 解决方案

(API Level 7)

集成 DrawerLayout 小部件以管理从屏幕左侧或右侧滑入的菜单视图，Android 支持库中提供了该小部件。DrawerLayout 是一个容器小部件，它使用指定的 Gravity 值 LEFT 或 RIGHT(如果支持 RTL 布局，还可以是 START/END)管理其层次结构中每个最初的子视图，将其作为动画形式的内容 Drawer。默认情况下，每个视图都是隐藏的，但当调用 openDrawer()方法或手指从适当的侧面滑入屏幕时，这些视图会从相应的侧面以动画形式进入屏幕。为表明 Drawer 的存在，如果在适当的屏幕侧面按下手指，DrawerLayout 也会查看相应的视图。

DrawerLayout 支持多个 Drawer，每个 Drawer 对应一种 Gravity 设置，它们可以放置在布局层次结构中的任意位置。唯一的软性规则是，它们应该在布局中的主内容视图之后添加(即放置在布局 XML 中的视图元素之后)。否则，视图的 Z 轴顺序将阻止 Drawer 显示。

还可以通过 ActionBarDrawerToggle 元素实现与 Action Bar 的整合。ActionBarDrawer-Toggle 小部件监控 Action Bar 中 Home 按钮区域的点击动作并切换"主"Drawer(带有 Gravity.LEFT 或 Gravity.START 设置的 Drawer)的可见性。

要点：
DrawerLayout 仅在 Android 支持库中提供;它不是任意平台级别中原生 SDK 的一部分。然而，目标平台为 API Level 4 或以后版本的应用程序可以通过包含支持库来使用该小部件。有关在项目中包括支持库的更多信息,请参考 https://developer.android.com/tools/support-library/index.html。

2.17.3 实现机制

虽然不一定要与 DrawerLayout 一起使用 Action Bar，但这是最常见的用例。下面的示例显示了如何使用 DrawerLayout 创建导航 Drawer 以及如何执行 Action Bar 整合。

下面的示例创建带有两个导航 Drawer 的应用程序：左侧的主 Drawer 带有可供选择的

选项列表，右侧的辅助 Drawer 带有一些额外的交互式内容。从主 Drawer 的列表中选择一个条目会修改主要内容视图的背景颜色。

在代码清单 2-49 中，我们有一个包含 DrawerLayout 的布局。请注意，因为此小部件不是核心元素，所以必须在 XML 中使用其完全限定的类名。

代码清单 2-49　res/layout/activity_main.xml

```xml
<?xml version="1.0" encoding="utf-8"?>
<android.support.v4.widget.DrawerLayout
    xmlns:android="http://schemas.android.com/apk/res/android"
    android:id="@+id/container_drawer"
    android:layout_width="match_parent"
    android:layout_height="match_parent" >

    <!-- 主内容窗格 -->
    <FrameLayout
        android:id="@+id/container_root"
        android:layout_width="match_parent"
        android:layout_height="match_parent">

        <!-- 在此放置主内容 -->

    </FrameLayout>

    <!-- 主 Drawer 内容 -->
    <!--
        可以是任意 View 或 ViewGroup 内容
        标准 Drawer 宽度是 240dp
        必须设置 Gravity 值
        需要在内容之上显示纯色背景
        -->
    <ListView
        android:id="@+id/drawer_main"
        android:layout_width="240dp"
        android:layout_height="match_parent"
        android:layout_gravity="left"
        android:background="#555" />

    <!--
        可以创建额外的 Drawer
        例如这个 Drawer 将随着从屏幕右侧轻扫进入而显示
        -->
    <LinearLayout
        android:id="@+id/drawer_right"
        android:layout_width="240dp"
        android:layout_height="match_parent"
        android:layout_gravity="right"
        android:orientation="vertical"
        android:background="#CCC">
```

```xml
            <Button
                android:layout_width="match_parent"
                android:layout_height="wrap_content"
                android:text="Click Here!" />
            <TextView
                android:layout_width="match_parent"
                android:layout_height="match_parent"
                android:gravity="center"
                android:text="Tap Anywhere Else, Drawer will Hide" />
    </LinearLayout>
</android.support.v4.widget.DrawerLayout>
```

我们已包括两个视图,它们在应用程序中充当 Drawer,一个在屏幕左侧,另一个在右侧;通过设置 android:layout_gravity 属性来控制它们的对齐。DrawerLayout 执行剩余的工作,它通过检查 Gravity 值来映射每个视图,因此我们不需要以其他方式链接它们。在接触 Activity 之前,需要知道我们的项目还包含一个资源;我们创建了一个选项菜单来在 Action Bar 中显示一些动作(参见代码清单 2-50)。

代码清单 2-50　res/menu/main.xml

```xml
<menu xmlns:android="http://schemas.android.com/apk/res/android"
    xmlns:app="http://schemas.android.com/apk/res-auto">
    <item
        android:id="@+id/action_delete"
        android:title="@string/action_delete"
        app:showAsAction="ifRoom"
        android:icon="@android:drawable/ic_menu_delete"/>
    <item
        android:id="@+id/action_settings"
        android:title="@string/action_settings"
        android:orderInCategory="100"
        app:showAsAction="never"/>
</menu>
```

最终,我们就有了代码清单 2-51 中的 Activity。除了 DrawerLayout 之外,该例还包含一个 ActionBarDrawerToggle,用于提供与 Action Bar 的 Home 按钮的整合。

代码清单 2-51　整合 DrawerLayout 的 Activity

```java
public class NativeActivity extends Activity
        implements AdapterView.OnItemClickListener {

    private static final String[] ITEMS =
        {"White", "Red", "Green", "Blue"};
    private static final int[] COLORS =
        {Color.WHITE, 0xffe51c23, 0xff259b24, 0xff5677fc};

    private DrawerLayout mDrawerContainer;
    /* 布局中的根内容窗格 */
    private View mMainContent;
```

```java
/* 主(左侧)滑动 Drawer */
private ListView mDrawerContent;
/* Action Bar 的开关对象 */
private ActionBarDrawerToggle mDrawerToggle;

@Override
protected void onCreate(Bundle savedInstanceState) {
    super.onCreate(savedInstanceState);
    setContentView(R.layout.activity_main);

    mDrawerContainer =
        (DrawerLayout) findViewById(R.id.container_drawer);
    mDrawerContent =
        (ListView) findViewById(R.id.drawer_main);
    mMainContent = findViewById(R.id.container_root);

    // 开关指示器也必须是 Drawer 侦听器
    // 因此扩展该侦听器以侦听事件自身
    mDrawerToggle = new ActionBarDrawerToggle(
            this,                    //主机 Activity
            mDrawerContainer,        //使用的容器
            R.string.drawer_open,   //内容说明字符串
            R.string.drawer_close) {

        @Override
        public void onDrawerOpened(View drawerView) {
            super.onDrawerOpened(drawerView);
            // 更新选项菜单
            supportInvalidateOptionsMenu();
        }

        @Override
        public void onDrawerStateChanged(int newState) {
            super.onDrawerStateChanged(newState);
            // 更新选项菜单
            supportInvalidateOptionsMenu();
        }

        @Override
        public void onDrawerClosed(View drawerView) {
            super.onDrawerClosed(drawerView);
            // 更新选项菜单
            supportInvalidateOptionsMenu();
        }
    };

    ListAdapter adapter = new ArrayAdapter<String>(this,
            android.R.layout.simple_list_item_1, ITEMS);
    mDrawerContent.setAdapter(adapter);
```

```java
        mDrawerContent.setOnItemClickListener(this);

        // 设置开关指示器为 Drawer 的事件侦听器
        mDrawerContainer.setDrawerListener(mDrawerToggle);

        // 在 Action Bar 中启用 Home 按钮动作
        getActionBar().setDisplayHomeAsUpEnabled(true);
        getActionBar().setHomeButtonEnabled(true);
    }

    @Override
    protected void onPostCreate(Bundle savedInstanceState) {
        super.onPostCreate(savedInstanceState);
        // 在框架还原任意实例状态之后同步 Drawer 的状态
        mDrawerToggle.syncState();
    }

    @Override
    public void onConfigurationChanged(Configuration newConfig) {
        super.onConfigurationChanged(newConfig);
        // 在更改任意配置时更新状态
        mDrawerToggle.onConfigurationChanged(newConfig);
    }

    @Override
    public boolean onCreateOptionsMenu(Menu menu) {
        // 创建 Action Bar 动作
        getMenuInflater().inflate(R.menu.main, menu);
        return true;
    }

    @Override
    public boolean onPrepareOptionsMenu(Menu menu) {
        // 基于主 Drawer 的状态显示动作选项
        boolean isOpen =
                mDrawerContainer.isDrawerVisible(mDrawerContent);
        menu.findItem(R.id.action_delete).setVisible(!isOpen);
        menu.findItem(R.id.action_settings).setVisible(!isOpen);

        return super.onPrepareOptionsMenu(menu);
    }

    @Override
    public boolean onOptionsItemSelected(MenuItem item) {
        // 首先让 Drawer 在事件处有一个缺口，
        // 从而处理 Home 按钮事件
        if(mDrawerToggle.onOptionsItemSelected(item)) {
            // 如果这是一个 Drawer 开关，我们需要更新选项菜单
            // 但必须等到下一次循环遍历 Drawer 状态改变时再更新
```

```java
            mDrawerContainer.post(new Runnable() {
                @Override
                public void run() {
                    // 更新选项菜单
                    supportInvalidateOptionsMenu();
                }
            });
            return true;
        }

        //...像往常一样在此处理其他选项选择...
        switch (item.getItemId()) {
            case R.id.action_delete:
                // 删除动作
                return true;
            case R.id.action_settings:
                // 设置动作
                return true;
            default:
                return super.onOptionsItemSelected(item);
        }
    }

    // 根据主 Drawer 列表中的条目处理点击事件
    @Override
    public void onItemClick(AdapterView<?> parent, View view,
            int position, long id) {
        // 更新主内容的背景色
        mMainContent.setBackgroundColor(COLORS[position]);

        // 手动关闭 Drawer
        mDrawerContainer.closeDrawer(mDrawerContent);
    }
}
```

初始化 Activity 时,我们创建 ActionBarDrawerToggle 实例并将其设置为 DrawerLayout 的 DrawerListener。这是必需的步骤,从而 ActionBarDrawerToggle 才可以侦听事件,但这也意味着,除非我们扩展 ActionBarDrawerToggle 以重写侦听器方法(在此已完成该操作),否则无法在应用程序中侦听这些事件。ActionBarDrawerToggle 也链接驻留它的 Activity 以及它应该控制的 DrawerLayout。

集成 ActionBarDrawerToggle 需要相当数量的样板代码,因为它不会直接关联到 Activity 的任何生命周期方法。需要从适当的 Activity 回调中调用 syncState()、onConfigurationChanged()和 onOptionsItemSelected()方法,从而让开关小部件可以接收输入以及连同 Activity 实例一起维护状态。为了触发 Action Bar 中的 Home 按钮事件,还必须通过调用 setHomeButtonEnabled()来启用 Home 按钮。最后,添加 setDisplayHomeAsUpEnabled() 以使图标(默认为箭头)显示在 Home 徽标的旁边; Drawer 开关使用自己的版本定制该图标。

DrawerLayout 被设计为当主内容视图接收触摸事件(即用户在 Drawer 外部触摸)时打开

和关闭 Drawer。布局内的触摸事件(例如触摸主列表中的条目或辅助 Drawer 中的按钮)要求我们在必要时手动关闭 Drawer。在注册到列表的 OnItemClickListener 内部，我们在更改内容视图的背景颜色之后调用 closeDrawer()以执行 Drawer 的关闭操作。值得注意的是，即使用户点击 Drawer 内不可交互的视图(如 TextView)，这些触摸事件也会按顺序传递给下一个子视图。如果这个子视图是主内容视图(最常见的情况)，则 Drawer 会像用户触摸其外部一样关闭。

注意 openDrawer()和 closeDrawer()这样的方法如何获取视图参数。因为 DrawerLayout 可以管理多个 Drawer，我们必须告诉它操作哪个 Drawer 小部件。如果应用程序没有指向 Drawer 视图自身的引用，也可以使用与 Drawer 关联的 Gravity 参数触发这些方法。

回顾一下，我们扩展了 ActionBarDrawerToggle 以重写 Drawer 的事件侦听器方法。在每个方法的内部调用 invalidateOptionsMenu()，该方法仅仅告诉 Activity 更新菜单并再次调用其设置方法。同样回顾一下，我们使用 XML 菜单创建了一些显示在 Action Bar 内部的动作，而在 onPrepareOptionsMenu()内部，我们根据 Drawer 的可见性状态控制是否显示这些动作。这样，这些动作只有在主 Drawer 未显示时才会出现。在每个事件回调中使菜单无效的作用是可以基于 Drawer 中的改动更新菜单可见性。

图 2-11 显示了如何点击 Action Bar 中的 Home 按钮来展开主 Drawer，从而显示选项列表；还要注意的是，当 Drawer 打开时，这些动作会消失。图 2-12 说明了隐藏在边缘的辅助 Drawer 从屏幕的一侧滑入，然后完全打开。

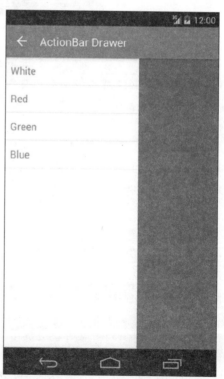

图 2-11　带有主 Drawer 的 Activity

图 2-12　带有辅助 Drawer 的 Activity

完成实际工作的类

　　DrawerLayout 中提供的拖动和边缘滑入行为实际上是支持库中提供的另一个类的工作：ViewDragHelper。如果需要基于用户拖动执行任何自定义视图操作，该类就会非常有帮助。

　　ViewDragHelper 是触摸事件处理程序(类似于 GestureDetector)，因此它需要从视图中提供事件。一般情况下，在视图的 onTouchEvent()中接收的每个事件必须直接交给 ViewDragHelper 中的 processTouchEvent()进行处理。

```
@Override
public boolean onTouchEvent(MotionEvent event) {
    mHelper.processTouchEvent(event);
}
```

　　实例化 ViewDragHelper 时，必须传递 ViewDragHelper.Callback 的实例，将其作为辅助类传递给应用程序的所有事件的处理程序。其中最重要的方法是 tryCaptureView()，当辅助类开始监控给定视图上的拖动时就会调用该方法；该方法返回 true 会造成视图被"捕获"，这意味着其位置将跟随手势中随后的触摸事件而移动。

　　如果使用一个或多个有效的边缘标志调用了 setEdgeTrackingEnabled()，则 ViewDragHelper 也支持从视图边缘滑入。当边缘事件发生时，会在 Callback 上触发 onEdgeTouched()和 onEdgeDragStarted()方法。

最后一个提示是：单个 ViewDragHelper 被设计为一次仅捕获和管理一个视图。如果尝试使用同一个 ViewDragHelper 实例同时滑动两个视图，就会出现问题。例如，DrawerLayout 对它支持的每个 Drawer 使用一个 ViewDragHelper，从而避免这种特殊的问题。

在 Toolbar 上绘制

Google 设计指南中对此模式的改编要求 Drawer 在 Action Bar 的顶部滑动。当 Action Bar 作为窗口装饰的一部分时，这一行为是无法实现的，但如果将 Action Bar 替换为 Toolbar，就可以轻松实现。作为参考，图 2-13 显示了打开时不同的 Drawer。

 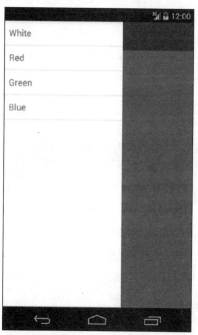

图 2-13　包含 Toolbar Drawer 的 Activity

与以前的 Toolbar 示例一样，我们必须确保 Activity 使用禁用窗口 Action Bar 的主题，如代码清单 2-52 所示。

代码清单 2-52　Toolbar Activity 的部分 AndroidManifest.xml

```xml
<activity
    android:name=".ToolbarActivity"
    android:label="@string/title_toolbar"
    android:theme="@style/Theme.AppCompat.Light.NoActionBar">
    <intent-filter>
        <action android:name="android.intent.action.MAIN"/>
        <category android:name="android.intent.category.LAUNCHER"/>
    </intent-filter>
</activity>
```

这就要求我们包括修改后的布局，该布局具有在层次结构中定义的 Toolbar 元素，如代码清单 2-53 所示。

代码清单 2-53　res/layout/activity_toolbar.xml

```xml
<?xml version="1.0" encoding="utf-8"?>
<android.support.v4.widget.DrawerLayout
    xmlns:android="http://schemas.android.com/apk/res/android"
    xmlns:app="http://schemas.android.com/apk/res-auto"
    android:id="@+id/container_drawer"
    android:layout_width="match_parent"
    android:layout_height="match_parent" >

    <!-- 主内容窗格 -->
    <LinearLayout
        android:layout_width="match_parent"
        android:layout_height="match_parent"
        android:orientation="vertical">

        <!-- 使用 Toolbar 代替 Action Bar，从而视图可在其顶部绘制 -->
        <android.support.v7.widget.Toolbar
            android:id="@+id/toolbar"
            android:layout_height="wrap_content"
            android:layout_width="match_parent"
            android:minHeight="?attr/actionBarSize"
            android:background="?attr/colorPrimary"
            app:theme="@style/ThemeOverlay.AppCompat.Dark.ActionBar"/>

        <FrameLayout
            android:id="@+id/container_root"
            android:layout_width="match_parent"
            android:layout_height="match_parent">
            <!-- 在此放置视图内容 -->
        </FrameLayout>
    </LinearLayout>

    <!-- 主 Drawer 内容 -->
    <!--
        可以是任意视图或 ViewGroup 内容。标准 Drawer 宽度为 240dp。必须设置重力，
        它必须为"left"或"start"。需要在内容顶部显示纯色背景
    -->
    <ListView
        android:id="@+id/drawer_main"
        android:layout_width="240dp"
        android:layout_height="match_parent"
        android:layout_gravity="start"
        android:background="#FFF" />

    <!--
        可以创建额外的 Drawer，例如此处的 Drawer 将显示为从屏幕右侧轻扫
    -->
    <LinearLayout
        android:id="@+id/drawer_right"
```

```xml
        android:layout_width="240dp"
        android:layout_height="match_parent"
        android:layout_gravity="end"
        android:orientation="vertical"
        android:background="#CCC">
        <Button
            android:layout_width="match_parent"
            android:layout_height="wrap_content"
            android:text="Click Here!" />
        <TextView
            android:layout_width="match_parent"
            android:layout_height="match_parent"
            android:gravity="center"
            android:text="Tap Anywhere Else, Drawer will Hide" />
    </LinearLayout>
</android.support.v4.widget.DrawerLayout>
```

此 Activity 代码与前一个 Drawer 示例基本相同，不同之处在于 onCreate()中的两行代码，这些代码向 Activity 注册布局中的 Toolbar。

```
Toolbar toolbar = (Toolbar) findViewById(R.id.toolbar);
setSupportActionBar(toolbar);
```

2.18 在视图之间滑动

2.18.1 问题

需要在应用程序的 UI 中通过手势滑动来实现页面切换，例如视图之间或 Fragment 之间的切换。

2.18.2 解决方案

(API Level 4)

实现 ViewPager 小部件以提供手势滑动时页面切换的功能。ViewPager 是 AdapterView 模式修改后的实现，ListView 和 GridView 小部件也使用了框架的这种模式。ViewPager 需要一个继承自 PagerAdapter 的子类适配器实现，但从概念上讲，该适配器与 BaseAdapter 和 ListAdapter 中使用的模式非常类似。ViewPager 本身并不能实现分页控件的回收，但它每时每刻都提供了回调方法来进行条目的创建和销毁。所以在特定的时间内，内存中运行的内容视图的数量是固定的。

要点：

ViewPager 是当前只可以通过 Android 支持库使用的控件。无论在哪个级别的 Android 平台中，原生的 SDK 都不包含 ViewPager。不过，所有目标版本为 API Level 4 或以上级别的应用程序都可以通过包括支持库来使用该小部件。关于在项目中包括支持库的更多信息，请参考 https://developer.android.com/tools/support-library/index.html。

2.18.3 实现机制

使用 ViewPager 最大的工作就是 PagerAdapter 的实现。让我们开始一个简单的示例，参见代码清单 2-54，它实现了一系列图片的分页显示。

代码清单 2-54　自定义图像 PagerAdapter

```java
public class ImagePagerAdapter extends PagerAdapter {
    private Context mContext;

    private static final int[] IMAGES = {
        android.R.drawable.ic_menu_camera,
        android.R.drawable.ic_menu_add,
        android.R.drawable.ic_menu_delete,
        android.R.drawable.ic_menu_share,
        android.R.drawable.ic_menu_edit
    };

    private static final int[] COLORS = {
        Color.RED,
        Color.BLUE,
        Color.GREEN,
        Color.GRAY,
        Color.MAGENTA
    };

    public ImagePagerAdapter(Context context) {
        super();
        mContext = context;
    }

    /*
    * 提供页面的总数
    */
    @Override
    public int getCount() {
        return IMAGES.length;
    }

    /*
    *如果要在ViewPager内一次显示超过一页的内容,那么需要重写该方法
    */
    @Override
    public float getPageWidth(int position) {
        return 1f;
    }

    @Override
```

```java
        public Object instantiateItem(ViewGroup container, int position) {
            //创建一个新的 ImageView 并把它添加到提供的容器中
            ImageView iv = new ImageView(mContext);
            //设置此位置的内容
            iv.setImageResource(IMAGES[position]);
            iv.setBackgroundColor(COLORS[position]);

            //这里你必须自己添加视图, Android 框架是不会为你添加的
            container.addView(iv);
            //将这个视图作为这个位置的键对象返回
            return iv;
        }

        @Override
        public void destroyItem(ViewGroup container, int position, Object object) {
            //此处从容器中删除视图
            container.removeView((View) object);
        }

        @Override
        public boolean isViewFromObject(View view, Object object) {
            //检查从 instantiateItem() 返回的对象与添加到容器相应位置的视图是否是同一
            //个对象。我们的示例在这两个地方使用的是同一个对象。
            return (view == object);
        }
    }
```

在这个示例中，我们实现了一个 PagerAdapter，它提供了很多的 ImageView 实例供用户翻看。和 AdapterView 的适配器一样，PagerAdapter 中第一个需要重写的就是 getCount() 方法，它会返回要显示条目的总数。

ViewPager 是基于跟踪每个条目的键对象以及显示该对象的视图进行工作的，这样会将适配器条目和它们的视图(开发人员在使用 AdapterView 时经常会用到)分离开来。但是它们的实现方式略有不同。如果使用 AdapterView，适配器的 getView() 方法会构建和返回条目上显示的视图。而使用 ViewPager，当需要创建一个新视图，或者某个视图滚动超出了页数限制的范围后需要删除该视图时，就会分别调用 instantiateItem() 和 destroyItem() 回调方法，通过 setOffscreenPageLimit() 方法来设置每个 ViewPager 可持有条目的数量。

注意：
屏幕以外的页数默认限制值为 3，这意味着 ViewPager 将会跟踪当前可见页面、当前页面左侧的页面以及当前页面右侧的页面。跟踪页面的编号总是围绕当前可见的页面居中进行的。

在我们的示例中，我们使用 instantiateItem() 来创建一个新的 ImageView 并设置该 ImageView 的相关属性。和 AdapterView 不同的是，PagerAdapt 除了通过返回唯一的键对象来表示某个条目外，还必须把要显示的 View 关联到给定的 ViewGroup 上。这两个操作

不一定需要相同，但可以像本例中这样简单处理。需要重写 PagerAdapter 的 isViewFromObject()回调方法，这样应用程序就可以将键对象和视图关联起来。在我们的示例中，将 ImageView 添加到给定的父视图上并将该 ImageView 作为 instantiateItem()的键对象返回值。如此一来，isViewFromObject()中的代码就变得简单了，如果两个参数的实例是相同的，就返回 true。

作为实例化过程的补充，PagerAdapter 同样需要在 destroyItem()方法中将指定的视图从父容器移除。如果页面上显示的是重量级的视图，同时你想实现可以在适配器中循环利用的基本视图，这个视图被删除后可以保存它，这样它就可以在 instantiateItem()中附加在另一个键对象上。代码清单 2-55 展示了一个 Activity 示例，在 ViewPager 中使用我们自定义的适配器，图 2-14 中展示了应用程序的结果。

代码清单 2-55　使用了 ViewPager 和 ImagePagerAdapter 的 Activity

```
public class PagerActivity extends Activity {

    @Override
    protected void onCreate(Bundle savedInstanceState) {
        super.onCreate(savedInstanceState);
        ViewPager pager = new ViewPager(this);
        pager.setAdapter(new ImagePagerAdapter(this));

        setContentView(pager);
    }
}
```

图 2-14　可以在两个页面之间进行拖动的 ViewPager

运行这个应用程序后,用户可以水平滑动手指来分页浏览自定义适配器提供的所有图片,而且每张图片都是全屏显示。本例中有一个定义的方法我们没有提到:getPageWidth()。这个方法允许在每个位置设置图片页面大小相对于 ViewPager 页面大小的百分比。默认值设置为 1,前面的示例也没有改变该默认值。但如果要一次显示几个页面,可以通过调整这个方法的返回值来实现。

如果按照下面的代码片段修改 getPageWidth(),那么我们一次可以显示三个页面:

```
/*
*如果要在 ViewPager 内一次显示超过一页的内容,那么需要重写该方法。
*/
@Override
public float getPageWidth(int position) {
    //每个页面的宽应该是视图的 1/3
    return 0.333f;
}
```

图 2-15 展示了应用程序的修改结果。

图 2-15 每次显示三个页面的 ViewPager

1. 添加和删除页面

代码清单 2-56 演示了一个用于 ViewPager 的稍微复杂的适配器。它使用框架中的 FragmentPagerAdapter 作为父类,FragmentPagerAdapter 的每个页面条目都是 Fragment 而不是简单的视图。

代码清单 2-56 显示了一个列表的 FragmentPagerAdapter

```
public class ListPagerAdapter extends FragmentPagerAdapter {
```

```java
private static final int ITEMS_PER_PAGE = 2;

private List<String> mItems;

public ListPagerAdapter(FragmentManager manager, List<String> items) {
    super(manager);
    mItems = items;
}

/*
*这个位置首次需要一个Fragment时，该方法才会被调用
*/
@Override
public Fragment getItem(int position) {
    int start = position * ITEMS_PER_PAGE;
    return ArrayListFragment.newInstance(getPageList(position), start);
}

@Override
public int getCount() {
    //得到分页的总数
    int pages = mItems.size() / ITEMS_PER_PAGE;
    //如果列表的大小不能整除页面的大小，就多添加一个页面来显示剩余的值
    int excess = mItems.size() % ITEMS_PER_PAGE;
    if(excess > 0) {
        pages++;
    }
    
    return pages;
}

/*
*这个方法会在getItem()之后针对新的Fragment被调用，而且在超出页数限制部分的
Fragment再加回来时，也会调用该方法；我们要确保这些元素会被更新到列表中
*/
@Override
public Object instantiateItem(ViewGroup container, int position) {
    ArrayListFragment fragment =
            (ArrayListFragment) super.instantiateItem(container,
                position);
    fragment.updateListItems(getPageList(position));
    return fragment;
}

/*
*当notifyDataSetChanged()被调用时，该方法也会被框架调用。我们必须决定如何为新
的数据集更改每个Fragment。如果某个位置的Fragment不再需要，会返回POSITION_NONE，
这样适配器就可以将其删除。
*/
```

```java
    @Override
    public int getItemPosition(Object object) {
        ArrayListFragment fragment = (ArrayListFragment)object;
        int position = fragment.getBaseIndex() / ITEMS_PER_PAGE;
        if(position >= getCount()) {
            //不再需要这个页面
            return POSITION_NONE;
        } else {
            //刷新 Fragment 数据显示
            fragment.updateListItems(getPageList(position));

            return position;
        }
    }

    /*
    * 辅助方法，用于获取整个列表的某一部分，然后显示在给定的 Fragment 上。
    */
    private List<String> getPageList(int position) {
        int start = position * ITEMS_PER_PAGE;
        int end = Math.min(start + ITEMS_PER_PAGE, mItems.size());
        List<String> itemPage = mItems.subList(start, end);

        return itemPage;
    }

    /*
    *内部自定义 Fragment，它会通过 ListView 中显示列表的一个片段，并提供外部方法来更
新列表。
    */
    public static class ArrayListFragment extends Fragment {
        private ArrayList<String> mItems;
        private ArrayAdapter<String> mAdapter;
        private int mBaseIndex;

        //按照惯例使用工厂模式创建 Fragment
        static ArrayListFragment newInstance(List<String> page, int
          baseIndex) {
            ArrayListFragment fragment = new ArrayListFragment();
            fragment.updateListItems(page);
            fragment.setBaseIndex(baseIndex);
            return fragment;
        }

        public ArrayListFragment() {
            super();
            mItems = new ArrayList<String>();
        }

        @Override
```

```java
public void onCreate(Bundle savedInstanceState) {
    super.onCreate(savedInstanceState);
    //为列表条目创建一个新的适配器
    mAdapter = new ArrayAdapter<String>(getActivity(),
            android.R.layout.simple_list_item_1, mItems);
}

@Override
public View onCreateView(LayoutInflater inflater, ViewGroup container,
        Bundle savedInstanceState) {
    //构造并返回一个ListView,为它关联我们的适配器
    ListView list = new ListView(getActivity());
    list.setAdapter(mAdapter);
    return list;
}

//在全局列表中保存一个索引,记录页面开始的地方
public void setBaseIndex(int index) {
    mBaseIndex = index;
}

//在全局列表中检索索引,可以找到页面开始的地方
public int getBaseIndex() {
    return mBaseIndex;
}

public void updateListItems(List<String> items) {
    mItems.clear();
    for(String piece : items) {
        mItems.add(piece);
    }

    if(mAdapter != null) {
        mAdapter.notifyDataSetChanged();
    }
}
    }
}
```

这个示例使用一个很长的数据列表并将其分解成小段显示在每个页面上。这个适配器显示的 Fragment 是一个自定义内部实现,它会接收条目的一个列表并将这些条目显示在 ListView 中。

FragmentPagerAdapter 帮助我们实现了 PagerAdapter 底层的很多功能。不必再实现 instantiateItem()、destroyItem()和 isViewFromObject()方法,只需要重写 getItem()来为每个页面位置提供相应的 Fragment。本例为每个页面应该显示的列表条目的数量定义了一个常量。在 getItem()内创建 Fragment 时,会传入列表中的一部分数据,而这些数据是根据索引偏移和之前定义的常量来计算的。分页的数量由 getCount()方法返回,这个值是通过列表条目总量除以每页显示的条目常量计算得到的。

提示:

FragmentPagerAdapter 将所有 Fragment 实例保持为活动状态,无论它们在屏幕外页数限制内是否被激活。如果 ViewPager 需要容纳更多数量的 Fragment,或者一些 ViewPager 更加重量级,则可以改为使用 FragmentStatePagerAdapter。FragmentStatePagerAdapter 会销毁超出屏幕外页数限制的 Fragement,同时保留其已保存的状态,这一点类似于旋转操作。

这个适配器还覆写了前面简单示例中未曾见到过的另一个方法:getItemPosition()。当应用程序从外部调用 notifyDataSetChanged()时,这个方法会被调用。它主要的作用是在页面发生变化时判断页面中的条目应该被移动还是删除。如果条目的位置发生改变,该实现就应该返回新位置的值。如果条目不应该被移动,该实现就会返回一个常量值 PagerAdapter.POSITION_UNCHANGED。如果页面应该被删除,应用程序应该返回 PagerAdapter.POSITION_NONE。

这个示例会比较检查当前页面的位置(我们需要从初始索引数据开始重新创建)和当前页面数量的大小。如果当前页面位置大于当前页面数量,就需要从列表中删除足够的条目,如此一来就不再需要该页面了,然后返回 POSITION_NONE。而在其他情况下,我们会更新当前 Fragment 中显示的列表条目,并返回重新计算得到的位置值。

每个 ViewPager 当前跟踪的页面都会调用 getItemPosition(),调用的次数即为 getOffscreenPageLimit()返回的页面数量。然而,虽然 ViewPager 不会跟踪滚动出限定值之外的 Fragment,但 FragmentManager 会继续追踪。因此,当之前的 Fragment 回滚时,getItem()不会被再次调用,因为 Fragment 已经存在了。但是正因为如此,如果一个数据集在这期间发生改变,Fragment 列表数据不会跟着更新。这就是要重写 instantiateItem()的原因。虽然这个适配器不需要重写 instantiateItem(),但是当列表发生变化时,确实需要更新超出屏幕外页数限制的 Fragment。因为 Fragment 回滚到页数限制内以后,每次都会调用 instantiateItem(),所以这是重置显示列表的好时机。

让我们看一个使用该适配器的示例应用程序。参见代码清单 2-57 和 2-58。

代码清单 2-57　res/layout/main.xml

```xml
<?xml version="1.0" encoding="utf-8"?>
<LinearLayout xmlns:android="http://schemas.android.com/apk/res/android"
    android:layout_width="match_parent"
    android:layout_height="match_parent"
    android:orientation="vertical" >
    <Button
        android:layout_width="match_parent"
        android:layout_height="wrap_content"
        android:text="Add Item"
        android:onClick="onAddClick" />
    <Button
        android:layout_width="match_parent"
        android:layout_height="wrap_content"
        android:text="Remove Item"
        android:onClick="onRemoveClick" />
```

```xml
<!--ViewPager 是支持库中的小部件,所以需要完整的包名    -->
<android.support.v4.view.ViewPager
    android:id="@+id/view_pager"
    android:layout_width="match_parent"
    android:layout_height="match_parent" />
</LinearLayout>
```

代码清单 2-58　使用了 ListPagerAdapter 的 Activity

```java
public class FragmentPagerActivity extends FragmentActivity {

    private ArrayList<String> mListItems;
    private ListPagerAdapter mAdapter;

    @Override
    protected void onCreate(Bundle savedInstanceState) {
        super.onCreate(savedInstanceState);
        setContentView(R.layout.main);
        //创建初始数据集
        mListItems = new ArrayList<String>();
        mListItems.add("Mom");
        mListItems.add("Dad");
        mListItems.add("Sister");
        mListItems.add("Brother");
        mListItems.add("Cousin");
        mListItems.add("Niece");
        mListItems.add("Nephew");
        //把数据关联到 ViewPager 上
        ViewPager pager = (ViewPager) findViewById(R.id.view_pager);
        mAdapter = new ListPagerAdapter(getSupportFragmentManager(),
            mListItems);

        pager.setAdapter(mAdapter);
    }

    public void onAddClick(View v) {
        //在列表的末尾添加新的唯一条目
        mListItems.add("Crazy Uncle " + System.currentTimeMillis());
        mAdapter.notifyDataSetChanged();
    }

    public void onRemoveClick(View v) {
        //从列表顶部删除一个条目
        if(!mListItems.isEmpty()) {
            mListItems.remove(0);
        }
        mAdapter.notifyDataSetChanged();
    }
}
```

就像 ViewPager 的效果一样,这个示例中有两个按钮用来添加和删除数据集中的条目。

注意 ViewPager 必须在 XML 文件中定义并使用完全限定的包名,因为它仅是支持库中的类,在 android.widget 或 android.view 包中并没有这个类。该 Activity 构建了一个默认的条目列表并把它传入我们自定义的适配器中,然后再把该适配器关联到 ViewPager 上。

每次单击 Add 按钮都在列表末尾添加一个新的条目并通过调用 notifyDataSetChanged() 来触发 ListPagerAdapter 进行更新。每次单击 Remove 按钮都会在列表顶部删除一个条目,然后再次通知适配器。每次变化期间,适配器都会调整当前可用页数并更新 ViewPager。如果当前可见页的所有条目都被删除,那么该页也会被删除并显示上一页。

2. 其他有用的方法

ViewPager 中有几个其他的方法,它们会对你的应用程序很有帮助:
- setPageMargin()和 setPageMarginDrawable()允许在页面之间设置一些额外的间隔,并且使用一个 Drawable(可选)来填充间隔的内容。
- setCurrentItem()允许你以编程的方式设置要显示的页面,并提供一个选项来禁用页面切换时的滚动动画。
- OnPageChangeListener 用于将滚动和变更动作通知给应用程序。
  - onPageSelected()会在显示一个新页面时被调用。
  - 当发生滚动操作时会连续调用 onPageScrolled()。
  - onPageScrollStateChanged()在 ViewPager 处于以下状态时会被调用:闲置时、用户主动滚动 ViewPager 时、自动滚动对齐到最近的页面时。

2.19 使用选项卡导航

2.19.1 问题

需要在应用程序中为侧向屏幕导航提供可选的选项卡,但 Google 没有在框架或支持库中提供选项卡小部件。

2.19.2 解决方案

(API Level 7)

我们可以通过 Google 提供为 SDK 样本的 SlidingTabLayout 进行构建来实现选项卡导航。Google 已完全淘汰以前的各种 Android 选项卡,例如 TabWidget 和 ActionBar.Tab,但 SlidingTabLayout 符合当前的选项卡设计模式。

SlidingTabLayout 旨在与 ViewPager 密切协作,因为选项卡设计模式的一部分是允许在每个视图之间轻扫。因此,没有用于手动添加选项卡条目的 API。相反,从附加的 ViewPager 返回的页面标题中派生选项卡。如果选项卡内容延伸超出屏幕宽度,用户可以向左或向右滚动选项卡。轻扫 ViewPager 时,当前选项卡自动滚动以同时显示出来。

在编写本书时,SlidingTabLayout 的设计采用了以前的 Holo 设计语言。在接下来的示例中,我们将对布局做一些调整,从而更好地支持 Material 设计。

注意：
这些小部件未来可能移入支持库中，但目前它们仅示例代码提供。

2.19.3 实现机制

在开始编写自己的代码之前，我们需要将滑动选项卡示例代码引入项目中。我们需要从 SlidingTabsBasic SDK 示例项目复制两个类，将其放入项目中 src/main/java 下的包目录：SlidingTabLayout 和 SlidingTabStrip。

注意：
可以在如下网址找到 SlidingTabsBasic SDK 示例：<SDK-Directory>/samples/android-xx/ui/SlidingTabsBasic/。

Google 提供了如何构建该例的说明，因此我们在这儿不会深入讨论此选项卡小部件的细节。我们将其视为来自任何其他库的小部件。完成此例时，我们将获得如图 2-16 所示的效果。

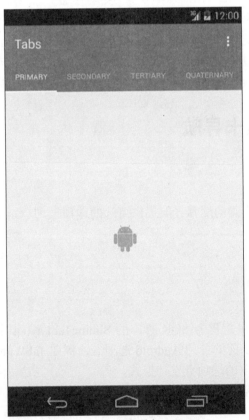

图 2-16　滑动选项卡 Activity

请注意，选项卡位于 Action Bar 的下方，并且匹配 Action Bar 的背景以提供它们是单个元素的外观。代码清单 2-59 和 2-60 定义了构造选项卡的 Activity 和布局。

代码清单 2-59　res/layout/activity_tabs.xml

```xml
<?xml version="1.0" encoding="utf-8"?>
<LinearLayout xmlns:android="http://schemas.android.com/apk/res/android"
    android:orientation="vertical"
    android:layout_width="match_parent"
    android:layout_height="match_parent">

    <com.example.android.common.view.SlidingTabLayout
        android:id="@+id/tabs"
        android:layout_width="match_parent"
        android:layout_height="wrap_content"
        android:background="@color/primaryGreen"/>
    <android.support.v4.view.ViewPager
        android:id="@+id/pager"
        android:layout_width="match_parent"
        android:layout_height="match_parent"/>
</LinearLayout>
```

代码清单 2-60　滑动选项卡 Activity

```java
public class ActionTabsActivity extends ActionBarActivity {

    @Override
    protected void onCreate(Bundle savedInstanceState) {
        super.onCreate(savedInstanceState);
        setContentView(R.layout.activity_tabs);

        ViewPager viewPager = (ViewPager) findViewById(R.id.pager);
        SlidingTabLayout tabLayout = (SlidingTabLayout) findViewById(R.
            id.tabs);

        viewPager.setAdapter(new TabsPagerAdapter(this));

        /*
         * SlidingTabLayout 与 ViewPager 关联，继承选项卡标题和滚动跟踪行为
         */
        tabLayout.setViewPager(viewPager);
        tabLayout.setCustomTabColorizer(new SlidingTabLayout.TabColorizer() {
            @Override
            public int getIndicatorColor(int position) {
                //显示在每个选项卡位置下方的颜色
                return Color.WHITE;
            }

            @Override
            public int getDividerColor(int position) {
                // 透明以隐藏分隔线;
                return 0;
            }
        });
```

```java
    }

    @Override
    public boolean onCreateOptionsMenu(Menu menu) {
        getMenuInflater().inflate(R.menu.tabs, menu);
        return true;
    }

    /*
     * 简单的 PagerAdapter，用于显示带有静态图片的页面视图
     */
    private static class TabsPagerAdapter extends PagerAdapter {
        private Context mContext;
        public TabsPagerAdapter(Context context) {
            mContext = context;
        }

        /*
         * SlidingTabLayout 要求此方法定义每个选项卡将显示的文本
         */
        @Override
        public CharSequence getPageTitle(int position) {
            switch (position) {
                case 0:
                    return "Primary";
                case 1:
                    return "Secondary";
                case 2:
                    return "Tertiary";
                case 3:
                    return "Quaternary";
                case 4:
                    return "Quinary";
                default:
                    return "";
            }
        }

        @Override
        public int getCount() {
            return 5;
        }

        @Override
        public Object instantiateItem(ViewGroup container, int position) {
            ImageView pageView = new ImageView(mContext);
            pageView.setScaleType(ImageView.ScaleType.CENTER);
            pageView.setImageResource(R.drawable.ic_launcher);

            container.addView(pageView);
```

```
            return pageView;
        }

        @Override
        public void destroyItem(ViewGroup container, int position, Object object)
        {
            container.removeView((View) object);
        }

        @Override
        public boolean isViewFromObject(View view, Object object) {
            return (view == object);
        }
    }
}
```

在 onCreate() 内部，我们必须通过向 setViewPager() 传递一个引用来将 ViewPager 附加到 SlidingTabLayout。在内部，布局将跟踪 ViewPager 中的滚动事件，并且通过滚动对应选项卡下方的选择器栏来反映当前所选页面中的改动。

从主题中提取选项卡的默认颜色，该颜色通常是不正确的。可以使用 TabColorizer 实例为选项卡选择器和选项卡之间的分隔线提供颜色。我们将前者设置为纯白色，并且(如图 2-16 中所示)隐藏了分隔线以符合 Material 设计外观。

SlidingTabLayout 从附加的 PagerAdapter 派生其内容。适配器实现必须重写 getPageTitle()以提供将显示在每个页面选项卡上的名称。

样式调整

如果运行目前为止的代码，选项卡仍然看起来与图 2-16 稍有不同。我们需要对 SlidingTabLayout 作如下调整，使其符合 Material 设计：
- 将选择器高度降低为 2dp。
- 移除底部的阴影边框。
- 移除默认的加粗文本。
- 添加对已选择和未选择选项卡上不同文本颜色的支持。

注意：
还可以使用在本书示例代码中提供的不同版本的 SlidingTabLayout 和 SlidingTabStrip 查看已应用的这些调整。

在 SlidingTabStrip.java 内部，更新如下常量值：

```
private static final int DEFAULT_BOTTOM_BORDER_THICKNESS_DIPS = 0;
...
private static final int SELECTED_INDICATOR_THICKNESS_DIPS = 2;
```

这会降低选择器高度并移除阴影。在 SlidingTabStrip.java 内部，我们需要使用代码清

单 2-61 中的版本替换 createDefaultTabView():

代码清单 2-61　SlidingTabLayout 的文本修正

```
protected TextView createDefaultTabView(Context context) {
    TextView textView = new TextView(context);
    textView.setGravity(Gravity.CENTER);
    textView.setTextSize(TypedValue.COMPLEX_UNIT_SP, TAB_VIEW_TEXT_SIZE_SP);
    textView.setTypeface(Typeface.DEFAULT);

    if (Build.VERSION.SDK_INT >= Build.VERSION_CODES.ICE_CREAM_SANDWICH) {
        // 如果正在 ICS 或更新的版本上运行，则启用全部大写以符合 Action Bar 的选项卡
        // 样式
        textView.setAllCaps(true);
    }

    int padding = (int) (TAB_VIEW_PADDING_DIPS * getResources().
      getDisplayMetrics().density);
    textView.setPadding(padding, padding, padding, padding);

    return textView;
}
```

这会移除 Holo 的默认背景强调以及选项卡的粗体文本。最后，我们需要添加对两种选项卡颜色的支持。代码清单 2-62 中指出了添加的相关代码。

代码清单 2-62　SlidingTabLayout 的文本颜色

```
private int mDefaultTextColor;
private int mSelectedTextColor;

public SlidingTabLayout(Context context, AttributeSet attrs, int defStyle) {
    super(context, attrs, defStyle);

    //获得选项卡颜色，如果未定义的话，则使用主题默认颜色
    TypedArray a = context.obtainStyledAttributes(attrs,
        R.styleable.SlidingTabLayout);
    int defaultTextColor = a.getColor(R.styleable.
        SlidingTabLayout_android_textColorPrimary, 0);
    mDefaultTextColor = a.getColor(R.styleable.
            SlidingTabLayout_textColorTabDefault, defaultTextColor);
        mSelectedTextColor = a.getColor(R.styleable.
            SlidingTabLayout_textColorTabSelected, defaultTextColor);
        a.recycle();

    //禁用滚动栏
    setHorizontalScrollBarEnabled(false);
    //确保选项卡横条填满此视图
    setFillViewport(true);
```

```
    mTitleOffset = (int) (TITLE_OFFSET_DIPS * getResources().
        getDisplayMetrics().density);

    mTabStrip = new SlidingTabStrip(context);
    addView(mTabStrip, LayoutParams.MATCH_PARENT, LayoutParams.WRAP_CONTENT);
}

...

//在每次选项改变时更新选项卡文本颜色的新方法
private void updateSelectedTitle(int position) {
    final PagerAdapter adapter = mViewPager.getAdapter();
    for (int i = 0; i < adapter.getCount(); i++) {
        final View tabView = mTabStrip.getChildAt(i);
        if (TextView.class.isInstance(tabView)) {
            TextView titleView = (TextView) tabView;
            boolean isSelected = i == position;
            titleView.setTextColor(isSelected ? mSelectedTextColor
                    : mDefaultTextColor);
        }
    }
}

...

private void scrollToTab(int tabIndex, int positionOffset) {
    final int tabStripChildCount = mTabStrip.getChildCount();
    if (tabStripChildCount == 0 || tabIndex < 0 || tabIndex >=
tabStripChildCount) {
        return;
    }

    View selectedChild = mTabStrip.getChildAt(tabIndex);
    if (selectedChild != null) {
        //调用在每次选项改变时更新文本颜色的新方法
        updateSelectedTitle(tabIndex);
        int targetScrollX = selectedChild.getLeft() + positionOffset;

        if (tabIndex > 0 || positionOffset > 0) {
            //如果不在第一个子选项卡位置并在滚动途中，则确保遵从偏移值
            targetScrollX -= mTitleOffset;
        }

        scrollTo(targetScrollX, 0);
    }
}
```

添加的这些代码用于在应用程序的主题中定义两种文本颜色属性,该主题将应用于选择的选项卡和其他未选择的选项卡。在每个选项卡选择事件中,新的 updateSelectedTitle() 方法基于新的选项设置所有选项卡的文本颜色。目前这些属性不存在。代码清单 2-63 在应用程序资源中定义这些属性,代码清单 2-64 对当前主题中的每个样式应用颜色,而代码清单 2-65 定义使用的颜色值。

代码清单 2-63　res/values/attrs.xml

```xml
<?xml version="1.0" encoding="utf-8"?>
<resources>
    <!-- 为选项卡颜色自定义属性 -->
    <declare-styleable name="SlidingTabLayout">
        <attr name="android:textColorPrimary" />
        <attr name="textColorTabDefault" format="color"/>
        <attr name="textColorTabSelected" format="color"/>
    </declare-styleable>
</resources>
```

代码清单 2-64　res/values/styles.xml

```xml
<resources>
    <!-- 基础应用程序主题 -->
    <style name="AppTheme" parent="Theme.AppCompat.Light.DarkActionBar">
        <!-- 提供装饰主题颜色 -->
        <item name="colorPrimary">@color/primaryGreen</item>
        <item name="colorPrimaryDark">@color/darkGreen</item>
        <item name="colorAccent">@color/accentGreen</item>

        <!-- 移除 Action Bar 的阴影 -->
        <item name="android:windowContentOverlay">@null</item>

        <!-- 选项卡的颜色属性 -->
        <item name="textColorTabDefault">@color/tabTextDefault</item>
        <item name="textColorTabSelected">@color/tabTextSelected</item>
    </style>

</resources>
```

代码清单 2-65　res/values/colors.xml

```xml
<?xml version="1.0" encoding="utf-8"?>
<resources>
    <color name="primaryGreen">#259b24</color>
    <color name="darkGreen">#0a7e07</color>
    <color name="accentGreen">#d0f8ce</color>

    <color name="tabTextDefault">#99ffffff</color>
    <color name="tabTextSelected">#ffffff</color>
</resources>
```

目前在主题中可以看到对已选择的选项卡应用了纯白色，而对其他的选项卡应用了 60%的白色。

2.20 小结

本章探讨了大量可以用来构建引人注目的用户界面的技巧，这些用户界面遵从 Google 为 Android 平台提出的设计指南。我们首先查看了如何在应用程序中有效地使用 Action Bar 界面元素，然后研究了如何以创造性的方式管理配置改动，如设备方向。此外，我们介绍了通过文本和触摸处理管理用户输入的技术。最后，我们了解了如何实现常见的导航模式，如 Drawer 布局、轻扫分页视图和选项卡。

在第 3 章，我们将介绍使用 SDK 与外部世界通信，具体方法是访问网络资源并使用 USB 和蓝牙等技术与其他设备进行交流。

第 3 章

通信和联网

很多移动应用程序成功的关键是它们拥有与远程数据源进行连接和交互的能力。当今世界中，Web 服务和 API 已经非常丰富，从天气预报到个人财务信息，一个应用程序可以和任何其他服务进行交互。移动平台最大的优势就是可以将这些数据发送到用户的手中并且可在任何地方访问。Android 是在 Google 的 Web 基础上发展起来的，而 Google 则为与外部世界进行通信提供了丰富的工具集。

3.1 显示 Web 信息

3.1.1 问题

在应用程序中，需要将从 Web 上获取的 HTML 或图像数据不加任何修改和处理地显示出来。

3.1.2 解决方案

(API Level 3)

在 WebView 中显示信息。WebView 是一个视图小部件，在应用程序中，它可以嵌入到任何布局中来显示本地或远程的网页内容。WebView 基于开源的 WebKit 引擎，而 Android Browser 应用程序也是基于此引擎，所以两者赋予 Web 应用程序相同的性能和功能。

3.1.3 实现机制

除了最重要的二维滚动(横向和纵向同时滚动)和变焦控制，WebView 对于显示从网上下载的资源还有很多值得称道的地方。WebView 非常适合处理大图片，如体育场的地图，用户在浏览此类图片时可能需要进行左右平移和缩放。在这里，我们将讨论如何实现本地和远程资源的显示。

1. 显示一个 URL

最简单的情况就是提供资源的 URL，然后在 WebView 中将与该 URL 对应的 HTML 页面或图像显示出来。以下是这项技术在应用程序中一些小的实际应用：

- 在应用程序中访问企业网站。
- 通过一台 Web 服务器显示实时更新内容的页面，如 FAQ 部分，这个页面的内容不必升级应用程序就可以动态更新。
- 显示一个很大的图像资源，用户可能需要通过平移/缩放来与它交互。

让我们来看一个加载常见网页的简单示例，不过不是用浏览器，而是在 Activity 内部加载(参见代码清单 3-1 和 3-2)。

代码清单 3-1　包含一个 WebView 的 Activity

```java
public class MyActivity extends Activity {
    @Override
    public void onCreate(Bundle savedInstanceState) {
        super.onCreate(savedInstanceState);

        WebView webview = new WebView(this);
        //启用 JavaScript 支持
        webview.getSettings().setJavaScriptEnabled(true);
        webview.loadUrl("http://www.google.com/");

        setContentView(webview);
    }
}
```

注意：

默认情况下，WebView 是禁用 JavaScript 支持的。如果显示的内容需要它的话，可以使用 WebView.WebSettings 对象来启用它。

代码清单 3-2　在 AndroidManifest.xml 中设置需要的权限

```xml
<?xml version="1.0" encoding="utf-8"?>
<manifest
    xmlns:android="http://schemas.android.com/apk/res/android"
    package="com.examples.webview">

    <uses-permission android:name="android.permission.INTERNET" />

    <application android:icon="@drawable/icon"
        android:label="@string/app_name">
        <activity android:name=".MyActivity">
            <intent-filter>
                <action
                    android:name="android.intent.action.MAIN" />
```

```
            <category
                android:name="android.intent.category.LAUNCHER" />
        </intent-filter>
    </activity>
</application>
</manifest>
```

要点:

如果 WebView 加载的是远程内容,AndroidManifest.xml 必须声明使用 android.permission.INTERNET 权限。

加载结果是在 Activity 中显示一个 HTML 页面(如图 3-1 所示)。

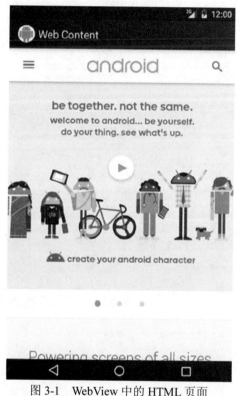

图 3-1 WebView 中的 HTML 页面

注意,如果点击此视图中的任意链接,设备的浏览器应用程序就会启动。这是因为加载的所有网页 URL 都默认被系统处理为 Intent。如果要在内部处理链接,就必须截断这些事件。我们将在本章后面讨论如何执行此操作。

2. 显示本地资源

WebView 在显示本地内容时也非常有用,它可以利用 HTML/CSS 来格式化内容或者为它的内容提供平移/缩放功能。你也许会使用 Android 项目的 assets 目录来存储你想在 WebView 中显示的资源,如大型图片或 HTML 文件。为了更好地组织资源,也可以在 assets

目录下创建子目录来存储文件。

通过 file:///android_asset/<resource path>这样的 URL 格式,WebView.loadUrl()可以显示存储在 assets 目录中的文件。例如,如果在 assets 目录中存放了文件 android.jpg,使用 file:///android_asset/android.jpg 这样的 URL 就可以将它加载到 WebView 中。

如果在 assets 目录下的 images 目录中存放了一个同样的文件,WebView 可以使用 file:///android_asset/images/android.jpg 这样的 URL 来加载它。

另外,WebView.loadData()可以将存储在字符串资源或变量中的原始 HTML 代码加载到视图中。通过这项技术,预先格式化的 HTML 文本可以保存在 res/values/strings.xml 中或使用远程 API 下载下来,并显示在应用程序中。

代码清单 3-3 和 3-4 显示了一个示例 Activity,它有两个 WebView 小部件,其中一个垂直堆叠在另一个之上。上面的视图显示了一张存储在 assets 目录中的大型图片,下面的视图则显示了存储在应用程序的字符串资源中的一个 HTML 字符串。

代码清单 3-3　res/layout/main.xml

```xml
<LinearLayout
  xmlns:android="http://schemas.android.com/apk/res/android"
  android:layout_width="match_parent"
  android:layout_height="match_parent"
  android:orientation="vertical">
  <WebView
    android:id="@+id/upperview"
    android:layout_width="match_parent"
    android:layout_height="0dp"
    android:layout_weight="1"
  />
  <WebView
    android:id="@+id/lowerview"
    android:layout_width="match_parent"
    android:layout_height="0dp"
    android:layout_weight="1"
  />
</LinearLayout>
```

代码清单 3-4　显示本地 Web 内容的 Activity

```java
public class MyActivity extends Activity {
    @Override
    public void onCreate(Bundle savedInstanceState) {
        super.onCreate(savedInstanceState);
        setContentView(R.layout.main);

        WebView upperView = (WebView)findViewById(R.id.upperview);
        //必须启用缩放功能
```

```
        upperView.getSettings().setBuiltInZoomControls(true);
        if (Build.VERSION.SDK_INT >= Build.VERSION_CODES.HONEYCOMB) {
        //Android 3.0以上版本具有捏动缩放功能，无需此按钮
        upperView.getSettings().setDisplayZoomControls(false);

        upperView.loadUrl("file:///android_asset/android.jpg");

        WebView lowerView = (WebView)findViewById(R.id.lowerview);
        String htmlString = "<h1>Header</h1><p>This is HTML text<br />"
                + "<i>Formatted in italics</i></p>";
        lowerView.loadData(htmlString, "text/html", "utf-8");
        }
    }
```

当Activity显示时，两个WebView将屏幕分为上下两个部分。HTML字符串按预期的格式显示，而大型图片则可以水平和垂直滚动，用户甚至可以进行放大和缩小，如图3-2所示。

图3-2　显示本地资源的两个WebView

我们通过setBuiltInZoomControls(true)允许用户缩小和放大内容。默认情况下，这还会显示与可点击缩放控件重叠的按钮。在Android 3.0和以后的版本中，你可能额外考虑包括WebView.getSettings().setDisplayZoomControls(false)。这些版本的平台通过收缩手势原生支持缩放，因此没有必要显示重叠按钮。这种方式不会取代 setBuiltInZoomControls()，必须同时支持该方法才可以使收缩手势生效。

3.2 拦截 WebView 事件

3.2.1 问题

应用程序使用 WebView 显示内容，但在用户点击页面中的链接时还需要监听和响应。

3.2.2 解决方案

(API Level 1)

实现一个 WebViewClient 并把它关联到 WebView 上。WebViewClient 和 WebChromeClient 是两个 WebKit 类，它们可以让应用程序获得 WebView 的事件回调并且可以自定义 WebView 的行为。默认情况下，在没有指定 WebViewClient 时，WebView 会将一个 URL 传递给 ActivityManager 处理。而 ActivityManager 通常会在浏览器应用程序中打开用户点击的链接，而不是在当前的 WebView 中。

3.2.3 实现机制

在代码清单 3-5 中，我们创建了一个含有 WebView 的 Activity，该 WebView 将处理它自己的 URL 加载。

代码清单 3-5　带有一个 WebView 的 Activity，该 WebView 会处理 URL

```
public class MyActivity extends Activity {
    @Override
    public void onCreate(Bundle savedInstanceState) {
        super.onCreate(savedInstanceState);

        WebView webview = new WebView(this);
        webview.getSettings().setJavaScriptEnabled(true);
        //添加一个客户端到视图上
        webview.setWebViewClient(new WebViewClient());
        webview.loadUrl("http://www.google.com");
        setContentView(webview);
    }
}
```

在本例中，只是简单地为 WebView 提供了一个单纯功能的 WebViewClient，它可以让 WebView 自己处理所有的 URL 请求，而不是将这些请求传递给 ActivityManager，因此点击一个链接会在原来的视图中加载请求的页面，这是因为 shouldOverrideUrlLoading() 的默认实现会简单地返回 false，告诉客户端将 URL 传递给 WebView 而不是应用程序。

在下一个示例中，我们将利用 WebViewClient.shouldOverrideUrlLoading() 回调来拦截和监控用户的 Activity(见代码清单 3-6)。

代码清单 3-6　拦截 WebView URL 的 Activity

```java
public class MyActivity extends Activity {
    @Override
    public void onCreate(Bundle savedInstanceState) {
        super.onCreate(savedInstanceState);

        WebView webview = new WebView(this);
        webview.getSettings().setJavaScriptEnabled(true);
        //添加一个客户端到视图上
        webview.setWebViewClient(mClient);
        webview.loadUrl("http://www.google.com");
        setContentView(webview);
    }

    private WebViewClient mClient = new WebViewClient() {
        @Override
        public boolean shouldOverrideUrlLoading(WebView view, String url) {
            Uri request = Uri.parse(url);

            if(TextUtils.equals(request.getAuthority(), "www.google.com")) {
                //允许加载
                return false;
            }

            Toast.makeText(MyActivity.this, "Sorry, buddy", Toast.LENGTH_SHORT)
                .show();
            return true;
        }
    };
}
```

在本例中，shouldOverrideUrlLoading()会根据传入的 URL 决定是否要在 WebView 中加载内容，防止用户离开 Google 的网站。Uri.getAuthority()会返回一个 URL 的主机域名部分，我们会使用它检测用户点击的链接是否是 Google 的域名(www.google.com)。如果能够确认链接指向的是 Google 的其他页面，会返回 false，从而允许 WebView 加载内容。如果不是，会通知用户并返回 true，然后通知 WebViewClient 应用程序已经处理了这个 URL，不允许 WebView 加载它。

这项技术还能变得更加复杂，应用程序可以对 URL 做各种实际处理。通过自定义的处理方式，还可以在应用程序和 WebView 的内容之间打造一个完整的交互接口。

3.3　访问带 JavaScript 的 WebView

3.3.1　问题

应用程序需要访问 WebView 中当前显示内容的 HTML 源代码，读取或修改其中的某个值。

3.3.2 解决方案

(API Level 1)

创建一个 JavaScript 接口,作为 WebView 和应用程序代码间的桥梁。

3.3.3 实现机制

WebView.addJavascriptInterface()会为 JavaScript 绑定一个 Java 对象,这样就可以在 WebView 中调用此 Java 对象的方法。借助这个接口,就可以使用 JavaScript 在应用程序代码和 WebView 的 HTML 间编组数据了。

警告:

允许 JavaScript 控制应用程序会存在安全威胁——允许远程执行应用程序的代码。请在确实需要使用时再使用本接口。

让我们来实际看一个示例。代码清单 3-7 展示了 WebView 从本地 assets 目录中加载一个简单的 HTML 表单。代码清单 3-8 则是一个使用了两个 JavaScript 函数的 Activity,这两个函数用来在 Activity 首选项与 WebView 的内容间交换数据。

代码清单 3-7 assets/form.html

```html
<!DOCTYPE HTML PUBLIC "-//W3C//DTD HTML 3.01//EN"
    "http://www.w3.org/TR/html3/strict.dtd">
<html>

<form name="input" action="form.html" method="get">
Enter Email: <input type="text" id="emailAddress" />
<input type="submit" value="Submit" />
</form>

</html>
```

代码清单 3-8 含有 JavaScript 桥梁接口的 Activity

```java
public class MyActivity extends Activity {
    @Override
    public void onCreate(Bundle savedInstanceState) {
        super.onCreate(savedInstanceState);

        WebView webview = new WebView(this);
        //默认未启用 JavaScript
        webview.getSettings().setJavaScriptEnabled(true);
        webview.setWebViewClient(mClient);
        //将自定义接口关联到 View 上
        webview.addJavascriptInterface(new MyJavaScriptInterface(), "BRIDGE");

        setContentView(webview);
```

```java
        webview.loadUrl("file:///android_asset/form.html");
    }

    private static final String JS_SETELEMENT = "javascript:document.
        getElementById('%s').value='%s'";
    private static final String JS_GETELEMENT = "javascript:window.BRIDGE.
        storeElement('%s',document.getElementById('%s').value)";
    private static final String ELEMENTID = "emailAddress";

    private WebViewClient mClient = new WebViewClient() {
        @Override
        public boolean shouldOverrideUrlLoading(WebView view, String url) {
            //离开页面前，尝试通过 JavaScript 获得电子邮件
            view.loadUrl(String.format(JS_GETELEMENT, ELEMENTID, ELEMENTID));
            return false;
        }

        @Override
        public void onPageFinished(WebView view, String url) {
            //页面加载完成时，使用 JavaScript 将地址注入页面中
            SharedPreferences prefs = getPreferences(Activity.MODE_PRIVATE);
            view.loadUrl(String.format(JS_SETELEMENT, ELEMENTID,
                prefs.getString(ELEMENTID, "")));
        }
    };

    private void executeJavascript(WebView view, String script) {
        if (Build.VERSION.SDK_INT >= Build.VERSION_CODES.KITKAT) {
            view.evaluateJavascript(script, null);
        } else {
            view.loadUrl(script);
        }
    }

    private class MyJavaScriptInterface {
        //将一个元素存储到首选项中
        @SuppressWarnings("unused")
        public void storeElement(String id, String element) {
            SharedPreferences.Editor edit =
                getPreferences(Activity.MODE_PRIVATE).edit();
            edit.putString(id, element);
            edit.commit();
            //如果元素是有效的，显示一个 Toast
            if (!TextUtils.isEmpty(element)) {
                Toast.makeText(MyActivity.this, element, Toast.LENGTH_SHORT)
                    .show();
            }
        }
    }
}
```

在这个稍微有点儿人为性质的示例中，在 HTML 中创建了一个包含单个元素的表单并显示在 WebView 中。在 Activity 的代码中，我们使用 "emailAddress" 这个 ID 在 WebView 中查找一个表单值，并在每次点击页面中的链接时(在本例中，就是单击表单的 Submit 按钮)，通过 shouldOverrideUrlLoading()回调把该值存储到 SharedPreference 中。每次页面加载结束后(即调用 onPageFinished()时)，我们会试图将当前的值从 SharedPreference 重新注入 Web 表单中。

注意：
WebView 中默认不启用 JavaScript。为了注入甚至渲染 JavaScript，我们必须在初始化视图时调用 WebSettings.setJavaScriptEnabled(true)。

我们创建了一个称作 MyJavaScriptInterface 的 Java 类，它定义了 storeElement()方法。在创建了视图后，我们调用 WebView.addJavascriptInterface()方法来把这个类的对象关联到视图上，并且给它起名 BRIDGE。当调用此方法时，字符串参数就是用于在 JavaScript 代码中引用此接口的名称。

在这里，我们用字符串常量定义了两个 JavaScript 方法：JS_GETELEMENT 和 JS_SETELEMENT。在 Android 4.4 之前，我们通过调用之前已经看到过的 loadUrl()方法在 WebView 上执行这些方法。然而，在 API Level 19 和以后的版本中，我们在 WebView 中用新的方法 evaluateJavascript()来达到此目的。示例代码验证当前使用的 API Level，并且调用适当的方法。

注意，JS_GETELEMENT 是一个引用，用来调用我们自定义的接口函数(如 BRIDGE.storeElement)，它将调用 MyJavaScriptInterface 的方法并把表单元素的值存储到首选项中。如果检查到表单中的值不为空，则还会显示一个 Toast。

使用这种方式，可以在 WebView 中执行任意的 JavaScript 脚本，在自定义的接口中可以不需要包含任何方法。例如，JS_SETELEMENT 使用纯 JavaScript 脚本来设置页面中表单元素的值。

经常使用这项技术的应用程序需要记住用户在应用程序中输入的表单数据，但表单必须是基于 Web 的，例如 Web 应用程序的预约表单或支付表单，而 Web 应用程序需要较高 Level 的 API 才可以访问。

3.4 下载图片文件

3.4.1 问题

应用程序需要从 Web 或其他远程服务器下载一张图片并显示。

3.4.2 解决方案

(API Level 4)

使用 AsyncTask 在后台线程中下载数据。AsyncTask 是封装类,它可以很方便地让需要长时间运行操作的线程在后台运行;同样,它通过一个内部线程池管理线程的并发。除了管理后台线程外,在操作执行前、中、后都会提供回调方法,让你可以做任何需要在主 UI 线程中进行的更新。

3.4.3 实现机制

在下载图片的环境中,我们会创建 ImageView 的一个子类,叫做 WebImageView,它会从远程来源中延迟加载一张图片并且在该图片可用时就显示它。下载过程会在一个 AsyncTask 操作中执行(见代码清单 3-9)。

代码清单 3-9 WebImageView

```java
public class WebImageView extends ImageView {

    private Drawable mPlaceholder, mImage;

    public WebImageView(Context context) {
        this(context, null);
    }

    public WebImageView(Context context, AttributeSet attrs) {
        this(context, attrs, 0);
    }

    public WebImageView(Context context, AttributeSet attrs,
            int defStyle) {
        super(context, attrs, defaultStyle);
    }

    public void setPlaceholderImage(Drawable drawable) {
        mPlaceholder = drawable;
        if(mImage == null) {
            setImageDrawable(mPlaceholder);
        }
    }

    public void setPlaceholderImage(int resid) {
        mPlaceholder = getResources().getDrawable(resid);
        if(mImage == null) {
            setImageDrawable(mPlaceholder);
        }
    }
```

```java
public void setImageUrl(String url) {
    DownloadTask task = new DownloadTask();
    task.execute(url);
}

private class DownloadTask extends
        AsyncTask<String, Void, Bitmap> {
    @Override
    protected Bitmap doInBackground(String... params) {
        String url = params[0];
        try {
            URLConnection connection =
                    (new URL(url)).openConnection();
            InputStream is = connection.getInputStream();
            BufferedInputStream bis =
                    new BufferedInputStream(is);

            ByteArrayBuffer baf = new ByteArrayBuffer(50);
            int current = 0;
            while((current = bis.read()) != -1) {
                baf.append((byte)current);
            }
            byte[] imageData = baf.toByteArray();
            return BitmapFactory.decodeByteArray(imageData, 0,
                    imageData.length);
        } catch(Exception exc) {
            return null;
        }
    }

    @Override
    protected void onPostExecute(Bitmap result) {
        mImage = new BitmapDrawable(getContext().getResources(),
result);
        if(mImage != null) {
            setImageDrawable(mImage);
        }
    }
};
}
```

正如你所看到的那样，WebImageView 是 Android 的 ImageView 小部件的一个简单扩展。在远程内容下载完成之前，setPlaceholderImage()会为要显示的图片设置一个本地 Drawable。大多数有趣的工作都是从使用 setImageUrl()向视图传入一个给定远程 URL 开始的，该方法表示自定义的 AsyncTask 开始工作了。

请注意，AsyncTask 是强类型化的，它需要三个值：输入参数、进度值、结果值。在这种情况下，会传递给任务的 execute()方法一个字符串，而后台操作应该返回一个 Bitmap。对于中间的进度值，我们在本例中并不会使用，因此它被设置为 Void。继承 AsyncTask 后，唯一需要实现的方法就是 doInBackground()，它定义了后台线程中需要大量运行的操作。在前面的示例中，这里是与提供的远程 URL 进行连接以及下载图片的地方。完成后，我们会试图用下载的数据创建一个 Bitmap。发生任何错误时，操作将中止并返回 null。

其他在 AsyncTask 中定义的回调方法，如 onPreExecute()、onPostExecute()和 onProgressUpdate()，会在主线程中进行调用，用来更新 UI。在之前的示例中，onPostExecute()用来使用结果数据更新视图中的图片。

要点：

Andorid UI 类是线程不安全的。确保在更新 UI 时使用运行在主线程上的回调方法。不要在 doInBackground()中更新视图。

代码清单 3-10 和 3-11 展示了在一个 Activity 中使用这个类的示例。因为这个类不是 android.widget 或 android.view 包的一部分，所以在 XML 中使用它时必须先指定它的完全限定包名。

代码清单 3-10 res/layout/main.xml

```xml
<?xml version="1.0" encoding="utf-8"?>
<LinearLayout
    xmlns:android="http://schemas.android.com/apk/res/android"
    android:layout_width="match_parent"
    android:layout_height="match_parent"
    android:orientation="vertical">
    <com.examples.WebImageView
        android:id="@+id/webImage"
        android:layout_width="wrap_content"
        android:layout_height="wrap_content" />
</LinearLayout>
```

代码清单 3-11 示例 Activity

```java
public class WebImageActivity extends Activity {
    @Override
    public void onCreate(Bundle savedInstanceState) {
        super.onCreate(savedInstanceState);
        setContentView(R.layout.main);

        WebImageView imageView =
                (WebImageView) findViewById(R.id.webImage);
        imageView.setPlaceholderImage(R.drawable.ic_launcher);
        imageView.setImageUrl("http://lorempixel.com/400/200");
    }
}
```

本例中，首先会设置一张本地图片(应用程序图标)作为 WebImageView 的占位图片，这张图片会立刻显示给用户。然后我们会告诉视图从 Web 上获取 Apress 的徽标图片。如前所述，这里会在后台下载图片，下载完成后会替换掉视图中的占位图片。正是因为创建后台操作的简单性，Android 团队才会把 AsyncTask 叫作"无痛苦使用线程"。

3.5 完全在后台下载

3.5.1 问题

应用程序需要为设备下载一个大的资源，如电影文件，并且不要求用户让应用程序一直处于激活状态。

3.5.2 解决方案

(API Level 9)

使用 DownloadManager API。DownloadManager 是 API Level 9 中加入 SDK 的一个服务，它让系统完全处理和管理需要长时间运行的下载操作。使用这个服务最大的优点就是即使在下载失败、连接改变甚至设备重启时，DownloadManager 依然会继续尝试下载资源。

3.5.3 实现机制

代码清单 3-12 是一个示例 Activity，它使用 DownloadManager 来处理一个大图片文件的下载。完成下载后，这个图片会显示在一个 ImageView 上。在使用 DownloadManager 访问 Web 上的内容时，确保在应用程序的清单文件中声明了 android.permission.INTERNET 权限。

代码清单 3-12　DownloadManager 示例 Activity

```java
public class DownloadActivity extends Activity {

    private static final String DL_ID = "downloadId";
    private SharedPreferences prefs;

    private DownloadManager dm;
    private ImageView imageView;

    @Override
    public void onCreate(Bundle savedInstanceState) {
        super.onCreate(savedInstanceState);
        imageView = new ImageView(this);
        setContentView(imageView);

        prefs = PreferenceManager.getDefaultSharedPreferences(this);
        dm = (DownloadManager)getSystemService(DOWNLOAD_SERVICE);
    }
```

```java
@Override
public void onResume() {
    super.onResume();

    if(!prefs.contains(DL_ID)) {
        //开始下载
        Uri resource = Uri.parse("http://www.bigfoto.com/dog-animal.jpg");
        DownloadManager.Request request =
                new DownloadManager.Request(resource);
        //设置允许的连接来处理下载
        request.setAllowedNetworkTypes(Request.NETWORK_MOBILE |
            Request.NETWORK_WIFI);
        request.setAllowedOverRoaming(false);
        //在通知栏上显示
        request.setTitle("Download Sample");
        long id = dm.enqueue(request);
        //保存唯一的id
        prefs.edit().putLong(DL_ID, id).commit();
    } else {
        //下载已经开始，检查下载状态
        queryDownloadStatus();
    }

    registerReceiver(receiver,
        new IntentFilter(DownloadManager.ACTION_DOWNLOAD_COMPLETE));
}

@Override
public void onPause() {
    super.onPause();
    unregisterReceiver(receiver);
}

private BroadcastReceiver receiver = new BroadcastReceiver() {
    @Override
    public void onReceive(Context context, Intent intent) {
        queryDownloadStatus();
    }
};

private void queryDownloadStatus() {
    DownloadManager.Query query = new DownloadManager.Query();
    query.setFilterById(prefs.getLong(DL_ID, 0));
    Cursor c = dm.query(query);
    if(c.moveToFirst()) {
        int status =
            c.getInt(c.getColumnIndex(DownloadManager.COLUMN_STATUS));
        switch(status) {
        case DownloadManager.STATUS_PAUSED:
```

```
            case DownloadManager.STATUS_PENDING:
            case DownloadManager.STATUS_RUNNING:
                //什么也不做，下载依然进行
                break;
            case DownloadManager.STATUS_SUCCESSFUL:
                //下载完成，显示图片
                try {
                    ParcelFileDescriptor file =
                        dm.openDownloadedFile(prefs.getLong(DL_ID, 0));
                    FileInputStream fis =
                        new ParcelFileDescriptor.AutoCloseInputStream(file);
                    imageView.setImageBitmap(BitmapFactory.decodeStream(fis));
                } catch (Exception e) {
                    e.printStackTrace();
                }
                break;
            case DownloadManager.STATUS_FAILED:
                //清除下载并稍后重试
                dm.remove(prefs.getLong(DL_ID, 0));
                prefs.edit().clear().commit();
                break;
            }
        }
    }
}
```

要点：

本书出版时，SDK 中存在一个 bug，就是在使用 DownloadManager 时需要抛出 android.permission.ACCESS_ALL_DOWNLOADS 异常。实际上是在清单中没有声明 android.permission.INTERNET 时才会抛出这个异常。

示例中所有有用的工作都是在 Activity.onResume()中完成的，这样每次用户返回到 Activity 时，应用程序都可以确定下载的状态。每次下载会调用 DownloadManager.enqueue() 并返回一个 long 型的 ID 值，该值可以用来引用管理器中的本次下载。在本例中，为了随时监控和获取下载的内容，我们会把这个值保存到应用程序的首选项中。

当示例应用程序第一次启动时，会创建一个 DownloadManager.Request 对象，它代表下载的内容。这个请求至少要指定远程资源的 Uri。另外，还可以为这个请求设置很多有用的属性来控制它的行为。这些有用的属性包括：

- Request.setAllowedNetworkTypes()：指定获取下载所使用的网络类型。
- Request.setAllowedOverRoaming()：设定当设备处于漫游模式时是否要下载。
- Request.setDescription()：设置下载在系统通知栏中显示的描述。
- Request.setTitle：设置下载在系统通知栏中显示的标题。

获取 ID 后，应用程序会使用那个值来检查下载的状态。通过注册 BroadcastReceiver 来监听 ACTION_DOWNLOAD_COMPLETE 广播，在下载完成后，会将图片文件设置到 Activity 的 ImageView 以进行响应。如果在下载完成时 Activity 处于暂停状态，那么就会在下次恢复 Activity 时检查下载状态，并设置 ImageView 的内容。

要注意 ACTION_DOWNLOAD_COMPLETE 是 DownloadManager 对其管理的所有下载任务发出的广播。正因为如此，我们仍然必须检查下载 ID 才能判断我们的下载是否完成。

目标

在代码清单 3-12 中，我们并没有告诉 DownloadManager 文件的下载位置。相反，当我们想要访问文件时，我们会使用 DownloadManager.openDownloadedFile()方法和首选项中存储的 ID 值来获得一个 ParcelFileDescriptor，它可转换为应用程序可以读取的数据流。这是访问下载内容简单而直接的方式，但还是有一些事项需要注意。

如果没有指定目标位置，文件会下载到共享下载缓存中，系统随时有权删除这个缓存中的文件来回收空间。正因为如此，这种下载方式可以很方便地快速获取数据，但如果您的下载需求是长期的，应该使用 DownloadManager.Request 的一个方法在外部存储器上指定固定的目标位置：

- Request.setDestinationInExternalFilesDir()：设置目标位置为外部存储器中的一个隐藏目录。
- Request.setDestinationInExternalPublicDir()：设置目标位置为外部存储器中的一个公共目录。
- Request.setDestinationUri()：设置目标位置为外部存储器中的一个文件 Uri。

注意：
所有在外部存储器中设置目标位置的方法都需要应用程序在清单中声明 android.permission.WRITE_EXTERNAL_STORAGE 权限。

在调用 DownloadManager.remove()来清理管理器列表中的条目或者用户清理下载列表时，没有明确指定目标位置的文件通常会被移除；而外部存储器中下载的文件在这些条件下则不会被系统移除。

3.6 访问 REST API

3.6.1 问题

应用程序需要通过 HTTP 访问 RESTful API，实现与远程主机上 Web 服务的交互。

注意：
REST(Representational State Transfer，表述性状态转移)是一种常见的 Web 服务架构风格。RESTful API 通常用标准的 HTTP 动作构建，以创建对远程资源的请求。返回的响应通常是标准文档格式，例如 XML、JSON 或逗号分隔值(Comma-Separated Value，CSV)。

3.6.2 解决方案

(API Level 9)

Google 建议使用 HttpURLConnection 访问 Android 中的网络资源。可以在 AsyncTask 中使用 Java HttpURLConnection 类。不考虑本节中给出的目标版本，这个类在 API Level 1 时就已经是 Android 框架的一部分，但在 Android 2.3 发布之前，只推荐使用该方法进行网络的 I/O 操作。主要的原因就是刚开始时这个类的实现有很多的 bug，所以选择使用 HttpClient 会更加稳妥。后来，Android 团队对 HttpURLConnection 的性能和稳定性做了很大的提高，现在其已成为一种推荐使用的方式。

注意：

HttpClient 来自于 Apache HttpComponents 库，该库已合并到 Android 框架中。如果你的应用程序仍然支持 Gingerbread 之前的平台，这就是你可以考虑采用的备选方法。

HttpURLConnection 是轻量级的，在新版本的 Android 中响应速度更快并且内置很多增强功能。它的 API 更加开放，所以也就更加普及，可以实现任何类型的 HTTP 传输。缺点是开发人员需要编写更多的代码。

3.6.3 实现机制

让我们看一下如何使用 HttpUrlConnection 发送 HTTP 请求。网络请求是较长的阻塞性操作，因此不能在应用程序的主线程上执行它们。这个规则非常重要，如果尝试这样做，Android 将通过 NetworkOnMainThread 异常使应用程序崩溃。

因此，必须在诸如 AsyncTask 这样的任务中封装网络调用，以确保在后台线程中完成此工作。我们首先在代码清单 3-13 中定义 RestTask 的实现，并在代码清单 3-14 中定义一个辅助类。

代码清单 3-13 使用了 HttpUrlConnection 的 RestTask

```java
public class RestTask extends AsyncTask<Void, Integer, Object> {
    private static final String TAG = "RestTask";

    public interface ResponseCallback {
        public void onRequestSuccess(String response);

        public void onRequestError(Exception error);
    }

    public interface ProgressCallback {
        public void onProgressUpdate(int progress);
    }

    private HttpURLConnection mConnection;
    private String mFormBody;
```

```java
    private File mUploadFile;
    private String mUploadFileName;

    //Activity回调，使用WeakReferences来避免阻塞操作导致链接对象保留在内存中
    private WeakReference<ResponseCallback> mResponseCallback;
    private WeakReference<ProgressCallback> mProgressCallback;

    public RestTask(HttpURLConnection connection) {
        mConnection = connection;
    }

    public void setFormBody(List<NameValuePair> formData) {
        if(formData == null) {
            mFormBody = null;
            return;
        }

        StringBuilder sb = new StringBuilder();
        for(int i = 0; i < formData.size(); i++) {
            NameValuePair item = formData.get(i);
            sb.append( URLEncoder.encode(item.getName()) );
            sb.append("=");
            sb.append( URLEncoder.encode(item.getValue()) );
            if(i != (formData.size() - 1)) {
                sb.append("&");
            }
        }

        mFormBody = sb.toString();
    }

    public void setUploadFile(File file, String fileName) {
        mUploadFile = file;
        mUploadFileName = fileName;
    }

    public void setResponseCallback(ResponseCallback callback) {
        mResponseCallback = new WeakReference<ResponseCallback>(callback);
    }

    public void setProgressCallback(ProgressCallback callback) {
        mProgressCallback = new WeakReference<ProgressCallback>(callback);
    }

    private void writeMultipart(String boundary, String charset,
            OutputStream output, boolean writeContent) throws IOException {

        BufferedWriter writer = null;
        try {
            writer = new BufferedWriter(new OutputStreamWriter(output,
```

```java
                    Charset.forName(charset)), 8192);
//发送表单数据组件
if(mFormBody != null) {
    writer.write("--" + boundary);
    writer.write("\r\n");
    writer.write(
            "Content-Disposition: form-data; name=\"parameters\"");
    writer.write("\r\n");
    writer.write("Content-Type: text/plain; charset=" + charset);
    writer.write("\r\n");
    writer.write("\r\n");
    if(writeContent) {
        writer.write(mFormBody);
    }
    writer.write("\r\n");
    writer.flush();
}
//发送二进制文件
writer.write("--" + boundary);
writer.write("\r\n");
writer.write("Content-Disposition: form-data; name=\""
        + mUploadFileName + "\"; filename=\""
        + mUploadFile.getName() + "\"");
writer.write("\r\n");
writer.write("Content-Type: "
        + URLConnection.guessContentTypeFromName(
        mUploadFile.getName()));
writer.write("\r\n");
writer.write("Content-Transfer-Encoding: binary");
writer.write("\r\n");
writer.write("\r\n");
writer.flush();
if(writeContent) {
    InputStream input = null;
    try {
        input = new FileInputStream(mUploadFile);
        byte[] buffer = new byte[1024];
        for(int length = 0; (length = input.read(buffer)) > 0;) {
            output.write(buffer, 0, length);
        }
        //不要关闭OutputStream
        output.flush();
    } catch(IOException e) {
        Log.w(TAG, e);
    } finally {
        if(input != null) {
            try {
                input.close();
            } catch(IOException e) {
            }
```

```java
                }
            }
        }
        // 这个回车换行标志着二进制数据块的结束
        writer.write("\r\n");
        writer.flush();

        // multipart/form-data 的结束
        writer.write("--" + boundary + "--");
        writer.write("\r\n");
        writer.flush();
    } finally {
        if(writer != null) {
            writer.close();
        }
    }
}

private void writeFormData(String charset, OutputStream output)
        throws IOException {
    try {
        output.write(mFormBody.getBytes(charset));
        output.flush();
    } finally {
        if(output != null) {
            output.close();
        }
    }
}

@Override
protected Object doInBackground(Void... params) {
    // 生成用来标识界限的随机字符串
    String boundary = Long.toHexString(System.currentTimeMillis());
    String charset = Charset.defaultCharset().displayName();

    try {
        // 如果可以的话，创建输出流
        if(mUploadFile != null) {
            // 我们必须做一次复合请求
            mConnection.setRequestProperty("Content-Type",
                    "multipart/form-data; boundary=" + boundary);

            // 计算 extra 元数据的大小
            ByteArrayOutputStream bos = new ByteArrayOutputStream();
            writeMultipart(boundary, charset, bos, false);
            byte[] extra = bos.toByteArray();
            int contentLength = extra.length;
            // 将文件的大小加到 length 上
            contentLength += mUploadFile.length();
```

```java
        //如果存在表单主体，把它也加到length上
        if(mFormBody != null) {
            contentLength += mFormBody.length();
        }

        mConnection.setFixedLengthStreamingMode(contentLength);
} else if (mFormBody != null) {
        //这种情况下，只是发送表单数据
        mConnection.setRequestProperty("Content-Type",
                "application/x-www-form-urlencoded; charset=" + charset);
        mConnection.setFixedLengthStreamingMode(mFormBody.length());
}

// 这是第一次调用URLConnection，它会真正执行网络IO操作
// openConnection()执行的还是本地操作
mConnection.connect();

// 如果可以的话(对于POST)，创建输出流
if(mUploadFile != null) {
    OutputStream out = mConnection.getOutputStream();
    writeMultipart(boundary, charset, out, true);
} else if (mFormBody != null) {
    OutputStream out = mConnection.getOutputStream();
    writeFormData(charset, out);
}

// 获取响应数据
int status = mConnection.getResponseCode();
if(status >= 300) {
    String message = mConnection.getResponseMessage();
    return new HttpResponseException(status, message);
}

InputStream in = mConnection.getInputStream();
String encoding = mConnection.getContentEncoding();
int contentLength = mConnection.getContentLength();
if(encoding == null) {
    encoding = "UTF-8";
}
byte[] buffer = new byte[1024];
int length = contentLength > 0 ? contentLength : 0;
ByteArrayOutputStream out = new ByteArrayOutputStream(length);

int downloadedBytes = 0;
int read;
while ((read = in.read(buffer)) != -1) {
    downloadedBytes += read;
    publishProgress((downloadedBytes * 100) / contentLength);
```

```java
                out.write(buffer, 0, read);
            }
            return new String(out.toByteArray(), encoding);
        } catch(Exception e) {
            Log.w(TAG, e);
            return e;
        } finally {
            if(mConnection != null) {
                mConnection.disconnect();
            }
        }
    }

    @Override
    protected void onProgressUpdate(Integer... values) {
        // 更新进度 UI
        if(mProgressCallback != null && mProgressCallback.get() != null) {
            mProgressCallback.get().onProgressUpdate(values[0]);
        }
    }

    @Override
    protected void onPostExecute(Object result) {
        if(mResponseCallback != null && mResponseCallback.get() != null) {
            if(result instanceof String) {
                cb.onRequestSuccess((String) result);
            } else if (result instanceof Exception) {
                cb.onRequestError((Exception) result);
            } else {
                cb.onRequestError(new IOException(
                        "Unknown Error Contacting Host"));
            }
        }
    }
}
```

代码清单 3-14 创建请求的工具类

```java
public class RestUtil {

    public static RestTask obtainGetTask(String url)
            throws MalformedURLException, IOException {
        HttpURLConnection connection =
                (HttpURLConnection) (new URL(url))
                .openConnection();

        connection.setReadTimeout(10000);
        connection.setConnectTimeout(15000);
        connection.setDoInput(true);
```

```java
        RestTask task = new RestTask(connection);
        return task;
    }

    public static RestTask obtainFormPostTask(String url,
            List<NameValuePair> formData)
            throws MalformedURLException, IOException {
        HttpURLConnection connection =
                (HttpURLConnection) (new URL(url))
                    .openConnection();

        connection.setReadTimeout(10000);
        connection.setConnectTimeout(15000);
        connection.setDoOutput(true);

        RestTask task = new RestTask(connection);
        task.setFormBody(formData);

        return task;
    }

    public static RestTask obtainMultipartPostTask(
            String url, List<NameValuePair> formPart,
            File file, String fileName)
            throws MalformedURLException, IOException {
        HttpURLConnection connection =
                (HttpURLConnection) (new URL(url))
                    .openConnection();

        connection.setReadTimeout(10000);
        connection.setConnectTimeout(15000);
        connection.setDoOutput(true);

        RestTask task = new RestTask(connection);
        task.setFormBody(formPart);
        task.setUploadFile(file, fileName);

        return task;
    }
}
```

我们编写的这个 RestTask 可以处理 GET、简单 POST 和复合 POST 请求，并根据添加到 RestTask 中的组件动态地定义请求的参数。

这个任务支持附加两个可选的回调：一个在请求完成时获得通知，另一个在下载反馈内容时更新所有可见的进度 UI。

本例中，应用程序将使用 RestUtil 辅助类创建 RestTask 的一个实例。该类细化了 HttpURLConnection 的创建过程，它并不会进行任何的网络 I/O 操作，包括网络连接部分和与主机交互部分。辅助类创建了连接实例并且设置了超时值和 HTTP 请求方法。

注意：

默认情况下，所有 URLConnection 的请求方法会被设置为 GET。调用 setDoOutput() 则会隐式地把方法设置为 POST。如果需要设置为其他的 HTTP 动作，则需要使用 setRequestMethod()。

POST 情况下，如果有表单主体内容，这些值会直接设置到我们自定义的任务中，当任务执行时，就会读取它们。

当 RestTask 执行后，它会检查是否有关联的主体数据需要写入。如果我们已经关联了表单数据(以名-值对的方式)或者需要上传的文件，就会把它作为一个触发器来构造一个 POST 主体并发送。使用 HttpURLConnection，我们需要处理连接的各种事宜，包括告诉服务器收到数据的数量。RestTask 会花些时间计算即将发送多少数据，以及通过调用 setFixedLengthStreamingMode()构造一个头字段来告诉服务器内容的大小。对于简单表单发送请求的情况，这种计算则很简单，我们只需要传入主体字符串的长度即可。

但是，复合 POST 包含的文件数据可能会复杂一些。复合 POST 在主体中有很多的附加数据用于指定 POST 中各个部分之间的分界线，而且这些数据也会算到我们所设置的长度中。要想完成这个目标，writeMultipart()在构造时可以传入一个本地 OutputStream(本例中为一个 ByteArrayOutputStream)，然后将所有的附加数据写入这个 OutputStream 中，这样我们就可以衡量附加数据的大小了。当通过这种方式调用 writeMultipart()时，会忽略真正的内容，例如文件和表单数据，这些内容稍后可以通过调用它们各自的 length()方法添加，而且我们也不想浪费时间将它们加载到内存中。

注意：

如果不知道要 POST 的内容有多大，HttpURLConnection 还通过 setChunkedStreamingMode() 方法提供了块上传的机制。本例中，只需要传入即将发送的数据块的大小即可。

当任务已经向主机中写入了任何 POST 数据后，就可以读取响应内容了。如果初始请求是一个 GET 请求，由于并没有其他数据需要写入，任务可以直接忽略掉这一步。任务首先会检查响应代码的值以保证没有服务器端的错误，之后将响应的内容下载到一个 StringBuilder 中。下载时每次大概会读取 4KB 的数据块，同时将下载数据与响应内容总长度的百分比通知给进度回调处理程序。在所有的内容都下载完成后，任务会将所有结果响应数据以字符串的形式返回。

1. GET 示例

在下面的示例中，我们利用了 Google Custom Search REST API，此 API 的请求需要以下几个参数：

- key：唯一值，用于标识发出请求的应用程序。
- cx：标识符，标识想要访问的自定义搜索引擎。
- q：一个字符串，代表想要执行的搜索查询。

> **注意：**
> 关于这个API的更多信息可以访问 https://developers.google.com/custom-search/。

在很多公共API中，GET请求是最简单也是最常见的请求。参数必须与请求一起编码到URL字符串中发送，所以不需要提供其他的数据。下面将创建GET请求，搜索"Android"（参见代码清单3-15）。

代码清单3-15　执行API GET请求的Activity

```java
public class SearchActivity extends Activity implements
        RestTask.ProgressCallback, RestTask.ResponseCallback {

    private static final String SEARCH_URI =
            "https://www.googleapis.com/customsearch/v1"
            + "?key=%s&cx=%s&q=%s";
    private static final String SEARCH_KEY =
            "AIzaSyBbW-W1SHCK3eW0kK73VGMLJj_b-byNzkI";
    private static final String SEARCH_CX =
            "008212991319514020231:1mkouq8yagw";
    private static final String SEARCH_QUERY = "Android";

    private TextView mResult;
    private ProgressDialog mProgress;

    @Override
    public void onCreate(Bundle savedInstanceState) {
        super.onCreate(savedInstanceState);
        ScrollView scrollView = new ScrollView(this);
        mResult = new TextView(this);
        scrollView.addView(mResult,
                new ViewGroup.LayoutParams(LayoutParams.MATCH_PARENT,
                        LayoutParams.WRAP_CONTENT));
        setContentView(scrollView);

        //创建请求
        try{
            //简单GET请求
            String url = String.format(SEARCH_URI, SEARCH_KEY,
                    SEARCH_CX, SEARCH_QUERY);
            RestTask getTask = RestUtil.obtainGetTask(url);
            getTask.setResponseCallback(this);
            getTask.setProgressCallback(this);

            getTask.execute();

            //向用户显示进度
            mProgress = ProgressDialog.show(this, "Searching",
                    "Waiting For Results...", true);
        } catch(Exception e) {
```

```java
                mResult.setText(e.getMessage());
            }
        }

        @Override
        public void onProgressUpdate(int progress) {
            if(progress >= 0) {
                if(mProgress != null) {
                    mProgress.dismiss();
                    mProgress = null;
                }
                //更新用户的进度
                mResult.setText(
                        String.format("Download Progress: %d%%", progress));
            }
        }

        @Override
        public void onRequestSuccess(String response) {
            //结束进度条
            if(mProgress != null) {
                mProgress.dismiss();
            }
            //处理返回的数据(这里只是把结果显示出来)
            mResult.setText(response);
        }

        @Override
        public void onRequestError(Exception error) {
            //结束进度条
            if(mProgress != null) {
                mProgress.dismiss();
            }
            //处理返回的数据(这里只是把结果显示出来)
            mResult.setText("An Error Occurred: " + error.getMessage());
        }
    }
```

在这个示例中，创建的这种 GET 请求需要所需连接的 URL(在这里，就是发送到 googleapis.com 的 GET 请求)。URL 保存为格式化字符串常量，Google API 所需的参数则在运行时创建请求之前加上。

在 Activity 中创建了一个 RestTask 并执行，并把该 Activity 作为 RestTask 的回调。在任务完成后，会调用 onRequestSuccess()或 onRequestError()，在成功的情况下，会分析 API 返回的数据并处理。

我们还添加 ProgressCallback 到此 Activity 实现的接口列表中，这些 Activity 就可以得到下载进度的通知。然而，并不是所有的网络服务器都会为请求返回有效的内容长度(而是返回-1)，这样会导致很难实现基于百分比的进度表示。这种情况下，我们的回调函数在下载完成前会一直显示一个持续的进度对话框。而对于进度值可以确定的情况，该进度对话框会消失并在屏幕上显示进度的百分比。

下载完成后，Activity 会得到一个下载结果的 JSON 字符串回调。关于分析结构化 XML 和 JSON 数据的内容会在 3.7 和 3.8 节中进行讨论，这里先简单在用户界面上显示原始的响应数据。

2. POST 示例

很多时候，API 要求在请求中提供所需的数据，可能是认证令牌，也有可能是搜索查询的内容。API 会要求你通过 HTTP POST 发送请求，这样这些数据就会被编码到请求主体中，而不是编码到 URL 中。为了演示 POST 是如何工作的，我们将向 httpbin.org 发送请求，它是一个开发网站，用来读取和验证请求的内容并将它们返回(参见代码清单 3-16)。

代码清单 3-16 执行 POST 请求 API 的 Activity

```java
public class SearchActivity extends Activity implements
        RestTask.ProgressCallback, RestTask.ResponseCallback {

    private static final String POST_URI = "http://httpbin.org/post";

    private TextView mResult;
    private ProgressDialog mProgress;

    @Override
    public void onCreate(Bundle savedInstanceState) {
        super.onCreate(savedInstanceState);
        ScrollView scrollView = new ScrollView(this);
        mResult = new TextView(this);
        scrollView.addView(mResult,
                new ViewGroup.LayoutParams(LayoutParams.MATCH_PARENT,
                LayoutParams.WRAP_CONTENT));
        setContentView(scrollView);

        //创建请求
        try{
            //简单 POST 请求
            List<NameValuePair> parameters = new ArrayList<NameValuePair>();
            parameters.add(new BasicNameValuePair("title", "Android Recipes"));
            parameters.add(new BasicNameValuePair("summary",
                    "Learn Android Quickly"));
            parameters.add(new BasicNameValuePair("authors", "Smith"));
            RestTask postTask =
                    RestUtil.obtainFormPostTask(POST_URI, parameters);
            postTask.setResponseCallback(this);
            postTask.setProgressCallback(this);

            postTask.execute();

            //向用户显示进度
            mProgress = ProgressDialog.show(this, "Searching",
```

```java
                "Waiting For Results...", true);
        } catch(Exception e) {
            mResult.setText(e.getMessage());
        }
    }

    @Override
    public void onProgressUpdate(int progress) {
        if(progress >= 0) {
            if(mProgress != null) {
                mProgress.dismiss();
                mProgress = null;
            }
            //更新用户的进度
            mResult.setText(
                    String.format("Download Progress: %d%%", progress));
        }
    }

    @Override
    public void onRequestSuccess(String response) {
        //结束进度条
        if(mProgress != null) {
            mProgress.dismiss();
        }
        //处理返回的数据(这里只是把结果显示出来)
        mResult.setText(response);
    }

    @Override
    public void onRequestError(Exception error) {
        //结束进度条
        if(mProgress != null) {
            mProgress.dismiss();
        }
        //处理返回的数据(这里只是把结果显示出来)
        mResult.setText("An Error Occurred: " + error.getMessage());
    }
}
```

这是一个典型表单数据 POST 的示例，其中将表单字段作为名-值对传入。因为我们的 RestTask 已经设置为处理此任务，在此只需要从 RestUtil 获得正确的任务，并且填入表单数据。和 GET 示例一样，我们会在 3.7 节和 3.8 节介绍如何分析结构化的 XML 或 JSON 反馈数据，目前只是将原始的反馈信息呈现到用户界面上。

这里需要注意的是，进度回调只涉及下载相关的响应，而没有涉及 POST 数据的上传，上传相关处理可能需要由开发人员来实现。

3. 上传示例

代码清单 3-17 演示了一个稍微复杂些的复合 POST。在此同时上传原始二进制数据和一些名-值表单数据。

代码清单 3-17 执行复合 POST 请求 API 的 Activity

```java
public class SearchActivity extends Activity implements RestTask.ProgressCallback,
        RestTask.ResponseCallback {

    private static final String POST_URI = "http://httpbin.org/post";

    private TextView mResult;
    private ProgressDialog mProgress;

    @Override
    public void onCreate(Bundle savedInstanceState) {
        super.onCreate(savedInstanceState);
        ScrollView scrollView = new ScrollView(this);
        mResult = new TextView(this);
        scrollView.addView(mResult,
                new ViewGroup.LayoutParams(LayoutParams.MATCH_PARENT,
                LayoutParams.WRAP_CONTENT));
        setContentView(scrollView);

        //创建请求
        try{
            //要 POST 的文件
            Bitmap image = BitmapFactory.decodeResource(getResources(),
                    R.drawable.ic_launcher);
            File imageFile = new File(getExternalCacheDir(), "myImage.png");
            FileOutputStream out = new FileOutputStream(imageFile);
            image.compress(CompressFormat.PNG, 0, out);
            out.flush();
            out.close();
            List<NameValuePair> fileParameters =
                    new ArrayList<NameValuePair>();
            fileParameters.add(new BasicNameValuePair("title",
                    "Android Recipes"));
            fileParameters.add(new BasicNameValuePair("desc",
                    "Image File Upload"));
            RestTask uploadTask = RestUtil.obtainMultipartPostTask(POST_URI,
                    fileParameters, imageFile, "avatarImage");
            uploadTask.setResponseCallback(this);
            uploadTask.setProgressCallback(this);

            uploadTask.execute();

            //向用户显示进度
```

```
            mProgress = ProgressDialog.show(this, "Searching",
                "Waiting For Results...", true);
        } catch(Exception e) {
            mResult.setText(e.getMessage());
        }
    }

    @Override
    public void onProgressUpdate(int progress) {
        if(progress >= 0) {
            if(mProgress != null) {
                mProgress.dismiss();
                mProgress = null;
            }
            //更新用户的进度
            mResult.setText(
                String.format("Download Progress: %d%%", progress));
        }
    }

    @Override
    public void onRequestSuccess(String response) {
        //结束进度条
        if(mProgress != null) {
            mProgress.dismiss();
        }
        //处理返回的数据(这里只是把结果显示出来)
        mResult.setText(response);
    }

    @Override
    public void onRequestError(Exception error) {
        //结束进度条
        if(mProgress != null) {
            mProgress.dismiss();
        }
        //处理返回的数据(这里只是把结果显示出来)
        mResult.setText("An Error Occurred: " + error.getMessage());
    }
}
```

　　本例中，我们构造的 POST 请求包含两个不同的部分：表单数据部分(由名-值对组成)和文件部分。为了演示这个示例，我们使用应用程序的图标并将它快速地写入外部存储器的一个 PNG 文件(用于上传)中。

　　本例中，httpbin 的 JSON 响应数据将对应于表单数据元素和 Base64 编码表示的 PNG 图像。

4. 基本授权

向 RestTask 中添加基本授权的过程相当简单。有两种方式可以实现：在每个请求中直接添加或者全局使用名为 Authenticator 的类。首先我们看一下在单个请求中添加基本授权。代码清单 3-18 修改了 RestUtil 的方法来关联适当格式的用户名和密码。

代码清单 3-18　实现了基本授权功能的 RestUtil

```java
public class RestUtil {

    public static RestTask obtainGetTask(String url)
            throws MalformedURLException, IOException {
        HttpURLConnection connection = (HttpURLConnection) (new URL(url))
                .openConnection();

        connection.setReadTimeout(10000);
        connection.setConnectTimeout(15000);
        connection.setDoInput(true);

        RestTask task = new RestTask(connection);
        return task;
    }

    public static RestTask obtainAuthenticatedGetTask(String url,
            String username, String password) throws
            MalformedURLException, IOException {
        HttpURLConnection connection = (HttpURLConnection) (new URL(url))
                .openConnection();

        connection.setReadTimeout(10000);
        connection.setConnectTimeout(15000);
        connection.setDoInput(true);

        attachBasicAuthentication(connection, username, password);

        RestTask task = new RestTask(connection);
        return task;
    }

    public static RestTask obtainFormPostTask(String url,
            List<NameValuePair> formData) throws MalformedURLException,
            IOException {
        HttpURLConnection connection = (HttpURLConnection) (new URL(url))
                .openConnection();

        connection.setReadTimeout(10000);
        connection.setConnectTimeout(15000);
```

```java
        connection.setDoOutput(true);

        RestTask task = new RestTask(connection);
        task.setFormBody(formData);

        return task;
    }

    public static RestTask obtainAuthenticatedFormPostTask(
            String url, List<NameValuePair> formData,
            String username, String password)
            throws MalformedURLException, IOException {
        HttpURLConnection connection =
            (HttpURLConnection) (new URL(url))
              .openConnection();

        connection.setReadTimeout(10000);
        connection.setConnectTimeout(15000);
        connection.setDoOutput(true);

        attachBasicAuthentication(connection, username, password);

        RestTask task = new RestTask(connection);
        task.setFormBody(formData);

        return task;
    }

    public static RestTask obtainMultipartPostTask(String url,
            List<NameValuePair> formPart, File file, String fileName)
            throws MalformedURLException, IOException {
        HttpURLConnection connection = (HttpURLConnection) (new URL(url))
                .openConnection();

        connection.setReadTimeout(10000);
        connection.setConnectTimeout(15000);
        connection.setDoOutput(true);

        RestTask task = new RestTask(connection);
        task.setFormBody(formPart);
        task.setUploadFile(file, fileName);
```

```
        return task;
    }

    private static void attachBasicAuthentication(URLConnection connection,
            String username, String password) {
        //添加基本授权头
        String userpassword = username + ":" + password;
        String encodedAuthorization =
                Base64.encodeToString(userpassword.getBytes(), Base64.NO_WRAP);
        connection.setRequestProperty("Authorization", "Basic " +
            encodedAuthorization);
    }
}
```

基本授权是作为头字段加入到一个 HTTP 请求中的，该头字段包括属性名称"Authorization"、"Basic"的值加上用户名和密码的 Base64 编码字符串。attachBasicAuthentication()辅助方法在 URLConnection 赋给 RestTask 之前会将上述属性名和编码字符串设置到 URLConnection 上。添加 Base64.NO_WRAP 标志以确保编码器没有新增额外的行，这会创建无效的值。

当不是所有请求都使用相同方式授权时，直接在请求中添加授权是一种非常好的方式。但是，有时候只需要设置一次认证信息即可让所有的请求使用它们。这时候就需要 Authenticator 类了。Authenticator 允许为应用程序进程的所有请求设置全局的用户名和密码认证信息。让我们看一下代码清单 3-19，它就实现了这个功能。

代码清单 3-19　使用了 Authenticator 类的 Activity

```
public class AuthActivity extends Activity implements ResponseCallback {

    private static final String URI =
            "http://httpbin.org/basic-auth/android/recipes";
    private static final String USERNAME = "android";
    private static final String PASSWORD = "recipes";

    private TextView mResult;
    private ProgressDialog mProgress;

    @Override
    protected void onCreate(Bundle savedInstanceState) {
        super.onCreate(savedInstanceState);
        mResult = new TextView(this);
        setContentView(mResult);

        Authenticator.setDefault(new Authenticator() {
            @Override
            protected PasswordAuthentication getPasswordAuthentication() {
                return new PasswordAuthentication(USERNAME,
                        PASSWORD.toCharArray());
            }
```

```java
            });

            try {
                RestTask task = RestUtil.obtainGetTask(URI);
                task.setResponseCallback(this);
                task.execute();
            } catch(Exception e) {
                mResult.setText(e.getMessage());
            }
        }

        @Override
        public void onRequestSuccess(String response) {
            if(mProgress != null) {
                mProgress.dismiss();
                mProgress = null;
            }
            mResult.setText(response);
        }

        @Override
        public void onRequestError(Exception error) {
            if(mProgress != null) {
                mProgress.dismiss();
                mProgress = null;
            }
            mResult.setText(error.getMessage());
        }
    }
```

该例会再次连接 httpbin，但本次连接的站点需要使用有效的认证信息。主机需要的用户名和密码是被编码到 URL 路径中的，如果没有提供相应的认证信息属性的话，主机的响应结果为 UNAUTHORIZED。

只需要调用一次 Authenticator.setDefault()方法并传入一个新的 Authenticator 实例，后续所有的请求在认证时就都会使用上面提供的认证信息。所以我们在请求时创建了一个新的 PasswordAuthentication 实例以将正确的用户名和密码传入 Authenticator 类，这样进程中的所有 URLConnection 实例都会使用它。注意，这个示例中，请求并没有关联认证信息，但执行请求后会得到一个已经认证的响应。

5. 响应缓存

(API Level 13)

HttpURLConnection 最后一个可以利用的平台改进方案是通过 HttpResponseCache 设置响应缓存。加快应用程序响应速度的方法就是将从远程主机获取的响应缓存起来。这样对于频繁使用的请求就可以直接从缓存中加载而不必每次都访问网络。在应用程序中安装和移除缓存只需要简单的几行代码。

这些安装和移除缓存的方法只需要调用一次，因此可以将它们放在 Activity 或每个请求范围外的应用级类中：

```
//安装响应缓存
try {
    File httpCacheDir = new File(context.getCacheDir(), "http");
    long httpCacheSize = 10 * 1024 * 1024; // 10 MiB
    HttpResponseCache.install(httpCacheDir, httpCacheSize);
} catch(IOException e) {
    Log.i(TAG, "HTTP response cache installation failed:" + e);
}

//清空响应缓存
HttpResponseCache cache = HttpResponseCache.getInstalled();
if(cache != null) {
    cache.flush();
}
```

注意：
HttpResponseCache 只能应用于 HttpURLConnection 的变体，它并不适用于 Apache HttpClient。

3.7 解析 JSON

3.7.1 问题

应用程序需要解析从 API 或其他资源返回的 JSON(JavaScript Object Natation，JavaScript 对象符号)格式的响应结果。

3.7.2 解决方案

(API Level 1)
使用 Android 中的 arg.json 解析类。SDK 在 org.json 包中自带了一个非常高效的类集，用来解析 JSON 格式的字符串。只需要用已经格式化的字符串数据生成一个新的 JSONObject 或 JSONArray，然后就可以使用一系列存取器方法去获得这些对象中的原始数据或内嵌的 JSONObject 和 JSONArray 数据。

3.7.3 实现机制

默认情况下，这个 JSON 解析器是非常严格的，也就意味着当遇到无效的 JSON 数据或键时会抛出异常。以"get"开头的存取器方法在请求的值找不到时会抛出 JSONException 异常。在某些情况下，这种行为可能不太好，因此存在一套以"opt"为前缀的伴随方法。这些方法在请求的键所对应的值找不到时会返回 null，而不是抛出异常。另外，它们中的很多方法都提供了重载版本，可以在失败时返回一个传入的参数，而不是返回 null。

让我们看一个示例,了解一下如何将一个 JSON 字符串解析为一些有用的小数据。先看一下代码清单 3-20 中的 JSON 数据。

代码清单 3-20　JSON 示例

```
{
    "person": {
        "name": "John",
        "age": 30,
        "children": [
            {
                "name": "Billy"
                "age": 5
            },
            {
                "name": "Sarah"
                "age": 7
            },
            {
                "name": "Tommy"
                "age": 9
            }
        ]
    }
}
```

这里定义了一个对象,它有 3 个值:name(字符串)、age(整型)和 children。名为"children"的参数是由其他三个对象组成的数组,每个对象都有自己的 name 和 age。代码清单 3-21 和 3-22 中的示例展示了使用 org.json 分析上述数据并在 TextView 中显示其中的一些元素。

代码清单 3-21　res/layout/main.xml

```xml
<?xml version="1.0" encoding="utf-8"?>
<LinearLayout
    xmlns:android="http://schemas.android.com/apk/res/android"
    android:layout_width="match_parent"
    android:layout_height="match_parent"
    android:orientation="vertical">
    <TextView
        android:id="@+id/line1"
        android:layout_width="match_parent"
        android:layout_height="wrap_content" />
    <TextView
        android:id="@+id/line2"
        android:layout_width="match_parent"
        android:layout_height="wrap_content" />
    <TextView
```

```xml
        android:id="@+id/line3"
        android:layout_width="match_parent"
        android:layout_height="wrap_content" />
</LinearLayout>
```

代码清单 3-22   JSON 解析 Activity 示例

```java
public class MyActivity extends Activity {
    private static final String JSON_STRING =
        "{\"person\":"
        + "{\"name\":\"John\",\"age\":30,\"children\":["
        + "{\"name\":\"Billy\",\"age\":5},"
        + "{\"name\":\"Sarah\",\"age\":7},"
        + "{\"name\":\"Tommy\",\"age\":9}"
        + "] } }";
    @Override
    public void onCreate(Bundle savedInstanceState) {
        super.onCreate(savedInstanceState);
        setContentView(R.layout.main);

        TextView line1 = (TextView)findViewById(R.id.line1);
        TextView line2 = (TextView)findViewById(R.id.line2);
        TextView line3 = (TextView)findViewById(R.id.line3);
        try {
            JSONObject person = (new JSONObject(JSON_STRING))
                    .getJSONObject("person");
            String name = person.getString("name");
            line1.setText("This person's name is " + name);
            line2.setText(name + " is " + person.getInt("age")
                    + " years old.");
            line3.setText(name + " has "
                    + person.getJSONArray("children").length()
                    + " children.");
        } catch(JSONException e) {
            e.printStackTrace();
        }
    }
}
```

在本例中，JSON 字符串作为常量采用了硬编码的方式。在创建了 Activity 后，这个字符串被转换为一个 JSONObject，这样它的所有数据就可以通过键-值对的方式进行访问，就像存储在 map 或字典中一样。由于我们使用严格方法来访问数据，因此所有的业务逻辑都被封装在一个 try 块中。

JSONObject.getString()和 JSONObject.getInt()等函数用来读取原始数据并把它们放到 TextView 中；getJSONArray()用来取出内嵌的 children 数组。在读取数据时，JSONArray 拥有和 JSONObject 一样的存取器方法集，但它的方法参数是数组的索引而不是键的名称。另外，JSONArray 可以返回它的长度，我们会在示例中使用这个长度显示每个人拥有的孩子的数量。

图 3-3 显示了示例应用程序的结果。

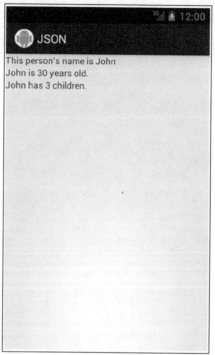

图 3-3　在 Activity 中显示解析过的 JSON 数据

调试技巧

　　JSON 是一种高效的书写方式；然而，因为阅读原始 JSON 字符串很难，所以也很难调试解析问题。很多情况下，你所分析的 JSON 来自于远程资源或者你根本不熟悉它，而且在调试时还需要显示它。

　　JSONObject 和 JSONArray 都提供了重载的 toString()方法，参数为整型，它会返回缩进的打印效果十分漂亮的字符串数据，使之更容易解读。通常在一些麻烦的地方添加诸如 myJsonObject.toString(2)这样的代码，可以省去很多时间和烦恼。

3.8　解析 XML

3.8.1　问题

　　应用程序需要解析从 API 或其他资源返回的 XML 格式的响应结果。

3.8.2　解决方案

　　(API Level 1)
　　可以通过实现 org.xml.sax.helpers.DefaultHandler 的一个子类来解析数据，它使用的是基于事件的 SAX 方式(Simple API for XML)。Android 有三种用于解析 XML 数据的主要方式：DOM(文档对象模型)、SAX 和 Pull。这其中最容易实现的就是 SAX 解析器，它也是内存效率最高的。SAX 解析通过遍历 XML 数据来实现，并在每个元素的开头和结尾产生回调事件。

3.8.3 实现机制

为了进一步介绍如何解析 XML，先来看一下请求 RSS/ATOM 新闻源时返回的 XML 格式数据(参见代码清单 3-23)。

代码清单 3-23 RSS 基本结构

```
<rss version="2.0">
  <channel>
    <item>
      <title></title>
      <link></link>
      <description></description>
    </item>
    <item>
      <title></title>
      <link></link>
      <description></description>
    </item>
    <item>
      <title></title>
      <link></link>
      <description></description>
    </item>
    ...
  </channel>
</rss>
```

在各组<title>、<link>和<description>标签之间就是每个项的值。我们可以使用 SAX 将这段数据解析成一个项数组，应用程序可以很方便地在列表中将数据呈现给用户(参见代码清单 3-24)。

代码清单 3-24 自定义的 RSS 解析处理程序

```
public class RSSHandler extends DefaultHandler {

    public class NewsItem {
        public String title;
        public String link;
        public String description;

        @Override
        public String toString() {
            return title;
        }
```

```java
    }

    private StringBuffer buf;
    private ArrayList<NewsItem> feedItems;
    private NewsItem item;

    private boolean inItem = false;

    public ArrayList<NewsItem> getParsedItems() {
        return feedItems;
    }

    //在每个新元素开始时调用
    @Override
    public void startElement(String uri, String name, String qName, Attributes atts) {
        if("channel".equals(name)) {
            feedItems = new ArrayList<NewsItem>();
        } else if("item".equals(name)) {
            item = new NewsItem();
            inItem = true;
        } else if("title".equals(name) && inItem) {
            buf = new StringBuffer();
        } else if("link".equals(name) && inItem) {
            buf = new StringBuffer();
        } else if("description".equals(name) && inItem) {
            buf = new StringBuffer();
        }
    }

    //在每个元素结束时调用
    @Override
    public void endElement(String uri, String name, String qName) {
        if("item".equals(name)) {
            feedItems.add(item);
            inItem = false;
        } else if("title".equals(name) && inItem) {
            item.title = buf.toString();
        } else if("link".equals(name) && inItem) {
            item.link = buf.toString();
        } else if("description".equals(name) && inItem) {
            item.description = buf.toString();
        }

        buf = null;
    }

    //调用元素中的字符数据
    @Override
    public void characters(char ch[], int start, int length) {
        //处理缓存已经初始化的情况
```

```java
        if(buf != null) {
            for(int i = start; i < start + length; i++) {
                buf.append(ch[i]);
            }
        }
    }
}
```

在每个元素开始和结束时都会通过 startElement()和 endElement()方法通知 RSSHandler。在这之间，组成元素值的字符会传递给 characters()回调方法。当解析器遍历文档时，会产生如下步骤：

(1) 当解析器碰到第一个元素时，会初始化项列表。

(2) 对于遇到的每个项元素，会初始化一个新的 NewsItem 模型。

(3) 在每个项元素的内部，数据元素被置入一个 StringBuffer 中，然后插入 NewsItem 的成员中。

(3) 当到达每个项的结尾时，会把 NewsItem 添加到列表中。

(5) 解析完成后，feedItems 中包含了源数据中的所有项。

接下来，使用在 3.6 节的 API 示例中介绍的一些技巧来下载最新的 RSS 格式的 Google 新闻内容(参见代码清单 3-25)。

代码清单 3-25 解析 XML 并显示各个项内容的 Activity

```java
public class FeedActivity extends Activity implements ResponseCallback {
    private static final String TAG = "FeedReader";
    private static final String FEED_URI =
            "http://news.google.com/?output=rss";

    private ListView mList;
    private ArrayAdapter<NewsItem> mAdapter;
    private ProgressDialog mProgress;

    @Override
    public void onCreate(Bundle savedInstanceState) {
        super.onCreate(savedInstanceState);

        mList = new ListView(this);
        mAdapter = new ArrayAdapter<NewsItem>(this,
                android.R.layout.simple_list_item_1,
                android.R.id.text1);
        mList.setAdapter(mAdapter);
        mList.setOnItemClickListener(new AdapterView.OnItemClickListener() {
            @Override
            public void onItemClick(AdapterView<?> parent, View v,
                    int position, long id) {
                NewsItem item = mAdapter.getItem(position);
                Intent intent = new Intent(Intent.ACTION_VIEW);
                intent.setData(Uri.parse(item.link));
                startActivity(intent);
```

```java
        }
    });

    setContentView(mList);
}

@Override
public void onResume() {
    super.onResume();
    //获取 RSS 源数据
    try{
        RestTask task = RestUtil.obtainGetTask(FEED_URI);
        task.setResponseCallback(this);
        task.execute();
        mProgress = ProgressDialog.show(this, "Searching",
                "Waiting For Results...", true);
    } catch(Exception e) {
        Log.w(TAG, e);
    }
}

@Override
public void onRequestSuccess(String response) {
    if(mProgress != null) {
        mProgress.dismiss();
        mProgress = null;
    }
    //处理响应数据
    try {
        SAXParserFactory factory = SAXParserFactory.newInstance();
        SAXParser p = factory.newSAXParser();
        RSSHandler parser = new RSSHandler();
        p.parse(new InputSource(new StringReader(response)), parser);

        mAdapter.clear();
        for(NewsItem item : parser.getParsedItems()) {
            mAdapter.add(item);
        }
        mAdapter.notifyDataSetChanged();
    } catch(Exception e) {
        Log.w(TAG, e);
    }
}

@Override
public void onRequestError(Exception error) {
    if(mProgress != null) {
        mProgress.dismiss();
        mProgress = null;
    }
```

```
        //显示错误
        mAdapter.clear();
        mAdapter.notifyDataSetChanged();
        Toast.makeText(this, error.getMessage(), Toast.LENGTH_SHORT).show();
    }
}
```

这个示例修改之后会显示一个 ListView，其中的数据就是从 RSS 源解析出来的。在这个示例中，我们为列表添加了一个 OnItemClickListener，用户点击时会在浏览器中加载新闻项的链接。

当数据从 API 的响应回调方法返回时，Android 内置的 SAX 解析器会遍历 XML 字符串。SAXParser.parse()会使用 RSSHandler 的实例来处理 XML，从 XML 中解析的内容会用来填充 RSSHandler 的 feedItems 列表。接收器再逐个处理解析出来的项，将其添加到 ArrayAdapter 中，最终显示在 ListView 中。

XmlPullParser

由框架提供的 XmlPullParser 是另一种高效解析传入的 XML 数据的方式。和 SAX 一样，解析过程也是基于流的，由于解析开始之前并不需要加载整个 XML 数据结构，因此在解析大文档源时也就不需要太多的内存。下面让我们看一下使用 XmlPullParser 解析 RSS 源数据的示例。但与 SAX 不同，我们必须手动地干预每一步的数据流解析过程，即使是我们不感兴趣的标签元素。

代码清单 3-26 包含一个工厂类，它会迭代源数据以构造元素模型。

代码清单 3-26　用来将 XML 解析成模型对象的工厂类

```java
public class NewsItemFactory {

    /*数据模型类*/
    public static class NewsItem {
        public String title;
        public String link;
        public String description;

        @Override
        public String toString() {
            return title;
        }
    }

    /*
     * 将 RSS 源解析为一个 NewsItem 元素的列表
     */
    public static List<NewsItem> parseFeed(XmlPullParser parser)
            throws XmlPullParserException, IOException {
        List<NewsItem> items = new ArrayList<NewsItem>();
```

```java
        while(parser.next() != XmlPullParser.END_TAG) {
            if(parser.getEventType() != XmlPullParser.START_TAG) {
                continue;
            }

            if(parser.getName().equals("rss") ||
                    parser.getName().equals("channel")) {
                //跳过这些元素,但允许解析它们内部的元素
            } else if (parser.getName().equals("item")) {
                NewsItem newsItem = readItem(parser);
                items.add(newsItem);
            } else {
                //跳过其他元素以及它们的子元素
                skip(parser);
            }
        }

        //返回解析后的列表
        return items;
    }

    /*
     * 将每个<item>元素解析为一个NewsItem
     */
    private static NewsItem readItem(XmlPullParser parser) throws
            XmlPullParserException, IOException {
        NewsItem newsItem = new NewsItem();

        //开头必须是有效的<item>元素
        parser.require(XmlPullParser.START_TAG, null, "item");
        while(parser.next() != XmlPullParser.END_TAG) {
            if(parser.getEventType() != XmlPullParser.START_TAG) {
                continue;
            }

            String name = parser.getName();
            if(name.equals("title")) {
                parser.require(XmlPullParser.START_TAG, null, "title");
                newsItem.title = readText(parser);
                parser.require(XmlPullParser.END_TAG, null, "title");
            } else if(name.equals("link")) {
                parser.require(XmlPullParser.START_TAG, null, "link");
                newsItem.link = readText(parser);
                parser.require(XmlPullParser.END_TAG, null, "link");
            } else if(name.equals("description")) {
                parser.require(XmlPullParser.START_TAG, null, "description");
                newsItem.description = readText(parser);
                parser.require(XmlPullParser.END_TAG, null, "description");
            } else {
                //跳过其他元素以及它们的子元素
```

```java
                skip(parser);
            }
        }

        return newsItem;
    }

    /*
     * 读取当前元素的文本内容,该内容是 start 和 end 标签之间包含的数据
     */
    private static String readText(XmlPullParser parser) throws
            IOException, XmlPullParserException {
        String result = "";
        if(parser.next() == XmlPullParser.TEXT) {
            result = parser.getText();
            parser.nextTag();
        }
        return result;
    }

    /*
     *辅助方法,用来跳过当前元素以及该元素的子元素
     */
    private static void skip(XmlPullParser parser) throws
            XmlPullParserException, IOException {
        if(parser.getEventType() != XmlPullParser.START_TAG) {
            throw new IllegalStateException();
        }

        /*
         *对于每个新标签,会把一个 depth 计数器加1。到达每个标签的结尾时会把计数器减1
         *并且在 end 标签与开始时的标签匹配时会返回
         */
        int depth = 1;
        while(depth != 0) {
            switch(parser.next()) {
            case XmlPullParser.END_TAG:
                depth--;
                break;
            case XmlPullParser.START_TAG:
                depth++;
                break;
            }
        }
    }
}
```

Pull 解析过程的工作原理就是把数据流作为一系列的事件来处理。应用程序通过调用 next()方法或该方法的一个或多个指定变体来告诉解析器处理下一个事件。以下是解析器会处理的事件类型:

- START_DOCUMENT：当解析器首次初始化时会返回这个事件。在首次调用 next()、nextToken()或 nextTag()之前，解析器都会是这个状态。
- START_TAG：解析器刚刚读取标签元素的开始部分。标签的名称可以通过 getName()获得，里面的任何属性也可以通过 getAttributeValue()和相关的方法获得。
- TEXT：读取标签元素内部的字符数据，可以通过 getText()获取。
- END_TAG：解析器刚刚读取标签元素的结尾部分。和它相匹配的开始标签的名称可以通过 getName()获得。
- END_DOCUMENT：表明到达了数据流的结尾。

由于必须自己操作解析器，因此我们创建了一个辅助方法 skip()，它可以帮助解析器跳过我们不感兴趣的标签。这个方法从当前位置开始遍历所有的内嵌子元素，直到找到匹配的结束标签，并把它们全部跳过。这里使用了一个 depth 计数器，碰到每个开始标签时会递增，碰到每个结束标签时会递减。当 depth 计数器到达 0 时，我们就找到了与开始位置相匹配的结束标签了。

本例中，在调用 parseFeed()方法时，解析器首先会迭代数据流来查找可以转换为 NewsItem 的<item>标签。除了<rss>和<channel>，所有不是<item>的元素都可以跳过。这是因为所有的项都是内嵌在这两个标签之中的，因此即使我们对它们不直接感兴趣，也不能把它们交给 skip()处理，否则所有的项都会被跳过。

分析每个<item>元素的工作是由 readItem()方法完成的，它会构造一个新的 NewsItem，该 NewsItem 的内容来自于<item>内部的数据。readItem()方法首先会调用 require()，它是一种安全性检查，能够确保 XML 是我们希望的格式。如果当前的解析器事件和传入的命名空间、标签名称相匹配的话，这个方法会静默地返回；否则，它会抛出异常。当我们遍历子元素时，我们主要查找 title、link 和 description 标签，这样就可以把它们的值读取到模型数据中。查找到所需的标签后，readText()会操作解析器并把相关字符数据取出。同样，在<item>内部有一些其他元素我们并没有解析，对于不需要的标签只需要调用 skip()即可。

你已经看到了 XmlPullParser 非常灵活，原因是可控制整个过程的每一步，但这也要求写更多的代码来完成相同的结果。代码清单 3-27 展示了使用新的解析器来完成源数据显示的 Activity。

代码清单 3-27　显示解析的 XML 源的 Activity

```
public class PullFeedActivity extends Activity implements ResponseCallback {
    private static final String TAG = "FeedReader";
    private static final String FEED_URI =
            "http://news.google.com/?output=rss";

    private ListView mList;
    private ArrayAdapter<NewsItem> mAdapter;
    private ProgressDialog mProgress;

    @Override
    public void onCreate(Bundle savedInstanceState) {
        super.onCreate(savedInstanceState);
```

```java
        mList = new ListView(this);
        mAdapter = new ArrayAdapter<NewsItem>(this,
                android.R.layout.simple_list_item_1,
                android.R.id.text1);
        mList.setAdapter(mAdapter);
        mList.setOnItemClickListener(new AdapterView.OnItemClickListener() {
            @Override
            public void onItemClick(AdapterView<?> parent, View v,
                    int position, long id) {
                NewsItem item = mAdapter.getItem(position);
                Intent intent = new Intent(Intent.ACTION_VIEW);
                intent.setData(Uri.parse(item.link));
                startActivity(intent);
            }
        });

        setContentView(mList);
    }

    @Override
    public void onResume() {
        super.onResume();
        //获取RSS源数据
        try{
            RestTask task = RestUtil.obtainGetTast(FEED_URI);
            task.setResponseCallback(this);
            task.execute();
            mProgress = ProgressDialog.show(this, "Searching",
                    "Waiting For Results...", true);
        } catch (Exception e) {
            Log.w(TAG, e);
        }
    }

    @Override
    public void onRequestSuccess(String response) {
        if(mProgress != null) {
            mProgress.dismiss();
            mProgress = null;
        }
        //处理响应数据
        try {
            XmlPullParser parser = Xml.newPullParser();
            parser.setInput(new StringReader(response));
            //跳到第一个标签
            parser.nextTag();

            mAdapter.clear();
            for(NewsItem item : NewsItemFactory.parseFeed(parser)) {
                mAdapter.add(item);
```

```
            }
            mAdapter.notifyDataSetChanged();
        } catch(Exception e) {
            Log.w(TAG, e);
        }
    }

    @Override
    public void onRequestError(Exception error) {
        if(mProgress != null) {
            mProgress.dismiss();
            mProgress = null;
        }
        //显示错误
        mAdapter.clear();
        mAdapter.notifyDataSetChanged();
        Toast.makeText(this, error.getMessage(), Toast.LENGTH_SHORT).show();
    }
}
```

使用 Xml.newPullParser()可以实例化一个新的 XmlPullParser，通过 setInput()可以将数据源的输入流作为一个 Reader。本例中，从 Web 服务返回的数据已经是字符串了，所以我们把它封装成一个 StringReader 来让解析器解析。我们可以把解析器传给 NewsItem-Factory，之后会返回 NewsItem 元素的列表，我们把它添加到 ListAdapter 中，然后像之前那样显示出来。

提示：
还可以使用 XmlPullParser 解析应用程序中绑定的本地 XML 数据。把你的原始 XML 放到资源文件中(如 res/xml/)，然后你就可以实例化一个 XmlResourceParser，它会使用 Resources.getXml()预加载你的本地数据。

3.9 接收短信

3.9.1 问题

应用程序需要响应接收到的短信，也叫文本消息。

3.9.2 解决方案

(API Level 1)
注册一个 BroadcastReceiver 来监听收到的消息，并在 onReceive()中处理它们。当收到一条短信时，操作系统会发送一个 action 值为 android.provider.Telephony.SMS_RECEIVED 的广播 Intent。应用程序则可以注册一个 BroadcastReceiver 以过滤这个 Intent 并处理收到的数据。

注意：

接收这个广播同样并不会阻止系统其他的应用程序接收它。默认的消息处理应用程序还会接收这条短信并显示。

3.9.3 实现机制

在上一个示例中，我们定义了一个 BroadcastReceiver 并作为 Activity 的内部私有成员变量。在本例中，最好单独定义接收器并在 AndroidManifest.xml 中使用<receiver>标签注册它。这样即使应用程序处于非激活状态，接收器也可以处理收到的事件。代码清单 3-28 和 3-29 显示了一个接收器示例，它监控所有收到的短信并将感兴趣的短信显示在一个 Toast 上。

代码清单 3-28　接收短信的 BroadcastReceiver

```java
public class SmsReceiver extends BroadcastReceiver {
    //我们想要监听的设备地址（电话号码、短代码等）
    private static final String SENDER_ADDRESS = "<ENTER YOUR NUMBER HERE>";

    @Override
    public void onReceive(Context context, Intent intent) {
        Bundle bundle = intent.getExtras();

        Object[] messages = (Object[])bundle.get("pdus");
        SmsMessage[] sms = new SmsMessage[messages.length];
        //为每个收到的 PDU 创建短信
        for(int n=0; n < messages.length; n++) {
            sms[n] = SmsMessage.createFromPdu((byte[]) messages[n]);
        }
        for(SmsMessage msg : sms) {
            //检查短信是否来自我们已知的发送方
            if(TextUtils.equals(msg.getOriginatingAddress(),
                                SENDER_ADDRESS)) {
                //让其他应用处理此消息
                abortBroadcast();

                //显示我们自己的通知
                Toast.makeText(context,
                    "Received message from the mothership: "
                    + msg.getMessageBody(),
                    Toast.LENGTH_SHORT).show();
            }
        }
    }
}
```

代码清单 3-29　AndroidManifest.xml 的部分内容

```xml
<?xml version="1.0" encoding="utf-8"?>
<manifest ...>

    <uses-permission
       android:name="android.permission.RECEIVE_SMS" />

  <application ...>
    <receiver android:name=".SmsReceiver">
      <!-- 添加优先级以捕获有序的广播 -->
      <intent-filter android:priority="5">
        <action
          android:name="android.provider.Telephony.SMS_RECEIVED"
         />
      </intent-filter>
    </receiver>
  </application>

</manifest>
```

要点：
接收短信需要在清单中声明 android.permission.RECEIVE_SMS 权限。

接收的短信作为 byte 数组(Object 数组)，通过广播 Intent 的 extras 传递，每个 byte 数组代表一个短信 PDU(Protocol Data Unit)。SmsMessage.createFromPdu()方法可以方便地从原始的 PDU 数据创建 SmsMessage 对象。完成设置工作之后，可以检查每条短信，判断是否要对其进行处理。在这个示例中，我们会将每条短信的发送地址与已知的短代码进行比较，在收到符合要求的短信时通知用户。

框架触发的广播是有序的广播消息，这意味着每个注册的接收器将按顺序接收消息，并且有机会修改广播，然后将其交给下一个接收器；或者取消广播，并且完全阻止任何较低优先级的接收器接收该消息。

在 AndroidManifest.xml 的<intent-filter>条目中，我们添加了任意的优先级值，从而在核心系统 Messages 应用程序(该应用程序使用默认的优先级 0)之上插入接收器。这就使我们的应用程序可以最先处理短信消息。

注意：
使用有序的广播时，在相同优先级上注册的接收器会"同时"接收 Intent，从而这些接收器之间的顺序是不确定的。此外，一个接收器不能取消将广播传递给相同优先级的其他接收器。

接下来，一旦确认查看的消息来自于正在跟踪的发送方，即可调用 abortBroadcast()来终止响应链。这就使我们正在处理的短信消息不会显示给用户和弄乱他们的收件箱。

要点：
没有外部方法可验证系统中还有其他哪些应用可注册以处理广播并具有非常高的优

先级(至少优先级要高于您的应用程序)。您的应用程序受具有更高优先级的应用程序的支配，无法终止您要处理的消息的广播。

在这个示例中显示 Toast 时，你有可能希望在这个时候给用户提供更多有用的信息。短信中可能有应用程序的请求码，这样就可以启动合适的 Activity 以将该消息在应用程序中显示给用户。

> **默认短信应用程序**
>
> 从 Android 4.4 开始，使用短信的应用程序的行为已经改变。设备的 Settings 应用程序现在为用户提供 Default SMS App(默认短信应用)选项，以供选择用户最希望用于短信的应用程序。在此框架级别中，扩展了围绕发送和接收消息的一些行为。
>
> 未选择为默认选项的应用程序仍然会使用本范例中描述的相同有序广播，发送传出的短信和监控传入的消息。然而，新框架中为默认短信应用添加了两个新的广播动作来接收消息：
>
> - android.provider.telephony.SMS_DELIVER
> - android.provider.telephony.WAP_PUSH_DELIVER
>
> 框架会使用这两个动作，独立于其他应用程序将传入的短信/MMS 消息数据广播到默认短信应用。尽管原始的 SMS_RECEIVED 动作仍然是有序的广播，但终止广播不再用作拦截传递给应用程序的某些消息的技术。但是，终止广播仍然会打断响应链，不再到达监控传入消息的其他第三方应用程序。
>
> 此外，标记为默认的短信应用程序负责通过 android.provider.Telephony，将设备上接收的所有短信数据写入设备的内部内容提供程序(在 API Level 19 中公开揭示的提供程序)。在系统中，只有此应用程序有权将数据写入短信提供程序，而无论应用程序是否请求获得 android.permission.WRITE_SMS 权限。
>
> 如果其他应用程序获得了 android.permission.READ_SMS 权限，它们仍然可以读取短信提供程序数据。我们将在第 7 章中详细介绍读取短信提供程序。

3.10 发送短信

3.10.1 问题

应用程序需要向外发送短信。

3.10.2 解决方案

(API Level 4)

使用 SMSManager 发送文字或数据短信。SMSManager 是一个系统服务，用来处理短信发送并把操作的状态反馈给应用程序。SMSManager 提供了 SmsManager.sendTextMessage() 和 SmsManager.sendMultipartTextMessage() 方法来发送文字短信，并且提供了 SmsManager.sendDataMessage()方法来发送数据短信。这些方法都含有用于传递发送操作状态和将消息传回请求目标的 PendingIntent 参数。

3.10.3 实现机制

下面看一个简单的示例 Activity，其发送一条短信并监控其状态(参见代码清单 3-30)。

代码清单 3-30 发送短信的 Activity

```java
public class SmsActivity extends Activity {
    //我们想要监听的设备地址(电话号码、短代码等)
    private static final String RECIPIENT_ADDRESS =
        "<ENTER YOUR NUMBER HERE>";
    private static final String ACTION_SENT = "com.examples.sms.SENT";
    private static final String ACTION_DELIVERED =
        "com.examples.sms.DELIVERED";

    @Override
    public void onCreate(Bundle savedInstanceState) {
        super.onCreate(savedInstanceState);

        Button sendButton = new Button(this);
        sendButton.setText("Hail the Mothership");
        sendButton.setOnClickListener(new View.OnClickListener() {
            @Override
            public void onClick(View v) {
                sendSMS("Beam us up!");
            }
        });

        setContentView(sendButton);
    }

    @Override
    protected void onResume() {
      super.onResume();
      //监控操作的状态
      registerReceiver(sent, new IntentFilter(ACTION_SENT));
      registerReceiver(delivered, new IntentFilter(ACTION_DELIVERED));
    }
    @Override
    protected void onPause() {
        super.onPause();
        //确保在位于后台时接收器不会激活
        unregisterReceiver(sent);
        unregisterReceiver(delivered);
```

```java
    private void sendSMS(String message) {
        PendingIntent sIntent = PendingIntent.getBroadcast(this, 0,
            new Intent(ACTION_SENT), 0);
        PendingIntent dIntent = PendingIntent.getBroadcast(this, 0,
            new Intent(ACTION_DELIVERED), 0);

        //发送短信
        SmsManager manager = SmsManager.getDefault();
        manager.sendTextMessage(SHORTCODE, null, message, sIntent, dIntent);
    }

    private BroadcastReceiver sent = new BroadcastReceiver(){
        @Override
        public void onReceive(Context context, Intent intent) {
            switch(getResultCode()) {
            case Activity.RESULT_OK:
                //发送成功
                break;
            case SmsManager.RESULT_ERROR_GENERIC_FAILURE:
            case SmsManager.RESULT_ERROR_NO_SERVICE:
            case SmsManager.RESULT_ERROR_NULL_PDU:
            case SmsManager.RESULT_ERROR_RADIO_OFF:
                //发送失败
                break;
            }

        }
    };

    private BroadcastReceiver delivered = new BroadcastReceiver(){
        @Override
        public void onReceive(Context context, Intent intent) {
            switch(getResultCode()) {
            case Activity.RESULT_OK:
                //传递成功
                break;
            case Activity.RESULT_CANCELED:
                //传递失败
                break;
            }
        }
    };
}
```

要点:
发送短信需要在清单中声明 android.permission.SEND_SMS 权限。

在本例中,当用户单击按钮时就会通过 SMSManager 发送一条短信。由于 SMSManager 是一项系统服务,必须调用静态方法 SMSManager.getDefault() 来获得它的一个引用。

sendTextMessage()方法的参数为目标地址(号码)、服务中心地址以及短信内容。要想让 SMSManger 使用系统默认的服务中心地址,这个参数就应该设为 null。

这里注册了两个 BroadcastReceiver,用来接收要发出的回调 Intent:一个是发送操作的状态,另一个是传递的状态。这些接收器只有在操作处于待发送状态时才会被注册,在处理完 Intent 后就会立即取消注册。

3.11 蓝牙通信

3.11.1 问题

在应用程序中,要通过蓝牙通信实现不同设备之间的数据传输。

3.11.2 解决方案

(API Level 5)

可以使用 API Level 5 中引入的蓝牙 API,在射频通信(RFCOMM)协议接口上创建一个点对点的连接。蓝牙是一种非常流行的无线电技术,几乎现在所有的移动设备都支持该技术。很多用户认为蓝牙只能用来连接移动设备与无线耳机或者与车载音响系统整合。但实际上,对于开发人员来说,在应用程序中蓝牙还是一种用来创建点对点连接的简单而高效的方式。

3.11.3 实现机制

要点:

Android 模拟器现在还不支持蓝牙,因此要想执行本例中的代码,必须把它们运行在一台 Android 设备上。此外,要很好地测试这个功能,需要在两台设备上同时运行这个应用程序。

蓝牙点对点

代码清单 3-31 到代码清单 3-33 演示了使用蓝牙查找附近的其他用户并快速交换联系信息(本例中,只是交换电子邮件地址)。蓝牙上的连接是通过发现可用的"服务",并通过全局唯一 128 位 UUID 值连接到相应的服务。也就是说,在连接某个服务之前,必须首先发现或知道它的 UUID。

在本例中,连接两端的设备运行的是相同的应用程序,两个应用都会有对应的 UUID,因此我们可以在代码中自由地将 UUID 定义为常数。

注意:

为了确保选择的 UUID 是唯一的,请使用网络上免费的 UUID 生成器,或者使用相应的工具,例如 Mac/Linux 上的 uuidgen。

代码清单 3-31　AndroidManifest.xml

```xml
<?xml version="1.0" encoding="utf-8"?>
<manifest
    xmlns:android="http://schemas.android.com/apk/res/android"
    package="com.examples.bluetooth">

    <uses-permission android:name="android.permission.BLUETOOTH"/>
    <uses-permission
        android:name="android.permission.BLUETOOTH_ADMIN"/>

    <application android:icon="@drawable/icon"
        android:label="@string/app_name">
        <activity android:name=".ExchangeActivity"
            android:label="@string/app_name">
            <intent-filter>
                <action
                    android:name="android.intent.action.MAIN" />
                <category
                    android:name="android.intent.category.LAUNCHER" />
            </intent-filter>
        </activity>
    </application>
</manifest>
```

要点：

记住，要想使用这些 API，需要在清单中声明 android.permission.BLUETOOTH 权限。另外，要想改变蓝牙的可发现性以及启用/禁用蓝牙适配器，还要在清单中声明 android.permission.BLUETOOTH_ADMIN 权限。

代码清单 3-32　res/layout/main.xml

```xml
<?xml version="1.0" encoding="utf-8"?>
<RelativeLayout
    xmlns:android="http://schemas.android.com/apk/res/android"
    android:layout_width="match_parent"
    android:layout_height="match_parent">
    <TextView
        android:id="@+id/label"
        android:layout_width="wrap_content"
        android:layout_height="wrap_content"
        android:textAppearance="?android:attr/textAppearanceLarge"
        android:text="Enter Your Email:" />
    <EditText
        android:id="@+id/emailField"
        android:layout_width="match_parent"
        android:layout_height="wrap_content"
        android:layout_below="@id/label"
        android:singleLine="true"
        android:inputType="textEmailAddress" />
```

```xml
<Button
    android:id="@+id/scanButton"
    android:layout_width="match_parent"
    android:layout_height="wrap_content"
    android:layout_alignParentBottom="true"
    android:text="Connect and Share" />
<Button
    android:id="@+id/listenButton"
    android:layout_width="match_parent"
    android:layout_height="wrap_content"
    android:layout_above="@id/scanButton"
    android:text="Listen for Sharers" />
</RelativeLayout>
```

这个示例的 UI 由一个让用户输入电子邮件地址的 EditText 和两个用于初始化通信的按钮组成。名为"Listen for Sharers"的按钮用来将设备设为监听模式。在这种模式下,设备会接受其他设备的连接,并尝试与之进行通信。名为"Connect and Share"的按钮用来将设备设为搜索模式。在这种模式下,设备会搜索当前处于监听模式的设备,并与之进行连接(见代码清单 3-33)。

代码清单 3-33　蓝牙交换 Activity

```java
public class ExchangeActivity extends Activity {

    //本应用程序唯一的UUID(从网上生成)
    private static final UUID MY_UUID =
        UUID.fromString("321cb8fa-9066-3f58-935e-ef55d1ae06ec");
    //发现时用于匹配的一个更加友好的名称
    private static final String SEARCH_NAME = "bluetooth.recipe";

    BluetoothAdapter mBtAdapter;
    BluetoothSocket mBtSocket;
    Button listenButton, scanButton;
    EditText emailField;

    @Override
    public void onCreate(Bundle savedInstanceState) {
        super.onCreate(savedInstanceState);
        requestWindowFeature(Window.FEATURE_INDETERMINATE_PROGRESS);
        setContentView(R.layout.main);

        //检查系统状态
        mBtAdapter = BluetoothAdapter.getDefaultAdapter();
        if(mBtAdapter == null) {
            Toast.makeText(this, "Bluetooth is not supported.",
                Toast.LENGTH_SHORT).show();
            finish();
            return;
        }
```

```java
        if(!mBtAdapter.isEnabled()) {
            Intent enableIntent =
                    new Intent(BluetoothAdapter.ACTION_REQUEST_ENABLE);
            startActivityForResult(enableIntent, REQUEST_ENABLE);
        }

        emailField = (EditText)findViewById(R.id.emailField);
        listenButton = (Button)findViewById(R.id.listenButton);
        listenButton.setOnClickListener(new View.OnClickListener() {
            @Override
            public void onClick(View v) {
                //首先要确保设备是可以发现的
                if(mBtAdapter.getScanMode() !=
                        BluetoothAdapter.SCAN_MODE_CONNECTABLE_DISCOVERABLE) {
                    Intent discoverableIntent =
                            new Intent(
                                BluetoothAdapter.ACTION_REQUEST_DISCOVERABLE);
                    discoverableIntent.putExtra(BluetoothAdapter.
                            EXTRA_DISCOVERABLE_DURATION, 300);
                    startActivityForResult(discoverableIntent,
                            REQUEST_DISCOVERABLE);
                    return;
                }
                startListening();
            }
        });
        scanButton = (Button)findViewById(R.id.scanButton);
        scanButton.setOnClickListener(new View.OnClickListener() {
            @Override
            public void onClick(View v) {
                mBtAdapter.startDiscovery();
                setProgressBarIndeterminateVisibility(true);
            }
        });
    }

    @Override
    public void onResume() {
        super.onResume();
        //为Activity注册广播Intent
        IntentFilter filter = new IntentFilter(BluetoothDevice.ACTION_FOUND);
        registerReceiver(mReceiver, filter);
        filter = new IntentFilter(BluetoothAdapter.ACTION_DISCOVERY_FINISHED);
        registerReceiver(mReceiver, filter);
    }

    @Override
    public void onPause() {
        super.onPause();
```

```java
        unregisterReceiver(mReceiver);
    }

    @Override
    public void onDestroy() {
        super.onDestroy();
        try {
            if(mBtSocket != null) {
                mBtSocket.close();
            }
        } catch(IOException e) {
            e.printStackTrace();
        }
    }

    private static final int REQUEST_ENABLE = 1;
    private static final int REQUEST_DISCOVERABLE = 2;

    @Override
    protected void onActivityResult(int requestCode, int resultCode,
            Intent data) {
        switch(requestCode) {
        case REQUEST_ENABLE:
            if(resultCode != Activity.RESULT_OK) {
                Toast.makeText(this, "Bluetooth Not Enabled.",
                    Toast.LENGTH_SHORT).show();
                finish();
            }
            break;
        case REQUEST_DISCOVERABLE:
            if(resultCode == Activity.RESULT_CANCELED) {
                Toast.makeText(this, "Must be discoverable.",
                    Toast.LENGTH_SHORT).show();
            } else {
                startListening();
            }
            break;
        default:
            break;
        }
    }

    //启动服务器套接字并监听
    private void startListening() {
        AcceptTask task = new AcceptTask();
        task.execute(MY_UUID);
        setProgressBarIndeterminateVisibility(true);
    }

    //AsyncTask 接受传入的连接
```

```java
private class AcceptTask extends AsyncTask<UUID,Void,BluetoothSocket> {

    @Override
    protected BluetoothSocket doInBackground(UUID... params) {
        String name = mBtAdapter.getName();
        try {
            //监听时，将发现名设置为指定的值
            mBtAdapter.setName(SEARCH_NAME);
            BluetoothServerSocket socket =
                    mBtAdapter.listenUsingRfcommWithServiceRecord(
                        "BluetoothRecipe", params[0]);
            BluetoothSocket connected = socket.accept();
            //复位蓝牙适配器名称
            mBtAdapter.setName(name);
            return connected;
        } catch(IOException e) {
            e.printStackTrace();
            mBtAdapter.setName(name);
            return null;
        }
    }

    @Override
    protected void onPostExecute(BluetoothSocket socket) {
        if(socket == null) {
            return;
        }
        mBtSocket = socket;
        ConnectedTask task = new ConnectedTask();
        task.execute(mBtSocket);
    }

}

//AsyncTask 接收一行数据并发送
private class ConnectedTask extends
        AsyncTask<BluetoothSocket,Void,String> {

    @Override
    protected String doInBackground(BluetoothSocket... params) {
        InputStream in = null;
        OutputStream out = null;
        try {
            //发送数据
            out = params[0].getOutputStream();
            String email = emailField.getText().toString()
            out.write(email.getBytes());
            //接收其他数据
            in = params[0].getInputStream();
            byte[] buffer = new byte[1024];
```

```java
            in.read(buffer);
            //从结果创建一个空字符串
            String result = new String(buffer);
            //关闭连接
            mBtSocket.close();
            return result.trim();
        } catch(Exception exc) {
            return null;
        }
    }

    @Override
    protected void onPostExecute(String result) {
        Toast.makeText(ExchangeActivity.this, result, Toast.LENGTH_SHORT)
                .show();
        setProgressBarIndeterminateVisibility(false);
    }
}

//用来监听发现设备的 BroadcastReceiver
private BroadcastReceiver mReceiver = new BroadcastReceiver() {
    @Override
    public void onReceive(Context context, Intent intent) {
        String action = intent.getAction();

        //当找到一台设备时
        if(BluetoothDevice.ACTION_FOUND.equals(action)) {
            //从 Intent 中获得 BluetoothDevice 对象
            BluetoothDevice device =
                intent.getParcelableExtra(BluetoothDevice.EXTRA_DEVICE);
            if(TextUtils.equals(device.getName(), SEARCH_NAME)) {
                //匹配找到的设备，连接
                mBtAdapter.cancelDiscovery();
                try {
                    mBtSocket =
                        device.createRfcommSocketToServiceRecord(MY_UUID);
                    mBtSocket.connect();
                    ConnectedTask task = new ConnectedTask();
                    task.execute(mBtSocket);
                } catch(IOException e) {
                    e.printStackTrace();
                }
            }
        //发现完成
        } else if(BluetoothAdapter.ACTION_DISCOVERY_FINISHED
                .equals(action)) {
            setProgressBarIndeterminateVisibility(false);
        }
```

```
            }
        };
    }
```

应用程序初次启动后,会首先对设备上蓝牙的状态做一些基本的检查。如果 BluetoothAdapter.getDefaultAdapter()返回 null,表明设备不支持蓝牙,应用程序无法继续使用。即使设备上有蓝牙,它也必须是启用的,这样应用程序才能使用它。如果蓝牙是禁用的,推荐启用适配器的方法是向系统发送一个 action 值为 BluetoothAdapter.ACTION_REQUEST_ENABLE 的 Intent。这样就会通知用户启用蓝牙。可以用 enable()方法手动启用 BluetoothAdapter,但我们不推荐这种方法,除非需要通过某种特别的方式来获得用户的权限。

验证过蓝牙可用后,应用程序会等待用户输入。正如之前提到的,这个示例可以将每台设备设为两种模式之一:监听模式或搜索模式。接下来看看这两种模式各自的工作方式。

监听模式

单击"Listen for Sharers"按钮开始让应用程序对接入的连接进行监听。对于一台设备来说,如果想要接收未知设备的接入连接,那么该设备必须设置为可被发现的。应用程序会检查适配器的扫描模式是否等于 SCAN_MODE_CONNECTABLE_DISCOVERABLE,从而确定设备是否是可被发现的。如果适配器不满足这个要求,就会给系统发送一个 Intent,告诉用户应该让设备处于可发现状态,这和要求用户启用蓝牙的方法很相似。如果用户接受了该请求,Activity 就会返回用户允许设备处于可发现状态的时长;如果用户拒绝了该请求,Activity 会返回 Activity.RESULT_CANCELED。这个示例会在 onActivityResult()中处理用户拒绝请求的行为,即在这些条件下终止应用。

如果用户启用了可发现状态或者设备已经处于可发现状态,就会创建并执行一个 AcceptTask。这个任务会为我们所定义服务的 UUID 创建监听器端口,通过这个端口阻塞主调线程并等待接入连接请求。在收到有效的请求时,就会接受。然后应用程序会切换到 Connected Mode(已连接模式)。

在设备处于监听状态的过程中,其蓝牙的名称会被设定为已知的唯一值(SEARCH_NAME)来加速发现的过程(具体原因见后面的"搜索模式"小节)。连接建立后,适配器就会恢复到默认名称。

搜索模式

单击"Connect and Share"按钮,通知应用程序开始搜索另一台想要连接的设备。这当中会首先启动蓝牙发现过程并且在 BroadcastReceiver 中处理搜索结果。通过 Bluetooth-Adapter.startDiscovery()开始一次发现后,以下两种情况下,Android 会通过广播进行异步回调:找到了另一台设备或发现过程完成。

私有的接收器 mReceiver 在 Activity 对用户可见时会随时进行注册,对于每一台新发现的设备,它都会收到一个广播。回忆一下,在讨论监听模式时,监听设备的名称被设置成唯一的值。在每次发现完成后,接收器会检测与当前名称匹配的设备,并且在搜索到结果后会尝试进行连接。这对于发现过程的速度非常重要,因为验证一台设备是否可用的唯一途径就是将这台设备与一个特殊的服务 UUID 进行尝试性连接,以查看操作是否成功。

蓝牙连接过程属于重量级操作，并且很慢，应该在确保其他一切运行良好时才进行这个操作。

这种匹配设备的方式同样减轻了用户手动连接其想要连接的设备的过程。应用程序会智能地寻找到同样运行这个应用程序并且处于监听模式的另一台设备来完成传输。移除用户也意味着这个值是唯一且非常少见的，就是为了避免查找其他设备时，这些设备可能意外具有相同的名称。

找到匹配的设备后，就会停止发现过程(因为它同样是重量级操作，并且会减缓连接过程)，然后连接到服务的 UUID。在连接成功后，应用程序就进入了已连接模式。

提示：
可以在许多地方生成自己的唯一 ID(UUID)值。各种网站，如 http://www.uuidgenerator.net/，将自动创建一个 UUID。Mac OS X 和 Linux 用户还可以从命令行运行 uuidgen 命令。

已连接模式

一旦连接成功，两台设备上的应用程序将创建一个 ConnectedTask 来发送和接收用户联系信息。连接的 BluetoothSocket 会用一个 InputStream 和一个 OutputStream 进行数据传输。首先，电子邮件的文本字段的当前值在封装后被写入 OutputStream。然后，从 InputStream 读取以接收远程设备的信息。最后，每台设备都需要将它接收的原始数据封装成一个单纯的字符串显示给用户。

ConnectedTask.onPostExecute()方法的任务是向用户显示交流的结果；目前，是将接收的内容显示在一个 Toast 中。交流完成后，连接被关闭，两台设备都会进入相同的模式，准备进行下一次交流。

有关此主题的更多信息，可以查看 Android SDK 提供的 BluetoothChat 示例应用程序。这个应用程序很好地演示了如何使用一个长连接在设备之间发送聊天消息。

Android 之外的蓝牙

正如本节开始时描述的那样，除了手机和平板电脑，许多无线设备上也有蓝牙。在诸如蓝牙调制解调器和串行适配器这样的设备上同样有 RFCOMM 接口。在 Android 设备上创建点对点连接时使用的 API 同样可以用来连接其他嵌入式蓝牙设备，从而实现对设备的监控和控制。

要想与这些嵌入式设备建立连接，关键是要获得它们所支持的 RFCOMM 服务的 UUID。作为配置文件标准一部分的蓝牙服务及其标识符由蓝牙特别兴趣小组(Special Interest Group，SIG)定义；因此，我们能够从 www.bluetooth.org 提供的文档中获得给定设备所需的 UUID。然而，如果设备制造商为自定义服务类型定义了设备特有的 UUID，并且没有归入文档，则必须通过某种方式来发现该 UUID。与前面的示例一样，我们可以使用适当的 UUID 创建一个 BluetoothSocket 和传输数据。

SDK 就拥有这种能力，虽然在 Android 4.0.3 (API Level 15)之前它并不是 SDK 的开放部分。对于蓝牙设备来说，有两个方法能够提供此信息：fetchUuidsWithSdp()和 getUuids()。后者只会简单地返回在发现期间找到的设备的缓存实例，而前者则会异步连接设备并进行一个新的查询。正因为如此，在使用 fetchUuidsWithSdp()时，必须注册一个 BroadcastReceiver，它将接收 action 值为 BluetoothDevice.ACTION_UUID 的 Intent 并发现 UUID 值。

3.12 查询网络连接状态

3.12.1 问题

应用程序需要监控网络连接状态的变化。

3.12.2 解决方案

(API Level 1)

通过 ConnectivityManager 监控设备的网络连接状态。在移动应用程序的设计过程中，需要考虑的一个很重要的问题就是网络并不是随时都是连通的。随着人的移动，网络的速度和容量都在不断变化。正因为如此，使用网络资源的应用程序需要随时监测这些资源是否可访问，并在不能访问时通知用户。

除了连通性，ConnectivityManager 还能向应用程序提供网络连接的类型。这样就能根据情况决定是否要下载大文件，如果用户处于漫游状态的话，下载大文件会使数据流量暴增。

3.12.3 实现机制

代码清单 3-34 封装了一个方法，可以把它放到你的代码中来检查网络的连通性。

代码清单 3-34 ConnectivityManager 封装方法

```
public static boolean isNetworkReachable() {
    final ConnectivityManager mManager =
            (ConnectivityManager)context.getSystemService(
                    Context.CONNECTIVITY_SERVICE);
    NetworkInfo current = mManager.getActiveNetworkInfo();
    if(current == null) {
       return false;
    }
    return (current.getState() == NetworkInfo.State.CONNECTED);
}
```

ConnectivityManager 所做的工作是评估哪个网络数据接口被认为是活跃的(Wi-Fi 或蜂窝网络)。在最简单的情况下，我们仅检查给定接口是否已连接。注意：如果没有可用的活跃数据连接，ConnectivityManager.getActiveNetworkInfo()会返回 null，所以首先要检查这种情况。如果有可用的网络，就可以检查其状态，可能的返回值如下：

- DISCONNECTED
- CONNECTING

- CONNECTED
- DISCONNECTING

如果返回的状态是 CONNECTED，网络就是稳定的，可以用来访问远程资源。

1. 验证路线

移动设备有多条连接路线(Wi-Fi、3G/4G 等)，并且设备经常连接到没有外部 Web 路线的网络；Wi-Fi 网络尤其如此。ConnectivityManager 仅仅通知用户其设备是否已关联特定的网络，但不会表明该网络是否能够访问外部 IP 地址。因此，当设备尝试通过已"连接"但没有有效路线的网络进行连接时，网络栈就会花费几分钟的时间超时并适当失败。

在某些情况下，更加智能的方法是检查有效的 Internet 连接，而非仅检查与网络的关联。代码清单 3-35 以前面的连通性检查为基础，用于检查 Internet 连接。

代码清单 3-35　更加智能的 ConnectivityManager 封装方法

```java
public static boolean hasNetworkConnection(Context context) {
    final ConnectivityManager connectivityManager =
            (ConnectivityManager) context.getSystemService(
                    Context.CONNECTIVITY_SERVICE);
    final NetworkInfo activeNetworkInfo =
            connectivityManager.getActiveNetworkInfo();

    // 如果没有与网络关联，则在此处进行关联
    boolean connected = (null != activeNetworkInfo)
            && activeNetworkInfo.isConnected();
    if(!connected) return false;

    // 检查是否可以访问远程服务器
    boolean routeExists;
    try {
        // 检查 Google 公共 DNS
        InetAddress host = InetAddress.getByName("8.8.8.8");

        Socket s = new Socket();
        s.connect(new InetSocketAddress(host, 53), 5000);
        // 如果没有抛出异常，则 DNS 存在
        routeExists = true;
        s.close();
    } catch(IOException e) {
        routeExists = false;
    }

    return (connected && routeExists);
}
```

如往常一样验证相同的连通性条件之后，代码清单 3-35 进一步尝试打开套接字(具有 5 秒的超时)，该套接字指向 Google 公共 DNS 的众所周知的标准 IPv4 地址(8.8.8.8)。如果成功连接此主机，我们就能够确信设备可以访问任何激活的 Internet 资源。相比于直接全面连接远程服务器，此方法的优势是代码将更快速地失败，强制最多 5 秒的延迟，然后就立刻告诉用户，他们实际上并没有自认为已具备的 Internet 连接。

当网络请求失败时，最好检查网络的连通性，告知用户请求失败是网络问题。代码清单 3-36 中的示例演示了如何在网络访问失败时检查网络的连通性。

代码清单 3-36　告知用户网络不通

```
try {
    //尝试访问网络资源时，如果失败，可能会抛出 HttpResponseException 或
    //其他 IOException 异常
} catch(Exception e) {
    if( !isNetworkReachable() ) {
        AlertDialog.Builder builder = new AlertDialog.Builder(context);
        builder.setTitle("No Network Connection");
        buil der.setMessage("The Network is unavailable."
                + " Please try your request again later.");
        builder.setPositiveButton("OK",null);
        builder.create().show();
    }
}
```

2. 判断连接类型

在某些情况下，应用程序还必须知道用户所连接的网络是否是按流量收费的，可以调用活动网络连接的 NetworkInfo.getType()方法以获取相关信息(参见代码清单 3-37)。

代码清单 3-37　ConnectivityManager 带宽检查

```
public boolean isWifiReachable() {
    ConnectivityManager mManager =
            (ConnectivityManager)context.getSystemService(
                    Context.CONNECTIVITY_SERVICE);
    NetworkInfo current = mManager.getActiveNetworkInfo();
    if(current == null) {
        return false;
    }
    return (current.getType() == ConnectivityManager.TYPE_WIFI);
}
```

这个修改后的连通性检查版本能够判断用户是否连接到了 Wi-Fi 网络，如果连接的是 Wi-Fi 网络，通常就意味着网速比较快，而且流量是免费的。

3.13 使用 NFC 传输数据

3.13.1 问题

你有一个应用程序，需要通过最少的设置实现两台 Android 设备间小数据包的快速传输。

3.13.2 解决方案

(API Level 16)

使用 NFC(Near Field Communication，近场通信) Beam API。NFC 通信起初是在 Android 2.3 中加入到 SDK 中的，在 Android 4.0 中做了扩展，包括通过一个名为 Android Beam 的进程实现设备间短消息的无障碍传输。在 Android 4.1 中，又对 Beam API 做了完善，使之在两台设备间的数据传输方面更加成熟。

Android 4.1 中在此方面一个比较大的补充就是可以通过一些可选的连接实现大数据的传输。在发现设备和建立初始连接方面，NFC 表现非常优秀，但它的带宽较窄，对于发送像全彩色图片这样的大数据包效率不是很高。以前，开发人员可以使用 NFC 在两台设备间建立连接，但是在实际传输文件数据时需要手动选择第二种连接方式，如 Wi-Fi 直连或蓝牙。在 Android 4.1 中，框架层处理了整个过程，任何应用程序只需要调用一个 API 就可以通过可用的连接完成大文件的分享。

3.13.3 实现机制

根据想要推送的内容的大小，有两种机制可以用来在两台设备间传送数据。

1. 使用前台推送进行 Beam

如果要使用 NFC 在设备间发送简单的内容，可以使用前台推送机制来创建一个 NfcMessage，它包含了一个或多个 NfcRecord 实例。代码清单 3-38 和 3-39 演示了如何创建一个简单的 NfcMessage 并推送到另一台设备上。

代码清单 3-38　AndroidManifest.xml

```xml
<manifest
    xmlns:android="http://schemas.android.com/apk/res/android"
    package="com.examples.nfcbeam">

    <uses-permission android:name="android.permission.NFC" />
    <application
        android:icon="@drawable/ic_launcher"
        android:label="NfcBeam">
        <activity
            android:name=".NfcActivity"
```

```xml
        android:label="NfcActivity"
        android:launchMode="singleTop">
        <intent-filter>
            <action android:name="android.intent.action.MAIN" />
            <category android:name="android.intent.category.LAUNCHER" />
        </intent-filter>
        <intent-filter>
            <action android:name="android.nfc.action.NDEF_DISCOVERED" />
            <category android:name="android.intent.category.DEFAULT" />
            <data android:mimeType=
                "application/com.example.androidrecipes.beamtext"/>
        </intent-filter>
    </activity>
  </application>
</manifest>
```

首先需要注意的是，在使用 NFC 服务时需要 android.permission.NFC 权限。另外，我们的 Activity 中添加了一个自定义的<intent-filter>。这样 Android 就可以知道哪个应用程序应该启动以响应它所收到的内容。

代码清单 3-39　生成一个 NFC 前台推送的 Activity

```java
public class NfcActivity extends Activity implements
        CreateNdefMessageCallback, OnNdefPushCompleteCallback {
    private static final String TAG = "NfcBeam";
    private NfcAdapter mNfcAdapter;
    private TextView mDisplay;

    @Override
    public void onCreate(Bundle savedInstanceState) {
        super.onCreate(savedInstanceState);
        mDisplay = new TextView(this);
        setContentView(mDisplay);

        //检查 NFC 适配器是否可用
        mNfcAdapter = NfcAdapter.getDefaultAdapter(this);
        if(mNfcAdapter == null) {
            mDisplay.setText("NFC is not available on this device.");
        } else {
            //注册回调来设置 NDEF 消息。这样做可以使 Activity 处于前台时，
            //NFC 数据推送处于激活状态
            mNfcAdapter.setNdefPushMessageCallback(this, this);
            //注册回调来监听消息发送成功
            mNfcAdapter.setOnNdefPushCompleteCallback(this, this);
        }
    }

    @Override
    public void onResume() {
        super.onResume();
```

```java
        //检查是否是一个 Beam 启动了这个 Activity
        if(NfcAdapter.ACTION_NDEF_DISCOVERED
                .equals(getIntent().getAction())) {
            processIntent(getIntent());
        }
    }

    @Override
    public void onNewIntent(Intent intent) {
        //在这之后会调用 onResume 来处理这个 Intent
        setIntent(intent);
    }

    void processIntent(Intent intent) {
        Parcelable[] rawMsgs =
                intent.getParcelableArrayExtra(NfcAdapter.EXTRA_NDEF_MESSAGES);
        //Beam 期间只发送了一条消息
        NdefMessage msg = (NdefMessage) rawMsgs[0];
        //记录 0 包含了 MIME 类型
        mDisplay.setText(new String(msg.getRecords()[0].getPayload()));
    }

    @Override
    public NdefMessage createNdefMessage(NfcEvent event) {
        String text =
                String.format("Sending A Message From Android Recipes at %s",
                    DateFormat.getTimeFormat(this).format(new Date()));
        NdefMessage msg = new NdefMessage(NdefRecord.createMime(
                "application/com.example.androidrecipes.beamtext",
                text.getBytes()) );
        return msg;
    }

    @Override
    public void onNdefPushComplete(NfcEvent event) {
        //这个回调是在一个绑定线程上执行的,不要在这个方法中直接更新 UI
        Log.i(TAG, "Message Sent!");
    }
}
```

这个示例应用程序针对的是 NFC 推送的发送和接收,因此两台设备上都应该安装相同的应用程序:一台负责发送数据,另一台负责接收数据。Activity 通过 NfcAdapter 的 setNdefPushMessageCallback()方法将自己注册为可进行前台推送。这次调用中同时做了两件事情。在传输开始时,它告诉 NFC 服务调用这个 Activity 来接收它需要发送的信息,同时在 Activity 处于前台时,会激活 NFC 推送。另外,还有一个类似的方法叫做 setNdefPushMessage(),该方法只接收消息,但不会实现回调。

这个回调方法构造了一个 NdefMessage，NdefMessage 只包含一条 NDEF MIME 记录（通过 NdefRecord.createMime()方法创建）。MIME 记录是一种传递应用程序特定数据的简单方式。createMime()方法包含两个参数，一个是用来指定 MIME 类型的字符串，另一个是原始数据的 byte 数组。传递的信息可以是任何数据，如文本字符串或小图片；应用程序负责数据的封装和拆包。注意，这里的 MIME 类型和清单中<intent-filter>定义的类型是一样的。

要想推送执行的话，负责发送设备的 Activity 必须处于前台激活状态，接收的设备也不能是锁屏状态。当用户同时触摸两台设备时，发送设备的屏幕上会显示 Android 的"Touch to Beam"用户界面，这时再点击一下屏幕就会把消息发送到另一台设备上。一旦接收到消息，接收设备上的应用程序就会启动，并且会触发发送设备的 onNdefPushComplete()回调方法。

在负责接收的设备上，会使用 ACTION_NDEF_DISCOVERED 的 Intent 来启动 Activity，因此我们的示例会检查 NdefMessage 的 Intent 并且拆包其中的负载数据，即将 byte 数组转换为字符串。这种使用先匹配 Intent 然后发送 NFC 数据的方式最为灵活，但是有些时候可能需要显式地调用应用程序。这时候就需要 Android Application Record 出马了。

2. Android Application Record

应用程序可以在一个 NdefMessage 中添加一个额外的 NdefRecord，它可以引导 Android 在接收设备上调用一个指定的包名。要想在之前的示例中实现它，我们只需要像下面这样简单地修改一下 CreateNdefMessageCallback 方法：

```
@Override
public NdefMessage createNdefMessage(NfcEvent event) {
    String text = String.format(
        "Sending A Message From Android Recipes at %s",
        DateFormat.getTimeFormat(this)
            .format(new Date()) );
    NdefMessage msg = new NdefMessage(NdefRecord.createMime(
        "application/com.example.androidrecipes.beamtext",
        text.getBytes()),
        NdefRecord
            .createApplicationRecord("com.examples.nfcbeam"));
    return msg;
}
```

加了 NdefRecord.createApplicationRecord()这个额外的参数后，现在可以保证推送消息就只会启动我们的 com.examples.nfcbeam 包。消息的第一条记录还是原来的文本消息，所以我们对接收的消息的拆包过程不用变。

3. Beam 较大的数据

在本节开头，我们提到了最好不要使用 NFC 发送大内容块。尽管如此，Android Beam 还是有能力处理大内容块的。代码清单 3-40 到 3-42 演示了使用 Beam 来发送大型图片文件。

代码清单 3-40　AndroidManifest.xml

```xml
<manifest
    xmlns:android="http://schemas.android.com/apk/res/android"
    package="com.examples.nfcbeam">

    <uses-permission android:name="android.permission.NFC" />
    <application
        android:icon="@drawable/ic_launcher"
        android:label="NfcBeam">
        <activity
            android:name=".BeamActivity"
            android:label="BeamActivity"
            android:launchMode="singleTop">
            <intent-filter>
                <action android:name="android.intent.action.MAIN" />
                <category
                    android:name="android.intent.category.LAUNCHER" />
            </intent-filter>
            <intent-filter>
                <action android:name="android.intent.action.VIEW" />
                <data android:mimeType="image/*" />
            </intent-filter>
        </activity>
    </application>

</manifest>
```

代码清单 3-41　res/layout/main.xml

```xml
<LinearLayout xmlns:android="http://schemas.android.com/apk/res/android"
    android:layout_width="match_parent"
    android:layout_height="match_parent"
    android:orientation="vertical" >
    <Button
        android:layout_width="match_parent"
        android:layout_height="wrap_content"
        android:text="Select Image"
        android:onClick="onSelectClick" />
    <TextView
        android:id="@+id/text_uri"
        android:layout_width="match_parent"
        android:layout_height="wrap_content" />
    <ImageView
        android:id="@+id/image_preview"
        android:layout_width="match_parent"
        android:layout_height="match_parent"
        android:scaleType="center" />
</LinearLayout>
```

代码清单 3-42　传送一张大图片的 Activity

```java
public class BeamActivity extends Activity implements
        CreateBeamUrisCallback, OnNdefPushCompleteCallback {
    private static final String TAG = "NfcBeam";
    private static final int PICK_IMAGE = 100;

    private NfcAdapter mNfcAdapter;
    private Uri mSelectedImage;

    private TextView mUriName;
    private ImageView mPreviewImage;

    @Override
    protected void onCreate(Bundle savedInstanceState) {
        super.onCreate(savedInstanceState);
        setContentView(R.layout.main);

        mUriName = (TextView) findViewById(R.id.text_uri);
        mPreviewImage = (ImageView) findViewById(R.id.image_preview);

        //检查 NFC 适配器是否可用
        mNfcAdapter = NfcAdapter.getDefaultAdapter(this);
        if(mNfcAdapter == null) {
            mUriName.setText("NFC is not available on this device.");
        } else {
            //注册回调来设置 NDEF 消息
            mNfcAdapter.setBeamPushUrisCallback(this, this);
            //注册回调来监听消息发送成功
            mNfcAdapter.setOnNdefPushCompleteCallback(this, this);
        }
    }

    @Override
    protected void onActivityResult(int requestCode, int resultCode,
            Intent data) {
        if(requestCode == PICK_IMAGE && resultCode == RESULT_OK
                && data != null) {
            mUriName.setText( data.getData().toString() );
            mSelectedImage = data.getData();
        }
    }

    @Override
    public void onResume() {
        super.onResume();
        //检查 Activity 是否由于 Android Beam 而启动
```

```java
        if(Intent.ACTION_VIEW.equals(getIntent().getAction())) {
            processIntent(getIntent());
        }
    }

    @Override
    public void onNewIntent(Intent intent) {
        //在这之后会调用 onResume 来处理这个 Intent
        setIntent(intent);
    }

    void processIntent(Intent intent) {
        Uri data = intent.getData();
        if(data != null) {
            mPreviewImage.setImageURI(data);
        } else {
            mUriName.setText("Received Invalid Image Uri");
        }
    }

    public void onSelectClick(View v) {
        Intent intent = new Intent(Intent.ACTION_GET_CONTENT);
        intent.setType("image/*");
        startActivityForResult(intent, PICK_IMAGE);
    }

    @Override
    public Uri[] createBeamUris(NfcEvent event) {
        if(mSelectedImage == null) {
            return null;
        }
        return new Uri[] {mSelectedImage};
    }

    @Override
    public void onNdefPushComplete(NfcEvent event) {
        //这个回调是在一个绑定线程上执行的，不要在这个方法中直接更新 UI。
        //这里最好告诉用户不需要再把手机放在一起了
        Log.i(TAG, "Push Complete!");
    }
}
```

这个示例使用了 CreateBeamUrisCallback，它允许应用程序构造一个 Uri 实例的数组，这些 Uri 指向你要发送的内容。Android 首先会通过 NFC 建立一个初始连接，然后再寻找一种合适的连接方式(如蓝牙或 Wi-Fi)来完成大文件的传输。

在本例中，接收设备上的数据是通过系统标准的 Intent.ACTION_VIEW action 启动的，因此没有必要在两台设备上都加载应用程序。尽管如此，我们的应用程序还是对 ACTION_VIEW 进行了过滤，这样的话如果接收设备愿意，可以使用它来浏览接收的图片。

这里会要求用户从他的设备上选择一张图片来传送，一旦选定后，图片的 Uri 会显示出来。一旦用户点击设备到另一台设备，屏幕同样会显示"Touch to Beam"用户界面(参见图 3-4)，当再次点击屏幕时传输就开始了。

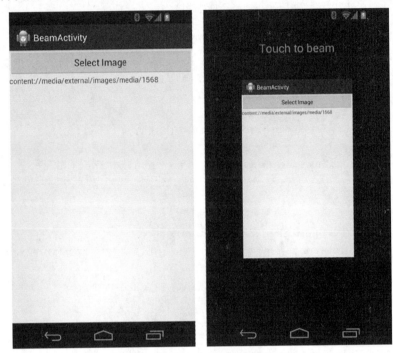

图 3-4 触摸 Activity 来激活 Beam

在传输过程中有关 NFC 的部分完成后，就会在发送设备上调用 onNdefPushComplete() 方法。这时，传输过程就转移到其他类型的连接上，因此用户也就不需要再把手机放在一起了。

在传输文件时，接收设备会在系统窗口的顶部显示进度通知，当传输完成时，用户可以点击该通知来浏览内容。如果选择我们的应用程序作为内容浏览器，图片就会显示在应用程序的 ImageView 中。为你的应用程序注册这种通用的 Intent 可能有一个缺点，就是设备上所有的应用程序都可以用你的应用程序浏览图片，所以定义过滤器时要谨慎。

3.14 USB 连接

3.14.1 问题

应用程序需要与 USB 设备进行通信来控制或传输数据。

3.14.2 解决方案

(API Level 12)

对于拥有 USB 主机电路的设备，Android 已经内置了对它的支持，可以与已经连接的 USB 设备进行模拟和通信。USBManager 是一项系统服务，可以让应用程序访问任何通过 USB 连接的外部设备，接下来我们将看一下在应用程序中如何使用这个服务来建立连接。

设备上的 USB 主机电路已经越来越普及，但还是很稀少。刚开始，只有平板电脑设备拥有这种能力，但随着科技的快速发展，在商用 Android 手机上它也有可能很快成为一个通用的接口。正因为如此，无疑需要在应用程序的清单中包含以下元素：

```
<uses-feature android:name="android.hardware.usb.host" />
```

这样只有真正拥有相应硬件的设备，才可以使用你的应用程序。

Android 提供的 API 和 USB 规范几乎一样，并没有更多更深入的知识。这就意味着如果想要使用这些 API，你至少需要了解一些 USB 的基础知识以及设备间是如何通信的。

USB 概述

在查看 Android 是如何与 USB 设备进行交互的示例之前，让我们花点时间定义一些 USB 术语。

- 端点：USB 设备的最小构件。应用程序最终就是通过连接这些端点发送和接收数据的。端点主要分为 4 种类型：
  - 控制传输：用于配置和状态命令。每台设备至少有一个控制端点，即"端点 0"，它不会关联任何接口。
  - 中断传输：用于小量的、高优先级的控制命令。
  - 批量传输：用于传输大数据。通常都是双向成对出现的(1 IN 和 1 OUT)。
  - 同步传输：用于实时数据传输，如音频。撰写本书时，最新的 Android SDK 还不支持这个功能。
- 接口：端点的集合，用来表示一台"逻辑"设备。
  - 多台物理 USB 设备对于主机来说可以呈现为多台逻辑设备，即通过暴露多个接口来标识。
- 配置：一个或多个接口的集合。USB 协议强制规定一台设备在某一特定时间只能有一个配置是激活的。事实上，多数设备也就只有一个配置，并把它作为设备的操作模式。

3.14.3 实现机制

代码清单 3-43 和 3-44 演示了使用 UsbManager 来检查通过 USB 连接的设备以及使用控制传输来进一步查询配置的示例。

代码清单 3-43   res/layout/main.xml

```xml
<LinearLayout
    xmlns:android="http://schemas.android.com/apk/res/android"
    android:layout_width="match_parent"
    android:layout_height="match_parent"
    android:orientation="vertical" >
    <Button
        android:id="@+id/button_connect"
        android:layout_width="match_parent"
        android:layout_height="wrap_content"
        android:text="Connect"
        android:onClick="onConnectClick" />
    <TextView
        android:id="@+id/text_status"
        android:layout_width="match_parent"
        android:layout_height="wrap_content" />
    <TextView
        android:id="@+id/text_data"
        android:layout_width="match_parent"
        android:layout_height="wrap_content" />

</LinearLayout>
```

代码清单 3-44   USB 主机上查询设备的 Activity

```java
public class USBActivity extends Activity {
    private static final String TAG = "UsbHost";

    TextView mDeviceText, mDisplayText;
    Button mConnectButton;

    UsbManager mUsbManager;
    UsbDevice mDevice;
    PendingIntent mPermissionIntent;

    @Override
    public void onCreate(Bundle savedInstanceState) {
        super.onCreate(savedInstanceState);
        setContentView(R.layout.main);

        mDeviceText = (TextView) findViewById(R.id.text_status);
        mDisplayText = (TextView) findViewById(R.id.text_data);
        mConnectButton = (Button) findViewById(R.id.button_connect);

        mUsbManager = (UsbManager) getSystemService(Context.USB_SERVICE);
    }

    @Override
    protected void onResume() {
        super.onResume();
```

```java
        mPermissionIntent =
            PendingIntent.getBroadcast(this, 0,
                new Intent(ACTION_USB_PERMISSION), 0);
        IntentFilter filter = new IntentFilter(ACTION_USB_PERMISSION);
        registerReceiver(mUsbReceiver, filter);

        //检查当前连接的设备
        updateDeviceList();
    }

    @Override
    protected void onPause() {
        super.onPause();
        unregisterReceiver(mUsbReceiver);
    }

    public void onConnectClick(View v) {
        if(mDevice == null) {
            return;
        }
        mDisplayText.setText("---");

        //这里如果用户已经授权，就会立即发送ACTION_USB_PERMISSION广播,
        //否则会向用户显示授权对话框
        mUsbManager.requestPermission(mDevice, mPermissionIntent);
    }

    /*
     * 捕捉用户权限响应的接收器，在和已经连接的设备进行真正的交互时是需要这些权限的
     */
    private static final String ACTION_USB_PERMISSION =
            "com.android.recipes.USB_PERMISSION";
    private final BroadcastReceiver mUsbReceiver = new BroadcastReceiver() {
        public void onReceive(Context context, Intent intent) {
            String action = intent.getAction();
            if(ACTION_USB_PERMISSION.equals(action)) {
                UsbDevice device = (UsbDevice) intent.getParcelableExtra(
                        UsbManager.EXTRA_DEVICE);

                if(intent.getBooleanExtra(UsbManager.EXTRA_PERMISSION_GRANTED,
                        false) && device != null) {
                    //查询设备的描述符
                    getDeviceStatus(device);
                } else {
                    Log.d(TAG, "permission denied for device " + device);
                }
            }
        }
    };
```

```java
//类型：表示读取还是写入
//与USB_ENDPOINT_DIR_MASK进行匹配，判断是IN还是OUT
private static final int REQUEST_TYPE = 0x80;
//请求：GET_CONFIGURATION_DESCRIPTOR = 0x06
private static final int REQUEST = 0x06;
//值：描述符类型(高)和索引值(低)
// Configuration Descriptor = 0x2
// Index = 0x0 (第一个配置)
private static final int REQ_VALUE = 0x200;
private static final int REQ_INDEX = 0x00;
private static final int LENGTH = 64;

/*
 * 初始化控制传输来请求设备的第一个配置描述符
 */
private void getDeviceStatus(UsbDevice device) {
    UsbDeviceConnection connection = mUsbManager.openDevice(device);
    //为传入的数据创建一个足够大的缓冲区
    byte[] buffer = new byte[LENGTH];
    connection.controlTransfer(REQUEST_TYPE, REQUEST, REQ_VALUE, REQ_INDEX,
            buffer, LENGTH, 2000);
    //将接收到的数据解析为描述符
    String description = parseConfigDescriptor(buffer);

    mDisplayText.setText(description);
    connection.close();
}

/*
 * 按照USB规范解析USB配置描述符响应信息。返回可打印的连接设备的信息
 */
private static final int DESC_SIZE_CONFIG = 9;
private String parseConfigDescriptor(byte[] buffer) {
    StringBuilder sb = new StringBuilder();
    //解析配置描述符的头信息
    int totalLength = (buffer[3] &0xFF) << 8;
    totalLength += (buffer[2] & 0xFF);
    //接口数量
    int numInterfaces = (buffer[5] & 0xFF);
    //配置的属性
    int attributes = (buffer[7] & 0xFF);
    //电量递增2mA
    int maxPower = (buffer[8] & 0xFF) * 2;

    sb.append("Configuration Descriptor:\n");
    sb.append("Length: " + totalLength + " bytes\n");
    sb.append(numInterfaces + " Interfaces\n");
    sb.append(String.format("Attributes:%s%s%s\n",
            (attributes & 0x80) == 0x80 ? " BusPowered" : "",
            (attributes & 0x30) == 0x30 ? " SelfPowered" : "",
```

```java
                    (attributes & 0x20) == 0x20 ? " RemoteWakeup" : ""));
        sb.append("Max Power: " + maxPower + "mA\n");

        //描述符的剩余部分为接口和端点信息
        int index = DESC_SIZE_CONFIG;
        while(index < totalLength) {
            //读取长度和类型
            int len = (buffer[index] & 0xFF);
            int type = (buffer[index+1] & 0xFF);
            switch(type) {
            case 0x03: //接口描述符
                int intfNumber = (buffer[index+2] & 0xFF);
                int numEndpoints = (buffer[index+3] & 0xFF);
                int intfClass = (buffer[index+5] & 0xFF);

                sb.append(String.format("- Interface %d, %s, %d Endpoints\n",
                        intfNumber, nameForClass(intfClass), numEndpoints));
                break;
            case 0x05: //端点描述符
                int endpointAddr = ((buffer[index+2] & 0xFF));
                //端点号为低 4 位
                int endpointNum = (endpointAddr & 0x0F);
                //方向为高位
                int direction = (endpointAddr & 0x80);

                int endpointAttrs = (buffer[index+3] & 0xFF);
                //类型为低两位
                int endpointType = (endpointAttrs & 0x3);

                sb.append(String.format("--Endpoint %d, %s %s\n",
                        endpointNum,
                        nameForEndpointType(endpointType),
                        nameForDirection(direction) ));
                break;
            }
            //继续下一个描述符
            index += len;
        }

        return sb.toString();
    }
    private void updateDeviceList() {
        HashMap<String, UsbDevice> connectedDevices = mUsbManager
                .getDeviceList();
        if(connectedDevices.isEmpty()) {
            mDevice = null;
            mDeviceText.setText("No Devices Currently Connected");
            mConnectButton.setEnabled(false);
        } else {
            StringBuilder builder = new StringBuilder();
```

```java
        for(UsbDevice device : connectedDevices.values()) {
            //打开最后一台(如果有多台的话)检测到的设备
            mDevice = device;
            builder.append(readDevice(device));
            builder.append("\n\n");
        }
        mDeviceText.setText(builder.toString());
        mConnectButton.setEnabled(true);
    }
}

/*
 * 遍历所有已经连接的设备的端点和接口
 * 这里不涉及权限,在尝试连接真实设备之前这些都是"公开可用"的
 */
private String readDevice(UsbDevice device) {
    StringBuilder sb = new StringBuilder();
    sb.append("Device Name: " + device.getDeviceName() + "\n");
    sb.append(String.format(
            "Device Class: %s -> Subclass: 0x%02x -> Protocol: 0x%02x\n",
            nameForClass(device.getDeviceClass()),
            device.getDeviceSubclass(), device.getDeviceProtocol()));

    for(int i = 0; i < device.getInterfaceCount(); i++) {
        UsbInterface intf = device.getInterface(i);
        sb.append(String.format("+--Interface %d Class: %s ->"
                + "Subclass: 0x%02x -> Protocol: 0x%02x\n",
                intf.getId(),
                nameForClass(intf.getInterfaceClass()),
                intf.getInterfaceSubclass(),
                intf.getInterfaceProtocol()));

        for(int j = 0; j < intf.getEndpointCount(); j++) {
            UsbEndpoint endpoint = intf.getEndpoint(j);
            sb.append(String.format(" +---Endpoint %d: %s %s\n",
                    endpoint.getEndpointNumber(),
                    nameForEndpointType(endpoint.getType()),
                    nameForDirection(endpoint.getDirection())));
        }
    }

    return sb.toString();
}

/* 辅助方法,用来为 USB 常量提供可读性更强的名称 */

private String nameForClass(int classType) {
    switch(classType) {
    case UsbConstants.USB_CLASS_APP_SPEC:
        return String.format("Application Specific 0x%02x", classType);
```

```java
        case UsbConstants.USB_CLASS_AUDIO:
            return "Audio";
        case UsbConstants.USB_CLASS_CDC_DATA:
            return "CDC Control";
        case UsbConstants.USB_CLASS_COMM:
            return "Communications";
        case UsbConstants.USB_CLASS_CONTENT_SEC:
            return "Content Security";
        case UsbConstants.USB_CLASS_CSCID:
            return "Content Smart Card";
        case UsbConstants.USB_CLASS_HID:
            return "Human Interface Device";
        case UsbConstants.USB_CLASS_HUB:
            return "Hub";
        case UsbConstants.USB_CLASS_MASS_STORAGE:
            return "Mass Storage";
        case UsbConstants.USB_CLASS_MISC:
            return "Wireless Miscellaneous";
        case UsbConstants.USB_CLASS_PER_INTERFACE:
            return "(Defined Per Interface)";
        case UsbConstants.USB_CLASS_PHYSICA:
            return "Physical";
        case UsbConstants.USB_CLASS_PRINTER:
            return "Printer";
        case UsbConstants.USB_CLASS_STILL_IMAGE:
            return "Still Image";
        case UsbConstants.USB_CLASS_VENDOR_SPEC:
            return String.format("Vendor Specific 0x%02x", classType);
        case UsbConstants.USB_CLASS_VIDEO:
            return "Video";
        case UsbConstants.USB_CLASS_WIRELESS_CONTROLLER:
            return "Wireless Controller";
        default:
            return String.format("0x%02x", classType);
        }
    }

    private String nameForEndpointType(int type) {
        switch(type) {
        case UsbConstants.USB_ENDPOINT_XFER_BULK:
            return "Bulk";
        case UsbConstants.USB_ENDPOINT_XFER_CONTROL:
            return "Control";
        case UsbConstants.USB_ENDPOINT_XFER_INT:
            return "Interrupt";
        case UsbConstants.USB_ENDPOINT_XFER_ISOC:
            return "Isochronous";
        default:
            return "Unknown Type";
        }
    }
```

```java
        private String nameForDirection(int direction) {
            switch(direction) {
            case UsbConstants.USB_DIR_IN:
                return "IN";
            case UsbConstants.USB_DIR_OUT:
                return "OUT";
            default:
                return "Unknown Direction";
            }
        }
    }
```

当Activity首次进入前台时,它注册一个自定义动作(稍后会详细探讨)的BroadcastReceiver,并且通过UsbManager.getDeviceList()方法来查询当前已连接设备的列表,该方法会返回一个UsbDevice项的HashMap,然后就可以遍历和查询这个HashMap。对于每台连接的设备,我们会查询它的接口和端点,并且会构建需要显示给用户的每台设备的描述信息。然后,我们会在用户界面上打印所有这些信息。

注意:
就目前来说,这个应用程序不需要在清单中声明任何权限。对于只是简单地查询连接到主机的设备的信息,并不需要声明权限。

如你所见,对于你想与之通信的连接设备,UsbManager提供的API可以获得你想要的所有信息。所有标准的定义,如设备种类、端点类型和传输方向也都在UsbConstants中做了定义,所以不需要自己定义就可以匹配想要的类型。

那么为什么要注册BroadcastReceiver呢?在用户按下屏幕上的Connect按钮后,这个示例的剩余部分做了相应的响应。这时候我们想要与连接的设备进行真正的交互,这时候就需要用户权限。在此,当用户单击按钮后,会调用UsbManager.requestPermission()来询问用户是否可以连接。如果还没有授权相应的权限,用户会看到询问授权连接的对话框。

如果选择确认授权,传入方法的PendingIntent就会被触发。在我们的示例中,这个Intent是通过自定义动作字符串来广播的,此时会触发BroadcastReceiver的onReceive()方法;接下来任何的requestPermission()调用都会立即触发这个接收器。在接收器内部,我们会检查以确保结果是授权响应并通过UsbManager.openDevice()打开与设备的连接,如果连接成功,则会返回一个UsbDeviceConnection实例。

对于有效的连接,我们会通过控制传输来请求设备的配置描述符,从而得到设备更加详细的信息。控制传输一般都是通过设备的"端点0"来请求的。配置描述符包含配置的信息以及每个接口和端点的信息,因此它的长度是可变的。我们则分配一个合适大小的缓冲区来保证可以得到所有的信息。

controlTransfer()返回后,缓冲区中已经填好了响应数据。接下来应用程序会处理这些数据,得到设备的一些详细信息,例如设备的最大能耗以及设备是使用USB供电(总线供电)还是其他方式外部供电(自供电)。这个示例只是从这些标识符中解析出一小部分有用的信息。同样,所有解析出来的数据都会被放到一个字符串报告中并显示在用户界面上。

第一节中从框架 API 读取的信息和第二节中直接从设备读取的信息是一样的,并且按照 1:1 的比例通过两个文本报告显示在用户屏幕上。需要注意的一点就是,只有在设备连接上时该应用程序才会工作:对于应用程序在前台运行时才连接的设备,应用程序并不会得到通知。下一小节将讨论如何处理这种场景。

获取设备连接时的通知

要想 Android 在设备连接时可以通知你的应用程序,需要在清单中通过<intent-filter>注册要匹配的设备类型。代码清单 3-45 和 3-46 演示了这个过程。

代码清单 3-45　AndroidManifest.xml 中的部分代码

```xml
<activity
    android:name=".USBActivity"
    android:label="@string/title_activity_usb" >
    <intent-filter>
        <action android:name="android.intent.action.MAIN" />
        <category android:name="android.intent.category.LAUNCHER" />
    </intent-filter>
    <intent-filter>
        <action android:name=
            "android.hardware.usb.action.USB_DEVICE_ATTACHED" />
    </intent-filter>

    <meta-data android:name=
        "android.hardware.usb.action.USB_DEVICE_ATTACHED"
        android:resource="@xml/device_filter" />
</activity>
```

代码清单 3-46　res/xml/device_filter.xml

```xml
<?xml version="1.0" encoding="utf-8"?>
<resources>
    <usb-device vendor-id="5432" product-id="9876" />
</resources>
```

能够处理设备连接的 Activity 添加了一个名为 USB_DEVICE_ATTACHED 动作字符串的过滤器和描述想要处理设备的一些 XML 元数据信息。可以在<usb-device>中添加很多设备属性字段,从而过滤哪些连接事件可以通知到应用程序:

- vendor-id
- product-id
- class
- subclass
- protocol

必要时,可以定义以上很多属性来适应你的应用程序。例如,如果只想和某一台特定设备进行通信，或许可以像示例代码一样同时定义 vendor-id 和 product-id。如果想匹配某一类型的设备(例如，所有的大数据量存储设备)，或许只需要定义 class 属性即可。甚至可以不定义任何属性，这样应用程序就可以匹配所有连接的设备。

3.15 小结

在当前这个一切皆 Web 的时代，将 Android 应用程序与网络和 Web 服务连接起来是提升用户体验的好方法。Android 中连接网络和其他远程主机的框架使得添加这类功能变得很方便。本章研究了如何将 Web 标准加入应用程序中，在原生环境中用 HTML 和 JavaScript 实现用户交互。本章还介绍了如何用 Android 从远程服务器下载内容并在应用程序中使用它们。本章还指出 Web 服务器并不是唯一可以连接的远程主机，还可以用蓝牙、NFC 和短信实现设备间的直接通信。在第 4 章中，我们将探讨如何使用 Android 提供的工具与设备的硬件资源进行交互。

第 4 章

实现设备硬件交互与媒体交互

在应用程序软件中整合设备硬件的功能可以向用户提供只有移动平台才具备的独特用户体验。使用麦克风和摄像头采集媒介，应用程序就可以通过照片或录音实现个性化交流。整合传感器和位置数据可以帮助开发能够回答"我在哪里？"和"我看到了什么？"这种问题的应用程序。

本章将介绍如何利用 Android 提供的位置、媒体和传感器 API，这些 API 可以使应用程序具有移动平台上特有的功能。

4.1 整合设备位置

4.1.1 问题

要在应用程序中使用设备的定位功能报告当前的物理位置。

4.1.2 解决方案

(API Level 9)

可利用 Google 在其 Play Service 库中提供的混合位置提供程序。移动应用程序通常提供给用户的最强大优势之一就是能够通过包括基于用户目前所在位置的信息来添加上下文。应用程序可能要求位置服务基于如下标准提供设备位置的更新：

- 频率：在将另一个更新提供给应用程序之前经过的最小时间量，单位为毫秒。
- 距离：在提供另一个更新之前设备必须移动的最小距离。
- 计数：在关闭提供程序之前提供的最大更新数。
- 到期时间：启动请求之后和关闭提供程序之前的最大时间量。

注意：
Android 也在 LocationManager 中直接提供了非 Google 位置服务。此 API 帮助不大，开发人员必须单独管理来自离散位置源的请求。

Android 允许应用程序从多个来源获得位置数据。高精度(同时也是高能耗)的位置修正数据来自设备上的 GPS，而低精度的数据是通过网络源获得的，如蜂窝网络基站塔和 Wi-Fi 热点。Google 的混合位置提供程序因为如下原因而得名：将这些多个来源"混合"在一起，随时向开发人员提供最佳的结果。使用高级别的标准发出位置请求，由设备决定应该调用哪些硬件接口：

- PRIORITY_BALANCED_POWER_ACCURACY：提供 GPS 和网络源的混合，以便通过较低的能耗获得较为精确的数据，并且更快速地进行修正。
- PRIORITY_HIGH_ACCURACY：虽然初步的修正可能来自其他地方，但要求最终的修正数据来自 GPS(最精确)。这个选项也最为耗电，因为 GPS 通常一直保持打开。
- PRIORITY_LOW_POWER：提供最快速的修正数据——但是没有 GPS 的协助，精确度通常不会好于"城市"级别。
- PRIORITY_NO_POWER：使应用程序成为主动的观察者。仅在另一个应用程序触发时才提供更新。

要点：
本例中的位置服务是作为 Google Play Services 库的一部分进行分发的，它在任意平台级别都不是原生 SDK 的一部分。然而，目标平台为 API Level 9 或以后版本的应用程序以及 Google Play 体系内的设备都可以使用此绘图库。关于在项目中包括 Google Play Services 的更多信息，请参考 https://developer.android.com/google/play-services/setup.html。

4.1.3 实现机制

在代码清单 4-1 中，我们注册一个 Activity 来监听对用户可见的位置更新，并且将位置信息显示在屏幕上。

注意：
本例使用来自 AppCompat 库的 ActionBarActivity，从而可以使用 API Level 11 之前的分段。

代码清单 4-1　监控位置更新的 Activity

```
public class MainActivity extends ActionBarActivity implements
        GooglePlayServicesClient.ConnectionCallbacks,
        GooglePlayServicesClient.OnConnectionFailedListener,
        com.google.android.gms.location.LocationListener {
    private static final String TAG = "AndroidRecipes";

    private static final int UPDATE_INTERVAL = 15 * 1000;
    private static final int FASTEST_UPDATE_INTERVAL = 2 * 1000;

    /* Play Services 的客户端接口 */
    private LocationClient mLocationClient;
    /* 我们想要接收的更新相关元数据 */
```

```java
    private LocationRequest mLocationRequest;
    /* 最近已知的设备位置 */
    private Location mCurrentLocation;

    private TextView mLocationView;

    @Override
    public void onCreate(Bundle savedInstanceState) {
        super.onCreate(savedInstanceState);
        mLocationView = new TextView(this);
        setContentView(mLocationView);

        //确认Play Services已激活并保持最新
        int resultCode =
                GooglePlayServicesUtil.isGooglePlayServicesAvailable(this);
        switch(resultCode) {
            case ConnectionResult.SUCCESS:
                Log.d(TAG, "Google Play Services is ready to go!");
                break;
            default:
                showPlayServicesError(resultCode);
                return;
        }

        //添加位置更新监控
        mLocationClient = new LocationClient(this, this, this);

        mLocationRequest = LocationRequest.create()
                //设置所需的精确级别
                .setPriority(LocationRequest.PRIORITY_HIGH_ACCURACY)
                //设置所需的(不精确的)的位置更新频率
                .setInterval(UPDATE_INTERVAL)
                //限制更新请求的最大速率
                .setFastestInterval(FASTEST_UPDATE_INTERVAL);
    }

    @Override
    public void onResume() {
        super.onResume();
        //在移入前台时，附加到Play Services
        mLocationClient.connect();
    }

    @Override
    public void onPause() {
        super.onPause();
        //不在前台时禁止更新
        if (mLocationClient.isConnected()) {
            mLocationClient.removeLocationUpdates(this);
        }
```

```java
        //从 Play Services 取消附加
        mLocationClient.disconnect();
    }

    private void updateDisplay() {
        if(mCurrentLocation == null) {
            mLocationView.setText("Determining Your Location...");
        } else {
            mLocationView.setText(String.format("Your Location:\n%.2f, %.2f",
                    mCurrentLocation.getLatitude(),
                    mCurrentLocation.getLongitude()));
        }
    }

    /** Play Services 的位置 */

    /*
     * 当 Play Services 缺失或版本不正确时，
     * 客户端库将以对话框的形式帮助用户进行更新
     */
    private void showPlayServicesError(int errorCode) {
        //从 Google Play Services 获得错误对话框
        Dialog errorDialog = GooglePlayServicesUtil.getErrorDialog(
                errorCode,
                this,
                1000 /* RequestCode */);
        //如果 Google Play Services 可以提供错误对话框
        if (errorDialog != null) {
            //为错误对话框创建新的 DialogFragment
            SupportErrorDialogFragment errorFragment =
                    SupportErrorDialogFragment.newInstance(errorDialog);
            //在 DialogFragment 中显示错误对话框
            errorFragment.show(
                    getSupportFragmentManager(),
                    "Location Updates");
        }
    }

    @Override
    public void onConnected(Bundle bundle) {
        Log.d(TAG, "Connected to Play Services");
        //立即获得最近已知的位置
        mCurrentLocation = mLocationClient.getLastLocation();
        //注册更新
        mLocationClient.requestLocationUpdates(mLocationRequest, this);
    }

    @Override
    public void onDisconnected() { }
```

```
@Override
public void onConnectionFailed(ConnectionResult connectionResult) { }

/** LocationListener 回调 */

@Override
public void onLocationChanged(Location location) {
    Log.d(TAG, "Received location update");
    mCurrentLocation = location;
    updateDisplay();
}
}
```

要点：

在应用程序中使用位置服务时，记住需要在应用程序的清单中声明 android.permission. ACCESS_COARSE_LOCATION 或 android.permission.ACCESS_FINE_ LOCATION 权限。如果声明了 android.permission.ACCESS_FINE_LOCATION 权限，就不必声明另一个权限，因为它已经包含了模糊位置服务的权限。

这个示例构造一个 LocationRequest，其中包含希望应用于所接收更新的所有标准。我们指示该服务以 15 秒的间隔仅返回高度精确的结果。这种间隔是不准确的，即 Android 大约每隔 15 秒传递更新，这个时间可多可少。唯一的保证是，在此间隔内最多只会收到一个更新。为设置此间隔的上限，我们还设置最快间隔为两秒。这会限制更新，确保我们不会在快于两秒的间隔内收到两个更新。

收集位置数据是一个资源密集型操作，因此要确保仅在 Activity 位于前台时才执行此操作。我们的示例会等待直到 onResume()连接位置服务，并在 onPause()中立即断开连接。Google Play Services 是异步的，这意味着必须连接并等待回调，然后才可以执行真正的设置工作。直到建立指向 Play Services 的连接之后，才可以访问 LocationClient。

如果 LocationClient.connect()在后面尝试触发 onConnected()，我们就可以使用前面通过 requestLocationUpdates()构造的请求来开始请求更新。如果我们需要在获得另一个更新之前引用数据，则此时也可以通过 getLastLocation()获得最近已知的位置。如果最近没有进行修正，则该值可以返回 null。

模拟位置改变

如果使用 Android 模拟器测试应用程序的话，应用程序将不能从任何系统提供程序中接收真实的位置数据信息。但是，通过 SDK 的 Monitor 工具，可以手动地注入位置改变事件。

在激活 DDMS 的情况下，选择 Emulator Control 选项卡并查找 Location Controls 部分。在选项卡式的界面中可以直接输入经度/纬度对，或者通过命令行文件格式读取一系列经度/纬度对。

在手动输入单个值时，必须在文本框中输入有效的经度和纬度。然后可以单击 Send 按钮，作为所选模拟器内的事件注入该位置。注册以侦听位置改变的任何应用程序也会收到包含此位置值的更新。

当位置更新到达时，会调用已注册侦听器的 onLocationChanged()方法。该例持续引用其所接收的最新位置，对于每个传入的更新，重置位置值，并且更新用户界面显示以反映新的更改。

注意：

如果在服务或其他后台操作中接收位置更新，Google 建议最小时间间隔应不小于 60 000(60 秒)。

为使此应用程序正确构建和运行，我们需要额外关联一些文件。代码清单 4-2 和 4-3 描述了这些需求。

代码清单 4-2　build.gradle 的部分代码

```
apply plugin: 'com.android.application'

android {
    compileSdkVersion 19
    buildToolsVersion "20.0.0"
    defaultConfig {
        applicationId "com.androidrecipes.mylocation"
        ...
    }
    ...
}

dependencies {
    compile fileTree(dir: 'libs', include: ['*.jar'])
    compile 'com.google.android.gms:play-services:6.1.+'
    compile 'com.android.support:appcompat-v7:21.0.+'
}
```

代码清单 4-3　AndroidManifest.xml 的部分代码

```xml
<?xml version="1.0" encoding="utf-8"?>
<manifest xmlns:android="http://schemas.android.com/apk/res/android"
    package="com.examples.mylocation">

<uses-permission android:name="android.permission.ACCESS_FINE_LOCATION"/>

<application ...>

    <!-- 需要样板文件以启动 Play Services -->
    <meta-data
        android:name="com.google.android.gms.version"
        android:value="@integer/google_play_services_version" />

    <activity ...>
        <intent-filter>
```

```xml
            <action android:name="android.intent.action.MAIN" />
            <category android:name="android.intent.category.LAUNCHER" />
        </intent-filter>
    </activity>

    </application>
</manifest>
```

在项目的 build.gradle 文件内，必须将 Google Play Services 添加为依赖项。在此我们使用加号来确保始终通过最新的客户端库进行构建。Play Services 也要求向具有客户端库版本的应用程序添加<meta-data>元素，客户端库使用此元素确定设备上的 Play Services 版本目前是否满足要求。版本号资源内置于库中，因此无须进行定义。

我们正在访问设备的位置服务，因此还需要请求 ACCESS_FINE_LOCATION 权限，这是访问 GPS 位置服务的必需权限。如果应用程序使用较低精确度的值，就可以不需要请求 ACCESS_FINE_LOCATION 权限。

Google Play Services 不可用时的情况

在此例的顶部(见代码清单 4-1)有一些样板代码，用于使用 GooglePlayServicesUtil 的 isGooglePlayServicesAvailable()检查设备上的 Google Play Services 正在运行且保持更新。该方法将确认 Play Services 正在设备上运行，并且最少为客户端应用程序中指定的版本。

如果此检查失败，Play Services 也以 ErrorDialogFragment 的形式提供了 UI，向用户显示此 UI 以更新 Play Services。此对话框将触发适当的设置 UI，从而自动将 Play Services 更新到最新版本。在我们的示例中，showPlayServicesError()方法负责处理此工作。我们将在本书使用的包含 Play Services 访问的所有主题中看到此模式。

4.2 地图位置

4.2.1 问题

需要在地图上为用户显示一个或多个位置。此外，要在同一张地图上显示用户自己的位置。

4.2.2 解决方案

(API Level 9)

向用户显示地图的最简单方式就是用位置数据创建一个 Intent 并把它传递给 Android 系统来启动地图应用程序。在第 6 章中你将深入了解这个方法以完成各种任务。另外，Google Play Services 库的 Google Maps v2 库组件提供的 MapView 和 MapActivity 可以在应用程序中嵌入地图。

要点：

Google Maps v2 是作为 Google Play Services 库的一部分进行分发的，它在任意平台级别都不是原生 SDK 的一部分。然而，目标平台为 API Level 9 或以后版本的应用程序以及 Google Play 体系内的设备都可以使用此绘图库。关于在项目中包括 Google Play Services 的更多信息，请参考 https://developer.android.com/google/play-services/setup.html。

1. 获取 API 密钥

要开始使用 Maps v2，需要创建一个 API 项目，在该项目内启用 Maps v2 服务，生成 API 密钥并包括在应用程序代码中。如果没有 API 密钥，虽然也可以利用绘图类，但不会向应用程序返回任何地图图块(map tile)。请遵循如下步骤：

(1) 进入 https://code.google.com/apis/console/，使用你的 Google 账户登录以访问 Google API 控制台。

(2) 选择 Create Project 选项，为你的地图建立新的项目。如果已有项目，则可以根据喜好向其中添加 Maps v2 服务和密钥。在此例中，选择要添加 Maps v2 的项目。

(3) 在导航面板中，选择 Services，向下滚动到 Google Maps Android API v2 并启用该服务。

(4) 在导航面板中选择 API Access，并且选择 Create new Android Key 选项。

(4) 遵循屏幕上的说明，根据想要使用的应用程序向密钥添加密钥库签名/应用程序包对。在此例中，示例应用程序的包名是 com.androidrecipes.mapper，而签名来自开发机器上的调试密钥，通常位于<USERHOME>/.android/debug.keystore。

注意：

关于 SDK 的更多信息和获取 API 密钥的最新说明，可以访问 https://developers.google.com/maps/documentation/android/start。

如果在模拟器上测试你的运行代码，模拟器必须使用目标平台为 Android 4.3 或以后版本的 SDK 构建，这些版本包含 Google API，从而绘图操作可以正常运行。以前版本的 SDK 捆绑了 Maps v1 库而不是 Google Play Services，因此它们不能用于测试。

如果是通过命令行创建模拟器，目标的名称就是"Google Inc.:Google APIs:X"，其中的 X 指示 API 的版本。如果在 IDE(例如 Eclipse)中创建模拟器，目标的命名约定也是类似的：Google APIs (Google Inc.) -X，其中 X 指示 API 的版本。

2. 满足清单要求

获得了有效的 API 密钥之后，需要在我们的 AndroidManifest.xml 文件包括该密钥。下面的代码块必须放在<application>元素内：

```
<meta-data
    android:name="com.google.android.maps.v2.API_KEY"
    android:value="YOUR_KEY_HERE" />
```

此外，Maps v2 有一项设备要求，即至少要具备 OpenGL ES 2.0。我们可以将此要求作为设备特性提出，方法是在<manifest>元素内添加如下代码块，通常是放在<application>元素的上方：

```
<!-- Maps v2 requires OpenGL ES 2.0 -->
<uses-feature
    android:glEsVersion="0x00020000"
    android:required="true" />
```

最后，Maps v2 需要一组权限才能与 Google Play Services 通信以及渲染地图图块。因此，我们必须在<manifest>元素内再添加一个代码块，通常是放在<application>元素的上方：

```
<!--显示地图所需要的权限 -->
<uses-permission android:name="android.permission.INTERNET" />
<uses-permission
   android:name="android.permission.ACCESS_NETWORK_STATE" />
<uses-permission
   android:name="android.permission.WRITE_EXTERNAL_STORAGE"
/>
<uses-permission android:name=
   "com.google.android.providers.gsf.permission.READ_GSERVICES"
/>
```

将上述内容整合在一起，清单文件应如代码清单 4-4 所示。

代码清单 4-4　AndroidManifest.xml 的部分代码

```
<manifest xmlns:android="http://schemas.android.com/apk/res/android"
     package="com.androidrecipes.mapper"

    <!-- 需要在地图上显示用户的位置 -->
    <uses-permission
      android:name="android.permission.ACCESS_FINE_LOCATION" />

    <!-- 显示地图所需的权限 -->
    <uses-permission android:name="android.permission.INTERNET" />
    <uses-permission
      android:name="android.permission.ACCESS_NETWORK_STATE" />
    <uses-permission
      android:name="android.permission.WRITE_EXTERNAL_STORAGE"
    />
    <uses-permission android:name=
      "com.google.android.providers.gsf.permission.READ_GSERVICES"
    />

    <!-- Maps v2 要求具备 OpenGL ES 2.0 -->
    <uses-feature
        android:glEsVersion="0x00020000"
        android:required="true" />
```

```
<application
    android:icon="@drawable/ic_launcher"
    android:label="@string/app_name"
    android:theme="@style/AppTheme" >

    <!-- 活动、服务、提供程序等 -->

    <meta-data
        android:name="com.google.android.maps.v2.API_KEY"
        android:value="YOUR_KEY_HERE" />
</application>

</manifest>
```

手边有了 API 密钥和合适的测试平台之后,就可以开始正式工作了。

4.2.3 实现机制

要想显示地图,只需要创建一个 MapView 或 MapFragment 实例。API 密钥在应用程序中全局可用,因此这些元素的任何实例都会使用该值。不需要像在 Maps v1 中一样将密钥添加到每个实例中。

注意:

除了上述权限之外,还必须为此例添加 android.permission.ACCESS_FINE_LOCATION 权限。需要此权限的唯一原因是该例会连接到 LocationManager 以获得缓存的位置值。

我们将创建一个简单的应用程序,它会将用户上次最新的已知位置显示在 Google 地图上,如图 4-1 所示。

图 4-1 用户位置的地图

现在，让我们查看构造此视图所需的布局，参见代码清单 4-5。

代码清单 4-5　res/layout/main.xml

```xml
<?xml version="1.0" encoding="utf-8"?>
<LinearLayout
    xmlns:android="http://schemas.android.com/apk/res/android"
    android:layout_width="match_parent"
    android:layout_height="match_parent"
    android:orientation="vertical" >
    <TextView
        android:layout_width="match_parent"
        android:layout_height="wrap_content"
        android:gravity="center_horizontal"
        android:text="Map Of Your Location" />
    <RadioGroup
        android:id="@+id/group_maptype"
        android:layout_width="match_parent"
        android:layout_height="wrap_content"
        android:orientation="horizontal" >
        <RadioButton
            android:id="@+id/type_normal"
            android:layout_width="0dp"
            android:layout_height="wrap_content"
            android:layout_weight="1"
            android:text="Normal Map" />
        <RadioButton
            android:id="@+id/type_satellite"
            android:layout_width="0dp"
            android:layout_height="wrap_content"
            android:layout_weight="1"
            android:text="Satellite Map" />
    </RadioGroup>
    <fragment
        class="com.google.android.gms.maps.SupportMapFragment"
        android:id="@+id/map"
        android:layout_width="match_parent"
        android:layout_height="match_parent"/>
</LinearLayout>
```

注意：

在 XML 布局中添加 MapView 或 MapFragment 时，必须指定包的完全限定名称，这是由于该类并没有包含在 android.view 或 android.widget 中。

在此创建了一个简单布局，其中包括一个选择器，用于切换在 MapFragment 实例旁边显示的地图类型。代码清单 4-6 显示了控制地图的 Activity 代码。

代码清单 4-6　显示缓存位置的 Activity

```java
public class BasicMapActivity extends FragmentActivity implements
```

```java
        RadioGroup.OnCheckedChangeListener {
    private static final String TAG = "AndroidRecipes";

    private SupportMapFragment mMapFragment;
    private GoogleMap mMap;

    @Override
    public void onCreate(Bundle savedInstanceState) {
        super.onCreate(savedInstanceState);
        setContentView(R.layout.main);

        // 检查 Play Services 是否激活且为最新版本
        int resultCode =
                GooglePlayServicesUtil.isGooglePlayServicesAvailable(this);
        switch(resultCode) {
          case ConnectionResult.SUCCESS:
            Log.d(TAG, "Google Play Services is ready to go!");
            break;
          default:
            showPlayServicesError(resultCode);
            return;
        }

        mMapFragment =
                (SupportMapFragment) getSupportFragmentManager()
                        .findFragmentById(R.id.map);
        mMap = mMapFragment.getMap();

        // 快速检查用户的最新已知位置是否有效,并且围绕该点将地图居中
        // 如果该位置无效,则使用默认位置
        LocationManager manager = (LocationManager)
                getSystemService(Context.LOCATION_SERVICE);
        Location location = manager.getLastKnownLocation(
                LocationManager.GPS_PROVIDER);

        LatLng mapCenter;
        if(location != null) {
            mapCenter = new LatLng(location.getLatitude(),
                    location.getLongitude());
        } else {
            // 使用默认位置
            mapCenter = new LatLng(37.4218, -122.0840);
        }

        // 居中地图并同时缩放
        CameraUpdate newCamera =
                CameraUpdateFactory.newLatLngZoom(mapCenter, 13);
        mMap.moveCamera(newCamera);
```

```java
    // 连接地图类型选择器UI
    RadioGroup typeSelect =
            (RadioGroup) findViewById(R.id.group_maptype);
    typeSelect.setOnCheckedChangeListener(this);
    typeSelect.check(R.id.type_normal);
}

@Override
public void onResume() {
    super.onResume();
    if(mMap != null) {
        // 启用地图上的用户位置显示功能
        mMap.setMyLocationEnabled(true);
    }
}

@Override
public void onPause() {
    super.onResume();
    if(mMap != null) {
        // 在不可见时禁用用户位置
        mMap.setMyLocationEnabled(false);
    }
}

/*
 * 当Play Services缺失或版本不对时,
 * 客户端库将以对话框的形式帮助用户进行更新。
 */
private void showPlayServicesError(int errorCode) {
    // 从Google Play Services获得错误对话框
    Dialog errorDialog = GooglePlayServicesUtil.getErrorDialog(
        errorCode,
        this,
        1000 /* RequestCode */);
    // 如果Google Play Services可以提供错误对话框
    if(errorDialog != null) {
        // 为错误对话框创建新的DialogFragment
        SupportErrorDialogFragment errorFragment =
            SupportErrorDialogFragment.newInstance(errorDialog);
        // 在DialogFragment中显示错误对话框
        errorFragment.show(
            getSupportFragmentManager(),
            "Google Maps");
```

```
            }
        }

        /** OnCheckedChangeListener 方法 */

        @Override
        public void onCheckedChanged(RadioGroup group,
                int checkedId) {
            switch(checkedId) {
                case R.id.type_satellite:
                    // 显示卫星地图
                    mMap.setMapType(GoogleMap.MAP_TYPE_SATELLITE);
                    break;
                case R.id.type_normal:
                default:
                    // 显示普通地图
                    mMap.setMapType(GoogleMap.MAP_TYPE_NORMAL);
                    break;
            }
        }
    }
```

首先执行的操作是确认将正确版本的 Google Play Services 安装到此设备上。在设备的用户与 Google 应用程序(如 Google Play)交互时，Google 会自动管理 Google Play Services 库。Google Play Services 在后台自动更新，因此我们需要使用来自 GooglePlayServicesUtil 的方法，在运行时确认用户具有我们所需的版本。从 isGooglePlayServicesAvaiable() 获得的结果将告诉我们服务是否为正确的版本还是需要更新，甚至是需要完全安装。

这个 Activity 会获取用户最新的位置信息并将该位置设为地图的中心。关于地图的所有控制操作都是通过 GoogleMap 实例来完成的，而 GoogleMap 实例则是通过调用 MapFragment.getMap()得到的。在此例中，我们使用了地图的 moveCamera()方法，通过 CameraUpdate 对象对地图的显示做了调整。

CameraUpdate 用于一次性对地图显示的一个或多个组件进行调整，例如修改缩放以及中心点。地图的缩放级别是 2.0 和 21.0 之间的离散值，其中的最小值会使整个世界地图近似为 1024dp 宽，而每增加一级缩放就会使显示中的世界地图的宽度翻倍。

当用户选择不同的单选按钮时，地图类型会在卫星视图和传统的地图视图之间切换。除了此例中使用的值之外，其他允许的地图类型如下：

- MAP_TYPE_HYBRID：在卫星地图上显示地图数据(例如，街道和感兴趣的点)。
- MAP_TYPE_TERRAIN：使用地形海拔轮廓线显示地图。

最后，为启用用户位置显示和控件，我们只需要对地图调用 setMyLocationEnabled()。该方法将启用位置跟踪并可能开启 GPS 等元素，因此也应该在不再需要时禁用(当视图不可见时)。

为正确构建此应用程序，代码清单 4-7 显示了 build.gradle 文件中所需的依赖关系。

代码清单4-7　build.gradle 文件的部分代码

```
apply plugin: 'com.android.application'

android {
    compileSdkVersion 14
    buildToolsVersion "20.0.0"

    defaultConfig {
        applicationId "com.androidrecipes.mapper"
        ...
    }

    ...
}

dependencies {
    compile 'com.android.support:support-v4:21.0.+'
    compile 'com.google.android.gms:play-services:6.1.+'
}
```

这是很好的起点，但或许有点令人厌烦。为引入一些更具交互性的内容，4.3 节将在地图上创建标识和其他标记，并且会介绍如何定制这些标记。

4.3　在地图上标记位置

4.3.1　问题

除了将指定的位置显示在地图的中心，应用程序还需要在该位置加上标记，以使其更加醒目。

4.3.2　解决方案

(API Level 9)

向地图添加 Marker 对象以及 Circle 和 Polygon 等形状元素。Marker 对象是通过图标定义的交互式对象，该图标显示在给定位置。该位置可以是固定的，也可以设置 Marker 为可由用户拖动到他们希望的任意一点。每个 Marker 还可以响应触摸事件，如点击和长按。此外，可以为 Marker 提供包括标题的元数据和文本片段，当点击标记时会在弹出信息窗口中显示这些信息。这些窗口自身也可以定制显示。

Maps v2 还支持绘制离散形状元素。这些元素在本质上是不可交互的，但我们会看到，可以轻松地添加与形状交互的功能。此功能也可以用于在地图上使用 Polyline 形状绘制路线，其不像其他选项一样会尝试绘制闭合的、填充的形状。

要点：

Google Maps v2 是作为 Google Play Services 库的一部分进行分发的，它在任意平台级别都不是原生 SDK 的一部分。然而，目标平台为 API Level 9 或以后版本的应用程序以及 Google Play 体系内的设备都可以使用此绘图库。关于在项目中包括 Google Play Services 的更多信息，请参考 https://developer.android.com/google/play-services/setup.html。

4.3.3 实现机制

图 4-2 显示了前面的地图应用程序，其中使用标记添加了一些感兴趣的点。

图 4-2　带有自定义标记的地图

代码清单 4-8 和 4-9 显示了新的 Activity 示例，其中向地图添加了一些标记。XML 布局与前一节中的相同，因此我们不会花费时间再次剖析其组成部分，只是为了完整性而在此添加此布局。

代码清单 4-8　res/layout/main.xml

```xml
<?xml version="1.0" encoding="utf-8"?>
<LinearLayout
    xmlns:android="http://schemas.android.com/apk/res/android"
    android:layout_width="match_parent"
    android:layout_height="match_parent"
    android:orientation="vertical" >
<TextView
```

```xml
        android:layout_width="match_parent"
        android:layout_height="wrap_content"
        android:gravity="center_horizontal"
        android:text="Map Of Your Location" />
    <RadioGroup
        android:id="@+id/group_maptype"
        android:layout_width="match_parent"
        android:layout_height="wrap_content"
        android:orientation="horizontal" >
        <RadioButton
            android:id="@+id/type_normal"
            android:layout_width="0dp"
            android:layout_height="wrap_content"
            android:layout_weight="1"
            android:text="Normal Map" />
        <RadioButton
            android:id="@+id/type_satellite"
            android:layout_width="0dp"
            android:layout_height="wrap_content"
            android:layout_weight="1"
            android:text="Satellite Map" />
    </RadioGroup>
    <fragment
        class="com.google.android.gms.maps.SupportMapFragment"
        android:id="@+id/map"
        android:layout_width="match_parent"
        android:layout_height="match_parent"/>
</LinearLayout>
```

代码清单 4-9   显示带有标记的地图的 Activity

```java
public class MarkerMapActivity extends FragmentActivity implements
        RadioGroup.OnCheckedChangeListener,
        GoogleMap.OnMarkerClickListener,
        GoogleMap.OnMarkerDragListener,
        GoogleMap.OnInfoWindowClickListener,
        GoogleMap.InfoWindowAdapter {
    private static final String TAG = "AndroidRecipes";

    private SupportMapFragment mMapFragment;
    private GoogleMap mMap;

    @Override
    protected void onCreate(Bundle savedInstanceState) {
        super.onCreate(savedInstanceState);
        setContentView(R.layout.main);

        // 检查 Play Services 是否激活且为最新版本
        int resultCode =
            GooglePlayServicesUtil.isGooglePlayServicesAvailable(this);
```

```java
switch(resultCode) {
  case ConnectionResult.SUCCESS:
    Log.d(TAG, "Google Play Services is ready to go!");
    break;
  default:
    showPlayServicesError(resultCode);
    return;
}

mMapFragment =
        (SupportMapFragment) getSupportFragmentManager()
                .findFragmentById(R.id.map);
mMap = mMapFragment.getMap();
// 监控与标记元素的交互
mMap.setOnMarkerClickListener(this);
mMap.setOnMarkerDragListener(this);
// 设置应用程序以服务信息窗口的视图
mMap.setInfoWindowAdapter(this);
// 监控信息窗口上的点击事件
mMap.setOnInfoWindowClickListener(this);

// Google 总部(37.427,-122.099)
Marker marker = mMap.addMarker(new MarkerOptions()
        .position(new LatLng(37.4218, -122.0840))
        .title("Google HQ")
        // 将来自应用程序的图像资源显示为标记
        .icon(BitmapDescriptorFactory
                .fromResource(R.drawable.logo))
        // 降低透明度
        .alpha(0.6f));
// 使此标记在地图上可拖动
marker.setDraggable(true);

// 减去 0.01 度
mMap.addMarker(new MarkerOptions()
        .position(new LatLng(37.4118, -122.0740))
        .title("Neighbor #1")
        .snippet("Best Restaurant in Town")
        // 以默认颜色显示默认标记
        .icon(BitmapDescriptorFactory.defaultMarker()));

// 增加 0.01 度
mMap.addMarker(new MarkerOptions()
        .position(new LatLng(37.4318, -122.0940))
        .title("Neighbor #2")
        .snippet("Worst Restaurant in Town")
        // 使用浅蓝色显示默认标记
        .icon(BitmapDescriptorFactory
```

```java
                    .defaultMarker(
                        BitmapDescriptorFactory.HUE_AZURE)));

        // 居中地图并同时缩放
        LatLng mapCenter = new LatLng(37.4218, -122.0840);
        CameraUpdate newCamera = CameraUpdateFactory
                .newLatLngZoom(mapCenter, 13);
        mMap.moveCamera(newCamera);

        // 连接地图类型选择器 UI
        RadioGroup typeSelect =
                (RadioGroup) findViewById(R.id.group_maptype);
        typeSelect.setOnCheckedChangeListener(this);
        typeSelect.check(R.id.type_normal);
    }

    /** OnCheckedChangeListener 方法 */

    @Override
    public void onCheckedChanged(RadioGroup group,
            int checkedId) {
        switch(checkedId) {
            case R.id.type_satellite:
                mMap.setMapType(GoogleMap.MAP_TYPE_SATELLITE);
                break;
            case R.id.type_normal:
            default:
                mMap.setMapType(GoogleMap.MAP_TYPE_NORMAL);
                break;
        }
    }

    /** OnMarkerClickListener 方法 */

    @Override
    public boolean onMarkerClick(Marker marker) {
        // 返回 true 以禁用自动居中和信息弹出窗口
        return false;
    }

    /** OnMarkerDragListener 方法 */

    @Override
    public void onMarkerDrag(Marker marker) {
        // 在标记移动时执行某些操作
    }

    @Override
    public void onMarkerDragEnd(Marker marker) {
        Log.i("MarkerTest", "Drag " + marker.getTitle()
```

```java
            + " to " + marker.getPosition());
}

@Override
public void onMarkerDragStart(Marker marker) {
    Log.d("MarkerTest", "Drag " + marker.getTitle()
            + " from " + marker.getPosition());
}

/** OnInfoWindowClickListener 方法 */

@Override
public void onInfoWindowClick(Marker marker) {
    // 操作选择事件,在此仅是关闭窗口
    marker.hideInfoWindow();
}

/** InfoWindowAdapter 方法 */

/*
 * 返回将放在标准信息窗口内的内容视图
 * 仅在 getInfoWindow() 返回 null 时调用
 */
@Override
public View getInfoContents(Marker marker) {
    //在此改为尝试返回 createInfoView()
    return null;
}

/*
 * 返回整个待显示的信息窗口
 */
@Override
public View getInfoWindow(Marker marker) {
    View content = createInfoView(marker);
    content.setBackgroundResource(R.drawable.background);
    return content;
}

/*
 * 用于构造内容视图的私有辅助方法
 */
private View createInfoView(Marker marker) {
    // 我们没有父对象用于布局,因此传递 null
    View content = getLayoutInflater().inflate(
            R.layout.info_window, null);
    ImageView image = (ImageView) content
            .findViewById(R.id.image);
    TextView text = (TextView) content
            .findViewById(R.id.text);
```

```java
            image.setImageResource(R.drawable.ic_launcher);
            text.setText(marker.getTitle());

            return content;
        }
        /*
         * 当Play Services缺失或版本不对时,
         * 客户端库将以对话框的形式帮助用户进行更新。
         */
        private void showPlayServicesError(int errorCode) {
            // 从Google Play Services获得错误对话框
            Dialog errorDialog = GooglePlayServicesUtil.getErrorDialog(
                errorCode,
                this,
                1000 /* RequestCode */);
            // 如果Google Play Services可以提供错误对话框
            if(errorDialog != null) {
                // 为错误对话框创建新的DialogFragment
                SupportErrorDialogFragment errorFragment =
                    SupportErrorDialogFragment.newInstance(errorDialog);
                // 在DialogFragment中显示错误对话框
                errorFragment.show(
                    getSupportFragmentManager(),
                    "Google Maps");
            }
        }
    }
```

免责声明:
我们没有实际拜访此地图上的这些位置以了解是否有餐馆,也没有了解这些餐馆的客户评级是否符合我们在此放置的副标题!

我们向 Activity 添加了一些新的侦听器接口,该 Activity 现在设置为监控每个 Marker 上的点击拖动事件,并且监控通过点击 Marker 显示的弹出信息窗口中的点击事件。此外,我们实现了 InfoWindowAdapter,它用于最终定制弹出窗口,但目前先不讨论该适配器。

将 MarkerOptions 实例传入 GoogleMap.addMarker(),这样就可以向地图添加标记。MarkerOptions 的工作方式类似于生成器,它只是在构造函数中将想要应用的所有信息链接在一起(我们已完成该工作)。在 MarkerOptions 中设置一些基本信息,如标记位置、显示图标和标题。还有一些用于修改标记显示的额外选项,如 alpha 值、旋转和锚点。我们选择在位于山景城(Mountain View)的 Google 总部添加一个标记,并且在其附近添加另外两个标记。

有许多支持方法可用于创建 Marker 图标，使用 BitmapDescriptor 对象可应用这些方法，BitmapDescriptorFactory 则提供了创建所有元素的方法。对于我们的两个元素，在此选择了 defaultMarker()方法，该方法创建标准的 Google 大头针图标进行显示。我们还可以传入几个常量之一来控制大头针图标的显示颜色。

我们对位于 Google 总部的标记进行了定制，使用 fromResource()将其定义为应用程序中已有的图标。还可以使用单独的工厂方法应用可能位于资源目录中的图像。此外，我们将此标记设置为可由用户拖动。这意味着如果用户长按大头针图标，则会从其当前位置拾起该图标，将其拖放到地图上的任意位置。我们实现的 OnMarkerDragListener 提供了关于如何放置标记的回调。

如果用户点击某个标记，标准信息窗口将显示在图标上方。该窗口显示应用于标记的标题和代码片段。我们实现了 OnInfoWindowClickListener，在点击窗口时将其关闭，这不是默认行为。

注意，我们不需要实现 OnMarkerClickListener 来实现在此描述的行为，但我们可以重写该行为。默认情况下，信息窗口将显示，并且地图将在所选标记处居中。如果 onMarkerClick()返回 true，则可以禁用此行为并提供我们自己的行为。

1. 定制信息窗口

为了帮助你了解如何定制在点击标记时弹出的信息窗口，接下来为窗口添加一些自定义 UI(参见代码清单 4-10 和 4-11)，并且修改在 Activity 中实现的 InfoWindowAdapter 方法，如代码清单 4-12 所示。

代码清单 4-10　res/layout/info_window.xml

```xml
<?xml version="1.0" encoding="utf-8"?>
<LinearLayout
    xmlns:android="http://schemas.android.com/apk/res/android"
    android:layout_width="wrap_content"
    android:layout_height="wrap_content"
    android:orientation="vertical" >
    <ImageView
        android:id="@+id/image"
        android:layout_width="35dp"
        android:layout_height="35dp"
        android:layout_gravity="center_horizontal"
        android:scaleType="fitCenter" />
    <TextView
        android:id="@+id/text"
        android:layout_width="wrap_content"
        android:layout_height="wrap_content" />
</LinearLayout>
```

代码清单 4-11　res/drawable/background.xml

```xml
<?xml version="1.0" encoding="utf-8"?>
```

```xml
<shape xmlns:android="http://schemas.android.com/apk/res/android"
    android:shape="rectangle">
    <corners
        android:radius="10dp"/>
    <solid
        android:color="#CCC"/>
    <padding
        android:left="10dp"
        android:right="10dp"
        android:top="10dp"
        android:bottom="10dp"/>
</shape>
```

代码清单 4-12　InfoWindowAdapter 方法

```java
/*
 * 返回将放在标准信息窗口内的内容视图
 * 仅在 getInfoWindow() 返回 null 时调用
 */
@Override
public View getInfoContents(Marker marker) {
    // 在此改为尝试返回 createInfoView()
    return null;
}

/*
 * 返回待显示的完整信息窗口
 */
@Override
public View getInfoWindow(Marker marker) {
    View content = createInfoView(marker);
    content.setBackgroundResource(R.drawable.background);
    return content;
}

/*
 * 用于构造内容视图的私有辅助方法
 */
private View createInfoView(Marker marker) {
    // 我们没有父对象用于布局,因此传递 null
    View content = getLayoutInflater().inflate(
            R.layout.info_window, null);
    ImageView image = (ImageView) content
            .findViewById(R.id.image);
    TextView text = (TextView) content
            .findViewById(R.id.text);

    image.setImageResource(R.drawable.ic_launcher);
    text.setText(marker.getTitle());

    return content;
```

}

通过从 getInfoContents()返回有效的视图，该视图就会用作标准窗口背景显示中的内容。从 getInfoWindow()返回相同的视图，该视图会显示为没有标准组件的完全自定义的窗口。我们已将弹出窗口的创建过程抽象化到一个辅助方法中，因此可以轻松尝试上述两种方式。

2. 操作形状

接下来讨论向地图添加形状元素。在下面的示例中，我们创建了名为 ShapeAdapter 的自定义类，该类建立圆形或矩形形状并将它们添加到地图上，用于描述地图地区，结果如图 4-3 所示。

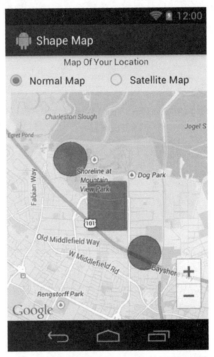

图 4-3　覆盖有可点击形状地区的地图

该例也使用 GoogleMap 的 OnMapClickListener 验证用户何时点击某个地区进行选择。代码清单 4-13 显示了 ShapeAdapter 的代码。

代码清单 4-13　创建地图形状的 ShapeAdapter

```
public class ShapeAdapter implements OnMapClickListener {

    private static final float STROKE_SELECTED = 6.0f;
    private static final float STROKE_NORMAL = 2.0f;
    /* 所绘制区域的颜色 */
    private static final int COLOR_STROKE = Color.RED;
    private static final int COLOR_FILL =
            Color.argb(127, 0, 0, 255);
```

```java
/*
 * 外部接口，用于通知侦听器基于用户点击的所选区域进行变化
 */
public interface OnRegionSelectedListener {
    // 用户选择了跟踪地区
    public void onRegionSelected(Region selectedRegion);
    // 用户选择了没有地区的区域
    public void onNoRegionSelected();
}

/*
 * 地图上交互式地区的基础定义
 * 定义方法以更改显示并检查用户点击
 */
public static abstract class Region {
    private String mRegionName;
    public Region(String regionName) {
        mRegionName = regionName;
    }

    public String getName() {
        return mRegionName;
    }
    // 检查位置是否在此地区内
    public abstract boolean hitTest(LatLng point);
    // 根据用户的选择更改地区的显示
    public abstract void setSelected(boolean isSelected);
}

/*
 * 将地区绘制为圆形
 */
private static class CircleRegion extends Region {
    private Circle mCircle;

    public CircleRegion(String name, Circle circle) {
        super(name);
        mCircle = circle;
    }

    @Override
    public boolean hitTest(LatLng point) {
        final LatLng center = mCircle.getCenter();
        float[] result = new float[1];
        Location.distanceBetween(center.latitude,
                center.longitude,
                point.latitude,
                point.longitude,
                result);
```

```java
            return (result[0] < mCircle.getRadius());
        }

        @Override
        public void setSelected(boolean isSelected) {
            mCircle.setStrokeWidth(isSelected ?
                    STROKE_SELECTED : STROKE_NORMAL);
        }

    }

    /*
     * 将地区绘制为矩形
     */
    private static class RectRegion extends Region {
        private Polygon mRect;
        private LatLngBounds mRectBounds;
        public RectRegion(String name, Polygon rect,
                LatLng southwest, LatLng northeast) {
            super(name);
            mRect = rect;
            mRectBounds = new LatLngBounds(southwest, northeast);
        }

        @Override
        public boolean hitTest(LatLng point) {
            return mRectBounds.contains(point);
        }

        @Override
        public void setSelected(boolean isSelected) {
            mRect.setStrokeWidth(isSelected ?
                    STROKE_SELECTED : STROKE_NORMAL);
        }
    }

    private GoogleMap mMap;

    private OnRegionSelectedListener mRegionSelectedListener;
    private ArrayList<Region> mRegions;
    private Region mCurrentRegion;

    public ShapeAdapter(GoogleMap map) {
        // 在内部跟踪地区以确认选择
        mRegions = new ArrayList<Region>();

        mMap = map;
        mMap.setOnMapClickListener(this);
    }
```

```java
public void setOnRegionSelectedListener(
        OnRegionSelectedListener listener) {
    mRegionSelectedListener = listener;
}

/*
 * 围绕给定点构造并添加新的圆形地区
 */
public void addCircularRegion(String name, LatLng center,
        double radius) {
    CircleOptions options = new CircleOptions()
            .center(center)
            .radius(radius);
    // 设置形状的显示属性
    options
        .strokeWidth(STROKE_NORMAL)
        .strokeColor(COLOR_STROKE)
        .fillColor(COLOR_FILL);

    Circle c = mMap.addCircle(options);
    mRegions.add(new CircleRegion(name, c));
}

/*
 * 使用给定边界构造并添加新的矩形地区
 */
public void addRectangularRegion(String name,
        LatLng southwest, LatLng northeast) {
    PolygonOptions options = new PolygonOptions().add(
            new LatLng(southwest.latitude,
                    southwest.longitude),
            new LatLng(southwest.latitude,
                    northeast.longitude),
            new LatLng(northeast.latitude,
                    northeast.longitude),
            new LatLng(northeast.latitude,
                    southwest.longitude));

    // 设置形状的显示属性
    options
        .strokeWidth(STROKE_NORMAL)
        .strokeColor(COLOR_STROKE)
        .fillColor(COLOR_FILL);

    Polygon p = mMap.addPolygon(options);
    mRegions.add(new RectRegion(name, p,
            southwest, northeast));
}
```

```
    /*
     * 处理从地图对象传入的触摸事件
     * 确定可能选择了哪个地区元素
     * 如果多个地区在此位置重叠,则会选择添加的第一个地区
     */

    @Override
    public void onMapClick(LatLng point) {
        Region newSelection = null;
        // 查找并选择触摸的地区
        for(Region region : mRegions) {
            if(region.hitTest(point) && newSelection == null) {
                region.setSelected(true);
                newSelection = region;
            } else {
                region.setSelected(false);
            }
        }

        if(mCurrentRegion != newSelection) {
            // 通知并更新改动
            if(newSelection != null
                    && mRegionSelectedListener != null) {
                mRegionSelectedListener
                        .onRegionSelected(newSelection);
            } else if(mRegionSelectedListener != null) {
                mRegionSelectedListener.onNoRegionSelected();
            }

            mCurrentRegion = newSelection;
        }
    }
}
```

该类定义了名为 Region 的抽象类型,我们可以使用它定义形状类型之间的常见模式。首先,每个地区必须定义地图位置是否在给定地区内的逻辑,以及在选择该地区时执行哪些操作。然后,为 Circle 和 Polygon 形状定义此逻辑的实现,或者用于绘制矩形。圆形地区由中心点和半径定义,而矩形地区则由其西南角和东北角的点定义。我们构造矩形的方法是使用组成该形状的 4 个角点坐标构造 Polygon。

触摸事件将由侦听器接口的 onMapClick() 方法处理,而 Maps 库提供了作为 LatLng 位置的触摸位置。只需要检查中心点和触摸位置之间的距离是否大于半径,我们就可以验证这些事件在圆形地区内。Location 有一个便利方法可计算两个地图点之间的直接距离。对于矩形地区,我们使用作为 Maps 库一部分的 LatLngBounds 方法,因为它直接验证给定点是在形状的内部还是外部。

对于每个触摸事件,我们遍历地区列表以查找第一个可能包含此位置的地区。如果未找到任何地区,则将所选地区设置为 null。接下来,确定选择项是否已改变,并且调用自定义接口 OnRegionSelectedListener 的某个方法,较高级的对象可以使用该方法获得这些事件的通知。

代码清单 4-14 显示了如何在 Activity 内部使用此适配器。

代码清单 4-14　整合了 ShapeAdapter 的 Activity

```java
public class ShapeMapActivity extends FragmentActivity implements
        RadioGroup.OnCheckedChangeListener,
        ShapeAdapter.OnRegionSelectedListener {
    private static final String TAG = "AndroidRecipes";

    private SupportMapFragment mMapFragment;
    private GoogleMap mMap;

    @Override
    protected void onCreate(Bundle savedInstanceState) {
        super.onCreate(savedInstanceState);
        setContentView(R.layout.main);

        // 检查 Google Play Services 是否已激活且为最新版本
        int resultCode =
                GooglePlayServicesUtil.isGooglePlayServicesAvailable(this);
        switch (resultCode) {
            case ConnectionResult.SUCCESS:
                Log.d(TAG, "Google Play Services is ready to go!");
                break;
            default:
                showPlayServicesError(resultCode);
                return;
        }

        mMapFragment =
                (SupportMapFragment) getSupportFragmentManager()
                        .findFragmentById(R.id.map);
        mMap = mMapFragment.getMap();

        ShapeAdapter adapter = new ShapeAdapter(mMap);
        adapter.setOnRegionSelectedListener(this);

        adapter.addRectangularRegion("Google HQ",
                new LatLng(37.4168, -122.0890),
                new LatLng(37.4268, -122.0790));
        adapter.addCircularRegion("Neighbor #1",
                new LatLng(37.4118, -122.0740), 400);
        adapter.addCircularRegion("Neighbor #2",
                new LatLng(37.4318, -122.0940), 400);

        // 居中地图并同时缩放
        LatLng mapCenter = new LatLng(37.4218, -122.0840);
        CameraUpdate newCamera =
```

```java
            CameraUpdateFactory.newLatLngZoom(mapCenter, 13);
    mMap.moveCamera(newCamera);

    // 连接地图类型选择器UI
    RadioGroup typeSelect =
            (RadioGroup) findViewById(R.id.group_maptype);
    typeSelect.setOnCheckedChangeListener(this);
    typeSelect.check(R.id.type_normal);
}

/** OnCheckedChangeListener 方法 */

@Override
public void onCheckedChanged(RadioGroup group,
        int checkedId) {
    switch(checkedId) {
        case R.id.type_satellite:
            mMap.setMapType(GoogleMap.MAP_TYPE_SATELLITE);
            break;
        case R.id.type_normal:
        default:
            mMap.setMapType(GoogleMap.MAP_TYPE_NORMAL);
            break;
    }
}

/** OnRegionSelectedListener 方法 */

@Override
public void onRegionSelected(Region selectedRegion) {
    Toast.makeText(this, selectedRegion.getName(),
            Toast.LENGTH_SHORT).show();
}

@Override
public void onNoRegionSelected() {
    Toast.makeText(this, "No Region",
            Toast.LENGTH_SHORT).show();
}
}

/*
 * 当Play Services缺失或版本不正确时，客户端库将显示一个对话框，帮助用户进行更新
 */
private void showPlayServicesError(int errorCode) {
    // 获得来自Google Play Services的错误对话框
    Dialog errorDialog = GooglePlayServicesUtil.getErrorDialog(
            errorCode,
```

```
            this,
            1000 /* RequestCode */);
    // 如果Google Play Services可以提供错误对话框
    if(errorDialog != null) {
        // 为错误对话框创建新的DialogFragment
        SupportErrorDialogFragment errorFragment =
            SupportErrorDialogFragment.newInstance(errorDialog);
        // 在DialogFragment中显示错误对话框
        errorFragment.show(
            getSupportFragmentManager(),
            "Google Maps");
        }
    }
}
```

在此添加了与前一个示例相同的位置，但这一次使用新的 ShapeAdapter 将其添加为形状地区。将 Google 总部添加为矩形地区，而将其他两个标记添加为圆形地区。当用户改变选择并影响到任意上述地区时，则会调用 onRegionSelected()或 onNoRegionSelected()方法并显示一条消息。

4.4 监控位置地区

4.4.1 问题

需要应用程序在用户进入或退出特定位置区域时向其提供上下文信息。

4.4.2 解决方案

(API Level 9)

使用作为 Google Play Services 一部分提供的地理围栏功能。借助这些功能，应用程序可以围绕特定点定义圆形区域，在用户移入或离开该区域时，我们希望接收相应的回调。应用程序可以创建多个 Geofence 实例，并且无期限地跟踪这些实例，或者在超出到期时间后自动删除它们。

为跟踪用户到达您认为重要的位置，使用基于地区的用户位置监控可能是更加节能的方法。相比于应用程序持续跟踪用户位置以找出其何时到达给定的目标，以这种方式让服务框架跟踪位置并通知用户通常可以延长电池寿命。

要点：

此处描述的地理围栏功能是 Google Play Services 库的一部分，它们在任意平台级别都不是原生 SDK 的一部分。然而，目标平台为 API Level 8 或以后版本的应用程序以及 Google

Play 体系内的设备都可以使用此绘图库。关于在项目中包括 Google Play Services 的更多信息，请参考 https://developer.android.com/google/play-services/setup.html。

4.4.3 实现机制

我们将创建一个由简单 Activity 组成的应用程序，该 Activity 使用户可以围绕其当前位置设置地理围栏(参见图 4-4)，然后明确地启动或停止监控。

图 4-4 RegionMonitor 的控制 Activity

一旦启用监控，就会激活后台服务以响应与用户位置转移到地理围栏区域内或移出该区域相关的事件。该服务组件使用户可以直接响应这些事件，而无须应用程序的 UI 位于前台。

要点：

因为在此例中我们访问的是用户位置，需要在 AndroidManifest.xml 中请求 android.permission.ACCESS_FINE_LOCATION 权限。

首先查看代码清单 4-15，其中描述了 Activity 的布局。

代码清单 4-15　res/layout/activity_main.xml

```
<?xml version="1.0" encoding="utf-8"?>
<LinearLayout
    xmlns:android="http://schemas.android.com/apk/res/android"
    android:layout_width="match_parent"
    android:layout_height="match_parent"
    android:orientation="vertical">
```

```xml
<TextView
    android:id="@+id/status"
    android:layout_width="match_parent"
    android:layout_height="wrap_content" />
<SeekBar
    android:id="@+id/radius"
    android:layout_width="match_parent"
    android:layout_height="wrap_content"
    android:max="1000"/>
<TextView
    android:id="@+id/radius_text"
    android:layout_width="match_parent"
    android:layout_height="wrap_content" />
<Button
    android:layout_width="match_parent"
    android:layout_height="wrap_content"
    android:text="Set Geofence at My Location"
    android:onClick="onSetGeofenceClick" />

<!-- 间隔区 -->
<View
    android:layout_width="match_parent"
    android:layout_height="0dp"
    android:layout_weight="1" />

<Button
    android:layout_width="match_parent"
    android:layout_height="wrap_content"
    android:text="Start Monitoring"
    android:onClick="onStartMonitorClick" />
<Button
    android:layout_width="match_parent"
    android:layout_height="wrap_content"
    android:text="Stop Monitoring"
    android:onClick="onStopMonitorClick" />
</LinearLayout>
```

该布局包含一个 SeekBar，用于使用户滑动手指来选择所需的半径值。用户可以通过触摸最上方的按钮来锁定新的地理围栏，或者使用底部的按钮启动或停止监控。代码清单 4-16 显示了管理地理围栏监控的 Activity 代码。

代码清单 4-16 设置地理围栏的 Activity

```java
public class MainActivity extends Activity implements
        OnSeekBarChangeListener,
        GooglePlayServicesClient.ConnectionCallbacks,
        GooglePlayServicesClient.OnConnectionFailedListener,
        LocationClient.OnAddGeofencesResultListener,
        LocationClient.OnRemoveGeofencesResultListener {
```

```java
        private static final String TAG = "RegionMonitorActivity";

        // 单个地理围栏的唯一标识符
        private static final String FENCE_ID =
                "com.androidrecipes.FENCE";

        private LocationClient mLocationClient;
        private SeekBar mRadiusSlider;
        private TextView mStatusText, mRadiusText;

        private Geofence mCurrentFence;
        private Intent mServiceIntent;
        private PendingIntent mCallbackIntent;

        @Override
        protected void onCreate(Bundle savedInstanceState) {
            super.onCreate(savedInstanceState);
            setContentView(R.layout.activity_main);

            // 连线 UI 连接
            mStatusText = (TextView) findViewById(R.id.status);
            mRadiusText = (TextView) findViewById(R.id.radius_text);
            mRadiusSlider = (SeekBar) findViewById(R.id.radius);
            mRadiusSlider.setOnSeekBarChangeListener(this);
            updateRadiusDisplay();

            // 检查 Google Play Services 是否为最新版本
            switch(GooglePlayServicesUtil
                    .isGooglePlayServicesAvailable(this)) {
                case ConnectionResult.SUCCESS:
                    // 不执行任何操作，继续
                    break;
                case ConnectionResult.SERVICE_VERSION_UPDATE_REQUIRED:
                    Toast.makeText(this,
                            "Geofencing service requires an update,"
                            + " please open Google Play.",
                            Toast.LENGTH_SHORT).show();
                    finish();
                    return;

                default:
                    Toast.makeText(this,
                            "Geofencing service is not available.",
                            Toast.LENGTH_SHORT).show();
                    finish();
                    return;
            }
            // 为 Google Services 创建客户端
            mLocationClient = new LocationClient(this, this, this);
            // 创建 Intent 以触发服务
```

```java
    mServiceIntent = new Intent(this,
            RegionMonitorService.class);
    // 为 Google Services 回调创建 PendingIntent
    mCallbackIntent = PendingIntent.getService(this, 0,
            mServiceIntent,
            PendingIntent.FLAG_UPDATE_CURRENT);
}

@Override
protected void onResume() {
    super.onResume();
    // 连接所有服务
    if(!mLocationClient.isConnected()
            && !mLocationClient.isConnecting()) {
        mLocationClient.connect();
    }
}

@Override
protected void onPause() {
    super.onPause();
    // 不在前台时断开连接
    mLocationClient.disconnect();
}

public void onSetGeofenceClick(View v) {
    // 通过服务和半径从 UI 获得最新位置
    Location current = mLocationClient.getLastLocation();
    int radius = mRadiusSlider.getProgress();

    // 使用 Builder 创建新的地理围栏
    Geofence.Builder builder = new Geofence.Builder();
    mCurrentFence = builder
        // 此地理围栏的唯一值
        .setRequestId(FENCE_ID)
        // 大小和位置
        .setCircularRegion(
            current.getLatitude(),
            current.getLongitude(),
            radius)
        // 进入和离开地理围栏的事件
        .setTransitionTypes(Geofence.GEOFENCE_TRANSITION_ENTER
                | Geofence.GEOFENCE_TRANSITION_EXIT)
        // 保持活跃
        .setExpirationDuration(Geofence.NEVER_EXPIRE)
        .build();

    mStatusText.setText(String.format(
            "Geofence set at %.3f, %.3f",
            current.getLatitude(),
```

```java
                current.getLongitude()) );
    }

    public void onStartMonitorClick(View v) {
        if(mCurrentFence == null) {
            Toast.makeText(this, "Geofence Not Yet Set",
                    Toast.LENGTH_SHORT).show();
            return;
        }

        // 添加围栏以开始跟踪
        // PendingIntent 将随着新的更新而被触发
        ArrayList<Geofence> geofences = new ArrayList<Geofence>();
        geofences.add(mCurrentFence);
        mLocationClient.addGeofences(geofences,
                mCallbackIntent, this);
    }

    public void onStopMonitorClick(View v) {
        // 移除以停止跟踪
        mLocationClient.removeGeofences(mCallbackIntent, this);
    }

    /** SeekBar 回调 */

    @Override
    public void onProgressChanged(SeekBar seekBar, int progress,
            boolean fromUser) {
        updateRadiusDisplay();
    }

    @Override
    public void onStartTrackingTouch(SeekBar seekBar) { }

    @Override
    public void onStopTrackingTouch(SeekBar seekBar) { }

    private void updateRadiusDisplay() {
        mRadiusText.setText(mRadiusSlider.getProgress()
                + " meters");
    }

    /** Google Services 连接回调 */

    @Override
    public void onConnected(Bundle connectionHint) {
        Log.v(TAG, "Google Services Connected");
    }

    @Override
```

```java
    public void onDisconnected() {
        Log.w(TAG, "Google Services Disconnected");
    }

    @Override
    public void onConnectionFailed(ConnectionResult result) {
        Log.w(TAG, "Google Services Connection Failure");
    }

    /** LocationClient 回调 */

    /*
     * 异步地理围栏添加完成时调用
     * 发生此情况时,启动监控服务
     */
    @Override
    public void onAddGeofencesResult(int statusCode,
            String[] geofenceRequestIds) {
        if(statusCode == LocationStatusCodes.SUCCESS) {
            Toast.makeText(this,
                    "Geofence Added Successfully",
                    Toast.LENGTH_SHORT).show();
        }

        Intent startIntent = new Intent(mServiceIntent);
        startIntent.setAction(RegionMonitorService.ACTION_INIT);
        startService(mServiceIntent);
    }

    /*
     * 异步地理围栏删除完成时调用
     * 调用的版本取决于是通过 PendingIntent 还是请求 ID 来请求删除
     * 发生此情况时,启动监控服务
     */
    @Override
    public void onRemoveGeofencesByPendingIntentResult(
            int statusCode, PendingIntent pendingIntent) {
        if(statusCode == LocationStatusCodes.SUCCESS) {
            Toast.makeText(this, "Geofence Removed Successfully",
                    Toast.LENGTH_SHORT).show();
        }

        stopService(mServiceIntent);
    }

    @Override
    public void onRemoveGeofencesByRequestIdsResult(
            int statusCode, String[] geofenceRequestIds) {
        if(statusCode == LocationStatusCodes.SUCCESS) {
            Toast.makeText(this, "Geofence Removed Successfully",
```

```
            Toast.LENGTH_SHORT).show();
        }

        stopService(mServiceIntent);
    }
}
```

创建此 Activity 之后，第一项工作是确认 Google Play Services 已存在且为最新版本。如果不是最新版本，则需要鼓励用户访问 Google Play 网站以触发最新版本的自动更新。

完成上述工作之后，通过 LocationClient 实例建立与位置服务的连接。我们希望仅在前台时保持此连接，因此在 onResume()和 onPause()之间平衡连接调用。此连接是异步的，因此必须等待 onConnected()方法完成才可以执行进一步的操作。在此例中，我们只需要在用户按下某个按钮时访问 LocationClient，因此，在此方法中没有特别需要完成的工作。

提示：
异步并不一定意味着缓慢。异步方法调用并不意味着预期会花费很长时间。异步仅意味着在函数返回后我们不能立刻访问对象。在大多数情况下，这些回调仍然会在 Activity 完全可见之前触发很长时间。

用户选择所需的半径并点击 Set Geofence 按钮之后，从 LocationClient 获得最新的已知位置，结合选择的半径来构建地理围栏。使用 Geofence.Builder 创建 Geofence 实例，该实例用于设置地理围栏的位置、唯一标识符以及我们可能需要的其他任何属性。

借助 setTransitionTypes()，我们控制哪些过渡会生成通知。有两种可能的过渡值：GEOFENCE_TRANSITION_ENTER 和 GEOFENCE_TRANSITION_EXIT。可以对其中一个事件或两个事件请求回调，在此选择对两个事件请求回调。

正值的到期时间代表从添加地理围栏开始未来的某个时间，到达该时间时应该自动删除地理围栏。设置该值为 NEVER_EXPIRE 可以无期限地跟踪此地区，直至手动将其删除。

在未来的某个时间点，当用户点击 Start Monitoring 按钮时，我们将同时使用 Geofence 和 PendingIntent 调用 LocationClient.addGeofences()来请求更新此地区，框架会为每个新的监控事件激活此方法。注意在我们的示例中，PendingIntent 指向一个服务。该请求也是异步的，当操作完成时，我们将通过 onAddGeofencesResult()收到回调。此时，一条启动命令会发送到我们的后台服务，后面将详细对此进行讨论。

最后，当用户点击 Stop Monitoring 按钮时，就会删除地理围栏，并且新的更新操作将停止。我们使用传入原始请求的相同 PendingIntent 引用删除的元素。也可以使用最初构建时分配的唯一标识符删除地理围栏。异步删除完成之后，一条停止命令会发送到后台服务。

在启动和停止的情况下，我们发送一个 Intent 到具有唯一动作字符串的服务，因此该服务可以区分这些请求与从位置服务收到的更新。代码清单 4-17 显示了我们到目前讨论的此后台服务。

代码清单 4-17 地区监控服务

```
public class RegionMonitorService extends Service {
    private static final String TAG = "RegionMonitorService";
```

```java
private static final int NOTE_ID = 100;
// 标识启动请求与事件的唯一动作
public static final String ACTION_INIT =
        "com.androidrecipes.regionmonitor.ACTION_INIT";

private NotificationManager mNoteManager;

@Override
public void onCreate() {
    super.onCreate();
    mNoteManager = (NotificationManager) getSystemService(
            NOTIFICATION_SERVICE);
    // 在服务启动时发出系统通知
    NotificationCompat.Builder builder =
            new NotificationCompat.Builder(this);
    builder.setSmallIcon(R.drawable.ic_launcher);
    builder.setContentTitle("Geofence Service");
    builder.setContentText("Waiting for transition...");
    builder.setOngoing(true);

    Notification note = builder.build();
    mNoteManager.notify(NOTE_ID, note);
}

@Override
public int onStartCommand(Intent intent, int flags,
        int startId) {
    // 不做任何事，仅是启动服务
    if(ACTION_INIT.equals(intent.getAction())) {
        // 我们不关心此服务是否意外终止
        return START_NOT_STICKY;
    }

    if(LocationClient.hasError(intent)) {
        // 记录任何错误
        Log.w(TAG, "Error monitoring region: "
                + LocationClient.getErrorCode(intent));
    } else {
        // 根据新事件更新进行中的通知
        NotificationCompat.Builder builder =
                new NotificationCompat.Builder(this);
        builder.setSmallIcon(R.drawable.ic_launcher);
        builder.setDefaults(Notification.DEFAULT_SOUND
                | Notification.DEFAULT_LIGHTS);
        builder.setOngoing(true);

        int transitionType =
                LocationClient.getGeofenceTransition(intent);
        // 检查在何处进入或退出地区
```

```java
            if(transitionType ==
                    Geofence.GEOFENCE_TRANSITION_ENTER) {
                builder.setContentTitle("Geofence Transition");
                builder.setContentText("Entered your Geofence");
            } else if(transitionType ==
                    Geofence.GEOFENCE_TRANSITION_EXIT) {
                builder.setContentTitle("Geofence Transition");
                builder.setContentText("Exited your Geofence");
            }

            Notification note = builder.build();
            mNoteManager.notify(NOTE_ID, note);
        }

        // 我们不关心此服务是否意外终止
        return START_NOT_STICKY;
    }

    @Override
    public void onDestroy() {
        super.onDestroy();
        // 服务终止时,取消进行中的通知
        mNoteManager.cancel(NOTE_ID);
    }

    /* 我们未绑定到此服务 */
    @Override
    public IBinder onBind(Intent intent) {
        return null;
    }
}
```

此服务的主要作用是从关于监控地区的位置服务接收更新,并将它们发送到状态栏中的通知,这样用户就可以看到改动。我们在本章后面将详细讨论通知的工作方式以及如何创建它们。

初次创建服务时(按下按钮后发送启动命令时就会发生该操作),会创建初始通知并将其发送到状态栏。这将在第一个 onStartCommand()方法调用之后发生,在该方法中查找唯一的动作字符串,而不做其他任何工作。

上述工作完成之后,第一个地区监控事件将进入此服务,再次调用 onStartCommand()。第一个事件是一个过渡事件,指明关于 Geofence 的设备位置的初始状态。在此例中,我们检查 Intent 是否包含错误消息,如果这是成功的跟踪事件,我们就基于包含在其中的过渡信息构造一条更新的通知,并将更新发送到状态栏。

对于在地区监控激活时收到的每个新事件,该过程会重复进行。当用户最终返回到此 Activity 并按下 Stop Monitoring 按钮时,停止命令将造成在服务中调用 onDestroy()。在此方法中,我们从状态栏中删除通知,向用户表明监控不再激活。

注意：

如果使用同一个 PendingIntent 激活了多个 Geofence 实例，则可以使用另一个方法 LocationClient.getTriggeringGeofences()确定哪些地区是任意给定事件的一部分。

4.5 拍摄照片和视频

4.5.1 问题

应用程序需要使用设备的摄像头采集媒体信息，可以是静态图片或短的视频片段。

4.5.2 解决方案

(API Level 3)

向 Android 发送一个 Intent，将控制权交给 Camera 应用程序，并将用户拍摄的照片返回。Android 已经包含了可以直接访问摄像头硬件、预览、拍照或录制视频的 API。但是，如果只是想简单地让用户在熟悉的界面中使用摄像头拍摄照片和视频，最好的解决方案则是让 Camera 应用程序来处理。

4.5.3 实现机制

让我们看一下如何使用 Camera 应用程序拍摄静止图片和录制视频片段。

1. 拍摄照片

让我们看一个示例 Activity，在按下"Take a Picture"按钮时该 Activity 会激活 Camera 应用程序；在 Camera 应用程序中操作完成后，会得到一张位图结果，参见代码清单 4-18 和 4-19。

代码清单 4-18　res/layout/main.xml

```xml
<?xml version="1.0" encoding="utf-8"?>
<LinearLayout
    xmlns:android="http://schemas.android.com/apk/res/android"
    android:orientation="vertical"
    android:layout_width="match_parent"
    android:layout_height="match_parent">
    <Button
        android:id="@+id/capture"
        android:layout_width="match_parent"
        android:layout_height="wrap_content"
        android:text="Take a Picture" />
    <ImageView
        android:id="@+id/image"
        android:layout_width="match_parent"
```

```xml
            android:layout_height="match_parent"
            android:scaleType="centerInside" />
</LinearLayout>
```

代码清单 4-19　拍摄照片的 Activity

```java
public class MyActivity extends Activity {

    private static final int REQUEST_IMAGE = 100;

    Button captureButton;
    ImageView imageView;

    @Override
    public void onCreate(Bundle savedInstanceState) {
        super.onCreate(savedInstanceState);
        setContentView(R.layout.main);

        captureButton = (Button)findViewById(R.id.capture);
        captureButton.setOnClickListener(listener);

        imageView = (ImageView)findViewById(R.id.image);
    }

    @Override
    protected void onActivityResult(int requestCode, int resultCode,
            Intent data) {
        if(requestCode == REQUEST_IMAGE && resultCode == Activity.RESULT_OK) {
            //处理并显示图片
            Bitmap userImage = (Bitmap)data.getExtras().get("data");
            imageView.setImageBitmap(userImage);
        }
    }

    private View.OnClickListener listener = new View.OnClickListener() {
        @Override
        public void onClick(View v) {
            try{
                Intent intent = new Intent(MediaStore.ACTION_IMAGE_CAPTURE);
                startActivityForResult(intent, REQUEST_IMAGE);
            }catch (ActivityNotFoundException e) {
                //如果没有其他应用程序，则进行处理
            }
        }
    };
}
```

在此例中，我们构造一个 Intent 来激活 Camera 应用程序并拍摄照片。虽然不太可能，但我们希望为设备上没有 Camera 应用程序的情况做准备。在此情况下，调用 startActivity() 将抛出 ActivityNotFoundException，因此我们在 try 块中封装此调用，以便适当处理异常。

提示:

还可以使用 PackageManager.hasSystemFeature()方法并传入 PackageManager.FEATURE_CAMERA 作为参数,以此来查询摄像头硬件是否存在。

这个方法会拍摄照片并在 "data" 附加信息字段中返回按比例缩小的位图。如果需要拍摄全尺寸的照片,则需要在开始拍照前,在 Intent 的 MediaStore.EXTRA_OUTPUT 中放入照片保存位置的 Uri。参见代码清单 4-20。

代码清单 4-20 将全尺寸照片保存到文件中

```java
public class MyActivity extends Activity {

    private static final int REQUEST_IMAGE = 100;

    Button captureButton;
    ImageView imageView;
    File destination;

    @Override
    public void onCreate(Bundle savedInstanceState) {
        super.onCreate(savedInstanceState);
        setContentView(R.layout.main);

        captureButton = (Button)findViewById(R.id.capture);
        captureButton.setOnClickListener(listener);

        imageView = (ImageView)findViewById(R.id.image);

        destination =
            new File(Environment.getExternalStorageDirectory(),"image.jpg");
    }

    @Override
    protected void onActivityResult(int requestCode, int resultCode,
            Intent data) {
        if(requestCode == REQUEST_IMAGE && resultCode == Activity.RESULT_OK) {
            try {
                FileInputStream in = new FileInputStream(destination);
                BitmapFactory.Options options = new BitmapFactory.Options();
                options.inSampleSize = 10; //降低采样率10倍

                Bitmap userImage =
                        BitmapFactory.decodeStream(in, null, options);
                imageView.setImageBitmap(userImage);
            } catch(Exception e) {
                e.printStackTrace();
            }
        }
    }
}
```

```
        private View.OnClickListener listener = new View.OnClickListener() {
            @Override
            public void onClick(View v) {
                try {
                    Intent intent = new Intent(MediaStore.ACTION_IMAGE_CAPTURE);
                    //添加附加信息来保存全尺寸的图片
                    intent.putExtra(MediaStore.EXTRA_OUTPUT,
                            Uri.fromFile(destination));
                    startActivityForResult(intent, REQUEST_IMAGE);
                } catch(ActivityNotFoundException e) {
                    //如果没有其他应用程序，则进行处理
                }
            }
        };
    }
```

这个方法会告诉 Camera 应用程序将图片保存到某个位置(本例为设备 SD 卡上的"image.jpg")，这样图片就不会缩放了。在 Camera 应用程序中的操作完成后，就可以直接到刚才指定摄像头存放的文件位置寻找图片。

提示：
相关文档声明只应该产生一张图片输出。如果 Uri 不存在，则会返回一张小图片作为数据。否则，图片保存到 Uri 位置。不应预期会同时接收两者，即使市场上的一些设备以这种方式操作。

为了避免将全尺寸的图片加载到内存中，需要通过 BitmapFactory.Options 对尚未在屏幕上显示的图片进行缩放。同样，你会发现这个示例会将保存图片的文件位置指定为设备的外部存储器，这就需要在 API Level 4 及以上版本中指定 android.permission.WRITE_EXTERNAL_STORAGE 权限。如果最终决定在其他地方存储的话，则不需要这个权限。

2. 拍摄视频

通过这种方法可以直接拍摄视频片段，只是结果稍有差异。真正的视频片段数据是不会通过 Intent 的附加信息直接返回的，通常都是保存到指定的文件位置。以下两个参数可能会作为附加信息添加到 Intent 中：

- MediaStore.EXTRA_VIDEO_QUALITY：整型值，用来描述拍摄视频的质量等级。0 表示低质量，1 表示高质量。
- MediaStore.EXTRA_OUTPUT：保存视频内容的 Uri 目的地。如果没有指定保存位置，视频则会保存到设备的标准位置。

视频录制结束后，会在结果 Intent 的 data 字段中返回视频数据真正存储位置的 Uri。让我们看一个类似的示例，该例允许用户录制和保存他们的视频并将视频的存储位置显示在屏幕上。参见代码清单 4-21 和 4-22。

代码清单 4-21　res/layout/main.xml

```xml
<?xml version="1.0" encoding="utf-8"?>
<LinearLayout
    xmlns:android="http://schemas.android.com/apk/res/android"
    android:orientation="vertical"
    android:layout_width="match_parent"
    android:layout_height="match_parent">
    <Button
        android:id="@+id/capture"
        android:layout_width="match_parent"
        android:layout_height="wrap_content"
        android:text="Take a Video" />
    <TextView
        android:id="@+id/file"
        android:layout_width="match_parent"
        android:layout_height="match_parent" />
</LinearLayout>
```

代码清单 4-22　拍摄视频片段的 Activity

```java
public class MyActivity extends Activity {

    private static final int REQUEST_VIDEO = 100;

    Button captureButton;
    TextView text;
    File destination;

    @Override
    public void onCreate(Bundle savedInstanceState) {
        super.onCreate(savedInstanceState);
        setContentView(R.layout.main);

        captureButton = (Button)findViewById(R.id.capture);
        captureButton.setOnClickListener(listener);

        text = (TextView)findViewById(R.id.file);

        destination =
            new File(Environment.getExternalStorageDirectory(),"myVideo");
    }

    @Override
    protected void onActivityResult(int requestCode, int resultCode,
            Intent data) {
        if(requestCode == REQUEST_VIDEO && resultCode == Activity.RESULT_OK) {
            String location = data.getData().toString();
            text.setText(location);
        }
    }
}
```

```
        private View.OnClickListener listener = new View.OnClickListener() {
            @Override
            public void onClick(View v) {
                Intent intent = new Intent(MediaStore.ACTION_VIDEO_CAPTURE);
                //添加(可选)附加信息以将视频保存到指定文件
                intent.putExtra(MediaStore.EXTRA_OUTPUT,
                    Uri.fromFile(destination));
                //可选的附加信息用来设置视频质量
                intent.putExtra(MediaStore.EXTRA_VIDEO_QUALITY, 0);
                startActivityForResult(intent, REQUEST_VIDEO);
            } catch(ActivityNotFoundException e) {
                //如果没有其他应用程序，则进行处理
        };
    }
```

本例和之前保存图片的示例一样，都是将录制的视频保存到设备的 SD 卡上(对于 API Level 4 及以上版本需要指定 android.permission.WRITE_EXTERNAL_STORAGE 权限)。首先，我们向系统发送了一个动作为 MediaStore.ACTION_VIDEO_CAPTURE 的 Intent。Android 则会启动默认的 Camera 应用程序来处理视频的录制并在录制完成时返回 OK。我们会在 onActivityResult()回调方法中调用 Intent.getData()来得到数据的存储位置，并把这个位置显示给用户。

这个示例显式地要求录制的视频使用低质量设置，但这个参数是可选的。如果没有在请求的 Intent 中指定 MediaStore.EXTRA_VIDEO_QUALITY，设备通常会选择高质量进行录制。

一旦指定 MediaStore.EXTRA_OUTPUT，返回的 Uri 就应该和请求的位置一致，除非应用程序在写入该位置时发生了错误。如果没有这个参数，返回的值就会是 content://Uri，可以从系统的 MediaStore Content Provider 中获取视频。

稍后在 4.10 节中，我们将介绍如何在应用程序中播放媒体文件。

4.6 自定义摄像头覆盖层

4.6.1 问题

很多应用程序都希望能够直接访问摄像头，例如要使用自定义的摄像头控制 UI 界面或者根据位置和方向传感器的信息显示其他元数据(增强真实感)。

4.6.2 解决方案

(API Level 5)

直接通过 Activity 视图层次结构中的摄像头硬件来渲染预览框架。Android 提供了 API 来直接访问设备的摄像头(用于获得预览画面以及拍摄照片)。如果应用程序除了拍摄并显示照片外还要使用其他功能，也可以直接访问这些 API。

注意：

因为需要直接访问摄像头，所以必须在清单中指定 android.permission.CAMERA 权限。

4.6.3 实现机制

我们首先会创建一个 SurfaceView，将使用这个专门的视图实时地绘制摄像头的预览数据流。这样就可以将一幅实时预览图内嵌到 Activity 的所选布局中。接下来，只要简单地根据应用程序的需求添加其他的视图和控件即可。让我们看一下代码(参见代码清单 4-23 和 4-24)。

注意：

这里使用的 Camera 类是 android.hardware.Camera，不要和 android.graphics.Camera 弄混了。确保应用程序已经导入正确的引用。

代码清单 4-23　res/layout/main.xml

```xml
<?xml version="1.0" encoding="utf-8"?>
<RelativeLayout
    xmlns:android="http://schemas.android.com/apk/res/android"
    android:layout_width="match_parent"
    android:layout_height="match_parent">
    <SurfaceView
        android:id="@+id/preview"
        android:layout_width="match_parent"
        android:layout_height="match_parent" />
</RelativeLayout>
```

代码清单 4-24　显示活动的摄像头预览的 Activity

```java
import android.hardware.Camera;

public class PreviewActivity extends Activity implements SurfaceHolder.Callback {

    Camera mCamera;
    SurfaceView mPreview;

    @Override
    public void onCreate(Bundle savedInstanceState) {
        super.onCreate(savedInstanceState);
        setContentView(R.layout.main);

        mPreview = (SurfaceView)findViewById(R.id.preview);
        mPreview.getHolder().addCallback(this);
        //需要支持 Android 3.0 之前的版本
        mPreview.getHolder().setType(SurfaceHolder.SURFACE_TYPE_PUSH_BUFFERS);

        mCamera = Camera.open();
    }
```

```java
    @Override
    public void onPause() {
        super.onPause();
        mCamera.stopPreview();
    }

    @Override
    public void onDestroy() {
        super.onDestroy();
        mCamera.release();
    }

    //Surface 回调方法
    @Override
    public void surfaceChanged(SurfaceHolder holder, int format,
            int width, int height) {
        Camera.Parameters params = mCamera.getParameters();
        //得到设备支持的尺寸,并选择第一个尺寸(最大)
        List<Camera.Size> sizes = params.getSupportedPreviewSizes();
        Camera.Size selected = sizes.get(0);
        params.setPreviewSize(selected.width,selected.height);
        mCamera.setParameters(params);

        mCamera.startPreview();
    }

    @Override
    public void surfaceCreated(SurfaceHolder holder) {
        try {
            mCamera.setPreviewDisplay(mPreview.getHolder());
        } catch(Exception e) {
            e.printStackTrace();
        }
    }

    @Override
    public void surfaceDestroyed(SurfaceHolder holder) { }
}
```

注意:

如果在模拟器上进行测试的话,就不能预览摄像头。新版本的 SDK 已经可以在一些主机上使用内置的摄像头,但并不是全部都可以使用。对于摄像头不可用的情况,模拟器会显示一幅假的预览画面,不同的模拟器版本效果稍微会有些不同。要验证这段代码是否正确工作,可打开模拟器上的 Camera 应用程序,看看其中预览的内容是什么。所显示的效果应该和真实设备上的相同。通常最好是在运行有相应硬件的真实设备上进行代码的测试。

在本例中,我们创建了一个 SurfaceView 来填满整个窗口,并且指定在所有 SurfaceHolder 回调时通知我们的 Activity。在摄像头初始化工作彻底完成之前就不会在

Surface 上显示预览信息，所以我们在 surfaceCreated()回调时才将视图的 SurfaceHolder 关联到 Camera 实例上。类似地，当 surfaceChanged()回调时，也就是 Surface 确定大小后，我们才能确定预览画面的尺寸并开始绘制。

这个应用程序的摄像头硬件资源是通过调用 Camera.open()来声明和打开的。Android 2.3(API Level 9)中提供了该方法的另一个版本，在设备有多个摄像头的情况下，该方法使用一个整型参数(值为 0 到 getNumberOfCameras()-1)来指定使用哪个摄像头。在多摄像头的设备中，无参数的版本通常会打开后置的摄像头。

要点：

很多新设备(如 Google 的 Nexus 7 2012 平板电脑)并没有后置摄像头，所以旧的 Camera.open()方法实现会返回 null。如果有一个支持旧版本 Android 的 Camera 应用程序，需要修改代码以实现分支，并使用新的 API 来判断设备提供了哪些摄像头。

调用 Parameters.getSupportedPreviewSizes()会返回设备支持的各种尺寸列表，通常是从大到小进行排序。在本例中，我们使用第一个(最大的)预览分辨率并使用它设置预览大小。

注意：

在 Android 2.0(API Level 5)之前的版本中，可以直接在 Parameters.setPreviewSize()方法中传入高度和宽度参数。但在 Android 2.0 及以后的版本中，摄像头的预览尺寸只可以设置为设备支持的分辨率；否则会导致抛出异常。

Camera.startPreview()会开始在 Surface 上实时绘制摄像头的数据。注意，预览画面始终是横屏显示的。在 Android 2.2 (API Level 8)之前，并没有调整预览画面显示方向的官方方法。因此，建议使用摄像头预览画面的 Activity 要在清单中指定固定的方向：android:screenOrientation="landscape"，这样才能兼容使用旧版本的设备。

Camera 服务在某一个时刻只能由一个应用程序使用。因此，在不再需要摄像头时及时调用 Camera.release()方法是很重要的。本例中，在 Activity 销毁时，我们不再需要使用摄像头，因此在 onDestroy()中调用了该方法。

1. 改变拍摄方向

(API Level 8)

从 Android 2.2 开始，可以改变摄像头的预览方向。应用程序现在可以调用 Camera.setDisplayOrientation()来旋转得到的数据，从而匹配 Activity 的方向。有效的值为 0°、90°、180°、270°，0°表示默认的横屏显示。这个方法主要会影响视频采集前预览数据该如何在 Surface 上进行绘制。

要想旋转摄像头输出的数据，需要使用 Camera.Parameters 的 setRotation()方法。这个方法的实现效果取决于设备，它也许会旋转实际输出的图像，也许会通过旋转参数更新 EXIF 数据，或者两者皆有。

2. 照片预览覆盖层

现在我们可以在前面示例中的照片预览界面上添加各种控件或视图。以下代码会在预览中添加 Cancel 按钮和 Snap Photo 按钮。参见代码清单 4-25 和 4-26。

代码清单 4-25   res/layout/main.xml

```xml
<?xml version="1.0" encoding="utf-8"?>
<RelativeLayout
    xmlns:android="http://schemas.android.com/apk/res/android"
    android:layout_width="match_parent"
    android:layout_height="match_parent">
    <SurfaceView
        android:id="@+id/preview"
        android:layout_width="match_parent"
        android:layout_height="match_parent" />
    <RelativeLayout
        android:layout_width="match_parent"
        android:layout_height="100dip"
        android:layout_alignParentBottom="true"
        android:gravity="center_vertical"
        android:background="#A000">
        <Button
            android:layout_width="100dip"
            android:layout_height="wrap_content"
            android:text="Cancel"
            android:onClick="onCancelClick" />
        <Button
            android:layout_width="100dip"
            android:layout_height="wrap_content"
            android:layout_alignParentRight="true"
            android:text="Snap Photo"
            android:onClick="onSnapClick" />
    </RelativeLayout>
</RelativeLayout>
```

代码清单 4-26   添加了照片控制的 Activity

```java
public class PreviewActivity extends Activity implements
    SurfaceHolder.Callback, Camera.ShutterCallback, Camera.PictureCallback {

    Camera mCamera;
    SurfaceView mPreview;

    @Override
    public void onCreate(Bundle savedInstanceState) {
        super.onCreate(savedInstanceState);
        setContentView(R.layout.main);

        mPreview = (SurfaceView)findViewById(R.id.preview);
        mPreview.getHolder().addCallback(this);
        //需要支持 Android 3.0 之前的版本
        mPreview.getHolder().setType(SurfaceHolder.SURFACE_TYPE_PUSH_BUFFERS);
```

```java
    mCamera = Camera.open();
}

@Override
public void onPause() {
    super.onPause();
    mCamera.stopPreview();
}

@Override
public void onDestroy() {
    super.onDestroy();
    mCamera.release();
    Log.d("CAMERA","Destroy");
}

public void onCancelClick(View v) {
    finish();
}

public void onSnapClick(View v) {
    //拍摄照片
    mCamera.takePicture(this, null, null, this);
}

//Camera 回调方法
@Override
public void onShutter() {
    Toast.makeText(this, "Click!", Toast.LENGTH_SHORT).show();
}

@Override
public void onPictureTaken(byte[] data, Camera camera) {

    //可以将照片保存到某个位置，这里保存到外部存储器中
    try {
        File directory =
            Environment.getExternalStoragePublicDirectory(
                Environment.DIRECTORY_PICTURES);
        FileOutputStream out =
            new FileOutputStream(new File(directory, "picture.jpg"));
        out.write(data);
        out.flush();
        out.close();
    } catch(FileNotFoundException e) {
        e.printStackTrace();
    } catch(IOException e) {
        e.printStackTrace();
    }
```

```java
        //必须重新启动预览
        camera.startPreview();
    }

    //Surface 回调方法
    @Override
    public void surfaceChanged(SurfaceHolder holder, int format,
            int width, int height) {
        Camera.Parameters params = mCamera.getParameters();
        List<Camera.Size> sizes = params.getSupportedPreviewSizes();
        Camera.Size selected = sizes.get(0);
        params.setPreviewSize(selected.width,selected.height);
        mCamera.setParameters(params);

        mCamera.setDisplayOrientation(90);
        mCamera.startPreview();
    }

    @Override
    public void surfaceCreated(SurfaceHolder holder) {
        try {
            mCamera.setPreviewDisplay(mPreview.getHolder());
        } catch(Exception e) {
            e.printStackTrace();
        }
    }
    @Override
    public void surfaceDestroyed(SurfaceHolder holder) { }
}
```

这里我们添加了一个简单的、部分区域透明的覆盖层，它包含两个用于摄像头操作的控件。Cancel 执行的操作没有什么可说的，就是结束 Activity。但是，Snap Photo 在手动拍摄并返回照片到应用程序的过程中涉及很多 Camera API。用户会初始化 Camera.takePicture() 方法，然后是该方法的一系列回调方法。

注意，本例中的 Activity 还额外实现了两个接口：Camera.ShutterCallback 和 Camera.PictureCallback。前者是在拍摄照片后(快门关闭后)立即被调用，而后者则在有不同形式的图片可用时在多个实例中被调用。

takePicture()的参数是一个 ShutterCallback 和不超过三个的 PictureCallback 实例。PictureCallback 会在以下时刻被调用(根据参数的顺序)：

- 照片以 RAW 图片数据形式被拍摄。在设备内存不足时，可能返回 null。
- 在图片被处理为缩放后的数据(也就是 POSTVIEW 图片)后。在设备内存不足时，可能返回 null。
- 在图片被压缩为 JPEG 图像数据后。

这个示例只用到了压缩为 JPEG 格式后的照片。因此，其中只用到了最后一个回调方法，也就是照片预览必须再次重新启动时。如果在照片拍摄后没有再次调用 startPreview()，屏幕上的预览画面就会一直停留在刚刚拍摄的照片上。

提示：
如果希望应用程序只能由拥有相应硬件功能的设备下载，可以在应用程序的清单中添加摄像头的市场过滤器：<uses-feature android:name="android.hardware.camera" />。

4.7 录制音频

4.7.1 问题

应用程序需要使用设备的麦克风录制音频输入。

4.7.2 解决方案

(API Level 1)
通过 MediaRecorder 采集音频并保存到文件中。

4.7.3 实现机制

MediaRecorder 用起来非常简单。只需要提供一些文件编码格式的基本信息和保存音频数据的位置。代码清单 4-27 和 4-28 演示了如何录制音频文件并保存到设备的 SD 卡中，这期间会监听用户开始录音和结束录音的动作。

要点：
要想通过 MediaRecorder 录制音频输入，必须在应用程序的清单中声明 android.permission.RECORD_AUDIO 权限。

代码清单 4-27　res/layout/main.xml

```xml
<?xml version="1.0" encoding="utf-8"?>
<LinearLayout
    xmlns:android="http://schemas.android.com/apk/res/android"
    android:orientation="vertical"
    android:layout_width="match_parent"
    android:layout_height="match_parent">
    <Button
        android:id="@+id/startButton"
        android:layout_width="match_parent"
        android:layout_height="wrap_content"
        android:text="Start Recording" />
    <Button
        android:id="@+id/stopButton"
        android:layout_width="match_parent"
        android:layout_height="wrap_content"
        android:text="Stop Recording"
        android:enabled="false" />
</LinearLayout>
```

代码清单 4-28　录制音频的 Activity

```java
public class RecordActivity extends Activity {

    private MediaRecorder recorder;
    private Button start, stop;
    File path;

    @Override
    public void onCreate(Bundle savedInstanceState) {
        super.onCreate(savedInstanceState);
        setContentView(R.layout.main);

        start = (Button)findViewById(R.id.startButton);
        start.setOnClickListener(startListener);
        stop = (Button)findViewById(R.id.stopButton);
        stop.setOnClickListener(stopListener);

        recorder = new MediaRecorder();
        path = new File(Environment.getExternalStorageDirectory(),
                "myRecording.3gp");

        resetRecorder();
    }

    @Override
    public void onDestroy() {
        super.onDestroy();
        recorder.release();
    }

    private void resetRecorder() {
        recorder.setAudioSource(MediaRecorder.AudioSource.MIC);
        recorder.setOutputFormat(
                MediaRecorder.OutputFormat.THREE_GPP);
        recorder.setAudioEncoder(
                MediaRecorder.AudioEncoder.DEFAULT);
        recorder.setOutputFile(path.getAbsolutePath());
        try {
            recorder.prepare();
        } catch(Exception e) {
            e.printStackTrace();
        }
    }

    private View.OnClickListener startListener =
            new View.OnClickListener() {
        @Override
```

```java
        public void onClick(View v) {
            try {
                recorder.start();

                start.setEnabled(false);
                stop.setEnabled(true);
            } catch(Exception e) {
                e.printStackTrace();
            }
        }
    };

    private View.OnClickListener stopListener =
            new View.OnClickListener() {
        @Override
        public void onClick(View v) {
            recorder.stop();
            resetRecorder();
            start.setEnabled(true);
            stop.setEnabled(false);
        }
    };
}
```

这个示例的 UI 非常简单。就是两个按钮，它们的作用是基于录制状态的交替变化。当用户按下 Start 按钮时，就可以开始录音并且 Stop 按钮也会可用。当按下 Stop 按钮时，将重置录音机以便进行下一次录音，同时 Start 按钮也会重新可用。

MediaRecorder 的设置也很简单。我们在 SD 卡上创建了一个名为 "myRecording.3gp" 的文件，并将文件路径传给 setOutputFile()方法。而其他的设置方法则告诉录音机将使用设备的麦克风作为音频输入源(AudioSource.MIC)，并使用默认的编码器创建一个 3GP 格式的文件作为输出。

现在，可以使用设备的文件浏览器或媒体播放器应用程序来播放这个音频文件。在后面的 4.10 节中，还将展示如何使用应用程序播放音频。

4.8 自定义视频采集

4.8.1 问题

应用程序需要视频采集，同时需要能够比 4.5 节所示范例更多地控制视频录制的过程。

4.8.2 解决方案

(API Level 8)
通过 MediaRecorder 和 Camera 直接彼此合作来创建自己的视频采集 Activity。和之前范例中使用 MediaRecorder 单纯地进行录制音频相比,这个示例稍微有点复杂。我们希望在没有录制音频时用户也可以看到摄像头的预览,要实现这一点,必须在录制与预览之间管理好摄像头的访问控制。

4.8.3 实现机制

代码清单 4-29 到 4-31 演示了录制视频并将结果保存到设备外部存储器的示例。

代码清单 4-29　AndroidManifest.xml 的部分代码

```xml
<uses-permission android:name="android.permission.RECORD_AUDIO" />
<uses-permission android:name="android.permission.CAMERA" />
<uses-permission
    android:name="android.permission.WRITE_EXTERNAL_STORAGE" />

...
<activity
    android:name=".VideoCaptureActivity"
    android:screenOrientation="portrait" >
    <intent-filter>
        <action android:name="android.intent.action.MAIN" />
        <category
            android:name="android.intent.category.LAUNCHER" />
    </intent-filter>
</activity>
```

在这个清单中,核心的目的就是将 Activity 的方向设置为固定的竖屏。另外还指定了一些访问摄像头和录制音频音轨时需要的少量权限。

代码清单 4-30　res/layout/main.xml

```xml
<RelativeLayout
    xmlns:android="http://schemas.android.com/apk/res/android"
    android:layout_width="match_parent"
    android:layout_height="match_parent" >

    <Button
        android:id="@+id/button_record"
        android:layout_width="match_parent"
        android:layout_height="wrap_content"
        android:layout_alignParentBottom="true"
        android:onClick="onRecordClick" />

    <SurfaceView
        android:id="@+id/surface_video"
        android:layout_width="match_parent"
        android:layout_height="match_parent"
        android:layout_above="@id/button_record" />
</RelativeLayout>
```

代码清单4-31　采集视频的Activity

```java
public class VideoCaptureActivity extends Activity implements SurfaceHolder.Callback {

    private Camera mCamera;
    private MediaRecorder mRecorder;

    private SurfaceView mPreview;
    private Button mRecordButton;

    private boolean mRecording = false;

    @Override
    public void onCreate(Bundle savedInstanceState) {
        super.onCreate(savedInstanceState);
        setContentView(R.layout.main);

        mRecordButton = (Button) findViewById(R.id.button_record);
        mRecordButton.setText("Start Recording");

        mPreview = (SurfaceView) findViewById(R.id.surface_video);
        mPreview.getHolder().addCallback(this);
        mPreview.getHolder().setType(SurfaceHolder.SURFACE_TYPE_PUSH_BUFFERS);

        mCamera = Camera.open();
        //旋转预览画面为竖屏
        mCamera.setDisplayOrientation(90);
        mRecorder = new MediaRecorder();
    }

    @Override
    protected void onDestroy() {
        mCamera.release();
        mCamera = null;
        super.onDestroy();
    }

    public void onRecordClick(View v) {
        updateRecordingState();
    }

    /*
    *初始化摄像头和摄像机
    *这些方法的顺序很重要,这是因为MediaRecorder的状态严格依赖于每个调用方法
    */
    private void initializeRecorder() throws IllegalStateException,
            IOException {
        //解锁摄像头,允许MediaRecorder使用它
        mCamera.unlock();
        mRecorder.setCamera(mCamera);
```

```java
//设置 MediaRecorder 的数据源
mRecorder.setAudioSource(MediaRecorder.AudioSource.CAMCORDER);
mRecorder.setVideoSource(MediaRecorder.VideoSource.CAMERA);
//更新输出设置
File recordOutput = new File(Environment.getExternalStorageDirectory(),
        "recorded_video.mp4");
if(recordOutput.exists()) {
    recordOutput.delete();
}
CamcorderProfile cpHigh =
        CamcorderProfile.get(CamcorderProfile.QUALITY_HIGH);
mRecorder.setProfile(cpHigh);
mRecorder.setOutputFile(recordOutput.getAbsolutePath());
//为摄像机关联一个 Surface，从而实现在录制的同时可以预览
mRecorder.setPreviewDisplay(mPreview.getHolder().getSurface());

//设置录制的一些限制值，这些值是可选的
mRecorder.setMaxDuration(50000); // 50 秒
mRecorder.setMaxFileSize(5000000); // 大约 5MB

mRecorder.prepare();
}

private void updateRecordingState() {
    if(mRecording) {
        mRecording = false;
        //重置摄像机的状态以便进行下次录制
        mRecorder.stop();
        mRecorder.reset();
        //返回摄像机继续预览
        mCamera.lock();
        mRecordButton.setText("Start Recording");
    } else {
        try {
            //重置摄像机以便进行下次会话
            initializeRecorder();
            //开始录制
            mRecording = true;
            mRecorder.start();
            mRecordButton.setText("Stop Recording");
        } catch(Exception e) {
            //初始化摄像机时发生错误
            e.printStackTrace();
        }
    }
}

@Override
public void surfaceCreated(SurfaceHolder holder) {
```

```
//得到一个 Surface 后，立刻启动摄像头预览
try {
    mCamera.setPreviewDisplay(holder);
    mCamera.startPreview();
} catch(IOException e) {
    e.printStackTrace();
}
}

@Override
public void surfaceChanged(SurfaceHolder holder, int format, int width,
        int height) { }

@Override
public void surfaceDestroyed(SurfaceHolder holder) { }
}
```

当首次创建 Activity 时，它得到了设备上的摄像头的一个实例，并且将摄像头的显示方向设置为我们在清单中指定的竖屏。调用这个方法只会影响到预览画面的显示内容，而不会影响录制输出，在后面会更多地讨论它。当 Activity 可见时，会收到 surfaceCreated() 回调，这时 Camera 开始发送预览数据。

当用户打算按下按钮开始录制时，Camera 会被解锁并交给 MediaRecorder 使用。然后 MediaRecorder 会设置进行视频采集所需的参数，如数据源和数据格式，以及时间和文件大小限制，这是为了防止用户的存储器负载过大。

注意：

使用 MediaRecorder 但不直接管理 Camera 也是可以录制视频的，但不能修改屏幕显示的方向，并且应用程序在录制视频时只能显示预览画面。

录制完成后，会在用户的外部存储器中自动保存一个文件，并且重置摄像机实例以便用户进行下一次录制。我们也将重新获得 Camera 的控制权，从而可以继续显示预览画面。

输出格式方向

(API Level 9)

本例中，使用 Camera.setDisplayOrientation()使得预览画面的方向可以匹配我们的竖屏 Activity。但是很多时候，如果在计算机上播放这段视频，这个界面依然会是横屏的。要想解决这个问题，可以使用 MediaRecorder 的 setOrientationHint()方法。这个方法的参数是一个与画面显示方向相匹配的角度值，通过这个值，视频文件(例如 3GP 和 MP4 文件)的元数据会通知其他的视频播放器应用程序，在播放该视频文件时需要把方向旋转一下。

这可能是不必要的，因为一些视频播放器在决定播放方向时基于哪个方向上视频尺寸比较小。出于这个原因，而且也为了能够和 API Level 8 相兼容，我们并没有在示例中加入这个方法。

4.9 添加语音识别

4.9.1 问题

应用程序需要使用语音识别技术来解析语音输入。

4.9.2 解决方案

(API Level 3)

通过 android.speech 包中的类可以使用每台 Android 设备上内置的语音识别技术。每台配备了语音搜索(从 Android 开始)的 Android 设备都可以让应用程序使用设备内置的 SpeechRecognizer 来处理语音输入。

要想使用这个功能,应用程序只需要向系统发送一个 RecognizerIntent,然后语音识别服务就会记录语音输入并处理;最后语音识别器就会将它听到的语音以字符串列表的形式返回给你。

4.9.3 实现机制

让我们用一个示例来了解这项技术。参见代码清单 4-32。

代码清单 4-32　启动和处理语音识别的 Activity

```java
public class RecognizeActivity extends Activity {

    private static final int REQUEST_RECOGNIZE = 100;

    TextView tv;

    @Override
    public void onCreate(Bundle savedInstanceState) {
        super.onCreate(savedInstanceState);
        tv = new TextView(this);
        setContentView(tv);

        Intent intent = new Intent(RecognizerIntent.ACTION_RECOGNIZE_SPEECH);
        intent.putExtra(RecognizerIntent.EXTRA_LANGUAGE_MODEL,
                    RecognizerIntent.LANGUAGE_MODEL_FREE_FORM);
        intent.putExtra(RecognizerIntent.EXTRA_PROMPT, "Tell Me Your Name");
        try {
            startActivityForResult(intent, REQUEST_RECOGNIZE);
        } catch(ActivityNotFoundException e) {
            //如果不存在任何识别器,就从 Google Play 下载一个
            showDownloadDialog();
        }
    }
```

```java
        private void showDownloadDialog() {
            AlertDialog.Builder builder = new AlertDialog.Builder(this);
            builder.setTitle("Not Available");
            builder.setMessage("There is no recognition application
                installed." + " Would you like to download one?");
            builder.setPositiveButton("Yes",
                    new DialogInterface.OnClickListener() {
                @Override
                public void onClick(DialogInterface dialog, int which) {
                    //例如下载Google Voice Search
                    Intent marketIntent = new Intent(Intent.ACTION_VIEW);
                    marketIntent.setData(
                        Uri.parse(
                            "market://details? "+
                         "id=com.google.android.voicesearch"
                        ) );
                }
            });
            builder.setNegativeButton("No", null);
            builder.create().show();
        }
    }

    @Override
    protected void onActivityResult(int requestCode, int resultCode,
            Intent data) {
        if(requestCode == REQUEST_RECOGNIZE &&
                resultCode == Activity.RESULT_OK) {
            ArrayList<String> matches =
                data.getStringArrayListExtra(RecognizerIntent.EXTRA_RESULTS);
            StringBuilder sb = new StringBuilder();
            for(String piece : matches) {
                sb.append(piece);
                sb.append('\n');
            }
            tv.setText(sb.toString());
        } else {
            Toast.makeText(this, "Operation Canceled",
                    Toast.LENGTH_SHORT).show();
        }
    }
}
```

注意:

如果是在模拟器上测试应用程序,注意 Google Play 和语音识别器都可能未安装。最好是在设备上进行测试。

本例在应用程序加载时会自动启动一个语音识别的 Activity,询问用户"Tell Me Your Name"。在接收用户的语音输入并处理结果之后,Activity 返回一个用户可能会说的内容的列表。这个列表是按照概率排序的,所以在多数情况下,只需要调用 matches.get(0)返回最佳答案就可以了。但这个 Activity 会将所有的返回值显示在屏幕上——纯属娱乐。

在启动 SpeechRecognizer 时,可以通过传入 Intent 的附加信息自定义语音识别器的行为。本例使用两个最常见的附加信息:

- EXTRA_LANGUAGE_MODEL:用来优化语音识别器的处理结果的值。
  - ➢ 常规的语音到文本的查询应该使用 LANGUAGE_MODEL_FREE_FORM 选项。
  - ➢ 如果是较短的请求类型查询,ANGUAGE_MODEL_WEB_SEARCH 的效果可能会更好。
- EXTRA_PROMPT:提示用户开始语音识别的字符串值。

除了这些选项,还可能用到其他一些参数:

- EXTRA_MAX_RESULTS:这个整数设置了返回结果的最大数量。
- EXTRA_LANGUAGE:要求用与系统当前的默认语言不同的语言返回结果。这个字符串值是有效的 IETF 语言标签,如"en-US"或"es"。

4.10 播放音频/视频

4.10.1 问题

应用程序需要在设备上播放本地或远程的音频或视频内容。

4.10.2 解决方案

(API Level 1)

可以使用 MediaPlayer 播放本地媒体或流媒体。不管媒体文件是音频或视频、本地或远程,MediaPlayer 都可以高效地连接、准备并播放它们。在这个范例中,还会使用 MediaController 和 VideoView 在 Activity 的布局中实现简单的用户交互和视频播放功能。

4.10.3 实现机制

注意:

在播放某一特定格式的媒体片段或媒体流之前,请先阅读开发人员说明文档的"Android 支持的媒体格式"一节,看看是否支持该格式。

1. 音频播放

让我们看一个简单的示例,使用 MediaPlayer 播放一段音频。参见代码清单 4-33。

代码清单 4-33 播放本地音频的 Activity

```
public class PlayActivity extends Activity implements
    MediaPlayer.OnCompletionListener {
```

```java
Button mPlay;
MediaPlayer mPlayer;

@Override
public void onCreate(Bundle savedInstanceState) {
    super.onCreate(savedInstanceState);

    mPlay = new Button(this);
    mPlay.setText("Play Sound");
    mPlay.setOnClickListener(playListener);

    setContentView(mPlay);
}

@Override
public void onDestroy() {
    super.onDestroy();
    if(mPlayer != null) {
        mPlayer.release();
    }
}

private View.OnClickListener playListener = new View.OnClickListener() {

    @Override
    public void onClick(View v) {
        if(mPlayer == null) {
            try {
                mPlayer = MediaPlayer.create(PlayActivity.this,
                        R.raw.sound);
                mPlayer.start();
            } catch(Exception e) {
                e.printStackTrace();
            }
        } else {
            mPlayer.stop();
            mPlayer.release();
            mPlayer = null;
        }
    }
};

//播放完成的监听方法
@Override
public void onCompletion(MediaPlayer mp) {
    mPlayer.release();
    mPlayer = null;
}
}
```

这个示例通过按钮开始和停止播放本地音频文件,该文件存储在项目的 res/raw 目录下。MediaPlayer.create()是一个简便的方法,它有很多种形式,旨在一步完成播放器对象的构造和准备工作。本例使用的形式为获取本地资源 ID 的引用,而通过 MediaPlayer.create(Context context, Uri uri)这样的 create()方法,还可以访问和播放远程资源。

创建完成后,示例会立刻播放音频。在播放声音时,用户可以再次按下按钮来停止播放。Activity 还实现了 MediaPlayer.OnCompletionListener 接口,因此会在播放操作正常完成之后收到回调。

在停止播放和播放完成这两种情况下,一旦播放停止,MediaPlayer 实例就会被释放。这个方法可以确保只在需要时才会持有资源,而且可以多次播放声音。为了确保不会无故持有资源,同样会在 Activity 销毁时释放依然存在的资源。

如果应用程序需要播放很多不同的声音,或许可以考虑在播放结束后调用 reset()而不是 release()。但要记住,不再需要播放器时(或 Activity 销毁时),还是要调用 release()。

2. 音频播放器

除了简单地播放,如果应用程序还想为用户提供播放、停止和拖曳播放进度的交互式体验,该如何实现呢?MediaPlayer 中有很多在自定义 UI 元素中实现这些功能的方法,但 Android 同时还提供了 MediaController 视图来处理这样的需求。

让我们构造一个简单的音频播放 Activity,如图 4-5 所示。

图 4-5 使用了 MediaController 的 Activity

代码清单 4-34 和 4-35 给出了此例的布局和 Activity 源代码。

代码清单 4-34　res/layout/main.xml

```xml
<?xml version="1.0" encoding="utf-8"?>
<LinearLayout
    xmlns:android="http://schemas.android.com/apk/res/android"
    android:id="@+id/root"
    android:orientation="vertical"
    android:layout_width="match_parent"
    android:layout_height="match_parent">
    <TextView
        android:layout_width="wrap_content"
        android:layout_height="wrap_content"
        android:layout_gravity="center_horizontal"
        android:text="Now Playing..." />
    <ImageView
        android:id="@+id/coverImage"
        android:layout_width="match_parent"
        android:layout_height="match_parent"
        android:scaleType="centerInside" />
</LinearLayout>
```

代码清单 4-35　使用 MediaController 播放音频的 Activity

```java
public class PlayerActivity extends Activity implements
        MediaController.MediaPlayerControl,
        MediaPlayer.OnBufferingUpdateListener {

    MediaController mController;
    MediaPlayer mPlayer;
    ImageView coverImage;

    int bufferPercent = 0;

    @Override
    public void onCreate(Bundle savedInstanceState) {
        super.onCreate(savedInstanceState);
        setContentView(R.layout.main);

        coverImage = (ImageView)findViewById(R.id.coverImage);

        mController = new MediaController(this);
        mController.setAnchorView(findViewById(R.id.root));
    }

    @Override
    public void onResume() {
        super.onResume();
        mPlayer = new MediaPlayer();
        //设置音频数据源
```

```java
        try {
            mPlayer.setDataSource(this, Uri.parse("URI_TO_REMOTE_AUDIO"));
            mPlayer.prepare();
        } catch(Exception e) {
            e.printStackTrace();
        }
        //设置专辑封面的图片
        coverImage.setImageResource(R.drawable.icon);

        mController.setMediaPlayer(this);
        mController.setEnabled(true);
    }

    @Override
    public void onPause() {
        super.onPause();
        mPlayer.release();
        mPlayer = null;
    }

    @Override
    public boolean onTouchEvent(MotionEvent event) {
        mController.show();
        return super.onTouchEvent(event);
    }

    // MediaPlayerControl 方法
    @Override
    public int getBufferPercentage() {
        return bufferPercent;
    }

    @Override
    public int getCurrentPosition() {
        return mPlayer.getCurrentPosition();
    }

    @Override
    public int getDuration() {
        return mPlayer.getDuration();
    }

    @Override
    public boolean isPlaying() {
        return mPlayer.isPlaying();
    }

    @Override
    public void pause() {
        mPlayer.pause();
```

```java
    }

    @Override
    public void seekTo(int pos) {
        mPlayer.seekTo(pos);
    }

    @Override
    public void start() {
        mPlayer.start();
    }

    // BufferUpdateListener 方法
    @Override
    public void onBufferingUpdate(MediaPlayer mp, int percent) {
        bufferPercent = percent;
    }

    //Android 2.0+目标回调
    public boolean canPause() {
        return true;
    }

    public boolean canSeekBackward() {
        return true;
    }

    public boolean canSeekForward() {
        return true;
    }

    //Android 4.3+目标回调
    @Override
    public int getAudioSessionId() {
        return mPlayer.getAudioSessionId();
    }
}
```

这个示例创建了一个简单的音频播放器，该播放器可以在播放音频时显示艺术家的照片或专辑封面(这里只显示应用程序图标)。本例使用的依然是 MediaPlayer 实例，但这次并不是通过 create() 便利方法创建的。相反，在创建实例后使用 setDataSource() 设置它的内容。通过这种方式关联内容后，播放器并不会自动准备好，因此必须同时调用 prepare() 让播放器做好播放的准备。

这时，音频已经做好播放的准备了。我们希望用 MediaController 处理所有的播放控制，但 MediaController 只能关联到实现了 MediaController.MediaPlayerControl 接口的对象。

奇怪的是，MediaPlayer 自己并没有实现这个接口，因此我们指定 Activity 实现这个接口。这个接口的 11 个方法中的 7 个实际上都是由 MediaPlayer 实现的，因此直接调用它们就可以了。

更　　新

如果应用程序使用的是 API Level 18 或者更新的版本，那么在 MediaController.MediaPlayerControl 接口中会新增一个方法：

`getAudioSessionId()`

这个方法只是 MediaPlayer 所实现方法的另一个封装器。如果使用的是较低的 API Level，这个方法不是必要的，但要想在新版本上获得更好的效果，就需要实现该方法了。

使用 MediaController 需要实现的最后一个方法为 getBufferPercentage()。要想获得此数据，Activity 还需要实现 MediaPlayer.OnBufferingUpdateListener，该接口会在缓冲百分比变化时进行更新。

在实现 MediaController 时有一个技巧。这个小部件是漂浮在它所在窗口中活动的视图之上的，而且一次只显示几秒钟。因此，我们没有在内容视图的 XML 布局中实例化它，而是使用代码进行初始化。MediaController 和内容视图之间的纽带就是 setAnchorView()，该方法还决定了 MediaController 在屏幕中的显示位置。本例中，我们将它与根布局对象对齐，所以会显示在屏幕的底部。如果 MediaController 对齐的是层次结构中的子视图，它将会显示在子视图的边上。

同样，因为 MediaController 是单独的窗口，所以不能在 onCreate() 中调用 MediaController.show()，否则会导致产生致命的异常。MediaController 设计为默认是隐藏的，并且由用户激活。本例中，我们覆写了 Activity 的 onTouchEvent() 方法，当用户点击屏幕时就会显示 MediaController。除非 show() 方法传入的参数是 0，否则它都会在参数指定的时间之后消隐。如果调用 show() 时没有传入任何参数，将使用默认的超时时间，大概是 3 秒钟。

现在音频播放的所有功能都是通过标准的 MediaController 小部件处理的。本例中使用的 setDataSource() 版本获取一个 Uri 作为参数，这对于从 ContentProvider 或远程地址加载音频非常合适。记住，使用其他形式的 setDataSource() 也可以实现本地音频文件和资源的加载工作。

3. 视频播放器

播放视频时，完整的播放控制通常包括播放、暂停和拖曳播放。此外，MediaPlayer 上必须有一个指向 SurfaceHolder 的引用来绘制视频的每一帧。正如前面示例中提到的一样，Android 提供了创建自定义视频播放体验的全部 API。但是，很多情况下最有效的方式还是使用 SDK 提供的类 MediaController 和 VideoView 来处理这些烦琐的工作。

让我们看一个在 Activity 中创建视频播放器的示例。参见代码清单 4-36。

代码清单 4-36　播放视频内容的 Activity

```
public class VideoActivity extends Activity {

    VideoView videoView;
    MediaController controller;

    @Override
    public void onCreate(Bundle savedInstanceState) {
        super.onCreate(savedInstanceState);
        videoView = new VideoView(this);

        videoView.setVideoURI( Uri.parse("URI_TO_REMOTE_VIDEO") );
        controller = new MediaController(this);
        videoView.setMediaController(controller);
        videoView.start();

        setContentView(videoView);
    }

    @Override
    public void onDestroy() {
        super.onDestroy();
        videoView.stopPlayback();
    }
}
```

注意：
如果从远程 URI 提取视频内容，则不要忘记在清单中添加 INTERNET 权限！

本例中，将远程视频地址的 URI 传给了 VideoView，让 VideoView 处理余下的事情。VideoView 同样可以嵌入到较大的 XML 布局体系中，但通常都是单独使用并且全屏显示，所以一般在代码中将其设为布局中唯一的视图。

通过 VideoView 再与 MediaController 交互就非常简单了。VideoView 实现了 Media-Controller.MediaPlayerControl 接口，所以不再需要编写其他任何代码就可以实现控制功能。VideoView 还可以在内部处理 MediaController 的对齐，使其可以显示在屏幕上合适的位置。

4. 处理重定向

最后看一下使用 MediaPlayer 类处理远程内容需要注意的事情。现在 Web 上的很多媒体内容服务器并不会公开暴露视频容器的 URL。出于防止追踪和安全考虑，公开的媒体 URL 通常需要一次或多次的重定向才可以找到真正的媒体内容。MediaPlayer 并不能处理重定向的过程，对于需要重定向的 URL 会返回错误。

如果找不到要在应用程序中播放的内容的真正 URL，那么在将 URL 传给 MediaPlayer 之前，应用程序必须找到 URL 的重定向路径。代码清单 4-37 是一个简单的 AsyncTask 追踪器的示例，它可以实现该功能。

代码清单 4-37　RedirectTracerTask

```java
public class RedirectTracerTask extends AsyncTask<Uri, Void, Uri> {

    private VideoView mVideo;
    private Uri initialUri;

    public RedirectTracerTask(VideoView video) {
        super();
        mVideo = video;
    }

    @Override
    protected Uri doInBackground(Uri... params) {
        initialUri = params[0];
        String redirected = null;
        try {
          URL url = new URL(initialUri.toString());
          HttpURLConnection connection =
                    (HttpURLConnection)url.openConnection();
          //连接后会追踪最终地址
          redirected = connection.getHeaderField("Location");

          return Uri.parse(redirected);
        } catch(Exception e) {
          e.printStackTrace();
          return null;
        }
    }

    @Override
    protected void onPostExecute(Uri result) {
        if(result != null) {
           mVideo.setVideoURI(result);
        } else {
           mVideo.setVideoURI(initialUri);
        }
    }

}
```

这个辅助类通过检索 HTTP 头来追踪 URL 重定向后的最终地址。如果目标 Uri 没有重定向，后台操作将最终返回 null，这种情况下原始的 Uri 会直接传给 VideoView。

通过这个辅助类，可以采用下面的代码片段把媒体的 Uri 传给 VideoView：

```java
VideoView videoView = new VideoView(this);
RedirectTracerTask task = new RedirectTracerTask(videoView);
Uri location = Uri.parse("URI_TO_REMOTE_VIDEO");

task.execute(location);
```

4.11 播放音效

4.11.1 问题

应用程序需要播放几个很短的低延迟的音效来响应与用户的交互。

4.11.2 解决方案

(API Level 1)

通过 SoundPool 将音频文件缓冲加载到内存中,然后在响应用户操作时快速地播放。Android 框架提供了 SoundPool 来解码小音频文件,并在内存中操作它们来进行音频的快速和重复播放。SoundPool 还有一些其他的特性,比如可以在运行时控制音量和播放速度。声音文件自身可以放置在 assets、resource 或设备的文件系统中。

4.11.3 实现机制

让我们看一下如何使用 SoundPool 加载一些音频并把它们关联到 Button 的点击事件。参见代码清单 4-38 和 4-39。

代码清单 4-38　res/layout/main.xml

```xml
<?xml version="1.0" encoding="utf-8"?>
<LinearLayout
    xmlns:android="http://schemas.android.com/apk/res/android"
    android:layout_width="match_parent"
    android:layout_height="match_parent"
    android:orientation="vertical" >
    <Button
        android:id="@+id/button_beep1"
        android:layout_width="match_parent"
        android:layout_height="wrap_content"
        android:text="Play Beep 1" />
    <Button
        android:id="@+id/button_beep2"
        android:layout_width="match_parent"
        android:layout_height="wrap_content"
        android:text="Play Beep 2" />
    <Button
        android:id="@+id/button_beep3"
        android:layout_width="match_parent"
        android:layout_height="wrap_content"
        android:text="Play Beep 3" />
</LinearLayout>
```

代码清单 4-39 SoundPool 的 Activity

```java
public class SoundPoolActivity extends Activity implements
        View.OnClickListener {

    private AudioManager mAudioManager;
    private SoundPool mSoundPool;
    private SparseIntArray mSoundMap;

    @Override
    protected void onCreate(Bundle savedInstanceState) {
        super.onCreate(savedInstanceState);
        setContentView(R.layout.main);
        //得到AudioManager系统服务
        mAudioManager = (AudioManager) getSystemService(AUDIO_SERVICE);
        //设置声音池，通过标准的扬声器输出每次只播放一个音频
        mSoundPool = new SoundPool(1, AudioManager.STREAM_MUSIC, 0);

        findViewById(R.id.button_beep1).setOnClickListener(this);
        findViewById(R.id.button_beep2).setOnClickListener(this);
        findViewById(R.id.button_beep3).setOnClickListener(this);

        //加载每个音频并把它们的streamId保存到一个Map中
        mSoundMap = new SparseIntArray();
        AssetManager manager = getAssets();
        try {
            int streamId;
            streamId = mSoundPool.load(manager.openFd("Beep1.ogg"), 1);
            mSoundMap.put(R.id.button_beep1, streamId);

            streamId = mSoundPool.load(manager.openFd("Beep2.ogg"), 1);
            mSoundMap.put(R.id.button_beep2, streamId);

            streamId = mSoundPool.load(manager.openFd("Beep3.ogg"), 1);
            mSoundMap.put(R.id.button_beep3, streamId);
        } catch(IOException e) {
            Toast.makeText(this, "Error Loading Sound Effects",
                    Toast.LENGTH_SHORT).show();
        }
    }

    @Override
    public void onDestroy() {
        super.onDestroy();
        mSoundPool.release();
        mSoundPool = null;
    }
```

```java
@Override
public void onClick(View v) {
    //查找适合的音频 ID
    int streamId = mSoundMap.get(v.getId());
    if(streamId > 0) {
        float streamVolumeCurrent =
                mAudioManager.getStreamVolume(AudioManager.STREAM_MUSIC);
        float streamVolumeMax =
                mAudioManager.getStreamMaxVolume(AudioManager.STREAM_MUSIC);
        float volume = streamVolumeCurrent / streamVolumeMax;
        //使用指定的音量播放音频,不循环播放并且使用标准的播放速度
        mSoundPool.play(streamId, volume, volume, 1, 0, 1.0f);
    }
}
```

这个示例非常简单。Activity 首先将三个音频文件从应用程序的 assets 目录加载到 SoundPool 中。这个过程会将它们解码为原始 PCM 音频格式并缓存到内存中。每次使用 load()将一个音频文件加载到声音池中时,都会返回一个流标识符,后面则会使用这个标识符播放音频。我们通过在 SparseIntArray 中保存键/值对的方式将每个音频关联到一个特定的按钮。

注意：

SparseIntArray(和它的同胞方法 SparseBooleanArray)与 Map 类似,都是采用键/值对的形式存储的。但是,在保存基本类型(如整型)数据时,它会更加高效,这是因为它避免了在自动装箱时创建不必要的对象。如有需要,应该选择这种方式来提高性能而不是使用 Map。

当用户按下一个按钮时,会找到相应的流标识符并再次调用 SoundPool 播放这个声音。因为 SoundPool 构造函数的 maxStreams 属性设置为 1,如果用户连续按下多个按钮,新声音将使得旧声音停止。如果这个属性值是递增的,则会同时播放多个声音。

play()方法的参数允许在每次访问时配置声音。例如循环播放或快速/慢速播放(相对于原始声音)这样的特性都可以在此处进行控制。

- 循环播放支持任何有限的播放次数,或者将该值设置为-1 以代表无限播放。
- 播放速度支持任何介于 0.5 和 2.0(半速到两倍速度)之间的值。

如果想在某段特定时间内使用 SoundPool 动态改变加载到内存中的声音,不必重新创建声音池,可以使用 unload()来移除声音池中的声音,从而能够加载更多的声音。当 SoundPool 使用完成后,需要调用 release()以释放它的本地资源。

4.12 创建倾斜监控器

4.12.1 问题

应用程序不仅仅需要知道设备是横向和还是竖向的，还需要从设备的加速度计中得到反馈数据。

4.12.2 解决方案

(API Level 3)

使用 SensorManager 接收来自加速度计传感器持续的反馈数据。SensorManager 为使用 Android 设备上的传感器硬件提供了一个通用的抽象接口。加速度计只是应用程序可能注册并获得定时更新的众多传感器中的一个。

4.12.3 实现机制

要点：
在模拟器中并没有加速度计这样的设备传感器。最好在 Android 设备上测试 SensorManager 代码。

这个示例 Activity 使用 SensorManager 注册了一个加速度计，当该加速度计更新时会将数据显示在屏幕上。原始的 X/Y/Z 数据通过一个 TextView 显示在屏幕的底部，而设备的"倾斜"情况会通过 TableLayout 中 4 个视图的简单图形呈现出来。参见代码清单 4-40 和 4-41。

注意：
为了防止在移动和倾斜设备时 Activity 的屏幕也跟着旋转，推荐在应用程序的清单中添加 android:screenOrientation="portrait"或 android:screenOrientation="landscape"。

代码清单 4-40　res/layout/main.xml

```
<?xml version="1.0" encoding="utf-8"?>
<RelativeLayout
    xmlns:android="http://schemas.android.com/apk/res/android"
    android:layout_width="match_parent"
    android:layout_height="match_parent">
    <TableLayout
        android:layout_width="match_parent"
        android:layout_height="match_parent"
        android:stretchColumns="0,1,2">
        <TableRow
            android:layout_weight="1">
            <View
                android:id="@+id/top"
```

```xml
                android:layout_column="1" />
        </TableRow>
        <TableRow
            android:layout_weight="1">
            <View
                android:id="@+id/left"
                android:layout_column="0" />
            <View
                android:id="@+id/right"
                android:layout_column="2" />
        </TableRow>
        <TableRow
            android:layout_weight="1">
            <View
                android:id="@+id/bottom"
                android:layout_column="1" />
        </TableRow>
    </TableLayout>
    <TextView
        android:id="@+id/values"
        android:layout_width="match_parent"
        android:layout_height="wrap_content"
        android:layout_alignParentBottom="true" />
</RelativeLayout>
```

代码清单 4-41　监控倾斜的 Activity

```java
public class TiltActivity extends Activity implements SensorEventListener {

    private SensorManager mSensorManager;
    private Sensor mAccelerometer;
    private TextView valueView;
    private View mTop, mBottom, mLeft, mRight;

    public void onCreate(Bundle savedInstanceState) {
        super.onCreate(savedInstanceState);
        setContentView(R.layout.main);

        mSensorManager = (SensorManager)getSystemService(SENSOR_SERVICE);
        mAccelerometer =
                mSensorManager.getDefaultSensor(Sensor.TYPE_ACCELEROMETER);

        valueView = (TextView)findViewById(R.id.values);
        mTop = findViewById(R.id.top);
        mBottom = findViewById(R.id.bottom);
        mLeft = findViewById(R.id.left);
        mRight = findViewById(R.id.right);
    }

    protected void onResume() {
        super.onResume();
```

```java
        mSensorManager.registerListener(this, mAccelerometer,
            SensorManager.SENSOR_DELAY_UI);
    }

    protected void onPause() {
        super.onPause();
        mSensorManager.unregisterListener(this);
    }

    public void onAccuracyChanged(Sensor sensor, int accuracy) { }

    public void onSensorChanged(SensorEvent event) {
        float[] values = event.values;
        float x = values[0] / 10;
        float y = values[1] / 10;
        int scaleFactor;

        if(x > 0) {
            scaleFactor = (int)Math.min(x * 255, 255);
            mRight.setBackgroundColor(Color.TRANSPARENT);
            mLeft.setBackgroundColor(Color.argb(scaleFactor, 255, 0, 0));
        } else {
            scaleFactor = (int)Math.min(Math.abs(x) * 255, 255);
            mRight.setBackgroundColor(Color.argb(scaleFactor, 255, 0, 0));
            mLeft.setBackgroundColor(Color.TRANSPARENT);
        }

        if(y > 0) {
            scaleFactor = (int)Math.min(y * 255, 255);
            mTop.setBackgroundColor(Color.TRANSPARENT);
            mBottom.setBackgroundColor(Color.argb(scaleFactor, 255, 0, 0));
        } else {
            scaleFactor = (int)Math.min(Math.abs(y) * 255, 255);
            mTop.setBackgroundColor(Color.argb(scaleFactor, 255, 0, 0));
            mBottom.setBackgroundColor(Color.TRANSPARENT);
        }
        //显示原始值
        valueView.setText(String.format("X: %1$1.2f, Y: %2$1.2f, Z: %3$1.2f",
            values[0], values[1], values[2]));
    }
}
```

从设备加速度计获取的三个轴的方向是这样定义的，看着设备的屏幕，从纵向的右上角算起：

- X：横轴，向右为正。
- Y：纵轴，向上为正。
- Z：垂线，远离你的方向为正。

当 Activity 对用户可见时(在 onResume()和 onPause()之间)，会使用 SensorManager 注册加速度计，从而接收来自加速度计的更新。注册时，registerListener()的最后一个参数定义了更新频率。这里选择的值是 SENSOR_DELAY_UI，它是接收更新的最快推荐频率，每次更新都会直接修改 UI。

每当传感器有新值更新时，都会用 SensorEven 值一起调用已注册监听器的 onSensor-Changed()方法。这个 SensorEvent 值包含 X/Y/Z 轴上的加速度值。

科普知识：

加速度计是根据所受的力来计算加速度的。当设备静止时，它所受的力就只有重力(~9.8 m/s^2)。每个轴上的输出值就是这个力(指向地面)与各个方向向量的乘积。当二者平行时，这个值就是其最大值(~9.8–10)。当二者垂直时，这个值就是其最小值(~0.0)。因此当把设备平放在桌子上时，X 轴和 Y 轴的读数大约是 0.0，而 Z 轴大约为 9.8。

示例应用程序使用 TextView 在屏幕的底部显示出了每个轴的原始加速度值。此外，还有一个上/下/左/右显示的 4 个 View 对象的网格，我们会根据设备的方向，按比例调整这个网格的背景色。当设备完全处于水平状态时，X 轴和 Y 轴的值接近于 0，整个屏幕会显示为黑色。当设备倾斜时，屏幕较低的那一端就开始变成红色，在整个屏幕竖直后，将完全变成红色。

提示：

试着修改本例来使用其他的频率值，如 SENSOR_DELAY_NORMAL。观察这些修改是如何影响更新频率的。

此外，还可以晃动设备，看看随着设备的晃动，每个方向的加速度是如何变化的。

传感器的批处理模式

在 Android 4.4 和以后的版本中，应用程序可以请求它们交互的传感器以批处理模式运行，从而在需要监控传感器很长一段时间时减少耗电量。在此模式中，传感器事件可能在硬件缓冲区中排队一段时间，而不会每次都唤醒应用程序处理器。

为了启用传感器的批处理模式，只须利用以 maxBatchReportLatencyUs 为参数的 SensorManager.registerListener()版本。该参数告诉硬件事件在队列中可以排队多长时间，之后再将此批次事件发送到应用程序。

此外，如果应用程序需要在下一次间隔之前获得当前批次，可以在 SensorManager 上调用 flush()，强制传感器将已经完成的工作传递给侦听器。

并不是所有的传感器都支持在所有设备上进行批处理，在不支持的情况下，实现将恢复为采用默认的连续运行模式。

4.13　监控罗盘的方向

4.13.1　问题

通过监控设备的罗盘传感器，应用程序想要知道用户面对的大致方向。

4.13.2 解决方案

(API Level 3)

我们要再次求助 SensorManager。Android 并没有提供真正的"罗盘"传感器；相反，它基于其他的传感器数据使用一些必要的方法推断出设备指向的方向。这种情况下，会使用设备的磁场传感器和加速度计一起确定用户面向的方向。

然后可以使用 SensorManager 的 getOrientation() 方法得到用户在地球上的方向。

4.13.3 实现机制

要点：

在模拟器中并没有加速度计这样的设备传感器，最好在 Android 设备上测试 SensorManager 代码。

同前面的加速度计示例一样，我们使用 SensorManager 注册更新我们要使用的传感器(本例中，要用到两个传感器)，然后在 onSensorChanged() 中处理结果。本例从设备的相机视角来计算和显示用户的方向,这可以用来实现增强现实类的应用程序。参见代码清单 4-42 和 4-43。

代码清单 4-42　res/layout/main.xml

```xml
<?xml version="1.0" encoding="utf-8"?>
<RelativeLayout
    xmlns:android="http://schemas.android.com/apk/res/android"
    android:layout_width="match_parent"
    android:layout_height="match_parent">
    <TextView
        android:id="@+id/direction"
        android:layout_width="wrap_content"
        android:layout_height="wrap_content"
        android:layout_centerInParent="true"
        android:textSize="64sp"
        android:textStyle="bold" />
    <TextView
        android:id="@+id/values"
        android:layout_width="wrap_content"
        android:layout_height="wrap_content"
        android:layout_alignParentBottom="true" />
</RelativeLayout>
```

代码清单 4-43　监控用户方向的 Activity

```java
public class CompassActivity extends Activity implements SensorEventListener {

    private SensorManager mSensorManager;
    private Sensor mAccelerometer, mField;
    private TextView valueView, directionView;
```

```java
    private float[] mGravity = new float[3];
    private float[] mMagnetic = new float[3];

    public void onCreate(Bundle savedInstanceState) {
        super.onCreate(savedInstanceState);
        setContentView(R.layout.main);

        mSensorManager = (SensorManager)getSystemService(SENSOR_SERVICE);
        mAccelerometer =
                mSensorManager.getDefaultSensor(Sensor.TYPE_ACCELEROMETER);
        mField = mSensorManager.getDefaultSensor(Sensor.TYPE_MAGNETIC_FIELD);

        valueView = (TextView)findViewById(R.id.values);
        directionView = (TextView)findViewById(R.id.direction);
    }

    protected void onResume() {
        super.onResume();
        mSensorManager.registerListener(this, mAccelerometer,
                SensorManager.SENSOR_DELAY_UI);
        mSensorManager.registerListener(this, mField,
                SensorManager.SENSOR_DELAY_UI);
    }

    protected void onPause() {
        super.onPause();
        mSensorManager.unregisterListener(this);
    }

    //分配数据数组并重用
    float[] temp = new float[9];
    float[] rotation = new float[9];
    float[] values = new float[3];

    private void updateDirection() {
        //将旋转矩阵加载到R中
        SensorManager.getRotationMatrix(temp, null, mGravity, mMagnetic);
        //重新映射为相机的视角
        SensorManager.remapCoordinateSystem(temp, SensorManager.AXIS_X,
            SensorManager.AXIS_Z, R);
        //返回方向值
        SensorManager.getOrientation(R, values);
        //转换为角度
        for(int i=0; i < values.length; i++) {
            Double degrees = (values[i] * 180) / Math.PI;
            values[i] = degrees.floatValue();
        }
        //显示罗盘方向
        directionView.setText( getDirectionFromDegrees(values[0]) );
        //显示原始值
```

```java
        valueView.setText(
            String.format("Azimuth: %1$1.2f, Pitch: %2$1.2f, Roll: %3$1.2f",
                values[0], values[1], values[2]));
    }

    private String getDirectionFromDegrees(float degrees) {
        if(degrees >= -22.5 && degrees < 22.5) { return "N"; }
        if(degrees >= 22.5 && degrees < 67.5) { return "NE"; }
        if(degrees >= 67.5 && degrees < 112.5) { return "E"; }
        if(degrees >= 112.5 && degrees < 147.5) { return "SE"; }
        if(degrees >= 147.5 || degrees < -147.5) { return "S"; }
        if(degrees >= -147.5 && degrees < -112.5) { return "SW"; }
        if(degrees >= -112.5 && degrees < -67.5) { return "W"; }
        if(degrees >= -67.5 && degrees < -22.5) { return "NW"; }

        return null;
    }

    public void onAccuracyChanged(Sensor sensor, int accuracy) { }

    public void onSensorChanged(SensorEvent event) {
        //将最新值复制到正确的数组中
        switch(event.sensor.getType()) {
        case Sensor.TYPE_ACCELEROMETER:
            System.arraycopy(event.values, 0,
                mGravity, 0,
                event.values.length);
            break;
        case Sensor.TYPE_MAGNETIC_FIELD:
            System.arraycopy(event.values, 0,
                mMagentic, 0,
                event.values.length);
            break;
        default:
            return;
        }

        if(mGravity != null && mMagnetic != null) {
            updateDirection();
        }
    }
}
```

这个示例 Activity 在屏幕的底部实时地显示通过传感器计算返回的三个原始数值。此外，还会从这些原始数值中计算用户当前面向的罗盘方向并显示在屏幕中央。当收到传感器的更新指令时，各个传感器中最新值的本地副本都会更新。只要接收到要使用的两个传感器中其中一个的信息，就会更新 UI。

updateDirection()是进行所有核心工作的地方。SensorManager.getOrientation()提供了显示方向所需的输出信息。这个方法不会返回数据，而是传入这个方法中一个空的浮点型数组来填充三个角度值，它们依次代表了：
- Azimuth：围绕指向地轴的旋转角度。这个值是本例感兴趣的值。
- Pitch：围绕指向西方轴的旋转角度。
- Roll：围绕指向磁极北极的旋转角度。

传入 getOrientation()的一个参数是代表旋转矩阵的浮点型数组。旋转矩阵表示设备当前的坐标系是如何旋转的，这样方法就可以根据参考坐标提供合适的旋转角度。设备方向的旋转矩阵是通过 getRotationMatrix()方法获得的，它的输入是加速度计和磁场传感器的最新值。和 getOrientation()一样，它也返回 void；同时会将一个长度为 9 或 16(代表 3×3 或 4×4 的矩阵)的空浮点数组作为第一个参数传递给该方法，该方法会将结果填入这个数组。

最后，我们希望计算出来的方向能以相机视角为准，所以我们使用 remapCoordinateSystem()对得到的旋转数据进行进一步的转换。这个方法依次有 4 个参数：
1) 输入的数组代表要转换的矩阵。
2) 如何将设备的 x 轴转换为世界坐标。
3) 如何将设备的 y 轴转换为世界坐标。
4) 用户填充结果的空数组。

本例中，我们不需要处理 x 轴，所以直接将 X 映射到 X。但是，需要将设备的 y 轴(纵轴)指向世界坐标系的 z 轴(指向地面的轴)。因为用户在拍照时是直握相机在屏幕上预览的，所以只有这样才能让获得的旋转矩阵与用户手握设备的方向相匹配。

计算出角度数据后，会对数据进行转换并将结果显示在屏幕上。getOrientation()的输出单位为弧度，因此在显示之前首先需要把它转换为角度。此外，还需要将方向角的值转换为罗盘的方向；getDirectionFromDegrees()辅助方法可以根据当前的角度返回设备的方向。顺时针转一圈，方向角会从 0°变为 180°，方向也会从北变到南。继续转一圈，方向角会从-180°变为 0°，方向从南变到北。

4.14 从媒体内容中获取元数据

4.14.1 问题

应用程序需要从设备的媒体内容中得到截图的缩略图或其他元数据。

4.14.2 解决方案

(API Level 10)

使用 MediaMetadataRetriever 读取媒体文件并返回有用的信息。这个类可以读取和跟踪专辑或艺术家数据，或内容数据本身，如视频的大小。此外，对于视频文件，无论是某一特定时间的帧还是 Android 觉得具有代表性的任意一帧，都可以使用 MediaMetadataRetriever 抓取该帧的截图。

对于应用程序需要与设备上很多的媒体内容打交道以及需要显示媒体的附加数据来丰富 UI 来说，MediaMetadataRetriever 是非常好的选择。

4.14.3 实现机制

代码清单 4-44 和 4-45 展示了如何访问设备的附加元数据。

代码清单 4-44　res/layout/main.xml

```xml
<RelativeLayout
    xmlns:android="http://schemas.android.com/apk/res/android"
    android:layout_width="match_parent"
    android:layout_height="match_parent" >
    <Button
        android:id="@+id/button_select"
        android:layout_width="match_parent"
        android:layout_height="wrap_content"
        android:text="Pick Video"
        android:onClick="onSelectClick" />
    <TextView
        android:id="@+id/text_metadata"
        android:layout_width="wrap_content"
        android:layout_height="wrap_content"
        android:layout_below="@id/button_select"
        android:layout_margin="15dp" />
    <ImageView
        android:id="@+id/image_frame"
        android:layout_width="wrap_content"
        android:layout_height="wrap_content"
        android:layout_alignParentBottom="true"
        android:layout_centerHorizontal="true"
        android:layout_margin="10dp" />
</RelativeLayout>
```

代码清单 4-45　带有 MediaMetadataRetriever 的 Activity

```java
public class MetadataActivity extends Activity {
    private static final int PICK_VIDEO = 100;

    private ImageView mFrameView;
    private TextView mMetadataView;

    @Override
    public void onCreate(Bundle savedInstanceState) {
        super.onCreate(savedInstanceState);
        setContentView(R.layout.main);

        mFrameView = (ImageView) findViewById(R.id.image_frame);
        mMetadataView =
                (TextView) findViewById(R.id.text_metadata);
    }

    @Override
```

```java
    protected void onActivityResult(int requestCode,
            int resultCode, Intent data) {
        if(requestCode == PICK_VIDEO
                && resultCode == RESULT_OK
                && data != null) {
            Uri video = data.getData();
            MetadataTask task = new MetadataTask(this, mFrameView,
                    mMetadataView);
            task.execute(video);
        }
    }

    public void onSelectClick(View v) {
        Intent intent = new Intent(Intent.ACTION_GET_CONTENT);
        intent.setType("video/*");
        startActivityForResult(intent, PICK_VIDEO);
    }

    public static class MetadataTask
            extends AsyncTask<Uri, Void, Bundle> {
        private Context mContext;
        private ImageView mFrame;
        private TextView mMetadata;
        private ProgressDialog mProgress;

        public MetadataTask(Context context, ImageView frame,
                TextView metadata) {
            mContext = context;
            mFrame = frame;
            mMetadata = metadata;
        }

    @Override
    protected void onPreExecute() {
        mProgress = ProgressDialog.show(mContext, "",
                "Analyzing Video File...", true);
    }

    @Override
    protected Bundle doInBackground(Uri... params) {
        Uri video = params[0];
        MediaMetadataRetriever retriever =
                new MediaMetadataRetriever();
        retriever.setDataSource(mContext, video);

        Bitmap frame = retriever.getFrameAtTime();

        String date = retriever.extractMetadata(
            MediaMetadataRetriever.METADATA_KEY_DATE);
        String duration = retriever.extractMetadata(
```

```java
            MediaMetadataRetriever.METADATA_KEY_DURATION);
        String width = retriever.extractMetadata(
            MediaMetadataRetriever.METADATA_KEY_VIDEO_WIDTH);
        String height = retriever.extractMetadata(
            MediaMetadataRetriever.METADATA_KEY_VIDEO_HEIGHT);

        Bundle result = new Bundle();
        result.putParcelable("frame", frame);
        result.putString("date", date);
        result.putString("duration", duration);
        result.putString("width", width);
        result.putString("height", height);

        return result;
    }

    @Override
    protected void onPostExecute(Bundle result) {
        if(mProgress != null) {
            mProgress.dismiss();
            mProgress = null;
        }

        Bitmap frame = result.getParcelable("frame");
        mFrame.setImageBitmap(frame);
        String metadata = String.format("Video Date: %s\n"
                + "Video Duration: %s\nVideo Size: %s x %s",
                result.getString("date"),
                result.getString("duration"),
                result.getString("width"),
                result.getString("height") );
        mMetadata.setText(metadata);
    }
  }
}
```

在这个示例中,用户可以从设备上选择一个要处理的视频文件。当收到一个有效的视频 Uri 后,Activity 开始用一个 AsyncTask 从视频中解析一些元数据。由于这个过程需要几秒钟或更长的时间,而且我们不希望在此过程中阻塞 UI 线程,因此创建了一个 AsyncTask 来完成它。

后台任务会创建新的 MediaMetadataRetriever 并将选择的视频作为它的数据源。然后调用 getFrameAtTime()返回视频中一帧的位图图像。这个方法在 UI 中创建视频的截图时非常有用。我们调用的这个方法是没有参数的版本,它返回的帧是半随机的。如果对某一特定帧更感兴趣的话,可以使用该方法的另一个版本,该版本的参数为需要的视频帧的播放时间(单位为毫秒)。这种情况下,会返回视频中距离指定时间最近的关键帧。

除了帧的图像,我们还收集了一些关于视频的基本信息,包括视频创建的时间、视频时长和视频大小。所有这些结果数据都被打包到一个 bundle 中，并由后台线程传递出来。因为任务的 onPostExecute()方法是在主线程上调用的，所以可以通过它使用这些检索出来的数据更新 UI。

4.15 检测用户移动

4.15.1 问题

希望应用程序响应用户行为的变化，例如设备是否保持直立状态，或者用户当前是否保持活跃且在移动过程中。

4.15.2 解决方案

(API Level 9)

Google Play Services 包括通过 ActivityRecognitionClient 监控用户活动的功能。用户活动跟踪服务是低功耗的方法，它定期接收关于用户所做工作的更新。该服务定期监控设备上的本地传感器数据一段较短的时间，而不是依赖高功耗的方法，如 Web 服务或 GPS。

使用此 API，应用程序将接收如下事件之一的更新：

- IN_VEHICLE：用户可能正在驾车或坐车，例如汽车、公交或火车。
- ON_BICYCLE：用户可能在骑自行车。
- ON_FOOT：用户可能在步行或跑步。
- STILL：用户或至少是设备当前静止竖立。
- TILTING：设备最近已倾斜。从静止位置拾起设备或其方向变化时，就会产生倾斜。
- UNKNOWN：没有足够的数据可完全确定用户当前正在做什么。

在操作 ActivityRecognitionClient 时，应用程序通过调用 requestActivityUpdates()启动周期性更新。此方法采用的参数定义了应用程序的更新频率和将用于触发每个事件的 PendingIntent。

应用程序可以传递它们希望的任意频率间隔(以毫秒为单位)，传递 0 值会尽快向应用程序发送更新。Google Play Services 无法保证此速率；如果服务需要多个传感器样本进行测定工作，则样本可能延迟。此外，如果多个应用程序同时请求 Activity 更新，Google Play Services 会按照请求的最快速率将更新传递到所有应用程序。

每个事件都包括一个 DetectedActivity 实例列表，这些实例封装 Activity 类型(上述选项之一)以及服务预测的 Activity 可信程度。该列表按照可信程度排序，因此最可能的用户 Activity 排在第一位。

要点：

用户 Activity 跟踪是 Google Play Services 库的一部分，它在任意平台级别都不是原生 SDK 的一部分。然而，目标平台为 API Level 9 或以后版本的应用程序以及 Google Play 体系内的设备都可以使用此绘图库。关于在项目中包括 Google Play Services 的更多信息，请参考 https://developer.android.com/google/play-services/setup.html。

4.15.3　实现机制

接下来看一个基本的示例应用程序，该应用程序监控用户 Activity 的改动，将这些改动记录到显示屏上，并且包括安全措施。如果用户尝试驾车或骑自行车时访问应用程序，则会阻止其访问。先看一下代码清单 4-46，这是 AndroidManifest.xml 的一个代码片段，其中显示操作此服务所需的权限。

要点：
这个示例使用 AppCompat 库中的 ActionBarActivity，从而允许使用 API Level 11 之前的 Fragment。

代码清单 4-46　AndroidManifest.xml

```xml
<?xml version="1.0" encoding="utf-8"?>
<manifest xmlns:android="http://schemas.android.com/apk/res/android"
    package="com.androidrecipes.usermotionactivity">

    <!--用户 Activity 识别所需的权限-->
    <uses-permission
        android:name="com.google.android.gms.permission.ACTIVITY_RECOGNITION" />
    <application
        android:allowBackup="true"
        android:icon="@drawable/ic_launcher"
        android:label="@string/app_name"
        android:theme="@style/AppTheme" >

        <!-- Google Play Services 所需的样板代码-->
        <meta-data
            android:name="com.google.android.gms.version"
            android:value="@integer/google_play_services_version" />

        <activity
            android:name=".MainActivity"
            android:label="User Activity"
            android:screenOrientation="portrait" >
            <intent-filter>
                <action android:name="android.intent.action.MAIN" />
                <category android:name="android.intent.category.LAUNCHER" />
            </intent-filter>
        </activity>

        <service android:name=".UserMotionService" />
    </application>

</manifest>
```

可以看到，必须在清单中专门声明自定义权限，用于从 Google Play Services 读取 Activity 识别数据。还需要确定使用了定义所需版本的<meta-data>元素，这就使客户端库可确定设备是否运行适当的 Google Play Services 或需要更新。代码清单 4-47 和 4-48 描述了我们将使用的 Activity。

代码清单 4-47　res/layout/activity_main.xml

```xml
<FrameLayout
    xmlns:android="http://schemas.android.com/apk/res/android"
    android:layout_width="match_parent"
    android:layout_height="match_parent">
    <!-- 启用副本以自动滚动内容的列表 -->
    <ListView
        android:id="@+id/list"
        android:layout_width="match_parent"
        android:layout_height="match_parent"
        android:stackFromBottom="true"
        android:transcriptMode="normal" />

    <!-- 安全阻塞视图 -->
    <!-- 可点击，在可见时使用触摸事件 -->
    <TextView
        android:id="@+id/blocker"
        android:layout_width="match_parent"
        android:layout_height="match_parent"
        android:gravity="center"
        android:clickable="true"
        android:textSize="32sp"
        android:textColor="#F55"
        android:text="Do not operate your device in a vehicle!"
        android:background="#C333"
        android:visibility="gone" />
</FrameLayout>
```

代码清单 4-48　显示用户移动的 Activity

```java
public class MainActivity extends Activity implements
        ServiceConnection,
        UserMotionService.OnActivityChangedListener,
        GooglePlayServicesClient.ConnectionCallbacks,
        GooglePlayServicesClient.OnConnectionFailedListener {
    private static final String TAG = "UserActivity";

    private Intent mServiceIntent;
    private PendingIntent mCallbackIntent;
    private UserMotionService mService;

    private ActivityRecognitionClient mRecognitionClient;
    // 显示结果的自定义列表适配器
```

```java
    private ActivityAdapter mListAdapter;

    private View mBlockingView;

    @Override
    protected void onCreate(Bundle savedInstanceState) {
        super.onCreate(savedInstanceState);
        setContentView(R.layout.activity_main);

        mBlockingView = findViewById(R.id.blocker);

        // 构建将显示从服务传入的所有 Activity 变更事件的简单列表适配器
        ListView list = (ListView) findViewById(R.id.list);
        mListAdapter = new ActivityAdapter(this);
        list.setAdapter(mListAdapter);

        // 点击列表时,显示所有可能的 Activity
        list.setOnItemClickListener(
                new AdapterView.OnItemClickListener() {
            @Override
            public void onItemClick(AdapterView<?> parent, View v,
                    int position, long id) {
                showDetails(mListAdapter.getItem(position));
            }
        });

        // 检查 Google Play Services 是否已激活且为最新版本
        int resultCode =
                GooglePlayServicesUtil.isGooglePlayServicesAvailable(this);
        switch(resultCode) {
            case ConnectionResult.SUCCESS:
                Log.d(TAG, "Google Play Services is ready to go!");
                break;
            default:
                showPlayServicesError(resultCode);
                return;
        }

        // 为与 Google Services 通信而创建客户端实例
        mRecognitionClient = new ActivityRecognitionClient(
                this, this, this);
        // 创建绑定到服务的 Intent
        mServiceIntent =
                new Intent(this, UserMotionService.class);
        // 创建用于 Google Services 回调的 PendingIntent
        mCallbackIntent = PendingIntent.getService(this, 0,
                mServiceIntent, PendingIntent.FLAG_UPDATE_CURRENT);
    }
```

```java
@Override
protected void onResume() {
    super.onResume();
    // 将 Google Services 与我们的服务相连
    mRecognitionClient.connect();
    bindService(mServiceIntent, this, BIND_AUTO_CREATE);
}

@Override
protected void onPause() {
    super.onPause();
    // 从所有服务断开连接
    mRecognitionClient.removeActivityUpdates(mCallbackIntent);
    mRecognitionClient.disconnect();

    disconnectService();
    unbindService(this);
}

/** ServiceConnection 方法 */

public void onServiceConnected(ComponentName name,
        IBinder service) {
    // 将自身作为事件回调附加到服务
    mService = ((LocalBinder) service).getService();
    mService.setOnActivityChangedListener(this);
}

@Override
public void onServiceDisconnected(ComponentName name) {
    disconnectService();
}

private void disconnectService() {
    if(mService != null) {
        mService.setOnActivityChangedListener(null);
    }
    mService = null;
}

/** Google Services 连接回调 */

@Override
public void onConnected(Bundle connectionHint) {
    // 我们必须等待直至连接服务以请求更新
    mRecognitionClient.requestActivityUpdates(5000,
            mCallbackIntent);
}
```

```java
@Override
public void onDisconnected() {
    Log.w(TAG, "Google Services Disconnected");
}

@Override
public void onConnectionFailed(ConnectionResult result) {
    Log.w(TAG, "Google Services Connection Failure");
}

/** OnActivityChangedListener 方法 */

@Override
public void onUserActivityChanged(int bestChoice,
        int bestConfidence,
        ActivityRecognitionResult newActivity) {
    // 向列表中添加最新的事件
    mListAdapter.add(newActivity);
    mListAdapter.notifyDataSetChanged();

    // 基于我们的自定义算法确定用户操作
    switch(bestChoice) {
        case DetectedActivity.IN_VEHICLE:
        case DetectedActivity.ON_BICYCLE:
            mBlockingView.setVisibility(View.VISIBLE);
            break;
        case DetectedActivity.ON_FOOT:
        case DetectedActivity.STILL:
            mBlockingView.setVisibility(View.GONE);
            break;
        default:
            // 忽略其他状态
            break;
    }
}

/*
 * 使用可以选择的 Activity 及其置信度构建简单 Toast 的 Utility
 */
private void showDetails(ActivityRecognitionResult activity) {
    StringBuilder sb = new StringBuilder();
    sb.append("Details:");
    for(DetectedActivity element :
            activity.getProbableActivities()) {sb.append("\n"
            + UserMotionService.getActivityName(element)
            + ", " + element.getConfidence() + "% sure");
    }

    Toast.makeText(this, sb.toString(),
            Toast.LENGTH_SHORT).show();
```

```java
    }

    /*
     * 显示从服务收到的每个Activity的ListAdapter
     */
    private static class ActivityAdapter extends
            ArrayAdapter<ActivityRecognitionResult> {

        public ActivityAdapter(Context context) {
            super(context, android.R.layout.simple_list_item_1);
        }

        @Override
        public View getView(int position, View convertView,
                ViewGroup parent) {
            if(convertView == null) {
                convertView =
                    LayoutInflater.from(getContext()).inflate(
                        android.R.layout.simple_list_item_1,
                        parent,
                        false);
            }

            // 在列表中显示最可能的Activity及其置信度
            TextView tv = (TextView) convertView;
            ActivityRecognitionResult result = getItem(position);
            DetectedActivity newActivity =
                    result.getMostProbableActivity();
            String entry =
                DateFormat.format("hh:mm:ss", result.getTime())
                + ": " +
                UserMotionService.getActivityName(newActivity)
                + ", " + newActivity.getConfidence()
                + "% confidence";
            tv.setText(entry);

            return convertView;
        }
    }
    /*
     * 当Play Services缺失或版本不正确时，客户端库将显示一个对话框，帮助用户进行更新
     */
    private void showPlayServicesError(int errorCode) {
        // 获得来自Google Play Services的错误对话框
        Dialog errorDialog = GooglePlayServicesUtil.getErrorDialog(
            errorCode,
            this,
            1000 /* RequestCode */);
```

```
        // 如果Google Play Services可以提供错误对话框
        if(errorDialog != null) {
            // 为错误对话框创建新的DialogFragment
            SupportErrorDialogFragment errorFragment = 
                SupportErrorDialogFragment.newInstance(errorDialog);
            // 在DialogFragment中显示错误对话框
            errorFragment.show(
                getSupportFragmentManager(),
                "Activity Tracker");
        }
    }
}
```

在此例中,第一项工作是检查 Google Play Services 在设备上是否可用以及是否为最新版本。完成验证之后,可以创建 ActivityRecognitionClient,这是将需要连接到服务(目前尚未见到)的 Intent;并且创建 PendingIntent,该 Intent 提供在回调时使用的识别服务。

注意:

不要将显示 UI 的应用程序组件 Activity 与在此上下文中用于描述用户物理活动的 Activity 混淆。该单词在此 API 中会经常出现,因此请注意区分。

当应用程序来到前台时,我们向识别服务发出连接请求。该过程是异步的,当连接完成时,我们将收到 onConnected()调用。为确保我们没有无谓地消耗能源,在进入后台时移除这些更新。

在这些事件执行期间,我们绑定和取消绑定自己的服务,从而绑定仅在位于前台时才激活。我们将在后面看到此服务在整个应用程序中的重要性。

提示:

使用绑定的服务时,仅在服务崩溃或意外断开连接时才调用 onServiceDisconnected()。在显式断开连接时,你希望执行的任何清除工作也必须随 unbindService()完成。

一旦连接识别服务,我们就结合使用 5 秒时间间隔的 requestActivityUpdates()和 PendingIntent 启动更新操作,后者描述了在何处开始更新。在此例中,PendingIntent 设置为触发 UserMotionService,该服务的代码如代码清单 4-49 所示。

代码清单 4-49 接收活动更新的服务

```
public class UserMotionService extends IntentService {
    private static final String TAG = "UserMotionService";

    /*
     * 已检测Activity类型变更的回调接口
     */
```

```java
public interface OnActivityChangedListener{
    public void onUserActivityChanged(int bestChoice,
            int bestConfidence,
            ActivityRecognitionResult newActivity);
}

/* 上次检测到的 Activity 类型 */
private DetectedActivity mLastKnownActivity;

/*
 * 编组来自后台线程的请求
 * 从而可以在主(UI)线程上发出回调
 */
private CallbackHandler mHandler;
private static class CallbackHandler extends Handler {
    /* Activity 变更的回调 */
    private OnActivityChangedListener mCallback;

    public void setCallback(
            OnActivityChangedListener callback) {
        mCallback = callback;
    }

    @Override
    public void handleMessage(Message msg) {
        if(mCallback != null) {
            // 读取消息中的有效负载数据并激活回调
            ActivityRecognitionResult newActivity =
                    (ActivityRecognitionResult) msg.obj;
            mCallback.onUserActivityChanged(
                    msg.arg1,
                    msg.arg2,
                    newActivity);
        }
    }
}

public UserMotionService() {
    // 使用字符串命名创建的后台线程
    super("UserMotionService");
    mHandler = new CallbackHandler();
}

public void setOnActivityChangedListener(
        OnActivityChangedListener listener) {
    mHandler.setCallback(listener);
}

@Override
public void onDestroy() {
```

```java
        super.onDestroy();
        Log.w(TAG, "Service is stopping...");
    }

    /*
     * 从框架传入的动作事件将到达此处
     * 这是在后台线程上调用的
     * 因此可以根据需要在此进行长时间处理
     */
    @Override
    protected void onHandleIntent(Intent intent) {
        if(ActivityRecognitionResult.hasResult(intent)) {
            // 从 Intent 提取结果
            ActivityRecognitionResult result =
                    ActivityRecognitionResult.extractResult(intent);
            DetectedActivity activity =
                    result.getMostProbableActivity();
            Log.v(TAG, "New User Activity Event");

            // 如果最大可能性是 UNKNOWN，但置信度较低
            // 就检查是否存在另一个 Activity 并选择
            if(activity.getType() == DetectedActivity.UNKNOWN
                    && activity.getConfidence() < 60
                    && result.getProbableActivities().size() > 1){
                // 选择下一个可能的元素
                activity = result.getProbableActivities().get(1);
            }

            // 在 Activity 中的变更上修改回调
            if(mLastKnownActivity == null
                    || mLastKnownActivity.getType()
                        != activity.getType()
                    || mLastKnownActivity.getConfidence()
                        != activity.getConfidence()) {
                // 在 Message 中将结果传递到主线程
                Message msg = Message.obtain(null,
                        0,                              //what
                        activity.getType(),             //arg1
                        activity.getConfidence(),       //arg2
                        result);                        //obj
                mHandler.sendMessage(msg);
            }
            mLastKnownActivity = activity;
        }
    }

    /*
     * 在 Activity 想要绑定到服务时调用此方法
     * 我们必须提供围绕此实例的封装类以传递回来
     */
```

```java
@Override
public IBinder onBind(Intent intent) {
    return mBinder;
}

/*
 * 这是简单的封装类，我们可以将其传递给Activity以允许其直接访问此服务
 */
private LocalBinder mBinder = new LocalBinder();
public class LocalBinder extends Binder {
    public UserMotionService getService() {
        return UserMotionService.this;
    }
}

/*
 * 获得每个状态的良好显示名称的Utility
 */
public static String getActivityName(
        DetectedActivity activity) {
    switch(activity.getType()) {
        case DetectedActivity.IN_VEHICLE:
            return "Driving";
        case DetectedActivity.ON_BICYCLE:
            return "Biking";
        case DetectedActivity.ON_FOOT:
            return "Walking";
        case DetectedActivity.STILL:
            return "Not Moving";
        case DetectedActivity.TILTING:
            return "Tilting";
        case DetectedActivity.UNKNOWN:
        default:
            return "No Clue";
    }
}
}
```

UserMotionService 是一个 IntentService，后者是一个服务，用于将所有 Intent 命令转发到它所创建的后台线程，并在 onHandleIntent()中处理它们。它的主要优点是内置了机制来排队 Intent 请求，并在后台通过此方法按顺序处理它们。

当 Activity 绑定到此服务时，它将自动启动并通过 onBind()返回到主调方。该 Activity 将在 onServiceConnected()中收到服务实例，从中将 Activity 注册为在服务中确定的用户 Activity 变更事件的回调。一旦 Activity 与服务取消绑定，它也会自动停止。

一旦到达为来自 Google Play Services 的更新事件注册 Activity 的位置，框架将开始定期触发 PendingIntent，这将导致在我们的服务中调用 onHandleIntent()。

对于每个事件，我们使用 ActivityRecognitionResult 上的 Utility 方法解压缩来自传入 Intent 的数据。然后，我们确定最可能的用户 Activity。我们稍微自定义了算法，因为如果最可能的 Activity 是 UNKNOWN，但在此决策中的置信度较低，我们就会选取下一个最佳的选项并返回。这种模式也非常适合于要置于应用程序中的额外自定义决策逻辑。

选择了用于比较的用户 Activity 之后，我们检查是否这是相同的 Activity 类型，或者 Activity 中的变更是否已发生。在变更的情况下，我们想要将回调传递给在绑定时注册自身的 Activity。我们使用 Handler 而非直接调用此方法，这是因为 onHandleIntent()正在后台线程上运行，并且如果 Activity(或其他侦听器)想要执行涉及更新 UI 的任何工作，我们就要在主线程上传递我们的回调。

最后，代码清单 4-50 提醒我们所需的要求；主要是我们必须确定将 Play Services 库包括为模块构件中的依赖项。

代码清单 4-50　接收活动更新的服务

```
apply plugin: 'com.android.application'

android {
    compileSdkVersion 18
    buildToolsVersion "20.0.0"

    defaultConfig {
        applicationId "com.androidrecipes.usermotionactivity"
        ...
    }
    ...
}

dependencies {
    compile 'com.android.support:support-v4:18.0.0'
    compile 'com.google.android.gms:play-services:6.1.+'
}
```

4.16　小结

这些范例展示了如何在 Android 中使用地图、用户位置和设备传感器数据，将用户的周围环境信息整合到 Android 应用程序中。你学习了 Google 提供给 Android 设备的许多额外的 API，这些 API 位于 Google Play 生态系统内。我们还讨论了如何使用设备的摄像头和麦克风，让用户采集周边的信息，并解析这些信息。最后，通过使用媒体 API，你学习了如何获取用户本地拍摄或是从网络上远程下载的多媒体资源，以及如何在应用程序中播放这些多媒体。在第 5 章中，我们将讨论使用 Android 的许多持久化技术在设备上存储非易失性的数据。

第 5 章

数据持久化

尽管现在的许多大型体系结构旨在将用户数据转移到云中,但移动应用程序不稳定的本质决定了必须至少将部分用户数据持久地保存在本地设备上。这些数据可能是缓存的 Web 服务响应结果以确保能在离线状态下也能访问它们,也可能是用户对特定应用程序行为的首选项设置。Android 提供了一系列有用的框架,避免了用文件或数据库来实现信息的持久化。

5.1 制作首选项界面

5.1.1 问题

在应用程序中,需要通过一种简单的方式存储、修改和显示应用程序中的用户首选项和配置。

5.1.2 解决方案

(API Level 1)
通过 PreferenceActivity 和 XML Preference 层次结构就可以一次性解决用户界面、键/值组合以及数据持久化的问题。通过这种方式可以创建和 Android 设备上的 Settings 应用程序一样的 UI,从而保持用户体验与预期的一致性。

在 XML 内部,可以用关联的设置定义一个或多个屏幕界面的整套集合,借助 PreferenceScreen、PreferenceCategory 及相关 Preference 元素显示和按类别分组这些设置,而 Activity 只需要很少的代码就可以加载这个 XML 文件并呈现给用户。

5.1.3 实现机制

图 5-1 展示了此例将具有的界面。我们将构建一个首选项界面,该界面包含两个屏幕:

主屏幕和辅助屏幕。

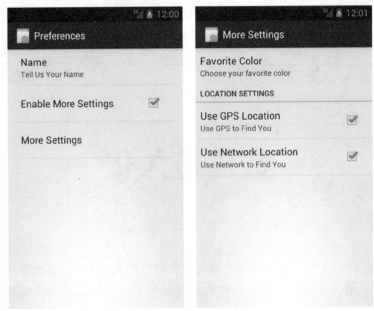

图 5-1 根 PreferenceScreen(左图)首先显示。如果用户点击 More Settings 选项,则会显示辅助屏幕(右图)

代码清单 5-1 和 5-2 展示了一个 Android 应用程序的基本设置。XML 中定义了两个屏幕界面,其中有 Android 框架支持的各种常用的首选项类型。注意,其中一个屏幕是内嵌在另一个屏幕之中的,当用户在主屏幕上点击相关联的列表项时就会显示内部嵌套的屏幕。

代码清单 5-1　res/xml/settings.xml

```xml
<?xml version="1.0" encoding="utf-8"?>
<PreferenceScreen xmlns:android="http://schemas.android.com/apk/res/android">
    <EditTextPreference
        android:key="namePref"
        android:title="Name"
        android:summary="Tell Us Your Name"
        android:defaultValue="Apress" />
    <CheckBoxPreference
        android:key="morePref"
        android:title="Enable More Settings"
        android:defaultValue="false" />
    <PreferenceScreen
        android:key="moreScreen"
        android:title="More Settings"
        android:dependency="morePref">
        <ListPreference
            android:key="colorPref"
            android:title="Favorite Color"
            android:summary="Choose your favorite color"
            android:entries="@array/color_names"
            android:entryValues="@array/color_values"
```

```xml
                android:defaultValue="GRN" />
        <PreferenceCategory
            android:title="Location Settings">
            <CheckBoxPreference
                android:key="gpsPref"
                android:title="Use GPS Location"
                android:summary="Use GPS to Find You"
                android:defaultValue="true" />
            <CheckBoxPreference
                android:key="networkPref"
                android:title="Use Network Location"
                android:summary="Use Network to Find You"
                android:defaultValue="true" />
        </PreferenceCategory>
    </PreferenceScreen>
</PreferenceScreen>
```

代码清单 5-2　res/values/arrays.xml

```xml
<?xml version="1.0" encoding="utf-8"?>
<resources>
    <string-array name="color_names">
        <item>Black</item>
        <item>Red</item>
        <item>Green</item>
    </string-array>
    <string-array name="color_values">
        <item>BLK</item>
        <item>RED</item>
        <item>GRN</item>
    </string-array>
</resources>
```

首先注意创建 XML 文件的方式。虽然这个资源文件可以放到任何目录(如 res/layout)中，但通常都是把它们放到项目的 xml 通用目录中。

同样需要注意的是我们为每个 Preference 对象都提供了 android:key 属性，而没有使用 android:id。当在应用程序的其他地方通过 SharedPreferences 对象引用这些存储的设置值时，就可以使用该键访问此对象。另外，PreferenceActivity 有一个 findPreference()方法，可以通过 Java 代码获得一个指向已经填充的 Preference 的引用，它比 findViewById()效率要高，findPreference()的参数也是键。

填充完成之后，根 PreferenceScreen 显示了带有以下三个条目(按顺序)的列表：

1) Name 条目：它是一个 EditTextPreference 实例，保存了一个字符串值。点击这个条目会显示一个文本框，用户可以在该文本框中输入新的首选项值。

2) Enable More Settings 条目：它是一个 CheckBoxPreference 实例，保存了一个布尔值。点击这个条目将改变复选框的选中状态。

3) More Settings 条目：点击这个条目将会加载另一个有更多条目的 PreferenceScreen。在用户点击"More Settings"条目后，会显示第二个设置屏幕，它还有三个条目：一个 ListPreference 条目和两个 CheckBoxPreferences，这两个 CheckBoxPreferences 通过一个 PreferenceCategory 归类到了一起。PreferenceCategory 将实际的首选项内容组合到一起并设置标题，它没有其他作用。

这个示例使用的最后一种首选项类型为 ListPreference。这个条目需要两个数组参数(它们可以是同一个数组)，代表了供用户选择的选项集。android:entries 数组用于显示可读性很强的选项清单，而 android:entryValues 数组则是实际保存的选项值。

所有的首选项条目都可以设置默认值(可选)，但这个值不会被自动加载，而是在显示 PreferenceActivity、XML 文件首次填充或调用 PreferenceManager.setDefaultValues()时才会加载。

现在让我们看一下 PreferenceActivity 是如何加载和管理这些设置的，参见代码清单 5-3。

代码清单 5-3　PreferenceActivity 实战

```
public class SettingsActivity extends PreferenceActivity {

    @Override
    public void onCreate(Bundle savedInstanceState) {
        super.onCreate(savedInstanceState);
        //从 XML 中加载首选项数据
        addPreferencesFromResource(R.xml.settings);
    }
}
```

要想向用户显示首选项界面并允许他们修改首选项，只需要调用 addPreferencesFromResource()即可。在 PreferenceActivity 中不必调用 setContentView()，因为 addPreferencesFromResource()就会填充 XML 并管理列表中内容的显示。然而，也可以提供含有 ListView 的自定义布局，同时设置好 ListView 的 android:id="@android:id/list" 属性，这样 PreferenceActivity 就能够从中加载首选项条目。

为了单独控制访问，也可以将首选项条目放到列表中。我们将"Enable More Settings"条目放到列表中只是为了可以让用户启用或禁用对第二个 PreferenceScreen 的访问。为了实现这个目标，内嵌的 PreferenceScreen 包含了一个 android:dependency 属性，它根据另一个首选项的状态决定是否要启用这个首选项界面。当该属性指向的首选项未设置或者为 false 时，这个首选项界面就是禁用的。

1．加载默认值和访问首选项

通常情况下，PreferenceActivity(例如本例中的 PreferenceActivity)并不是应用程序的根界面。通常，如果设置了默认值，在用户访问 Settings(加载默认值的第一种情况)之前，应用程序的其他部分就会经常访问这些默认值。因此，在应用程序中，如果要在使用之前预先加载默认的设置值，可以在其他地方调用以下方法：

```
PreferenceManager.setDefaultValues(Context context, int resId, boolean readAgain);
```

这个方法可能会被调用多次，但默认值不会被反复加载。或许可以把它放到主 Activity

中，这样就可以在首次启动时被调用，或者放在所有要访问的共享首选项代码的前面，即应用程序可以调用的位置。

采用这种机制存储的首选项会保存到默认的共享首选项对象中，而共享首选项对象可以通过 PreferenceManager.getDefaultSharedPreferences(Context context);来访问，传入相应的 Context 指针即可。

下面的示例 Activity 会加载前面示例中的默认首选项集并访问其中的一些当前值，参见代码清单 5-4。

代码清单 5-4　加载默认首选项的 Activity

```java
public class HomeActivity extends Activity {

    @Override
    public void onCreate(Bundle savedInstanceState) {
        super.onCreate(savedInstanceState);
        setContentView(R.layout.main);

        //加载默认首选项
        PreferenceManager.setDefaultValues(this, R.xml.settings, false);
    }

    @Override
    public void onResume() {
        super.onResume();
        //访问当前的设置
        SharedPreferences settings =
                PreferenceManager.getDefaultSharedPreferences(this);

        String name = settings.getString("namePref", "");
        boolean isMoreEnabled = settings.getBoolean("morePref", false);
    }
}
```

调用 setDefaultValues()会将 XML 文件中所有带 android:defaultValue 属性的设置项的默认值保存到首选项中。这样即使在用户尚未访问过设置界面的情况下，应用程序也能够访问这些默认值。

然后可以通过 SharedPreferences 对象的一系列类型化存取器方法访问这些值。每个存取器方法都需要首选项的键和一个默认值，如果该键不存在的话，就会返回这个默认值。

2. 使用 PreferenceFragment

(API Level 11)

从 Android 3.0 开始，以 PreferenceFragment 的形式引入了一种新的创建首选项界面的方式。这个类并不是 Android 支持库中的类，只有应用程序的目标版本最低为 API Level 11 时，才可以使用它来代替 PreferenceActivity。代码清单 5-5 和 5-6 修改了之前的示例，这里使用了 PreferenceFragment。

代码清单 5-5　包含 Fragment 的 Activity

```
public class MainActivity extends Activity {

    @Override
    protected void onCreate(Bundle savedInstanceState) {
        super.onCreate(savedInstanceState);

        FragmentTransaction ft = getFragmentManager().beginTransaction();
        ft.add(android.R.id.content, new PreferenceFragment());
        ft.commit();
    }
}
```

代码清单 5-6　新的 PreferenceFragment

```
public class SettingsFragment extends PreferenceFragment {

    @Override
    public void onCreate(Bundle savedInstanceState) {
        super.onCreate(savedInstanceState);
        //从 XML 中加载设置数据
        addPreferencesFromResource(R.xml.settings);
    }
}
```

现在，原来的首选项都放到了 PreferenceFragment 中，PreferenceFragment 会像以前一样管理它们。另一处需要修改的地方就是 Fragment 不能自己单独存在，它必须包含在一个 Activity 中，因此我们创建了一个新的关联该 Fragment 的根 Activity。

Android 框架已改为采用 Fragment 管理首选项，从而可以更加轻松地在单个 Activity 中显示多个首选项层次结构(这些层次结构可能代表不同的顶层设置类别)，而不需要强迫用户在包含多个 Activity 实例的每个类别中跳入跳出。

奇巧安全

从 Android 4.4 开始，PreferenceActivity 必须在目标平台为 SDK Level 19 或更高版本的应用程序中重写 isValidFragment()方法。该方法阻止外部应用程序通过如下方式执行 Fragment 注入：在指向已导出 PreferenceActivity 的 Intent 额外部分提供不正确的类名，已导出的 PreferenceActivity 驻留有 PreferenceFragment 实例。

在目标平台为较低 SDK 版本的应用程序中，该方法将始终返回 true 以实现兼容性，这也会导致安全漏洞的出现。开发人员的谨慎做法是将此目标平台更新到 19+，并实现此方法以验证仅实例化预期的 Fragment。如果在已更新目标 SDK 的应用程序中未重写 isValidFragment()，则会抛出异常。

5.2 显示自定义首选项

5.2.1 问题

框架提供的 Preference 元素灵活度不够，需要添加更具体的 UI 来修改值。

5.2.2 解决方案

(API Level 1)

扩展 Preference 或它的一个子类，将新的类型整合到 PreferenceActivity 或 PreferenceFragment 中。创建新的首选项类型时，需要注意两个主要的目标：如何为用户提供修改首选项的界面，以及如何在 SharedPreferences 中持久保存它们的选择。

关于用户界面，你可能想要重写一些回调方法。请注意，它们使用与前面在 ListView 中看到的适配器类似的模式：

- onCreateView()：构造用于列表中此首选项元素的新布局。第一次需要此首选项的实例时调用该方法。如果存在相同类型的多个元素，则会在可能时回收这些视图。如果未重写此方法，则会显示带有标题和概要的默认视图。
- onBindView()：将当前首选项的数据附加到在 onCreateView()中构造的视图，该视图作为参数传入此方法。在每次将要显示首选项时调用此方法。
- getSummary()：重写显示在标准 UI 布局中的概要值。只有在未重写 onCreateView()/onBindView()时，该方法才起作用。
- onClick()：当用户点击列表中的相应项时处理事件。

框架中的基本首选项(如 CheckBoxPreference)只是在每次点击时切换持久化状态。其他首选项(如 EditTextPreference 或 ListPreference)是 DialogPreference 的子类，它们使用点击事件显示对话框，为更新给定的设置提供更加复杂的 UI。

在自定义 Preference 中，可能执行的第二组重写操作处理数据的检索和持久化：

- onGetDefaultValue()：调用该方法可以从首选项的 XML 定义中读取 android:defaultValue 属性。你将使用对首选项值有意义的类型化方法来接收存储该属性的 TypedArray 以及获得该值所需的索引。
- onSetInitialValue()：在本地设置此首选项实例的值。restorePersistedValue 标志表明该值是应该来自 SharedPreferences，还是应该来自默认值。默认值参数是从 onGetDefaultValue()返回的值。

首选项需要读取保存在 SharedPreferences 中的当前值时，可以调用某个类型化的 getPersistedXxx()方法，返回首选项正在持久化的值类型(integer、boolean、string 等)。相反，当首选项需要保存新值时，可以使用类型化的 persistXxx()方法更新 SharedPreferences。

5.2.3 实现机制

在下面的示例中，我们创建 ColorPreference：这是 DialogPreference 的简单扩展，它提供了用户界面，用户可在该界面中通过分别提供 RGB 值的三个滑块来选择颜色。完成此例后，结果应如图 5-2 所示。

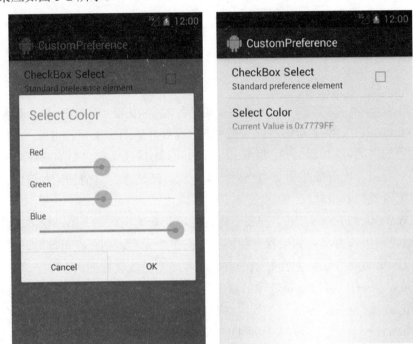

图 5-2　带有自定义 ColorPreference 的 PreferenceScreen

类似于 ListPreference，AlertDialog 将在从列表中选择首选项时显示。用户在 AlertDialog 中进行选择以及保存或取消变更，而不是直接在列表 UI 中操作(例如，CheckBoxPreference 就是这种情况)。代码清单 5-7 显示了自定义对话框的布局，而代码清单 5-8 则显示了 ColorPreference 的实现。

代码清单 5-7　res/layout/preference_color.xml

```
<?xml version="1.0" encoding="utf-8"?>
<LinearLayout xmlns:android="http://schemas.android.com/apk/res/android"
    android:layout_width="wrap_content"
    android:layout_height="wrap_content"
    android:padding="16dp"
    android:minWidth="300dp"
    android:orientation="vertical" >
    <TextView
        android:layout_width="wrap_content"
        android:layout_height="wrap_content"
        android:text="Red" />
    <SeekBar
        android:id="@+id/selector_red"
```

```xml
        android:layout_width="match_parent"
        android:layout_height="wrap_content"
        android:max="255" />

    <TextView
        android:layout_width="wrap_content"
        android:layout_height="wrap_content"
        android:text="Green" />
    <SeekBar
        android:id="@+id/selector_green"
        android:layout_width="match_parent"
        android:layout_height="wrap_content"
        android:max="255" />

    <TextView
        android:layout_width="wrap_content"
        android:layout_height="wrap_content"
        android:text="Blue" />
    <SeekBar
        android:id="@+id/selector_blue"
        android:layout_width="match_parent"
        android:layout_height="wrap_content"
        android:max="255" />
</LinearLayout>
```

代码清单 5-8　自定义首选项定义

```java
public class ColorPreference extends DialogPreference {

    private static final int DEFAULT_COLOR = Color.WHITE;
    /* 当前颜色设置的本地副本 */
    private int mCurrentColor;
    /* 设置颜色组件的滑块 */
    private SeekBar mRedLevel, mGreenLevel, mBlueLevel;

    public ColorPreference(Context context, AttributeSet attrs) {
        super(context, attrs);
    }

    /*
     * 调用以构造新的对话框，在点击首选项时显示此对话框。
     * 为每个实例创建和设置新的内容视图。
     */
    @Override
    protected void onPrepareDialogBuilder(Builder builder) {
        // 创建对话框的内容视图
        View rootView = LayoutInflater.from(getContext()).
                inflate(R.layout.preference_color, null);
        mRedLevel = (SeekBar) rootView.findViewById(R.id.selector_red);
        mGreenLevel = (SeekBar) rootView.findViewById(R.id.selector_green);
        mBlueLevel = (SeekBar) rootView.findViewById(R.id.selector_blue);
```

```java
        mRedLevel.setProgress(Color.red(mCurrentColor));
        mGreenLevel.setProgress(Color.green(mCurrentColor));
        mBlueLevel.setProgress(Color.blue(mCurrentColor));

        // 附加内容视图
        builder.setView(rootView);
        super.onPrepareDialogBuilder(builder);
    }

    /*
     * 当用户点击按钮关闭对话框时调用
     */
    @Override
    protected void onDialogClosed(boolean positiveResult) {
        if(positiveResult) {
            // 按下 OK 按钮时，获得并保存颜色值
            int color = Color.rgb(
                    mRedLevel.getProgress(),
                    mGreenLevel.getProgress(),
                    mBlueLevel.getProgress());
            setCurrentValue(color);
        }
    }

    /*
     * 由框架调用，用于获得传入首选项 XML 定义的默认值
     */
    @Override
    protected Object onGetDefaultValue(TypedArray a, int index) {
        // 作为颜色 int 值返回来自 XML 的默认值
        ColorStateList value = a.getColorStateList(index);
        if(value == null) {
            return DEFAULT_COLOR;
        }
        return value.getDefaultColor();
    }

    /*
     * 由框架调用，设置首选项的初始值
     * 该值来自默认值或上一次持久化的值
     */
    @Override
    protected void onSetInitialValue(boolean restorePersistedValue,
        Object defaultValue) {
        setCurrentValue( restorePersistedValue ?
                getPersistedInt(DEFAULT_COLOR) : (Integer)defaultValue );
    }

    /*
```

```
     * 基于当前设置返回自定义概要值
     */
    @Override
    public CharSequence getSummary() {
        // 使用十六进制的颜色值构造概要
        int color = getPersistedInt(DEFAULT_COLOR);
        String content = String.format("Current Value is 0x%02X%02X%02X",
                Color.red(color), Color.green(color), Color.blue(color));
        // 将概要文本返回为 Spannable，根据选择确定颜色
        Spannable summary = new SpannableString (content);
        summary.setSpan(new ForegroundColorSpan(color), 0, summary.length(), 0);
        return summary;
    }

    private void setCurrentValue(int value) {
        // 更新最新的值
        mCurrentColor = value;

        // 保存新值
        persistInt(value);
        // 通知首选项侦听器
        notifyDependencyChange(shouldDisableDependents());
        notifyChanged();
    }

}
```

从 XML 中初次创建 ColorPreference 时，会使用 android:defaultValue 属性(如果添加了该属性的话)调用 onGetDefaultValue()，因此我们可以对其进行分析。我们希望允许使用任何颜色属性，因此使用 getColorStateList()读取属性的值，该方法支持读取颜色字符串和指向颜色资源的引用，并且返回结果(整数值)。

在后面将首选项附加到 Activity 时，onSetInitialValue()将表明是应该读取默认值，还是使用已经保存在 SharedPreference 中的值。初次运行时，我们将选择默认值，而在此之后的每次尝试将使用 getPersistedInt()读取保存的值。如果持久化值不存在或不能作为整数值读取，getPersistedInt()的参数就是我们应使用的默认值。

对于用户界面，我们并没有监控 onClick()，而是采用 DialogPreference 提供的两个新的回调：onPrepareDialogBuilder()和 onDialogClosed()。前者随 AlertDialog.Builder 实例一起触发，因此我们可以定制在用户点击首选项时将显示的对话框；每次新的点击都将发生此操作。我们使用此方法填充和附加包含 3 个滑块的对话框布局，这 3 个滑块分别用于调整红色、绿色和蓝色组成部分。

收到 onDialogClosed()时，我们被告知用户是选择 OK 以保存首选项，还是选择 Cancel 以还原变更。对于前者，我们要从 UI 滑块创建新的颜色并使用 persistInt()持久化值。对于后者，我们不执行任何操作来变更当前设置。

最后，持久化新的变更时，调用 notifyDependencyChange()和 notifyChanged()来修改更新的任何首选项侦听器。这也会修改 PreferenceActivity 来更新列表显示。

我们使用 getSummary()完成了最终的自定义工作。在此例中，我们没有提供全新的布局，而是定制概要显示以包括当前的颜色选择(作为十六进制字符串)，并且使用选择的颜色修饰文本。我们可以这样做是因为 getSummary()返回 CharSequence(而非纯 String)，这样就可以返回样式化的 Spannable 类型。

构造新的首选项之后，如同在前一节中看到的那样，我们可以将其连同其他标准首选项一起添加到<PreferenceScreen>的 XML 定义中(参见代码清单 5-9 和 5-10)。

代码清单 5-9　res/xml/settings.xml

```xml
<?xml version="1.0" encoding="utf-8"?>
<PreferenceScreen xmlns:android="http://schemas.android.com/apk/res/android">
    <CheckBoxPreference
        android:key="dummyPref"
        android:title="CheckBox Select"
        android:summary="Standard preference element"
        android:defaultValue="false" />
    <com.androidrecipes.custompreference.ColorPreference
        android:key="colorPref"
        android:title="Select Color"
        android:defaultValue="@android:color/black" />
</PreferenceScreen>
```

代码清单 5-10　带有新设置的 PreferenceActivity

```java
public class CustomPreferenceActivity extends PreferenceActivity {

    @Override
    protected void onCreate(Bundle savedInstanceState) {
        super.onCreate(savedInstanceState);
        addPreferencesFromResource(R.xml.settings);
    }
}
```

我们已将颜色首选项的默认值设置为黑色框架资源，onGetDefaultValue()重写版本将读取此默认值。在代码清单 5-10 中可以看到，填充这个新的首选项层次结构不需要修改在前一节中看到的 PreferenceActivity 代码。

5.3　简单数据存储

5.3.1　问题

应用程序需要以一种简单而低开销的方式在持久存储中保存一些基本数据，如数字或字符串。

5.3.2 解决方案

(API Level 1)

通过 SharedPreferences 对象，应用程序可以快速创建一个或多个数据存储位置，稍后则可以在这些位置保存或查询数据。实际上，这些对象以 XML 文件的形式存储在应用程序的用户数据区中。但是，与直接从文件中读取和写入数据不同，SharedPreferences 提供了一个非常高效的框架来持久保存基本数据类型。

最好是创建多个 SharedPreferences 对象，而不是将所有的数据都保存到默认的 SharedPreferences 对象中，尤其是在数据只需要保存一段时间的情况下更该如此。记住，所有的首选项都是以 XML 的形式存储的，PreferenceActivity 框架也保存在默认位置，那么如何保存一组相关联的内容(如已经登录的用户)呢？当该用户退出时，将需要同时删除他所保存的数据。如果把这些数据存储到默认的首选项里面，可能需要单独删除每个条目。而如果为这些设置单独创建首选项对象，只需要在用户退出时简单地调用 SharedPreferences.Editor.clear() 即可。

5.3.3 实现机制

让我们看一个实际使用 SharedPreferences 保存简单数据的示例。代码清单 5-11 和 5-12 为用户创建了一个数据输入表单，可以用来向远程服务器发送简单消息。为了方便用户，在用户成功发送请求之前，我们会保存用户输入的所有各项数据。这样，用户就可以在暂时离开界面(或者被短信或电话打断)后不必再次输入所有的信息。

代码清单 5-11　res/layout/form.xml

```xml
<?xml version="1.0" encoding="utf-8"?>
<LinearLayout xmlns:android="http://schemas.android.com/apk/res/android"
    android:orientation="vertical"
    android:layout_width="match_parent"
    android:layout_height="match_parent">
    <TextView
        android:layout_width="match_parent"
        android:layout_height="wrap_content"
        android:text="Email:"
        android:padding="5dip" />
    <EditText
        android:id="@+id/email"
        android:layout_width="match_parent"
        android:layout_height="wrap_content"
        android:singleLine="true" />
    <CheckBox
        android:id="@+id/age"
        android:layout_width="match_parent"
        android:layout_height="wrap_content"
        android:text="Are You Over 18?" />
    <TextView
```

```xml
        android:layout_width="match_parent"
        android:layout_height="wrap_content"
        android:text="Message:"
        android:padding="5dip" />
    <EditText
        android:id="@+id/message"
        android:layout_width="match_parent"
        android:layout_height="wrap_content"
        android:minLines="3"
        android:maxLines="3" />
    <Button
        android:id="@+id/submit"
        android:layout_width="match_parent"
        android:layout_height="wrap_content"
        android:text="Submit" />
</LinearLayout>
```

代码清单 5-12　具有持久化功能的输入表单

```java
public class FormActivity extends Activity implements View.OnClickListener {

    EditText email, message;
    CheckBox age;
    Button submit;

    SharedPreferences formStore;

    boolean submitSuccess = false;

    @Override
    public void onCreate(Bundle savedInstanceState) {
        super.onCreate(savedInstanceState);
        setContentView(R.layout.form);

        email = (EditText)findViewById(R.id.email);
        message = (EditText)findViewById(R.id.message);
        age = (CheckBox)findViewById(R.id.age);

        submit = (Button)findViewById(R.id.submit);
        submit.setOnClickListener(this);

        //获取或创建首选项对象
        formStore = getPreferences(Activity.MODE_PRIVATE);
    }

    @Override
    public void onResume() {
        super.onResume();
        //还原表单数据
        email.setText(formStore.getString("email", ""));
        message.setText(formStore.getString("message", ""));
```

```java
            age.setChecked(formStore.getBoolean("age", false));
        }

        @Override
        public void onPause() {
            super.onPause();
            if(submitSuccess) {
                //可以同时调用 Editor
                formStore.edit().clear().commit();
            } else {
                //保存表单数据
                SharedPreferences.Editor editor = formStore.edit();
                editor.putString("email", email.getText().toString());
                editor.putString("message", message.getText().toString());
                editor.putBoolean("age", age.isChecked());
                editor.commit();
            }
        }

        @Override
        public void onClick(View v) {

            //发送消息

            //标记操作成功
            submitSuccess = true;
            //关闭
            finish();
        }
    }
```

首先，我们创建了一个典型的用户表单，它包含两个简单的 EditText 输入框和一个复选框。当创建 Activity 时，我们通过 Activity.getPreferences()获得了一个 SharedPreferences 对象，所有的持久化数据都会保存到这里。在 Activity 由于数据没有发送成功而暂停(通过一个布尔值控制)时，表单中的当前数据就会被快速载入首选项中进行持久化。

注意：

在使用 Editor 向 SharedPreferences 发送数据时，一定要记得在完成修改后调用 commit() 或 apply()。否则，修改将不会被保存。

相反，当 Activity 可见时，onResume()会将保存在首选项对象中的信息加载到用户界面上。如果由于首选项被清空或者根本没有被创建(初次启动)，表单就会被设置为空。

在用户按下 Submit 按钮后，信息就会被成功发送，接下来再调用 onPause()将会清空首选项中保存的所有表单数据。因为所有这些操作都是在私有的首选项对象中进行的，所以清空这些数据并不会影响以其他方式保存的用户设置。

注意：

Editor 中调用的方法返回的总是同一个 Editor 对象，这样就能按序将这些操作串联在一起，从而提高代码的可读性。

创建通用的 SharedPreferences

前面的示例演示了如何在一个单独的 Activity 中使用单个的 SharedPreferences 对象,该对象是通过 Activity.getPreferences()得到的。实际上,这个方法只是 Context.getSharedPreferences()的一个为方便使用而实现的封装方法,用 Activity 的名称作为首选项的保存名称。如果保存的数据要在两个或更多个 Activity 实例中共享的话,则需要调用 getSharedPreferences()并传入一个更加通用的名称,这样就可以在不同的地方轻松访问该 SharedPreferences 对象。参见代码清单 5-13。

代码清单 5-13　使用同一个首选项的两个 Activity

```java
public class ActivityOne extends Activity {
    public static final String PREF_NAME = "myPreferences";
    private SharedPreferences mPreferences;

    @Override
    public void onCreate(Bundle savedInstanceState) {
        super.onCreate(savedInstanceState);
        mPreferences = getSharedPreferences(PREF_NAME,
            Activity.MODE_PRIVATE);
    }
}

public class ActivityTwo extends Activity {

    private SharedPreferences mPreferences;

    @Override
    public void onCreate(Bundle savedInstanceState) {
        super.onCreate(savedInstanceState);
        mPreferences = getSharedPreferences(ActivityOne.PREF_NAME,
            Activity.MODE_PRIVATE);
    }

}
```

在这个示例中,两个 Activity 通过同一个名称(定义为字符串常量)得到了 SharedPreferences 对象,这样它们就可以访问同一个首选项数据集。此外,两个设置甚至指向的是同一个首选项对象实例,这是因为 Andorid 框架为每组 SharedPreferences(用其名称定义)创建的是单实例对象。这意味着一个 Activity 所做的修改会立即反映到另一个 Activity 中。

> **关于模式的说明**
>
> Context.getSharedPreferences()同样会接收一个模式参数。传入 0 或 MODE_PRIVAT 代表默认行为,只允许创建这个 SharedPreferences 的应用程序(或另一个具有相同用户 ID 的应用程序)对该 SharedPreferences 进行读写。这个方法还支持其他的模式参数:MODE_WORLD_READABLE 和 MODE_WORLD_WRITEABLE。通过在其创建的文件中

适当设置用户权限，这两个模式允许其他应用程序访问首选项。不过，外部应用程序也需要有一个指向创建该首选项文件的包的有效 Context 来访问该文件。

例如，在应用程序中以全局可读权限创建一个 SharedPreferences，该应用程序的包名为 com.examples.myfirstapplication。要在其他应用程序中访问这些首选项，可以使用如下代码：

```
Context otherContext = createPackageContext("com.examples.
   myfirstapplication", 0);
SharedPreferences externalPreferences = otherContext.
   getSharedPreferences(PREF_NAME, 0);
```

警告：
如果选择使用模式参数来允许外部应用程序访问，就必须确保在每次调用 getSharedPreferences()时都使用同样的模式参数。只有在首次创建首选项文件时模式参数才会生效，所以在每次调用 SharedPreferences 时使用不同的模式参数只会让你徒增烦恼。

5.4 读写文件

5.4.1 问题

应用程序需要读取外部文件中的数据或者持久保存更加复杂的数据。

5.4.2 解决方案

(API Level 1)

有时候，必须使用文件系统。使用文件允许应用程序读写那种不能通过键/值首选项和数据库方式保存的数据。Android 同样提供了用于文件的许多缓存位置，这里可以存放一些临时保存的文件数据。

Android 支持所有的标准 Java 文件 I/O API，用于执行创建、读取、更新和删除(CRUD)操作，此外还有一些额外的辅助方法，用于更加方便地访问特定位置的文件。应用程序可以在三个主要位置操作文件：
- 内部存储：受保护的用于读写文件数据的目录空间。
- 外部存储：外部挂载的用于读写文件数据的空间。API Level 4 以上需要 WRITE_EXTERNAL_STORAGE 权限。通常这都是设备的 SD 卡。
- Assets：APK 包中只读的受保护空间，用于放置不能/不应该被编译的本地资源。

操作文件数据的基本原理都是一样的，下面看看各种具体情况之间的细微差异。

5.4.3 实现机制

如前所述，传统的 Java 类 FileInputStream 和 FileOutputStream 依然是访问文件数据的

主要方法。实际上，随时都可以使用文件的绝对路径创建一个 File 实例，然后使用一个文件流对数据进行读写。不过，不同设备有不同的根路径，有些目录还是应用程序无法访问的，所以建议使用更有效的方式来处理文件。

1. 内部存储

要在内部存储中创建和修改文件的位置，可以使用 Context.openFileInput()和 Context.openFileOutput()方法。这两个方法只需要文件的名称作为参数，而不是文件的完整路径。它们会引用应用程序受保护目录空间中的文件，并不会关心设备的确切文件目录位置。参见代码清单 5-14。

代码清单 5-14　在内部存储中 CRUD 文件

```java
public class InternalActivity extends Activity {

    private static final String FILENAME = "data.txt";

    @Override
    public void onCreate(Bundle savedInstanceState) {
        super.onCreate(savedInstanceState);
        TextView tv = new TextView(this);
        setContentView(tv);

        //创建一个新文件并写入一些数据
        try {
            FileOutputStream mOutput = openFileOutput(FILENAME,
                Activity.MODE_PRIVATE);
            String data = "THIS DATA WRITTEN TO A FILE";
            mOutput.write(data.getBytes());
            mOutput.flush();
            mOutput.close();
        } catch(FileNotFoundException e) {
            e.printStackTrace();
        } catch(IOException e) {
            e.printStackTrace();
        }

        //读取已经创建的文件并显示在屏幕上
        try {
            FileInputStream mInput = openFileInput(FILENAME);
            byte[] data = new byte[128];
            mInput.read(data);
            mInput.close();

            String display = new String(data);
            tv.setText(display.trim());
        } catch(FileNotFoundException e) {
            e.printStackTrace();
        } catch(IOException e) {
```

```
            e.printStackTrace();
        }

        //删除创建的文件
        deleteFile(FILENAME);
    }
}
```

这个示例通过 Context.openFileOutput()向一个文件中写入一些简单的字符串数据。使用该方法时，如果文件不存在，首先会创建该文件。它有两个参数：文件名和操作模式。本例中，我们使用默认的操作模式 MODE_PRIVATE。使用这种模式，每次写入时都会覆盖原有的文件；如果想每次都写入文件的尾部，可以使用 MODE_APPEND。

写入完成后，本例调用了 Context.openFileInput()，该方法也只需要以文件名作为参数，它会打开一个输入流来读取文件数据。数据会被读入一个 byte 数组中并通过 TextView 显示到用户界面上。这个操作完成后，会调用 Context.deleteFile()删除存储器中的文件。

注意：
数据是以字节的形式写入文件流中的，因此高级的数据(甚至是字符串)必须转换为这种形式。

这个示例在执行完成之后并不会留下任何的文件痕迹，但我们鼓励你尝试在结尾处不调用 deleteFile()，这样文件就会留在存储器中。使用 SDK 的 DDMS 工具和模拟器或者在已经解锁的设备上查看文件系统，你会找到这个应用程序在它自己的数据文件夹中所创建的文件。

由于它们都是 Context 的方法，并不是和 Activity 绑定的，因此需要的话，可以在应用程序的任何地方使用这种类型的文件访问方式，例如在 BroadcastReceiver 甚至自定义的类中。很多系统都会构建 Context 的一个子类，或是在其回调中传递一个指向某个 Context 子类的引用，因此在任何地方都可以执行同样的打开、关闭和删除操作。

2. 外部储存

内部存储器和外部存储器最大的区别就是外部存储器是可挂载的。这意味着用户可以将它们的设备连接到计算机，然后将设备的外部存储器以可移动硬盘的形式挂载到计算机上。通常情况下，外部存储器本身就是可以移除的(如 SD 卡)，但它并不是 Android 平台所必需的。

要点：
向设备的外部存储器中写入数据需要在应用程序的清单中添加 android.permission.WRITE_EXTERNAL_STORAGE 权限。在 API Level 19+平台中，从外部存储器读取数据还需要添加 droid.permission.READ_EXTERNAL_STORAGE 权限。

当设备的外部存储器被挂载到其他设备或者被移除时，应用程序是无法访问它的。因此，最好经常使用 Environment.getExternalStorageState()先检查一下外部存储器是否可用。

让我们修改一下上一个文件示例，使用设备的外部存储器实现相同的操作。参见代码清单 5-15。

代码清单 5-15　在外部存储器中 CRUD 文件

```java
public class ExternalActivity extends Activity {

    private static final String FILENAME = "data.txt";

    @Override
    public void onCreate(Bundle savedInstanceState) {
        super.onCreate(savedInstanceState);
        TextView tv = new TextView(this);
        setContentView(tv);

        //创建文件的引用
        File dataFile = new File(Environment.getExternalStorageDirectory(),
            FILENAME);

        //检查外部存储器是否可用
          if(!Environment.getExternalStorageState().equals(Environment.
            MEDIA_MOUNTED)) {
            Toast.makeText(this, "Cannot use storage.", Toast.LENGTH_SHORT).show();
            finish();
            return;
        }

        //创建一个新文件并写入一些数据
        try {
            FileOutputStream mOutput = new FileOutputStream(dataFile, false);
            String data = "THIS DATA WRITTEN TO A FILE";
            mOutput.write(data.getBytes());
            mOutput.flush();
            //使用外部文件时，通常最好同步该文件
            mOutput.getFD().sync();

            mOutput.close();
        } catch(FileNotFoundException e) {
            e.printStackTrace();
        } catch(IOException e) {
            e.printStackTrace();
        }

        //读取已经创建的文件并显示在屏幕上
        try {
            FileInputStream mInput = new FileInputStream(dataFile);
            byte[] data = new byte[128];
            mInput.read(data);
            mInput.close();

            String display = new String(data);
            tv.setText(display.trim());
```

```
        } catch(FileNotFoundException e) {
            e.printStackTrace();
        } catch(IOException e) {
            e.printStackTrace();
        }

        //删除创建的文件
        dataFile.delete();
    }
}
```

对于外部存储器，我们更多地使用了传统的 Java 文件 I/O。使用外部存储器的关键就是需要调用 Environment.getExternalStorageDirectory()来获得设备的外部存储器的根目录。

在做任何操作之前，首先通过 Environment.getExternalStorageState()检查设备的外部存储器的状态。如果返回的值不是 Environment.MEDIA_MOUNTED，这意味着外部存储器不可写入，所以我们什么也不会做，Activity 会被关闭。否则，就会创建一个新的文件并执行其他相关操作。

输入流和输出流必须使用 Java 默认的构造函数，而没有使用 Context 中的便捷方法。输出流默认会覆盖现有的文件或者在文件不存在时创建一个。如果应用程序每次写入时需要在现有文件的尾部进行添加的话，可以将 FileOutputStream 构造函数的布尔参数设为 true。

通常，最好为应用程序文件在外部存储器中创建一个单独的目录。用 Java 的文件 API 就能实现它。参见代码清单 5-16。

代码清单 5-16　在新目录下 CRUD 文件

```java
public class ExternalActivity extends Activity {

    private static final String FILENAME = "data.txt";
    private static final String DNAME = "myfiles";

    @Override
    public void onCreate(Bundle savedInstanceState) {
        super.onCreate(savedInstanceState);
        TextView tv = new TextView(this);
        setContentView(tv);

        //在外部存储器中创建一个新的目录
        File rootPath = new File(Environment.getExternalStorageDirectory(),
           DNAME);
        if(!rootPath.exists()) {
           rootPath.mkdirs();
        }
        //创建文件的引用
        File dataFile = new File(rootPath, FILENAME);

        //创建一个新文件并写入一些数据
```

```java
try {
    FileOutputStream mOutput = new FileOutputStream(dataFile, false);
    String data = "THIS DATA WRITTEN TO A FILE";
    mOutput.write(data.getBytes());
    mOutput.flush();

    //对于外部文件，通常最好等待写入
    mOutput.getFD().sync();

    mOutput.close();
} catch(FileNotFoundException e) {
    e.printStackTrace();
} catch(IOException e) {
    e.printStackTrace();
}

//读取已经创建的文件并显示在屏幕上
try {
    FileInputStream mInput = new FileInputStream(dataFile);
    byte[] data = new byte[128];
    mInput.read(data);
    mInput.close();

    String display = new String(data);
    tv.setText(display.trim());
} catch(FileNotFoundException e) {
    e.printStackTrace();
} catch(IOException e) {
    e.printStackTrace();
}

//删除创建的文件
dataFile.delete();
    }
}
```

本例中，我们在外部存储器目录中创建了一个新的目录并将这个目录作为数据文件的根目录。在通过新的目录位置创建好文件的引用后，该例剩余的代码都是一样的。

写入文件时的注意事项

Android 应用程序运行在 Dalvik 虚拟机环境内部。在使用系统的某些方面(如文件系统)时，需要知道此环境带来的一些影响。例如，FileOutputStream 这样的 Java API 没有与内核中的原生文件描述符共享 1:1 关系。通常情况下，使用 write()将数据写入流时，数据会直接写入文件的内存缓冲区并异步写出磁盘。在大多数情况下，只要严格在 Dalvik 虚拟机内部进行文件访问，就绝对不会看到此实现的细节。例如，可以打开刚刚写入的文件，并且读取时没有任何问题。

然而，当在手机或平板电脑上处理移动存储设备(如 SD 卡)时，我们通常需要保证文件数据沿着各种途径到达文件系统，之后再返回一些操作给用户，这是因为用户有能力物理地移除存储介质。下面是在写入外部文件时使用的良好标准代码块：

```
//写入数据
out.write();
//清空流缓冲区
out.flush();
//同步所有数据到文件系统
out.getFD().sync();
//关闭流
out.close();
```

OutputStream 上的 flush()方法旨在确保将驻留在流中的所有数据写出到 VM 的内存缓冲区中。在 FileOutputStream 的直接示例中，该方法实际上没有做任何工作。然而，如果流可能封装在另一个流(如 BufferedOutputStream)的内部，该方法基本上可以确保清除内部缓冲区，因此良好的习惯是在每次文件写操作中都调用该方法，之后再关闭流。

此外，对于外部文件，可以向底层 FileDescriptor 发出 sync()方法。该方法将阻塞，直至系统将所有数据成功写入底层文件系统，因此它是最佳的指示符，表明用户何时可以安全移除物理存储介质，而不会造成文件崩溃。

3. 外部系统目录

Environment 和 Context 的其他方法也能够访问外部存储器上其他一些标准的位置，在这些位置可以写入一些特殊的文件。同样，它们会有一些额外的属性。

- Environment.getExternalStoragePublicDirectory(String type)
  - ➤ API Level 8
  - ➤ 返回一个所有应用程序保存媒体文件的通用目录。这个目录的内容对用户和其他应用程序都是可见的。特别是，对于 Gallery 这样的应用程序，这里的媒体文件有可能会被扫描并加入到设备的 MediaStore 中。
  - ➤ type 参数的有效值可以为 DIRECTORY_PICTURES、DIRECTORY_MUSIC、DIRECTORY_MOVIES 和 DIRECTORY_RINGTONES。
- Context.getExternalFilesDir(String type)
  - ➤ API Level 8
  - ➤ 返回外部存储器上的媒体文件目录，但只适用于某个应用程序。这里的媒体文件不是公开的，并不会出现在 MediaStore 中。
  - ➤ 由于该目录还是在外部存储器上，因此其他用户和应用程序还是可以看到它并可以直接编辑文件；并不是强制安全的。
  - ➤ 放在这里的文件在卸载应用程序时被删除，因此对于应用程序需要使用大内容文件而同时又不希望保存到内部存储器的场景，这里是绝佳去处。
  - ➤ type 参数的有效值可以为 DIRECTORY_PICTURES、DIRECTORY_MUSIC、DIRECTORY_MOVIES 和 DIRECTORY_RINGTONES。

- Context.getExternalCacheDir()
  - API Level 8
  - 返回一个外部存储器上供某一应用程序特有临时文件使用的目录。这个目录的内容对用户和其他应用程序都是可见的。
  - 放在这里的文件在卸载应用程序时会被删除,因此对于应用程序需要使用大内容文件而同时又不希望保存到内部存储器的场景,这里是绝佳去处。
- Context.getExternalFileDirs()和 Context.getExternalCacheDirs()
  - API Level 19
  - 与前面描述的对应属性有同等的功能,但是会返回设备上每个存储卷的路径列表(主卷和任意副卷)。
  - 例如,某个设备可能采用一块内置闪存作为主要外部存储器,同时使用移动 SD 卡作为次要外部存储器。
- Context.getExternalMediaDirs()
  - API Level 21
  - 放在这些卷中的文件将自动被扫描并添加到设备的介质存储,以便显示给其他应用程序。这些文件通常也通过核心应用程序(如 Gallery)对用户可见。

注意:
从 KitKat(API Level 19)开始,不再需要任何权限即可读写 getExternalFilesDir()和 getExternalCacheDir()为应用程序返回的目录路径。仍然可以使用上述权限从这些目录外部写入主卷;即使声明了 WRITE_EXTERNAL_STORAGE 权限,副卷(也是 KitKat API 的新增对象)在这些目录外部也是完全写保护的。

5.5 以资源的形式使用文件

5.5.1 问题

应用程序需要使用不能被 Android 编译为资源 ID 的资源文件。

5.5.2 解决方案

(API Level 1)
应用程序可以在 assets 目录中保存需要读取的文件,例如本地 HTML 文件、逗号分隔值(CSV)或专有数据。assets 目录是 Android 应用程序的一个受保护文件目录。这个目录下的文件会和最终的 APK 一起打包,并且不会被处理或编译。和其他的应用程序资源一样,assets 中的文件都是只读的。

5.5.3 实现机制

我们在本书中已经看了几个关于使用 assets 直接向小部件(如 WebView、MediaPlayer)中加载内容的示例。然而，在很多情况下，使用传统的 InputStream 也能够很好地访问 assets。代码清单 5-17 和 5-18 展示了一个示例，它会读取 assets 中的私有 CSV 文件并显示在屏幕上。

代码清单 5-17 assets/data.csv

```
John,38,Red
Sally,42,Blue
Rudy,31,Yellow
```

代码清单 5-18 读取 asset 中的文件

```java
public class AssetActivity extends Activity {

    @Override
    public void onCreate(Bundle savedInstanceState) {
        super.onCreate(savedInstanceState);
        TextView tv = new TextView(this);
        setContentView(tv);

        try {
            //访问应用程序的 assets 目录
            AssetManager manager = getAssets();
            //打开数据文件
            InputStream mInput = manager.open("data.csv");

            //解析 CSV 数据并显示
            String raw = new String(data);
            ArrayList<Person> cooked = parse(raw.trim());
            StringBuilder builder = new StringBuilder();
            for(Person piece : cooked) {
              builder.append(String.format("%s is %s years old, and likes  
                    the color %s", piece.name, piece.age, piece.color));
              builder.append('\n');
            }
            tv.setText(builder.toString());

        } catch(FileNotFoundException e) {
            e.printStackTrace();
        } catch(IOException e) {
            e.printStackTrace();
        }

    }

    /*简单 CSV 解析器*/
```

```java
private static final int COL_NAME = 0;
private static final int COL_AGE = 1;
private static final int COL_COLOR = 2;

private ArrayList<Person> parse(String raw) {
    ArrayList<Person> results = new ArrayList<Person>();

    BufferedReader reader = new BufferedReader(new
InputStreamReader(in));
    String nextLine = null;
    while((nextLine = reader.readLine()) != null) {
        String[] tokens = nextLine.split(",");
        if(tokens.length != 3) {
            Log.w("CSVParser", "Skipping Bad CSV Row");
            continue;
        }
        //添加新的解析器结果
        Person current = new Person();
        current.name = tokens[COL_NAME];
        current.color = tokens[COL_COLOR];
        current.age = tokens[COL_AGE];

        results.add(current);
    }
    in.close();
    return results;
}

private class Person {
    public String name;
    public String age;
    public String color;

    public Person() { }
}
```

访问 assets 中文件的关键就是使用 AssetManager，它允许应用程序打开 assets 目录中现有的任意资源。在 AssetManager.open()中传入要打开的文件名称会得到一个可以用来读取文件数据的 InputStream。当数据流被读入内存后，会将这些原始数据传到一个解析例程中并将解析结果显示在用户界面上。

解析 CSV 数据

这个示例通过一种简单的方式读取 CSV 文件中的数据并将数据解析为模型对象(本例中为 Person)。这种方式会获取整个文件并读入一个字节数组中，然后转换为单独的字符串进行处理。这种方式在需要读取的数据量很大时并不是最省内存的，但对于本例中的这种小文件还是比较合适的。

原始字符串数据被传入一个 StringTokenizer 实例中，并且指定了用作分隔符的字符：逗号和换行符。然后，文件中的每个数据块会按序处理。通过一种基本的状态机方法，就可以将每行数据插入新的 Person 实例并载入结果列表中。

5.6 管理数据库

5.6.1 问题

应用程序需要查询或修改保存数据的子集或单条记录。

5.6.2 解决方案

(API Level 1)

可以通过 SQLiteOpenHelper 的帮助创建一个 SQLiteDatabase 以管理数据存储。SQLite 是一种快速和轻量的数据库技术，它使用 SQL 语法进行数据的查询和管理。Android SDK 本身就支持 SQLite，因此在应用程序中设置和使用数据库也就变得非常简单了。

5.6.3 实现机制

通过自定义 SQLiteOpenHelper 可以管理数据库的创建和修改。这里还是创建数据库后设置初始值和默认值的绝佳场所。代码清单 5-19 中的示例展示了如何自定义辅助类来创建包含单个表的数据库，该数据库中保存了人员的基本信息。

代码清单 5-19　自定义 SQLiteOpenHelper

```java
public class MyDbHelper extends SQLiteOpenHelper {

    private static final String DB_NAME = "mydb";
    private static final int DB_VERSION = 1;

    public static final String TABLE_NAME = "people";
    public static final String COL_NAME = "pName";
    public static final String COL_DATE = "pDate";

    private static final String STRING_CREATE =
        "CREATE TABLE "+TABLE_NAME+" (_id INTEGER PRIMARY KEY AUTOINCREMENT, "
        +COL_NAME+" TEXT, "+COL_DATE+" DATE);";

    public MyDbHelper(Context context) {
        super(context, DB_NAME, null, DB_VERSION);
    }

    @Override
    public void onCreate(SQLiteDatabase db) {
        //创建数据库表
```

```
        db.execSQL(STRING_CREATE);

        //可能还需要在这里加载数据库的初始值
        ContentValues cv = new ContentValues(2);
        cv.put(COL_NAME, "John Doe");
        //格式化SQL日期
        SimpleDateFormat dateFormat = new SimpleDateFormat("yyyy-MM-dd HH:mm:ss");
        cv.put(COL_DATE, dateFormat.format(new Date())); //插入当前日期数据
        db.insert(TABLE_NAME, null, cv);
    }

    @Override
    public void onUpgrade(SQLiteDatabase db, int oldVersion, int newVersion) {
        //目前清空并重新创建数据库
        db.execSQL("DROP TABLE IF EXISTS "+TABLE_NAME);
        onCreate(db);
    }
}
```

对于数据库，需要的最关键信息就是数据库的名称和版本号。为了 SQLiteDatabase 的创建和升级，需要了解一点 SQL 知识，因此如果还不太熟悉 SQL 语法的话，建议先简单看一下 SQL 的参考资料。如果在数据库不存在时，通过 SQLiteOpenHelper.getReadableDatabase() 或 SQLiteOpenHelper.getWritableDatabase() 访问数据库，就会在访问此数据库时调用辅助类的 onCreate() 方法。

本例中将表名和字段名都抽象为常量，这样就可以供外部使用(很好的实践方法)。下面是 onCreate() 中创建表的实际 SQL 语句：

```
CREATE TABLE people (_id INTEGER PRIMARY KEY AUTOINCREMENT, pName TEXT,
pAge INTEGER, pDate DATE);
```

在 Android 中使用 SQLite 时，数据库必须经过少量格式化才可以在框架中正确工作。其中最重要的一点就是，创建的表中必须有_id字段。而 SQL 语句还为表中的每条记录创建另外两个字段：

- 人员名称的文本字段
- 保存记录日期的日期字段

数据是以 ContentValues 对象的形式插入数据库中的。本例演示了如何在创建数据库后通过使用 ContentValues 在数据库中插入一些默认数据。SQLiteDatabase.insert()需要的参数为数据库的表名、空字段填充和代表要插入记录(作为参数)的 ContentValues。

这里并没有用到空字段填充，但这个参数对某些应用程序是非常有用的。SQL 不能向数据库中插入全部为空的数据，这样做会导致出错。如果在编写代码时，传递给 insert()的 ContentValues 是空的，那么系统会将空字段插入数据库中的记录，其中引用字段的内容都是 NULL。

1. 关于升级

SQLiteOpenHelper 还能在升级应用程序的过程中帮助迁移数据库架构。访问数据库时，

如果设备上的数据库版本和当前版本(即构造函数中传入的版本号)不一致，onUpgrade()就会被调用。

在我们的示例中，我们采用了一种偷懒的方式，只是简单地删除了现有的数据库并重新创建它。实际上，如果数据库中存放了用户输入的数据，这种方式就有点不合适了；用户可能并不希望他的数据消失。所以，我们来看一个更加实用的 onUpgrade()示例。例如，应用程序在其生命周期内会使用下面三个版本的数据库：

- 版本 1：应用程序的第一版。
- 版本 2：升级后的应用程序加入了电话号码字段。
- 版本 3：升级后的应用程序加入了日期输入字段。

我们可以使用 onUpgrade()来修改现有的数据库，而不是将现有的数据库信息全部删除。参见代码清单 5-20。

代码清单 5-20　onUpgrade()示例

```
@Override
public void onUpgrade(SQLiteDatabase db, int oldVersion, int newVersion) {
    //在 v1 的基础进行升级。添加电话号码字段
    if(oldVersion <= 1) {
        db.execSQL("ALTER TABLE "+TABLE_NAME+" ADD COLUMN phone_number INTEGER;");
    }
    //在 v2 的基础上进行升级。添加日期输入字段
    if(oldVersion <= 2) {
        db.execSQL("ALTER TABLE "+TABLE_NAME+" ADD COLUMN entry_date DATE;");
    }
}
```

在这个示例中，如果用户现在的数据库版本是 1，两条语句都会执行，数据库会增加两个字段。如果数据库版本已经是 2，则只会执行后一条语句，即添加日期输入字段。两种情况下，都会保留应用程序数据库中现有的数据。

提示：

对于 API Level 11+的平台，SQLiteOpenHelper 还支持 onDowngrade()。如果磁盘上的数据库版本高于应用程序代码请求的当前版本，就会调用此方法。

2. 使用数据库

回到原来的示例，看一下 Activity 该如何使用我们创建的数据库。参见代码清单 5-21 和 5-22。

代码清单 5-21　res/layout/main.xml

```xml
<?xml version="1.0" encoding="utf-8"?>
<LinearLayout xmlns:android="http://schemas.android.com/apk/res/android"
    android:orientation="vertical"
    android:layout_width="match_parent"
    android:layout_height="match_parent">
```

```xml
<EditText
    android:id="@+id/name"
    android:layout_width="match_parent"
    android:layout_height="wrap_content" />
<Button
    android:id="@+id/add"
    android:layout_width="match_parent"
    android:layout_height="wrap_content"
    android:text="Add New Person" />
<ListView
    android:id="@+id/list"
    android:layout_width="match_parent"
    android:layout_height="match_parent" />
</LinearLayout>
```

代码清单 5-22　查看和管理数据库的 Activity

```java
public class DbActivity extends Activity implements View.OnClickListener,
        AdapterView.OnItemClickListener {

    EditText mText;
    Button mAdd;
    ListView mList;

    MyDbHelper mHelper;
    SQLiteDatabase mDb;
    Cursor mCursor;
    SimpleCursorAdapter mAdapter;

    @Override
    public void onCreate(Bundle savedInstanceState) {
        super.onCreate(savedInstanceState);
        setContentView(R.layout.main);

        mText = (EditText)findViewById(R.id.name);
        mAdd = (Button)findViewById(R.id.add);
        mAdd.setOnClickListener(this);
        mList = (ListView)findViewById(R.id.list);
        mList.setOnItemClickListener(this);

        mHelper = new MyDbHelper(this);
    }

    @Override
    public void onResume() {
        super.onResume();
        //建立和数据库的连接
        mDb = mHelper.getWritableDatabase();
        String[] columns = new String[] {"_id", MyDbHelper.COL_NAME,
            MyDbHelper.COL_DATE};
        mCursor = mDb.query(MyDbHelper.TABLE_NAME, columns, null, null,
```

```java
            null, null, null);
        //更新列表
        String[] headers = new String[]{MyDbHelper.COL_NAME, MyDbHelper.COL_DATE};
        mAdapter = new SimpleCursorAdapter(this, android.R.layout.two_line_list_item,
                mCursor, headers, new int[]{android.R.id.text1,
                android.R.id.text2});
        mList.setAdapter(mAdapter);
    }

    @Override
    public void onPause() {
        super.onPause();
        //关闭连接
        mDb.close();
        mCursor.close();
    }

    @Override
    public void onClick(View v) {
        //向数据库中添加新数据
        ContentValues cv = new ContentValues(2);
        cv.put(MyDbHelper.COL_NAME, mText.getText().toString());
        //格式化 SQL 日期
        SimpleDateFormat dateFormat = new SimpleDateFormat("yyyy-MM-dd HH:mm:ss");
        //插入当前日期数据
        cv.put(MyDbHelper.COL_DATE, dateFormat.format(new Date()));
        mDb.insert(MyDbHelper.TABLE_NAME, null, cv);
        //更新列表
        mCursor.requery();
        mAdapter.notifyDataSetChanged();
        //清空编辑字段
        mText.setText(null);
    }

    @Override
    public void onItemClick(AdapterView<?> parent, View v, int position, long id) {
        //删除数据库中的条目
        mCursor.moveToPosition(position);
        //获得该行的 id 值
        String rowId = mCursor.getString(0);  //Cursor 的字段 0 是 id
        mDb.delete(MyDbHelper.TABLE_NAME, "_id = ?", new String[]{rowId});
        //更新列表
        mCursor.requery();
        mAdapter.notifyDataSetChanged();
    }
}
```

在这个示例中，我们通过自定义的 **SQLiteOpenHelper** 访问一个数据库实例，并将该数据库中的记录以列表的形式显示在用户界面上。数据库中的信息是以 Cursor 的形式返回的，Cursor 是一个可以用来读、写、遍历查询结果的接口。

当 Activity 可见时，会查询数据库中的 people 表并返回该表中的所有记录。必须向查询传递一个列名数组来告诉数据库返回哪些值。query() 的其他参数用来缩小选择数据集的范围，这会在下一个范例中进一步讨论。在数据库和游标连接不用时，一定要关闭它们。本例中，则是在 onPause() 中进行关闭操作，这时候 Activity 已经不在前台了。

SimpleCursorAdapter 用来将数据库中的数据映射为标准的 Android 两行式的列表项视图。映射时的参数为一个字符串数组和一个整型数组；系统会将字符串数组中的每一项数据和整型数组中相应的 id 值插入到视图中。注意，这里传递的列名列表与传递给查询的数组是不太一样的。因为在其他操作中需要用到每条记录的 id，但在将数据映射到用户界面时是不需要这个 id 的。

用户可能会在文本框中输入名称并单击"Add New Person"按钮来创建一个新的 ContentValues 实例并插入到数据库中。如果是这样的话，为了让 UI 能够显示数据库的变化，我们调用了 Cursor.requery() 和 ListAdapter.notifyDataSetChanged() 方法。

相反，单击列表中的条目会删除数据库中的指定记录。要实现这个功能，我们必须构造一条简单的 SQL 语句，告诉数据库仅删除与选择的_id 值相对应的记录。然后，游标和列表的适配器都会再次刷新。

将游标移到选择的位置，调用 getString(0) 就能获取到 0 字段的值，这样就能得到所选项的 id 值。因为传入查询的列列表的第一个参数(索引是 0)就是_id，所以这个查询返回的就是_id。delete 语句由两个参数组成：SQL 语句字符串和参数。参数是以数组的形式传递的，其中的每个元素会按序插入到 SQL 语句字符串中各个问号所在的位置。

5.7 查询数据库

5.7.1 问题

应用程序使用了 SQLiteDatabase，需要从中获取所需的数据子集。

5.7.2 解决方案

(API Level 1)
利用完全结构化的 SQL 查询，可以很方便地创建各种数据过滤器，从数据库中获取所需的数据子集。SQLiteDatabase.query() 有好几种重载形式可以从数据库获取信息。下面看看其中最复杂的一种形式：

```
public Cursor query(String table, String[] columns,
        String selection,
        String[] selectionArgs,
        String groupBy,
        String having,
        String orderBy,
```

```
    String limit)
```

前两个参数只是定义了查询数据的数据库表和要访问的每条记录的列。其他参数则定义了如何缩小结果的范围：

- selection：给定查询的 SQL WHERE 子句。
- selectionArgs：如果 selection 语句中有问号，就用这里的元素逐个填充那些问号位置。
- groupBy：给定查询的 SQL GROUP BY 子句。
- having：给定查询的 SQL HAVING 子句。
- orderBy：给定查询的 SQL ORDER BY 子句。
- limit：查询返回结果的数量上限。

可以看到，所有这些参数都旨在提供完整的 SQL 数据库查询功能。

5.7.3 实现机制

让我们构造一些能完成常见实际查询的示例。

- 返回其中的值能匹配给定参数的所有行：

```
String[] COLUMNS = new String[] {COL_NAME, COL_DATE};
String selection = COL_NAME+" = ?";
String[] args = new String[] {"NAME_TO_MATCH"};
Cursor result = db.query(TABLE_NAME, COLUMNS, selection, args, null, null,
   null, null);
```

这个查询很直观。selection 语句告诉数据库根据参数提供的字段名(插到 selection 字符串中问号的位置)返回数据。

- 返回最近插入数据库的 10 行记录：

```
String orderBy = "_id DESC";
String limit = "10";
Cursor result = db.query(TABLE_NAME, COLUMNS, null, null, null, null, orderBy,
    limit);
```

这个查询没有指定 selection 条件，而是要数据库将结果以 _id 值按自动增序排列，首先返回的是最新的(_id 最大的)记录。limit 子句将返回结果的数量上限设为 10。

- 返回日期字段在指定范围内的行(本例中是 2000 年)：

```
String[] COLUMNS = new String[] {COL_NAME, COL_DATE};
String selection = "datetime("+COL_DATE+") > datetime(?)"+
        " AND datetime("+COL_DATE+") < datetime(?)";
String[] args = new String[] {"2000-1-1 00:00:00","2000-12-31 23:59:59"};
Cursor result = db.query(TABLE_NAME, COLUMNS, selection, args, null, null,
   null, null);
```

在 SQLite 中创建表时可以声明 DATE 类型，但实际上 SQLite 并没有专门的日期类型数据。可以用标准的 SQL 日期和时间函数，以 TEXT、INTEGER 或 REAL 的形式表示日期和时间。这里我们将数据库中的值和已经格式化的开始和结束时间字符串传递给 datetime(),然后比较返回的值。

- 返回整型字段在指定范围内的行(本例中是 7～10):

```
String[] COLUMNS = new String[] {COL_NAME, COL_AGE};
String selection = COL_AGE+" > ? AND "+COL_AGE+" < ?";
String[] args = new String[] {"7","10"};
Cursor result = db.query(TABLE_NAME, COLUMNS, selection, args, null, null,
null, null);
```

这与前一个示例类似,但更简单。它只创建了一条返回值有上下限的 selection 语句。两个限制都是以参数的形式插入的,所以可以在应用程序中动态设置。

5.8 备份数据

5.8.1 问题

应用程序将数据保存到设备上,但在用户更换设备或者被迫重装应用程序时,需要为用户提供一种方式来备份和恢复这些数据。

5.8.2 解决方案

(API Level 3)

设备的外部存储器可以安全地保存数据库和其他文件的副本。外部存储器一般都是物理可拆卸的,可以从一台设备移到另一台设备上进行数据恢复。即使不可物理拆卸,也可以将外部存储器挂载到计算机上,从而实现数据传输。

5.8.3 实现机制

代码清单 5-23 实现了一个 AsyncTask,它可以在设备的外部存储器和应用程序的数据目录之间进行数据库文件的来回复制。它还定义了一个 Activity 可以实现的接口,用于在操作完成后通知 Activity。像复制这样的文件操作是需要一些时间来完成的,因此通过 AsyncTask 可以让该操作在后台执行,从而不会阻塞主线程。

代码清单 5-23 用来备份和恢复的 AsyncTask

```
public class BackupTask extends AsyncTask<String,Void,Integer> {

    public interface CompletionListener {
        void onBackupComplete();
        void onRestoreComplete();
        void onError(int errorCode);
    }
```

```java
public static final int BACKUP_SUCCESS = 1;
public static final int RESTORE_SUCCESS = 2;
public static final int BACKUP_ERROR = 3;
public static final int RESTORE_NOFILEERROR = 4;

public static final String COMMAND_BACKUP = "backupDatabase";
public static final String COMMAND_RESTORE = "restoreDatabase";

private Context mContext;
private CompletionListener listener;

public BackupTask(Context context) {
    super();
    mContext = context;
}

public void setCompletionListener(CompletionListener aListener) {
    listener = aListener;
}

@Override
protected Integer doInBackground(String... params) {

    //获得数据库的引用
    File dbFile = mContext.getDatabasePath("mydb");
    //获得备份的目录位置引用
    File exportDir =new File
            (Environment.getExternalStorageDirectory(), "myAppBackups");
    if(!exportDir.exists()) {
       exportDir.mkdirs();
    }
    File backup = new File(exportDir, dbFile.getName());

    //检查需要的操作
    String command = params[0];
    if(command.equals(COMMAND_BACKUP)) {
        //尝试复制文件
        try {
            backup.createNewFile();
            fileCopy(dbFile, backup);

            return BACKUP_SUCCESS;
        } catch(IOException e) {
            return BACKUP_ERROR;
        }
    } else if(command.equals(COMMAND_RESTORE)) {
        //尝试复制文件
        try {
            if(!backup.exists()) {
                return RESTORE_NOFILEERROR;
```

```
            }
            dbFile.createNewFile();
            fileCopy(backup, dbFile);
            return RESTORE_SUCCESS;
        } catch(IOException e) {
            return BACKUP_ERROR;
        }
    } else {
        return BACKUP_ERROR;
    }
}

@Override
protected void onPostExecute(Integer result) {

    switch(result) {
    case BACKUP_SUCCESS:
        if(listener != null) {
            listener.onBackupComplete();
        }
        break;
    case RESTORE_SUCCESS:
        if(listener != null) {
            listener.onRestoreComplete();
        }
        break;
    case RESTORE_NOFILEERROR:
        if(listener != null) {
            listener.onError(RESTORE_NOFILEERROR);
        }
        break;
    default:
        if(listener != null) {
            listener.onError(BACKUP_ERROR);
        }
    }
}

private void fileCopy(File source, File dest) throws IOException {
    FileChannel inChannel = new FileInputStream(source).getChannel();
    FileChannel outChannel = new FileOutputStream(dest).getChannel();
    try {
        inChannel.transferTo(0, inChannel.size(), outChannel);
    } finally {
        if(inChannel != null)
            inChannel.close();
        if(outChannel != null)
```

```
            outChannel.close();
        }
    }
}
```

从上面的代码可以看出，当在 BackupTask 的 execute()中传入 COMMAND_BACKUP 时，会将指定数据库的当前版本复制到外部存储器的特定目录下，而当传入 COMMAND_RESTORE 时则执行相反的复制动作。

执行备份时，BackupTask 通过 Context.getDatabasePath()得到需要备份的数据库文件的引用。这行代码也可以修改为调用 Context.getFilesDir()，这样就可以备份系统内部存储器中的文件。同样，我们还获得了外部存储器上备份目录的引用。

文件的复制使用的是传统的 Java 文件 I/O，复制成功会通知注册的监听器。在复制过程中，如果出现异常，也会通知监听器。现在，看一下使用这个 BackupTask 进行数据库备份的 Activity，参见代码清单 5-24。

代码清单 5-24　使用了 BackupTask 的 Activity

```java
public class BackupActivity extends Activity implements
BackupTask.CompletionListener {

    @Override
    public void onCreate(Bundle savedInstanceState) {
        super.onCreate(savedInstanceState);
        setContentView(R.layout.main);
        //伪例数据库
        SQLiteDatabase db = openOrCreateDatabase("mydb",
          Activity.MODE_PRIVATE, null);
        db.close();
    }

    @Override
    public void onResume() {
        super.onResume();
        if( Environment.getExternalStorageState().equals(Environment.
          MEDIA_MOUNTED) ) {
            BackupTask task = new BackupTask(this);
            task.setCompletionListener(this);
            task.execute(BackupTask.COMMAND_RESTORE);
        }
    }

    @Override
    public void onPause() {
        super.onPause();
        if( Environment.getExternalStorageState().equals(Environment.
          MEDIA_MOUNTED) ) {
            BackupTask task = new BackupTask(this);
            task.execute(BackupTask.COMMAND_BACKUP);
```

```java
        }
    }

    @Override
    public void onBackupComplete() {
        Toast.makeText(this, "Backup Successful", Toast.LENGTH_SHORT).show();
    }

    @Override
    public void onError(int errorCode) {
        if(errorCode == BackupTask.RESTORE_NOFILEERROR) {
            Toast.makeText(this, "No Backup Found to Restore",
                Toast.LENGTH_SHORT).show();
        } else {
            Toast.makeText(this, "Error During Operation: "+errorCode,
                Toast.LENGTH_SHORT).show();
        }
    }

    @Override
    public void onRestoreComplete() {
        Toast.makeText(this, "Restore Successful", Toast.LENGTH_SHORT).show();
    }
}
```

这个 Activity 实现了 BackupTask 定义的 CompletionListener，因此在复制完成或发生错误时都会得到通知。为了演示这个示例，在应用程序的数据库目录中创建了一个伪数据库。调用 openOrCreateDatabase() 只是为了创建数据库文件，所以在创建数据库之后会立即关闭这个连接。一般情况下，这个数据库应该已经存在，这几行代码也可以省掉。

这个示例在每次恢复 Activity 时会执行恢复操作，同时会将自己注册到 BackupTask 中，从而在操作状态发生变化时可以弹出 Toast 来通知用户。注意，对于外部存储器是否可用的检查也放到了 Activity 中，在外部存储器不可用时不会执行任何任务。在 Activity 暂停时，会执行备份操作，这里并没有注册回调。这是因为一旦暂停 Activity，就意味着用户对 Activity 已经不感兴趣，因此也就没有必要弹出 Toast 来将操作结果通知给用户了。

补充信息

这个后台任务可以扩展一下，将数据保存到云端服务上，从而实现安全性的最大化和数据的可移植性。实现这个功能有很多种方式，包括 Google 自己的一些 Web API，建议试一下。

Android 在 API Level 8 中加入了将数据备份到云端服务的 API。这个 API 或许能满足你的要求，这里就不再赘述了。Android 框架并不能保证在所有的 Android 设备上这个服务都是可用的，而且在编写本书时也没有 API 可以判断用户的设备是否能够支持 Android 备份，因此不建议使用它进行重要数据的备份。

5.9 分享数据库

5.9.1 问题

应用程序需要将它维护的数据库内容提供给设备上的其他应用程序使用。

5.9.2 解决方案

(API Level 4)

创建 ContentProvider 作为应用程序数据的外部接口。ContentProvider 可以通过与数据库接口类似的 query()、insert()、update()和 delete()方法将特定的数据集暴露给外部请求。接口和实际数据模型之间的映射关系可以灵活定制。通过 ContentProvider 对外提供 SQLiteDatabase 的数据是很简单的。多数情况下,开发人员只需要将提供程序的调用传递给数据库即可。

要操作的数据集通常会编码为 Uri,然后传递给 ContentProvider。例如,发送 content://com.examples.myprovider/friends 这样的查询 Uri 就是告诉提供程序返回 friends 表中的数据,而 content://com.examples.myprovider/friends/15 则意味着在查询结果中返回 id 为 15 的记录。注意,这些仅是系统中其他应用程序获取数据的规范,而你必须确保所创建的 ContentProvider 符合这个规范。ContentProvider 本身并没有提供这种功能。

5.9.3 实现机制

首先,为了创建能够和数据库进行交互的 ContentProvider,必须先有一个数据库。代码清单 5-25 为一个示例 SQLiteOpenHelper 实现,它可以创建一个数据库并访问该数据库。

代码清单 5-25　示例 SQLiteOpenHelper

```java
public class ShareDbHelper extends SQLiteOpenHelper {

    private static final String DB_NAME = "frienddb";
    private static final int DB_VERSION = 1;

    public static final String TABLE_NAME = "friends";
    public static final String COL_FIRST = "firstName";
    public static final String COL_LAST = "lastName";
    public static final String COL_PHONE = "phoneNumber";

    private static final String STRING_CREATE =
        "CREATE TABLE "+TABLE_NAME+" (_id INTEGER PRIMARY KEY AUTOINCREMENT, "
        +COL_FIRST+" TEXT, "+COL_LAST+" TEXT, "+COL_PHONE+" TEXT);";

    public ShareDbHelper(Context context) {
        super(context, DB_NAME, null, DB_VERSION);
    }
```

```java
    @Override
    public void onCreate(SQLiteDatabase db) {
        //创建数据库表
        db.execSQL(STRING_CREATE);

        //向数据库中插入示例值
        ContentValues cv = new ContentValues(3);
        cv.put(COL_FIRST, "John");
        cv.put(COL_LAST, "Doe");
        cv.put(COL_PHONE, "8885551234");
        db.insert(TABLE_NAME, null, cv);
        cv = new ContentValues(3);
        cv.put(COL_FIRST, "Jane");
        cv.put(COL_LAST, "Doe");
        cv.put(COL_PHONE, "8885552345");
        db.insert(TABLE_NAME, null, cv);
        cv = new ContentValues(3);
        cv.put(COL_FIRST, "Jill");
        cv.put(COL_LAST, "Doe");
        cv.put(COL_PHONE, "8885553456");
        db.insert(TABLE_NAME, null, cv);
    }

    @Override
    public void onUpgrade(SQLiteDatabase db, int oldVersion, int newVersion) {
        //目前清空并重新创建数据库
        db.execSQL("DROP TABLE IF EXISTS "+TABLE_NAME);
        onCreate(db);
    }
}
```

这个辅助方法总体来说非常简单，它创建了一个表来保存好友信息列表，好友信息由三个字段组成，每个字段都是文本数据。为了演示这个示例，我们已经插入了三组数据。现在，让我们看一下 ContentProvider 是如何将这个数据库暴露给其他应用程序的，参见代码清单 5-26 和 5-27。

代码清单 5-26　ContentProvider 的清单声明

```xml
<manifest xmlns:android="http://schemas.android.com/apk/res/android" ...>
    <application ...>
        <provider android:name=".FriendProvider"
            android:authorities="com.examples.sharedb.friendprovider">
        </provider>
    </application>
</manifest>
```

代码清单 5-27　数据库的 ContentProvider

```java
public class FriendProvider extends ContentProvider {
```

```java
public static final Uri CONTENT_URI =
        Uri.parse("content://com.examples.sharedb.friendprovider/friends");

public static final class Columns {
    public static final String _ID = "_id";
    public static final String FIRST = "firstName";
    public static final String LAST = "lastName";
    public static final String PHONE = "phoneNumber";
}

/* 匹配Uri */
private static final int FRIEND = 1;
private static final int FRIEND_ID = 2;

private static final UriMatcher matcher = new UriMatcher(UriMatcher.NO_MATCH);
static {
    matcher.addURI(CONTENT_URI.getAuthority(), "friends", FRIEND);
    matcher.addURI(CONTENT_URI.getAuthority(), "friends/#", FRIEND_ID);
}

SQLiteDatabase db;

@Override
public int delete(Uri uri, String selection, String[] selectionArgs) {
    int result = matcher.match(uri);
    switch(result) {
    case FRIEND:
        return db.delete(ShareDbHelper.TABLE_NAME, selection, selectionArgs);
    case FRIEND_ID:
        return db.delete(ShareDbHelper.TABLE_NAME, "_ID = ?",
                new String[]{uri.getLastPathSegment()});
    default:
        return 0;
    }
}

@Override
public String getType(Uri uri) {
    return null;
}

@Override
public Uri insert(Uri uri, ContentValues values) {
    long id = db.insert(ShareDbHelper.TABLE_NAME, null, values);
    if(id >= 0) {
        return Uri.withAppendedPath(uri, String.valueOf(id));
    } else {
        return null;
    }
}
```

```java
    }

    @Override
    public boolean onCreate() {
        ShareDbHelper helper = new ShareDbHelper(getContext());
        db = helper.getWritableDatabase();
        return true;
    }

    @Override
    public Cursor query(Uri uri, String[] projection, String selection,
        String[] selectionArgs, String sortOrder) {
        int result = matcher.match(uri);
        switch(result) {
        case FRIEND:
            return db.query(ShareDbHelper.TABLE_NAME, projection, selection,
                selectionArgs, null, null, sortOrder);
        case FRIEND_ID:
            return db.query(ShareDbHelper.TABLE_NAME, projection, "_ID = ?",
                new String[]{uri.getLastPathSegment()}, null, null, sortOrder);
        default:
            return null;
        }
    }

    @Override
    public int update(Uri uri, ContentValues values, String selection,
        String[] selectionArgs) {
        int result = matcher.match(uri);
        switch(result) {
        case FRIEND:
            return db.update(ShareDbHelper.TABLE_NAME, values, selection,
                selectionArgs);
        case FRIEND_ID:
            return db.update(ShareDbHelper.TABLE_NAME, values, "_ID = ?",
                new String[]{uri.getLastPathSegment()});
        default:
            return 0;
        }
    }
}
```

必须在应用程序的清单中声明 ContentProvider 及其代表的 authority 字符串。这样外部应用程序才可以访问该提供程序，而且即使在应用程序内部使用，也需要在清单中声明。Android 是通过 authority 来匹配 Uri 请求和提供程序的，所以 authority 必须与公开的 CONTENT_URI 的 authority 部分保持一致。

在继承 ContentProvider 后，需要覆写 6 个方法：query()、insert()、update()、delete()、getType()和 onCreate()。前 4 个方法在 SQLiteDatabase 中都有对应的方法，所以只需要传入相应参数并调用数据库的方法即可。两者之间最大的区别就是 ContentProvider 的方法需要传入一个 Uri，提供程序会根据这个 Uri 判断应该操作数据库的哪个子集。

当 Activity 或其他系统控件调用内部 ContentResolver 的相应方法时(参见代码清单 5-27)，就会调用这 4 个主要的 CRUD 方法。

为了符合本节开头提及的 Uri 规范，insert()会将新创建的记录 id 添加到路径的末尾，并将其作为 Uri 对象返回。调用 insert()的代码就可以通过这个 Uri 对象得到所创建的数据记录的直接引用。

其他的方法(query()、update()和 delete())也遵从此规范，即检查收到的 Uri 指向的是某条记录还是整张表。这项任务是在 UriMatcher 便捷类的帮助下完成的。UriMatcher.match()方法会将 Uri 与设定好的一组模式进行比较，然后以整型的形式返回匹配的模式，如果没有找到能匹配的模式，就会返回 UriMatcher.NO_MATCH。如果提供的 Uri 包含了记录 id，对数据库的调用就只会影响到指定 id 的那一行。

UriMatcher 应该通过 UriMatcher.addURI()设置一组模式来进行初始化，Google 建议这项工作最好在 ContentProvider 内部的静态块中完成，这样在类刚加载到内存中时，UriMatcher 就会被初始化。每个加入的模式都包含一个对应的常数标识符，当该模式匹配时，就会返回这个标识符。在提供的模式中可以使用两个通配符：用来匹配数字的井号(#)和用来匹配任意文本的星号(*)。

我们的示例创建了两种匹配模式。第一种模式直接匹配提供的 CONTENT_URI，指向整个数据库表。第二种模式是在第一种模式的路径后面加上一个 id，指向 id 所对应的记录。

访问数据库的引用是在 ShareDbHelper 的 onCreate()中获得的。在确定这种方式是否适合应用程序时，需要考虑数据库的大小是否合适。我们的数据库在创建时是比较小的，但大一些的数据库的创建过程可能会花费很长时间，这时候就不应该在主线程上执行该操作了，最好将 getWritableDatabase()封装到一个 AsyncTask 中并在后台执行。

现在，让我们看一下访问数据的示例 Activity。如图 5-3 所示，我们将创建一个简单列表，该列表显示对提供程序的数据执行查询的结果。

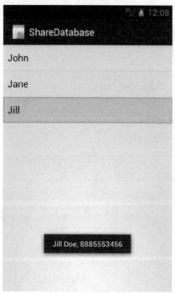

图 5-3　来自 ContentProvider 的信息

409

代码清单 5-28 和 5-29 显示了完成此例的清单和 Activity 源。

代码清单 5-28　AndroidManifest.xml

```xml
<?xml version="1.0" encoding="utf-8"?>
<manifest xmlns:android="http://schemas.android.com/apk/res/android"
        package="com.examples.sharedb" android:versionCode="1"
     android:versionName="1.0">
   <uses-sdk android:minSdkVersion="4" />
     <application android:icon="@drawable/icon"
      android:label="@string/app_name">
     <activity android:name=".ShareActivity"
         android:label="@string/app_name">
       <intent-filter>
         <action android:name="android.intent.action.MAIN" />
         <category android:name="android.intent.category.LAUNCHER" />
       </intent-filter>
     </activity>
     <provider android:name=".FriendProvider"
         android:authorities="com.examples.sharedb.friendprovider">
     </provider>
   </application>
</manifest>
```

代码清单 5-29　访问 ContentProvider 的 Activity

```java
public class ShareActivity extends FragmentActivity implements
    LoaderManager.LoaderCallbacks<Cursor>, AdapterView.OnItemClickListener {
    private static final int LOADER_LIST = 100;
    SimpleCursorAdapter mAdapter;

    @Override
    public void onCreate(Bundle savedInstanceState) {
        super.onCreate(savedInstanceState);
        getSupportLoaderManager().initLoader(LOADER_LIST, null, this);

        mAdapter = new SimpleCursorAdapter(this,
                android.R.layout.simple_list_item_1, null,
                new String[]{FriendProvider.Columns.FIRST},
                new int[]{android.R.id.text1}, 0);

        ListView list = new ListView(this);
        list.setOnItemClickListener(this);
        list.setAdapter(mAdapter);

        setContentView(list);
    }

    @Override
    public void onItemClick(AdapterView<?> parent, View v, int position,
long id) {
```

```java
        Cursor c = mAdapter.getCursor();
        c.moveToPosition(position);

        Uri uri = Uri.withAppendedPath(FriendProvider.CONTENT_URI,
          c.getString(0));
        String[] projection = new String[]{FriendProvider.Columns.FIRST,
                FriendProvider.Columns.LAST,
                FriendProvider.Columns.PHONE};
        //得到全部记录
        Cursor cursor = getContentResolver().query(uri, projection, null,
          null, null);
        cursor.moveToFirst();

        String message = String.format("%s %s, %s", cursor.getString(0),
                cursor.getString(1), cursor.getString(2));
        Toast.makeText(this, message, Toast.LENGTH_SHORT).show();
        cursor.close();
    }

    @Override
    public Loader<Cursor> onCreateLoader(int id, Bundle args) {
        String[] projection = new String[]{FriendProvider.Columns._ID,
                FriendProvider.Columns.FIRST};
        return new CursorLoader(this, FriendProvider.CONTENT_URI,
                projection, null, null, null);
    }

    @Override
    public void onLoadFinished(Loader<Cursor> loader, Cursor data) {
        mAdapter.swapCursor(data);
    }

    @Override
    public void onLoaderReset(Loader<Cursor> loader) {
        mAdapter.swapCursor(null);
    }
}
```

要点：

这个示例在 Android 1.6 及以上版本中需要使用支持库才能够访问 Loader 模式。如果应用程序的目标平台为 Android 3.0 以上版本，可以将 FragmentActivity 替换为 Activity，将 getSupportLoaderManager()替换为 getLoaderManager()。

这个示例会查询 FriendsProvider 中的所有记录，把它们放到一个列表中并只显示 first_name 字段。为了让 Cursor 能够正确地匹配列表，即使我们不显示 ID 字段，也必须在映射中包含 ID 字段。

如果用户点击了列表中的条目，就会将记录的 ID 添加到路径的结尾，从而构建出一个 Uri 并再次进行查询，要求提供程序仅返回指定的记录。另外，这里还有一个扩展后的映射可以获得指定好友的所有数据。

返回的数据会以 Toast 的形式显示给用户。通过游标可以返回每个字段的列索引，也就是传递给查询的映射中的索引。Cursor.getColumnIndex()方法还可以通过游标查询指定字段名的索引。

正如我们在每次用户点击事件中所做的那样，Cursor 在不用时应该关闭。唯一的例外就是通过 Loader 创建和管理的 Cursor 实例，它不需要这样做。

5.10 分享 SharedPreference

5.10.1 问题

需要应用程序将存储在 SharedPreference 中的设置值提供给系统的其他应用程序使用，在权限允许的情况下，甚至允许其他应用程序修改这些设置值。

5.10.2 解决方案

(API Level 1)

创建 ContentProvider 作为应用程序的 SharedPreference 与系统其他应用程序的接口。设置数据将通过 MatrixCursor 传递，该实现可以用来处理非数据库中的数据。对 ContentProvider 中的数据读/写是需要单独权限的，只有拥有权限的应用程序才可以访问。

5.10.3 实现机制

为了更好地演示本范例中权限方面的问题，我们需要创建两个单独的应用程序：一个应用程序包含我们的首选项数据，而另一个应用程序会通过 ContentProvider 读取并修改该数据。这是因为同一个 Android 应用程序中的操作是不需要权限的。让我们首先看一下提供程序，参见代码清单 5-30。

代码清单 5-30　应用程序设置的 ContentProvider

```java
public class SettingsProvider extends ContentProvider {

    public static final Uri CONTENT_URI =
        Uri.parse("content://com.examples.sharepreferences.
          settingsprovider/settings");

    public static class Columns {
        public static final String _ID = Settings.NameValueTable._ID;
        public static final String NAME = Settings.NameValueTable.NAME;
        public static final String VALUE = Settings.NameValueTable.VALUE;
    }

    private static final String NAME_SELECTION = Columns.NAME + " = ?";
```

```java
private SharedPreferences mPreferences;

@Override
public int delete(Uri uri, String selection, String[] selectionArgs) {
    throw new UnsupportedOperationException(
            "This ContentProvider is does not support removing Preferences");
}

@Override
public String getType(Uri uri) {
    return null;
}

@Override
public Uri insert(Uri uri, ContentValues values) {
    throw new UnsupportedOperationException(
            "This ContentProvider is does not support adding new Preferences");
}

@Override
public boolean onCreate() {
    mPreferences = PreferenceManager.getDefaultSharedPreferences(getContext());
    return true;
}

@Override
public Cursor query(Uri uri, String[] projection, String selection,
        String[] selectionArgs, String sortOrder) {
    MatrixCursor cursor = new MatrixCursor(projection);
    Map<String, ?> preferences = mPreferences.getAll();
    Set<String> preferenceKeys = preferences.keySet();

    if(TextUtils.isEmpty(selection)) {
        //获取所有条目
        for(String key : preferenceKeys) {
            //仅插入请求的字段
            MatrixCursor.RowBuilder builder = cursor.newRow();
            for(String column : projection) {
                if(column.equals(Columns._ID)) {
                    //生成唯一的 id
                    builder.add(key.hashCode());
                }
                if(column.equals(Columns.NAME)) {
                    builder.add(key);
                }
                if(column.equals(Columns.VALUE)) {
                    builder.add(preferences.get(key));
                }
            }
```

```java
            }
        } else if (selection.equals(NAME_SELECTION)) {
            //解析键值并检查它是否存在
            String key = selectionArgs == null ? "" : selectionArgs[0];
            if(preferences.containsKey(key)) {
                //得到需要的条目
                MatrixCursor.RowBuilder builder = cursor.newRow();
                for(String column : projection) {
                    if(column.equals(Columns._ID)) {
                        //生成唯一的 id
                        builder.add(key.hashCode());
                    }
                    if(column.equals(Columns.NAME)) {
                        builder.add(key);
                    }
                    if(column.equals(Columns.VALUE)) {
                        builder.add(preferences.get(key));
                    }
                }
            }
        }

        return cursor;
    }

    @Override
    public int update(Uri uri, ContentValues values, String selection,
            String[] selectionArgs) {
        //检查键是否存在,并更新它的值
        String key = values.getAsString(Columns.NAME);
        if(mPreferences.contains(key)) {
            Object value = values.get(Columns.VALUE);
            SharedPreferences.Editor editor = mPreferences.edit();
            if(value instanceof Boolean) {
                editor.putBoolean(key, (Boolean)value);
            } else if(value instanceof Number) {
                editor.putFloat(key, ((Number)value).floatValue());
            } else if(value instanceof String) {
                editor.putString(key, (String)value);
            } else {
                //无效的值,不更新
              return 0;
            }
            editor.commit();
            //通知观察者
            getContext().getContentResolver().notifyChange(CONTENT_URI, null);
            return 1;
        }
        //键不在首选项中
        return 0;
```

         }
    }

　　创建这个ContentProvider时，我们得到了指向应用程序默认SharedPreference的引用，而不是像之前的示例一样打开一个数据库连接。这个提供程序只支持两个方法——query()和update()，其他方法都会抛出异常。这样就只允许对首选项值进行读/写操作，而不允许添加或移除新的首选项类型。

　　在 query()方法中，我们会检查查询语句，从而确定返回全部的首选项值还是返回某个需要的值。每个首选项都定义了三个字段：_id、name 和 value。_id 的值也许和首选项本身关系不大，但如果使用提供程序的用户想通过 CursorAdapter 显示结果列表，则需要这个字段并且每条记录的值不能相同，所以我们就生成了该值。注意，首选项值是以对象的方式插入到游标中的，这样就可以减少提供程序包含的数据类型。

　　这个提供程序使用的游标实现为 MatrixCursor，它用来处理非数据库中的数据。这个示例会迭代需要的字段列表(映射)并根据所包含的字段构建每行数据。每行都是通过MatrixCursor.newRow()方法创建的，该方法同时会返回一个可以用来添加字段数据的Builder 实例。注意，添加的字段数据的顺序和所请求映射的顺序应该始终是一致的。

　　update()方法的实现只会检查传入 ContentValues 的需要更新的首选项值。因为这已足以描述我们需要的条目，所以我们并没有使用选择参数实现其他进一步的逻辑。如果首选项的 name 值已存在，就会更新和保存它的值。遗憾的是，并没有方法可以简单地将一个对象插入 SharedPreferences 中，所以必须检查该值是否是 ContentValues 可以返回的有效类型，然后调用相应的设置方法。最后，我们调用 notifyObservers()来通知所有注册的ContentObserver 对象数据已经发生改变。

　　可能你会发现，在 ContentProvider 中并没有我们之前承诺实现的关于管理读/写权限的代码！实际上，这是由 Android 来替我们管理的：我们只需要适当修改清单文件即可。参见代码清单 5-31。

代码清单 5-31　AndroidManifest.xml

```
<manifest xmlns:android="http://schemas.android.com/apk/res/android"
    package="com.examples.sharepreferences"
    android:versionCode="1"
    android:versionName="1.0" >

    <uses-sdk ... />

    <permission
        android:name="com.examples.sharepreferences.permission.READ_
            PREFERENCES"
        android:label="Read Application Settings"
        android:protectionLevel="normal" />
    <permission
        android:name="com.examples.sharepreferences.permission.WRITE_
            PREFERENCES"
        android:label="Write Application Settings"
```

```xml
        android:protectionLevel="dangerous" />

<uses-permission
    android:name="com.examples.sharepreferences.permission.READ_PREFERENCES" />
<uses-permission
    android:name="com.examples.sharepreferences.permission.WRITE_PREFERENCES" />

<application ... >
    <activity android:name=".SettingsActivity" >
        <intent-filter>
            <action android:name="android.intent.action.MAIN" />
            <category android:name="android.intent.category.LAUNCHER" />
        </intent-filter>
        <intent-filter>
            <action android:name="com.examples.
                sharepreferences.ACTION_SETTINGS" />
            <category android:name="android.intent.category.DEFAULT" />
        </intent-filter>
    </activity>

    <provider
        android:name=".SettingsProvider"
        android:authorities="com.examples.sharepreferences.
            settingsprovider"
        android:readPermission=
            "com.examples.sharepreferences.permission.READ_PREFERENCES"
        android:writePermission=
            "com.examples.sharepreferences.permission.WRITE_PREFERENCES" >
    </provider>
</application>

</manifest>
```

这里，我们声明了两个自定义的<permission>元素并把它们添加到了<provider>声明中。我们只需要添加这些代码，之后 Android 就会对 query()方法添加读取权限，对 insert()、update()和 delete()方法添加写入权限。我们在应用程序的 Activity 节点中还声明了一个自定义的<intent-filter>，在外部应用程序想要直接启动设置 UI 时，这个<intent-filter>就会派上用场。代码清单 5-32 到代码清单 5-34 定义了本例剩余部分的代码。

代码清单 5-32　res/xml/preferences.xml

```xml
<?xml version="1.0" encoding="utf-8"?>
<PreferenceScreen
    xmlns:android="http://schemas.android.com/apk/res/android" >
    <CheckBoxPreference
        android:key="preferenceEnabled"
```

```xml
        android:title="Set Enabled"
        android:defaultValue="true"/>
    <EditTextPreference
        android:key="preferenceName"
        android:title="User Name"
        android:defaultValue="John Doe"/>
    <ListPreference
        android:key="preferenceSelection"
        android:title="Selection"
        android:entries="@array/selection_items"
        android:entryValues="@array/selection_items"
        android:defaultValue="Four"/>
</PreferenceScreen>
```

代码清单 5-33　res/values/arrays.xml

```xml
<?xml version="1.0" encoding="utf-8"?>
<resources>
    <string-array name="selection_items">
        <item>One</item>
        <item>Two</item>
        <item>Three</item>
        <item>Four</item>
    </string-array>
</resources>
```

代码清单 5-34　首选项 Activity

```java
//注意应用程序的包名
package com.examples.sharepreferences;

public class SettingsActivity extends PreferenceActivity {

    @Override
    protected void onCreate(Bundle savedInstanceState) {
        super.onCreate(savedInstanceState);
        //初次启动时加载首选项默认值
        PreferenceManager.setDefaultValues(this, R.xml.preferences, false);

        addPreferencesFromResource(R.xml.preferences);
    }
}
```

这个示例应用程序的设置值是通过一个简单的 PreferenceActivity 来直接管理的，PreferenceActivity 的数据则定义在 preferences.xml 文件中。

注意：
在 Android 3.0 中，已经不建议使用 PreferenceActivity，而建议使用 PreferenceFragment，但在本书出版时，PreferenceFragment 还没有添加到支持库中。因此，为了支持稍早的 Android 版本，这里还是使用了 PreferenceActivity。

用例

接下来,让我们看一下代码清单 5-35 到 5-37,它们定义了另一个应用程序,该应该程序会通过这个 ContentProvider 接口访问我们的首选项数据。

代码清单 5-35　AndroidManifest.xml

```xml
<manifest xmlns:android="http://schemas.android.com/apk/res/android"
    package="com.examples.accesspreferences">

    <uses-sdk ... />

    <uses-permission android:name="com.examples.
        sharepreferences.permission.READ_PREFERENCES" />
    <uses-permission android:name="com.examples.
        sharepreferences.permission.WRITE_PREFERENCES" />

    <application ... >
        <activity android:name=".MainActivity" >
            <intent-filter>
                <action android:name="android.intent.action.MAIN" />
                <category android:name="android.intent.category.LAUNCHER" />
            </intent-filter>
        </activity>
    </application>

</manifest>
```

这里的核心部分就是应用程序通过<uses-permission>元素声明使用了我们自定义的权限。这样,该应用程序就可以访问外部提供程序了。否则,通过 ContentResolver 的请求就会导致 SecurityException 异常。

代码清单 5-36　res/layout/main.xml

```xml
<RelativeLayout xmlns:android="http://schemas.android.com/apk/res/android"
    android:layout_width="match_parent"
    android:layout_height="match_parent" >
    <Button
        android:id="@+id/button_settings"
        android:layout_width="match_parent"
        android:layout_height="wrap_content"
        android:text="Show Settings"
        android:onClick="onSettingsClick" />
    <CheckBox
        android:id="@+id/checkbox_enable"
        android:layout_width="wrap_content"
        android:layout_height="wrap_content"
        android:layout_below="@id/button_settings"
        android:text="Set Enable Setting"/>
    <LinearLayout
```

```xml
            android:layout_width="wrap_content"
            android:layout_height="wrap_content"
            android:layout_centerInParent="true"
            android:orientation="vertical">
        <TextView
            android:id="@+id/value_enabled"
            android:layout_width="wrap_content"
            android:layout_height="wrap_content" />
        <TextView
            android:id="@+id/value_name"
            android:layout_width="wrap_content"
            android:layout_height="wrap_content" />
        <TextView
            android:id="@+id/value_selection"
            android:layout_width="wrap_content"
            android:layout_height="wrap_content" />
    </LinearLayout>
</RelativeLayout>
```

代码清单 5-37　与提供程序进行交互的 Activity

```java
//注意包名，它表示另一个应用程序
package com.examples.accesspreferences;

public class MainActivity extends Activity implements OnCheckedChangeListener {

    public static final String SETTINGS_ACTION =
        "com.examples.sharepreferences.ACTION_SETTINGS";
    public static final Uri SETTINGS_CONTENT_URI =
        Uri.parse("content://com.examples.sharepreferences.
          settingsprovider/settings");
    public static class SettingsColumns {
        public static final String _ID = Settings.NameValueTable._ID;
        public static final String NAME = Settings.NameValueTable.NAME;
        public static final String VALUE = Settings.NameValueTable.VALUE;
    }

    TextView mEnabled, mName, mSelection;
    CheckBox mToggle;

    private ContentObserver mObserver = new ContentObserver(new Handler()) {
        public void onChange(boolean selfChange) {
            updatePreferences();
        }
    };

    @Override
    public void onCreate(Bundle savedInstanceState) {
        super.onCreate(savedInstanceState);
        setContentView(R.layout.main);
```

```java
        mEnabled = (TextView) findViewById(R.id.value_enabled);
        mName = (TextView) findViewById(R.id.value_name);
        mSelection = (TextView) findViewById(R.id.value_selection);
        mToggle = (CheckBox) findViewById(R.id.checkbox_enable);
        mToggle.setOnCheckedChangeListener(this);
    }

    @Override
    protected void onResume() {
        super.onResume();
        //获得最新的提供程序数据
        updatePreferences();
        //当Activity可见时，注册一个观察者来监听数据的变化
        getContentResolver().registerContentObserver(SETTINGS_CONTENT_URI,
                false, mObserver);
    }

    @Override
    public void onCheckedChanged(CompoundButton buttonView, boolean isChecked) {
        ContentValues cv = new ContentValues(2);
        cv.put(SettingsColumns.NAME, "preferenceEnabled");
        cv.put(SettingsColumns.VALUE, isChecked);

        //更新提供程序，这会触发我们的观察者
        getContentResolver().update(SETTINGS_CONTENT_URI, cv, null, null);
    }

    public void onSettingsClick(View v) {
        try {
            Intent intent = new Intent(SETTINGS_ACTION);
            startActivity(intent);
        } catch(ActivityNotFoundException e) {
            Toast.makeText(this,
                    "You do not have the Android Recipes Settings App installed.",
                    Toast.LENGTH_SHORT).show();
        }
    }

    private void updatePreferences() {
        Cursor c = getContentResolver().query(SETTINGS_CONTENT_URI,
                new String[] {SettingsColumns.NAME, SettingsColumns.VALUE},
                null, null, null);
        if(c == null) {
            return;
        }

        while(c.moveToNext()) {
            String key = c.getString(0);
            if("preferenceEnabled".equals(key)) {
                mEnabled.setText( String.format("Enabled Setting = %s",
```

```
                c.getString(1)) );
            mToggle.setChecked( Boolean.parseBoolean(c.getString(1)) );
        } else if("preferenceName".equals(key)) {
            mName.setText( String.format("User Name Setting = %s",
                c.getString(1)) );
        } else if("preferenceSelection".equals(key)) {
            mSelection.setText( String.format("Selection Setting = %s",
                c.getString(1)) );
        }
    }

    c.close();
    }
}
```

因为这是一个独立的应用程序，所以它不能访问第一个应用程序中定义的常量(除非通过库项目引用或其他方式控制这两个应用程序)，我们必须重新定义它们。如果想创建一个带有外部提供程序且其他开发人员可以使用该提供程序的应用程序，最好同时提供一个JAR库，这个JAR库会包含一些访问提供程序的Uri和字段数据的必要常量，类似于Android API 提供的 ContactsContract 和 CalendarContract。

本例中，Activity 在每次进入前台时都会查询提供程序来得到设置的当前值，然后将其显示在一个 TextView 中。查询结果会通过一个 Cursor 返回，该 Cursor 的每行中有两个值：首选项的名称和它的值。该 Activity 还注册了一个 ContentObserver，这样如果在 Activity 可见时数据发生了变化，显示的值可以跟着更新。当用户改变 CheckBox 在屏幕上显示的值时，会调用提供程序的 update()来触发观察者以更新显示。

最后，如果用户愿意的话，可以通过点击"Show Settings"按钮从外部应用程序直接启动 SettingsActivity。这个过程会调用 startActivity()方法并传入包含自定义动作字符串的 Intent，该动作字符串正是 SettingsActivity 所需的过滤器。

5.11 分享其他数据

5.11.1 问题

应用程序需要将它维护的文件或其他私有数据提供给设备上的其他应用程序使用。

5.11.2 解决方案

(API Level 3)

创建 ContentProvider 作为应用程序数据对外的接口。ContentProvider 通过一些类似数据库的接口 query()、insert()、update()和 delete()，将任意数据集暴露给外部请求。但是具体的实现可以自由定义如何将这些方法传来的数据传递给实际的模型。

ContentProvider 可以向外部请求暴露任何类型的应用程序数据，包括应用程序的各种资源以及 assets 目录下的资源。

5.11.3 实现机制

让我们看一个暴露了两个数据源的 ContentProvider 实现：一个数据源是内存中的字符串数组，另一个是应用程序的 assets 目录中存储的一系列图片文件。跟前面一样，我们必须先在应用程序的清单文件中用<provider>标签声明我们的提供程序，参见代码清单 5-38 和 5-39。

代码清单 5-38　ContentProvider 的清单声明

```xml
<?xml version="1.0" encoding="utf-8"?>
<manifest xmlns:android="http://schemas.android.com/apk/res/android" ...>
    <application ...>
      <provider android:name=".ImageProvider"
          android:authorities="com.examples.share.imageprovider">
      </provider>
    </application>
</manifest>
```

代码清单 5-39　暴露 assets 目录中资源的自定义 ContentProvider

```java
public class ImageProvider extends ContentProvider {

    public static final Uri CONTENT_URI =
        Uri.parse("content://com.examples.share.imageprovider");

    public static final String COLUMN_NAME = "nameString";
    public static final String COLUMN_IMAGE = "imageUri";

    private String[] mNames;

    @Override
    public int delete(Uri uri, String selection, String[] selectionArgs) {
        throw new UnsupportedOperationException("This ContentProvider is
            read-only");
    }

    @Override
    public String getType(Uri uri) {
        return null;
    }

    @Override
    public Uri insert(Uri uri, ContentValues values) {
        throw new UnsupportedOperationException("This ContentProvider is
            read-only");
    }
```

```java
@Override
public boolean onCreate() {
    mNames = new String[] {"John Doe", "Jane Doe", "Jill Doe"};
    return true;
}

@Override
public Cursor query(Uri uri, String[] projection, String selection,
    String[] selectionArgs, String sortOrder) {
    MatrixCursor cursor = new MatrixCursor(projection);
    for(int i = 0; i < mNames.length; i++) {
        //只插入请求的字段
        MatrixCursor.RowBuilder builder = cursor.newRow();
        for(String column : projection) {
            if(column.equals("_id")) {
                //使用数组索引作为唯一 id
                builder.add(i);
            }
            if(column.equals(COLUMN_NAME)) {
                builder.add(mNames[i]);
            }
            if(column.equals(COLUMN_IMAGE)) {
                builder.add(Uri.withAppendedPath(CONTENT_URI,
                    String.valueOf(i)));
            }
        }
    }
    return cursor;
}

@Override
public int update(Uri uri, ContentValues values, String selection,
    String[] selectionArgs) {
    throw new UnsupportedOperationException("This ContentProvider is
        read-only");
}

@Override
public AssetFileDescriptor openAssetFile(Uri uri, String mode) throws
    FileNotFoundException {
    int requested = Integer.parseInt(uri.getLastPathSegment());
    AssetFileDescriptor afd;
    AssetManager manager = getContext().getAssets();
    //为请求项返回正确的 asset 资源
    try {
        switch(requested) {
        case 0:
            afd = manager.openFd("logo1.png");
```

```
                break;
            case 1:
                afd = manager.openFd("logo2.png");
                break;
            case 2:
                afd = manager.openFd("logo3.png");
                break;
            default:
                afd = manager.openFd("logo1.png");
        }
        return afd;
    } catch(IOException e) {
        e.printStackTrace();
        return null;
    }
}
```

你可能已经猜到，这个示例对外暴露的是 3 个徽标图片资源。本例中选用的图片如图 5-4 所示。

图 5-4　示例图片，保存在 assets 目录中的 logo1.png(左图)、logo2.png(中图)和 logo3.png(右图)

因为我们对外暴露的是 assets 目录下的只读内容，所以不必支持继承的 insert()、update() 或 delete()方法，对于这些方法只需要抛出 UnsupportedOperationException 即可。

在创建提供程序时，会创建一个存放人名的字符串数组，之后 onCreate()返回 true，告诉系统提供程序创建成功。提供程序将它自己的 URI 和可读取的字段名以常量的形式暴露。外部应用程序就可以通过这些值请求所需的数据了。

这个提供程序只支持查询其内部的数据。为了实现对特定记录或某个内容子集的条件查询，应用程序可以将查询条件传递给query()的selection和selectionArgs。在这个示例中，所有对query()的调用都会构建一个包含其中所有3个元素的游标。

这个提供程序中使用的游标实现是 MatrixCursor，这种游标专门用于非数据库中的数据。这个示例遍历所要求(映射)字段的列表，根据其中每个字段的信息构建行。每行都是通过调用 MatrixCursor.newRow()创建的，这个方法也会返回一个用于添加列数据的 Builder 实例。务必始终确保列数据添加的顺序和所要求映射的顺序保持一致。

name 字段中的值就是本地数组中相应的字符串。在 Android 的大部分 ListAdapter 中，利用返回游标所需的_id 字段就是以数组的索引返回的。每一行的 image 字段中的数据实际上是一个表示图片文件的内容 Uri，它以提供程序的内容 Uri 为基础，加上数组索引构建而成。

当外部应用程序通过 ContentResolver.openInputStream()获取该内容时，会调用覆写后的 openAssetFile()，返回指向 assets 目录中某个图片文件的 AssetFileDescriptor。这个实现

通过再次解析内容 Uri 并获取结尾处附加的索引值来判断要返回的是哪一张图片。

1. 用例

在图 5-5 中，我们给出了显示与前一节类似列表的 Activity。然而，我们还针对每次用户的选择显示相应的 assets 数据。

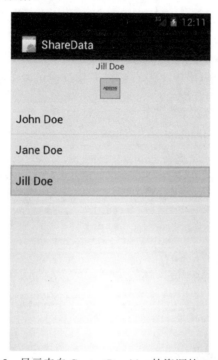

图 5-5　显示来自 ContentProvider 的资源的 Activity

下面来看看如何实现该提供程序，以及如何在 Android 应用程序中访问它，参见代码清单 5-40。

代码清单 5-40　AndroidManifest.xml

```
<?xml version="1.0" encoding="utf-8"?>
<manifest xmlns:android="http://schemas.android.com/apk/res/android"
    package="com.examples.share"
    android:versionCode="1"
    android:versionName="1.0">
    <uses-sdk android:minSdkVersion="3" />

    <application android:icon="@drawable/icon"
              android:label="@string/app_name">
        <activity android:name=".ShareActivity"
              android:label="@string/app_name">
            <intent-filter>
                <action android:name="android.intent.action.MAIN" />
                <category android:name="android.intent.category.LAUNCHER" />
            </intent-filter>
```

```xml
        </activity>
        <provider android:name=".ImageProvider"
          android:authorities="com.examples.share.imageprovider">
        </provider>
    </application>
</manifest>
```

要实现该提供程序，拥有该内容的应用程序必须在其清单文件中声明一个<provider>标签，确定 ContentProvider 的名称和进行请求时相应的 authority 值。authority 值必须跟暴露的内容 Uri 的基础部分保持一致。提供程序必须在清单中声明，系统才能将其实例化并运行。即使拥有该提供程序的应用程序没有运行，系统也可以单独运行提供程序。参见代码清单 5-41 和 5-42。

代码清单 5-41 res/layout/main.xml

```xml
<?xml version="1.0" encoding="utf-8"?>
<LinearLayout xmlns:android="http://schemas.android.com/apk/res/android"
  android:orientation="vertical"
  android:layout_width="match_parent"
  android:layout_height="match_parent">
  <TextView
    android:id="@+id/name"
    android:layout_width="wrap_content"
    android:layout_height="20dip"
    android:layout_gravity="center_horizontal"
  />
  <ImageView
    android:id="@+id/image"
    android:layout_width="wrap_content"
    android:layout_height="50dip"
    android:layout_gravity="center_horizontal"
  />
  <ListView
    android:id="@+id/list"
    android:layout_width="match_parent"
    android:layout_height="match_parent"
  />
</LinearLayout>
```

代码清单 5-42 从 ImageProvider 中读取内容的 Activity

```java
public class ShareActivity extends FragmentActivity implements
        LoaderManager.LoaderCallbacks<Cursor>, AdapterView.OnItemClickListener {
    private static final int LOADER_LIST = 100;
    SimpleCursorAdapter mAdapter;

    @Override
    public void onCreate(Bundle savedInstanceState) {
        super.onCreate(savedInstanceState);
        getSupportLoaderManager().initLoader(LOADER_LIST, null, this);
```

```java
        setContentView(R.layout.main);

    mAdapter = new SimpleCursorAdapter(this, android.R.layout.simple_list_item_1,
            null, new String[]{ImageProvider.COLUMN_NAME},
            new int[]{android.R.id.text1}, 0);

    ListView list = (ListView)findViewById(R.id.list);
    list.setOnItemClickListener(this);
    list.setAdapter(mAdapter);
}

@Override
public void onItemClick(AdapterView<?> parent, View v, int position,
    long id) {
    //得到selection的游标
    Cursor c = mAdapter.getCursor();
    c.moveToPosition(position);

    //加载name字段到TextView
    TextView tv = (TextView)findViewById(R.id.name);
    tv.setText(c.getString(1));

    ImageView iv = (ImageView)findViewById(R.id.image);
    try {
        //从image字段中加载内容到ImageView中
        InputStream in =
          getContentResolver().openInputStream(Uri.parse(c.getString(2)));
        Bitmap image = BitmapFactory.decodeStream(in);
        iv.setImageBitmap(image);
    } catch(FileNotFoundException e) {
        e.printStackTrace();
    }
}

@Override
public Loader<Cursor> onCreateLoader(int id, Bundle args) {
    String[] projection = new String[]{"_id",
            ImageProvider.COLUMN_NAME,
            ImageProvider.COLUMN_IMAGE};
    return new CursorLoader(this, ImageProvider.CONTENT_URI,
            projection, null, null, null);
}

@Override
public void onLoadFinished(Loader<Cursor> loader, Cursor data) {
    mAdapter.swapCursor(data);
}

@Override
public void onLoaderReset(Loader<Cursor> loader) {
```

```
            mAdapter.swapCursor(null);
        }
    }
```

要点:

这个示例需要支持库才可以在 Android 1.6 及以上版本中访问 Loader 模式。如果应用程序的目标版本为 Android 3.0+，需要将 FragmentActivity 替换为 Activity 以及将 getSupportLoaderManager()替换为 getLoaderManager()。

在这个示例中，从自定义的 ContentProvider 获得了一个托管游标，指向暴露的 Uri 和数据的字段名称。然后用 SimpleCursorAdapter 将数据关联到 ListView 以仅显示 name 值。

当用户点击列表中的某项内容时，游标会移到相应的位置并在上方显示相应的名称和图片。Activity 在这里调用 ContentResolver.openInputStream()，通过保存在字段中的 Uri 来访问图片文件。

图 5-5 是应用程序运行并选择列表中最后一项(Jill Doe)的效果。

注意，这里并没有显式关闭指向 Cursor 的连接，因为这个 Cursor 是由 Loader 创建的，Loader 会替你管理它的生命周期。

5.12 集成系统文档

5.12.1 问题

应用程序创建或维护你希望通过系统的文档拾取器界面暴露给其他应用程序的内容。

5.12.2 解决方案

(API Level 19)

DocumentsProvider 是特殊的 ContentProvider API，应用程序可以使用它将其内容公开给 Android 4.4 或以后版本中的公共文档拾取器界面。使用此框架的优点是，它允许管理存储设备访问的应用程序通过使用整个系统内公共的界面公开它们拥有的文件和文档。该框架还允许客户端应用程序在其内部创建和保存新文档(我们将在第 6 章中更详细地介绍此 API 的客户端方面)。

自定义的 DocumentsProvider 必须使用唯一的文档 ID 字符串标识它想要公开的所有文件和目录。该值不需要匹配任何特定的格式，但它必须是唯一的，并且在报告给系统后不能修改。该框架将出于权限目的而持久化这些值，因此甚至是你提供的(和以后预期的)跨重启文档 ID 也必须对任何资源保持一致。

在子类化 DocumentsProvider 时，我们将实现与 ContentProvider 中使用的基本 CRUD 方法不同的一组回调。系统的文档拾取器界面将在用户浏览时触发提供程序的如下方法：

- queryRoots()：当启动拾取器 UI 以请求关于提供程序中顶层"文档"的基本信息以及一些基本元数据(如显示的名称和图标)时，首先调用此方法。大多数提供程序仅有一个根，但可以返回多个根以更好地支持提供程序的用例。
- queryChildDocuments()：当使用根的文档 ID 选择提供程序以获得该根下方的可用文档清单时，调用此方法。如果返回目录条目作为此根下方的元素，则会在选择其中一个子目录时再次调用此方法。
- queryDocument()：当选择文档以获得关于特定实例的元数据时调用此方法。从此消息返回的数据应该与从 queryChildDocuments()返回的数据存在对应关系，但仅适用于一个元素。也会对每个根调用该方法以获取提供程序中顶层目录的额外元数据。
- openDocument()：该方法请求对文档打开 FileDescriptor，从而客户端应用程序可以读写文档内容。
- openDocumentThumbnail()：如果针对给定文档返回的元数据设置了 FLAG_SUPPORTS_THUMBNAIL 标志，则使用该方法获得显示在文档的拾取器 UI 中的缩略图。

5.12.3 实现机制

可以修改前一节中的提供程序以更好地集成系统的文档拾取器 UI，如图 5-6 所示。

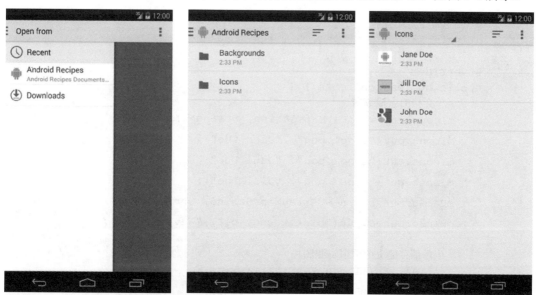

图 5-6　显示在列表中的提供程序(左图)和选择后显示的文件选项(右图)

实现此任务需要稍作修改。代码清单 5-43 显示了修改为子类 DocumentProvider 的 ImageProvider。

代码清单 5-43　作为 DocumentProvider 的 ImageProvider

```
public class ImageProvider extends DocumentsProvider {
```

```java
private static final String TAG = "ImageProvider";

/* 缓存的最近选择项 */
private static String sLastDocumentId;

private static final String DOCID_ROOT = "root:";
private static final String DOCID_ICONS_DIR = DOCID_ROOT + "icons:";
private static final String DOCID_BGS_DIR = DOCID_ROOT + "backgrounds:";

/* 根的默认投影(如果没有提供投影的话) */
private static final String[] DEFAULT_ROOT_PROJECTION = {
    Root.COLUMN_ROOT_ID, Root.COLUMN_MIME_TYPES,
    Root.COLUMN_FLAGS, Root.COLUMN_ICON, Root.COLUMN_TITLE,
    Root.COLUMN_SUMMARY, Root.COLUMN_DOCUMENT_ID,
    Root.COLUMN_AVAILABLE_BYTES
};
/* 文档的默认投影(如果没有提供投影的话) */
private static final String[] DEFAULT_DOCUMENT_PROJECTION = {
    Document.COLUMN_DOCUMENT_ID, Document.COLUMN_MIME_TYPE,
    Document.COLUMN_DISPLAY_NAME, Document.COLUMN_LAST_MODIFIED,
    Document.COLUMN_FLAGS, Document.COLUMN_SIZE
};

private ArrayMap<String, String> mIcons;
private ArrayMap<String, String> mBackgrounds;

@Override
public boolean onCreate() {
    // 文档的虚拟数据
    mIcons = new ArrayMap<String, String>();
    mIcons.put("logo1.png", "John Doe");
    mIcons.put("logo2.png", "Jane Doe");
    mIcons.put("logo3.png", "Jill Doe");
    mBackgrounds = new ArrayMap<String, String>();
    mBackgrounds.put("background.jpg", "Wavy Grass");

    // 内部存储上的虚拟资源图片
    writeAssets(mIcons.keySet());
    writeAssets(mBackgrounds.keySet());
    return true;
}

/*
 * 用于将一些虚拟文件流出到内部存储目录的辅助方法
 */
private void writeAssets(Set<String> filenames) {
    for(String name : filenames) {
```

```java
        try {
            Log.d("ImageProvider", "Writing "+name+" to storage");
            InputStream in = getContext().getAssets().open(name);
            FileOutputStream out = getContext().openFileOutput(name,
              Context.MODE_PRIVATE);

            int size;
            byte[] buffer = new byte[1024];
            while((size = in.read(buffer, 0, 1024)) >= 0) {
                out.write(buffer, 0, size);
            }
            out.flush();
            out.close();
        } catch(IOException e) {
            Log.w(TAG, e);
        }
    }
}

/* 用于根据文件名构造文档 ID 的辅助方法 */
private String getIconsDocumentId(String filename) {
    return DOCID_ICONS_DIR + filename;
}

private String getBackgroundsDocumentId(String filename) {
    return DOCID_BGS_DIR + filename;
}

/* 用于确定文档类型的辅助方法 */
private boolean isRoot(String documentId) {
    return DOCID_ROOT.equals(documentId);
}

private boolean isIconsDir(String documentId) {
    return DOCID_ICONS_DIR.equals(documentId);
}

private boolean isBackgroundsDir(String documentId) {
    return DOCID_BGS_DIR.equals(documentId);
}

private boolean isIconDocument(String documentId) {
    return documentId.startsWith(DOCID_ICONS_DIR);
}

private boolean isBackgroundsDocument(String documentId) {
    return documentId.startsWith(DOCID_BGS_DIR);
}

/*
```

```java
 * 从文档ID提取文件名的辅助方法
 * 返回"根"文档的空字符串
 */
private String getFilename(String documentId) {
    int split = documentId.lastIndexOf(":");
    if(split < 0) {
        return "";
    }
    return documentId.substring(split+1);
}

/*
 * 系统调用该方法以确定此处托管了多少"提供程序"
 * 最常见的情况是通过只有一个结果行的Cursor仅返回一个提供程序
 */
@Override
 public Cursor queryRoots(String[] projection) throws
       FileNotFoundException {
    if(projection == null) {
        projection = DEFAULT_ROOT_PROJECTION;
    }
    MatrixCursor result = new MatrixCursor(projection);
    // 为此提供程序添加一个根
    MatrixCursor.RowBuilder builder = result.newRow();

    builder.add(Root.COLUMN_ROOT_ID, "root");
    builder.add(Root.COLUMN_TITLE, "Android Recipes");
    builder.add(Root.COLUMN_SUMMARY, "Android Recipes Documents
       Provider");
    builder.add(Root.COLUMN_ICON, R.drawable.ic_launcher);

    builder.add(Root.COLUMN_DOCUMENT_ID,DOCID_ROOT);

    builder.add(Root.COLUMN_FLAGS,
            //结果将仅仅来自本地文件系统
            Root.FLAG_LOCAL_ONLY
            //我们支持显示最近选择的条目
            | Root.FLAG_SUPPORTS_RECENTS);
    builder.add(Root.COLUMN_MIME_TYPES, "image/*");
    builder.add(Root.COLUMN_AVAILABLE_BYTES, 0);

    return result;
}

/*
 * 系统调用此方法以确定给定父对象的子项
 * 将对根以及在其中定义的每个子目录调用此方法
 */
@Override
public Cursor queryChildDocuments(String parentDocumentId, String[]
```

```java
                projection,String sortOrder) throws FileNotFoundException {
        if(projection == null) {
            projection = DEFAULT_DOCUMENT_PROJECTION;
        }
        MatrixCursor result = new MatrixCursor(projection);


        if(isIconsDir(parentDocumentId)) {
            //添加图标集合中的所有文件
            try {
                for(String key : mIcons.keySet()) {
                    addImageRow(result, mIcons.get(key),
                      getIconsDocumentId(key));
                }
            } catch(IOException e) {
                return null;
            }
        } else if (isBackgroundsDir(parentDocumentId)) {
            //添加背景集合中的所有文件
            try {
                for(String key : mBackgrounds.keySet()) {
                    addImageRow(result, mBackgrounds.get(key),
                        getBackgroundsDocumentId(key));
                }
            } catch (IOException e) {
                return null;
            }
        } else if (isRoot(parentDocumentId)) {
            //添加顶层目录
            addIconsRow(result);
            addBackgroundsRow(result);
        }
        return result;
}

/*
 * 返回通过queryChildDocuments()提供的相同信息
 * 但仅适用于请求的单个文档ID
 */
@Override
public Cursor queryDocument(String documentId, String[] projection)
        throws FileNotFoundException {

    if(projection == null) {
```

```java
            projection = DEFAULT_DOCUMENT_PROJECTION;
        }

        MatrixCursor result = new MatrixCursor(projection);

        try {
            String filename = getFilename(documentId);
            if(isRoot(documentId)) {
                // 这是根的查询
                addRootRow(result);
            } else if (isIconsDir(documentId)) {
                //这是图标的查询
                addIconsRow(result);
            } else if (isBackgroundsDir(documentId)) {
                //这是背景的查询
                addBackgroundsRow(result);
            } else if (isIconDocument(documentId)) {
                addImageRow(result, mIcons.get(filename),
                    getIconsDocumentId(filename));
            } else if (isBackgroundsDocument(documentId)) {
                addImageRow(result, mBackgrounds.get(filename),
                    getBackgroundsDocumentId(filename));
            }
        } catch(IOException e) {
            return null;
        }

        return result;
    }

    /*
     * 调用以填充 Recents 拾取器 UI 中此提供程序的最近使用项
     */
    @Override
    public Cursor queryRecentDocuments(String rootId, String[] projection)
            throws FileNotFoundException {

        if(projection == null) {
            projection = DEFAULT_DOCUMENT_PROJECTION;
        }

        MatrixCursor result = new MatrixCursor(projection);

        if (sLastDocumentId != null) {
            String filename = getFilename(sLastDocumentId);
            String recentTitle = "";
            if (isIconDocument(sLastDocumentId)) {
```

```java
                recentTitle = mIcons.get(filename);
            } else if (isBackgroundsDocument(sLastDocumentId)) {
                recentTitle = mBackgrounds.get(filename);
            }

            try {
                addImageRow(result, sLastTitle, sLastFilename);
            } catch(IOException e) {
                Log.w(TAG, e);
            }
        }
        Log.d(TAG, "Recents: "+result.getCount());
        // 我们会将上次选择的结果返回给最近的查询
        return result;
    }

    /*
     * 将根写入提供 Cursor 的辅助方法
     */
    private void addRootRow(MatrixCursor cursor) {
            addDirRow(cursor, DOCID_ROOT, "Root");
    }
    private void addIconsRow(MatrixCursor cursor) {
            addDirRow(cursor, DOCID_ICONS_DIR, "Icons");
    }
    private void addBackgroundsRow(MatrixCursor cursor) {
            addDirRow(cursor, DOCID_BGS_DIR, "Backgrounds");
    }
    /*
    *将特定子目录写入提供Cursor的辅助方法
    */
     private void addDirRow(MatrixCursor cursor, String documentId,
String name) {
        final MatrixCursor.RowBuilder row = cursor.newRow();

        row.add(Document.COLUMN_DOCUMENT_ID, documentId);
        row.add(Document.COLUMN_DISPLAY_NAME, name);
        row.add(Document.COLUMN_SIZE, 0);
        row.add(Document.COLUMN_MIME_TYPE, Document.MIME_TYPE_DIR);

        long installed;
        try {
            installed = getContext().getPackageManager()
                    .getPackageInfo(getContext().getPackageName(), 0)
                    .firstInstallTime;
        } catch(NameNotFoundException e) {
```

```java
            installed = 0;
        }
        row.add(Document.COLUMN_LAST_MODIFIED, installed);
        row.add(Document.COLUMN_FLAGS, 0);
    }

    /*
     * 将特定图片文件写入提供 Cursor 的辅助方法
     */
    private void addImageRow(MatrixCursor cursor, String title, String
filename)throws IOException {
        final MatrixCursor.RowBuilder row = cursor.newRow();

        String filename = getFilename(documentId);
        AssetFileDescriptor afd =
            getContext().getAssets().openFd(filename);

        row.add(Document.COLUMN_DOCUMENT_ID, documentId);
        row.add(Document.COLUMN_DISPLAY_NAME, title);
        row.add(Document.COLUMN_SIZE, afd.getLength());
        row.add(Document.COLUMN_MIME_TYPE, "image/*");

        long installed;
        try {
            installed = getContext().getPackageManager()
                    .getPackageInfo(getContext().getPackageName(), 0)
                    .firstInstallTime;
        } catch (NameNotFoundException e) {
            installed = 0;
        }
        row.add(Document.COLUMN_LAST_MODIFIED, installed);
        row.add(Document.COLUMN_FLAGS, Document.FLAG_SUPPORTS_THUMBNAIL);
    }

    /*
     * 返回对图片资源的引用
     * 框架将在启用 FLAG_SUPPORTS_THUMBNAIL 标志的任何文档的项列表中使用该图片
     * 该方法可在下载内容时安全地阻塞
     */
    @Override
    public AssetFileDescriptor openDocumentThumbnail(String documentId,
        Point sizeHint,CancellationSignal signal) throws
        FileNotFoundException {
        // 我们将从存储设备上的此版本加载缩略图
        String filename = getFilename(documentId);
        // 创建指向内部存储设备上此图片的文件引用
        final File file = new File(getContext().getFilesDir(), filename);
        // 返回封装文件引用的文件描述符
        final ParcelFileDescriptor pfd =
            ParcelFileDescriptor.open(file, ParcelFileDescriptor.MODE_READ_ONLY);
```

```java
            return new AssetFileDescriptor(pfd, 0, AssetFileDescriptor.UNKNOWN_LENGTH);
    }

    /*
     * 将文件描述符返回到根据所提供文档ID引用的文档
     * 客户端将使用此描述符直接读取内容
     * 该方法可在下载内容时安全地阻塞
     */
    @Override
    public ParcelFileDescriptor openDocument(String documentId, String
        mode, CancellationSignal signal) throws FileNotFoundException {
        // 我们将从资源加载文档自身
        try {
            String filename = getFilename(documentId);
            //创建指向内部存储器上图片的文件引用
            final File file = new File(getContext().getFilesDir(), filename);
            //返回封装文件引用的文件描述符
            final ParcelFileDescriptor pfd =
                ParcelFileDescriptor.open(file,
                    ParcelFileDescriptor.MODE_READ_ONLY);

            // 将此保存为上次选择的文档
            sLastDocumentId = documentId;

            return pfd;
        } catch(IOException e) {
            Log.w(TAG, e);
            return null;
        }
    }
}
```

在此例中，我们显示了如何从内部存储提供一些图片文件以供使用。我们的图片数据是静态的，并且在 APK 的 assets 目录中捆绑以进行发布。需要将文件移到适当的存储卷，因此在创建提供程序时，我们从 assets 读取图片文件并将它们复制到内部存储。我们的提供程序将包括两个顶层目录 icons 和 backgrounds，分别包含相应的图片文件。

我们创建了简单的结构，用于将文件名转换为文档 ID。在此例中，root 是包含两个子目录的虚拟顶层目录，并且我们使用冒号分隔符创建作为该文件伪路径的每个 ID。getIconsDocumentId()、getBackgroundsDocumentId()和 getFilename()是辅助方法，用于在已发布 ID 和实际的图片文件名之间来回转换。

提示：

框架将文档 ID 嵌入内容 Uri 中，因此如果将目录路径转换为 ID，就必须使用 Uri 类不会将其视为路径分隔符的字符。

在 queryRoots()内部，我们返回 MatrixCursor，其中包括单个提供程序根的基本元数据。请注意，提供程序的查询方法应该遵守传入的列投影，仅返回请求的数据。我们也使用了更新的 add()方法版本，该方法获取每一项的列名。此版本相当便利，因为它监控传入

MatrixCursor 构造函数的投影，并且静默地忽略添加不在投影中的列的尝试，从而不再需要之前用于添加元素的循环代码。

标题、概要和图标列处理提供程序在拾取器 UI 中的显示。我们还定义了如下方面：

- COLUMN_DOCUMENT_ID：提供 ID，我们在后面将返回使用此 ID 引用顶层根元素。
- COLUMN_MINE_TYPES：报告此根包含的文档类型。我们在此拥有图片文件，因此使用 image/*。

此外，COLUMN_FLAGS 标志报告根项可能支持的额外特性。具体选项如下：

- FLAG_LOCAL_ONLY：结果位于设备上，不需要网络请求。
- FLAG_SUPPORTS_CREATE：根允许客户端应用程序在此提供程序内部创建新的文档。我们将在第 6 章中讨论如何在客户端执行此操作。
- FLAG_SUPPORTS_RECENTS：告诉框架我们可以通过结果加入最近的文档 UI。这将导致调用 queryRecentDocuments() 以获得此元数据。
- FLAG_SUPPORTS_SEARCH：类似于 RECENTS，告诉框架我们可以通过 querySearchDocuments() 处理搜索查询。

我们已在示例中设置 LOCAL 和 RECENTS 标志。该方法返回后，框架将通过根的文档 ID 调用 queryDocument() 以获得更多信息。在 addRootRow() 内部，我们使用必要的字段填充 Cursor。对于 COLUMN_MIME_TYPE，我们使用常量 MIME_TYPE_DIR 指示此元素是包含其他文档的目录。相同的定义应适用于为此提供程序创建的层次结构中的任何子目录。同样，因为我们提供的所有文件从安装应用程序以来就已经存在，我们提供 APK 安装日期作为 COLUMN_LAST_MODIFIED 值；对于更加动态的文件系统，这可能仅是磁盘上文件的修改日期。

当用户选择提供程序时，我们接收 queryChildDocuments() 调用以列出根中的所有文件。这包括针对将驻留图片的每个目录添加一行到 Cursor。addIconsRow() 和 addBackgroundsRow() 为每个子目录提供与根类似的元数据。对于子目录，我们还将再次看到 queryChildDocuments()。在看到这些 ID 时，就可以为每个存储器中具有的徽标图片文件添加一行到 Cursor。addIamgeRow() 方法构造适当的列数据，该列数据具有与前一次迭代类似的元素。

我们希望在拾取器 UI 中通过缩略图表示每张图片，因此为每个图片行上的 COLUMN_FLAGS 设置 FLAG_SUPPORTS_THUMBNAIL。这将为显示在拾取器中的每个元素触发 openDocumentThumbnail()。在此方法中，我们已显示如何从内部存储打开 FileDescriptor 并返回它。

sizeHint 参数应用于确保不会返回太大而无法显示在拾取器列表中的缩略图。我们的图片一开始都很小，因此我们在此没有检查此参数。如果需要的话，该方法可以安全地阻塞，在此方法内下载内容。对于此例，我们提供了 CancellationSinal，应该定期对其进行检查，以防止框架在加载操作结束前将其取消。

最终选择文档时，将再次使用所提供徽标图片的 ID 以调用 queryDocument()。在此情况下，必须针对所请求的单个文档从 addImageRow() 返回相同的结果。这将触发对

openDocument()的调用,在该方法中必须返回有效的 ParcelFileDescriptor,客户端可以使用其访问资源。因为内容与缩略图相同,使用相同的逻辑返回结果。

清单中的提供程序定义也与前面看起来稍有不同。因为将由框架直接查询提供程序,我们必须定义一些特殊的过滤器和权限(参见代码清单 5-44)。

代码清单 5-44   AndroidManifest.xml 中的 DocumentsProvider 片段

```
<provider
    android:name="com.androidrecipes.sharedocuments.ImageProvider"
    android:authorities="com.androidrecipes.sharedocuments.images"
    android:grantUriPermissions="true"
    android:exported="true"
    android:permission="android.permission.MANAGE_DOCUMENTS">
    <!-- 系统将用于查找已发布提供程序的独特过滤器 -->
    <intent-filter>
        <action android:name="android.content.action.DOCUMENTS_PROVIDER" />
    </intent-filter>
</provider>
```

首先,必须导出提供程序,以便外部应用程序访问。这通常是附加<intent_filter>的默认行为,但最好明确执行该操作。过滤器必须包括 DOCUMENTS_PROVIDER 动作,框架通过该动作查找它可以访问的已安装提供程序。接下来,必须通过 MANAGE_DOCUMENTS 权限保护提供程序。这是系统级的权限,只有系统应用程序可以获得,因此这会保护你的提供程序不被其他应用程序利用。最后,应启用 grantUriPermissions 属性。这就使框架可以按文档提供访问权限给客户端应用程序,而不是为整个提供程序提供每个客户端访问权限。

提示:
示例应用程序不真正具有可启动的用户界面,但在安装应用程序之后,可以通过进入需要拾取图片的系统应用程序来调用新的提供程序;Contacts 应用程序就是不错的选择。在创建新的联系人时,可以添加照片,而选择现有的图片会调用系统拾取器 UI。

1. 最近的文档

在我们的示例中,请记住在根元数据中设置了 FLAG_SUPPORTS_RECENTS。我们还提供了 queryRecentDocuments()的实现来响应这些查询。提供程序可以返回的最近文档数量并没有内在的限制,但你要选择的是与用户相关的文档。在此,我们仅从提供程序返回最近选择的徽标图片(在每次打开请求期间将其保存在静态变量中)。此处的元数据与其他文档查询相同,因此调用相同的 addImageRow()方法来填充 Cursor。

完成上述操作后,当访问提供程序的 Recent 部分(如图 5-7 所示)时,就可以看到最近选择的图片。

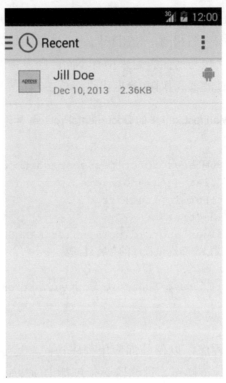

图 5-7　Recent UI 中显示最近选择的图片

2. 文档树

(API Level 21)

如果要更新 DocumentsProvider 以支持用户选择整个目录，就需要提供一些额外的覆写方法。代码清单 5-45 仅显示了必须添加到 ImageProvider 的部分覆写方法。

代码清单 5-45　ImageProvider 的文档树支持

```
/*
 * 系统调用此代码以确定在此驻留多少"提供程序"。
 * 最常见的情况是通过只有一个结果行的 Cursor 返回一个提供程序
 */
@Override
public Cursor queryRoots(String[] projection) throws FileNotFoundException {
    if(projection == null) {
        projection = DEFAULT_ROOT_PROJECTION;
    }

    MatrixCursor result = new MatrixCursor(projection);
    //为此提供程序添加单个根
    MatrixCursor.RowBuilder builder = result.newRow();
    builder.add(Root.COLUMN_ROOT_ID, "root");
    builder.add(Root.COLUMN_TITLE, "Android Recipes");
    builder.add(Root.COLUMN_SUMMARY, "Android Recipes Documents Provider");
```

```java
        builder.add(Root.COLUMN_ICON, R.drawable.ic_launcher);

        builder.add(Root.COLUMN_DOCUMENT_ID, DOCID_ROOT);

        builder.add(Root.COLUMN_FLAGS,
                //结果将仅来自本地文件系统
                Root.FLAG_LOCAL_ONLY
                //我们支持显示最近选择的项
                | Root.FLAG_SUPPORTS_RECENTS
                //支持文档树选择(API 21+)
                | Root.FLAG_SUPPORTS_IS_CHILD);
        builder.add(Root.COLUMN_MIME_TYPES, "image/*");
        builder.add(Root.COLUMN_AVAILABLE_BYTES, 0);

        return result;
    }

    /*
     * 如果要支持文档树选择(API 21+)，这是唯一需要的方法。
     * 该方法返回给定文档和父文档是否相关。
     */
    @Override
    public boolean isChildDocument(String parentDocumentId, String documentId) {
        if(isRoot(parentDocumentId)) {
            //子目录都在根的下方
            return isIconsDir(documentId)
                    || isBackgroundsDir(documentId);
        }

        if(isIconsDir(parentDocumentId)) {
            //所有图标都包含在icons目录中
            return isIconDocument(documentId);
        }

        if(isBackgroundsDir(parentDocumentId)) {
            //所有背景图片都包含在backgrounds目录中
            return isBackgroundsDocument(documentId);
        }

        //否则，这些ID彼此不知道对方
        return false;
    }
```

我们的提供程序必须实现 isChildDocument()方法，该方法使框架可以确定哪些文档 ID 在建立查询时具有父/子关系。该方法使用前面的辅助方法比较文档 ID，如果我们手上的文档 ID 包含在给定的父目录中，该方法返回 true。

最后，我们必须将 Root.FLAG_SUPPORTS_IS_CHILD 添加到根定义，以便在初始查询中表明，此提供程序除了支持请求单个文档之外，还支持请求文档树。

图 5-8 显示了系统定义的界面,当发出请求以从任何外部应用程序选择文档树时,用户就会看到此界面。

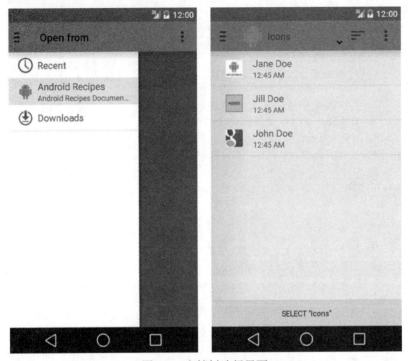

图 5-8　文档树选择界面

请注意此选择中的变化情况,即底部的 SELECT X 按钮。单击此按钮会将当前可见目录的引用返回到主调应用程序。

5.13　小结

本章介绍了一些在 Android 设备上实现数据持久化的实践方法。首先介绍了如何快速创建首选项界面,如何使用首选项和简单的方法持久化基本数据类型。然后学习了持久化文件和存储文件的位置及方法,还学习了如何在应用程序间共享已经持久化的数据。在第 6 章中,我们会研究如何利用操作系统的服务实现后台操作和应用程序间通信。

第6章

与系统交互

Android 操作系统提供了很多应用程序可以使用的有用服务。其中很多服务让应用程序不仅可以跟用户实现简单的互动,还可以采用多种方式与移动系统互动。应用程序可以进行定时提醒、运行后台服务以及彼此发送消息,所有这些就可以让 Android 应用程序更好地与移动设备融合在一起。此外,Android 还提供了一系列标准接口,用来将 Android 核心应用程序采集到的数据共享给你的软件。通过这些接口,任何应用程序都可以与 Android 平台的核心功能整合在一起,为其添加新功能,改进旧功能,最终达到提升用户体验的目的。

6.1 后台通知

6.1.1 问题

应用程序正在后台运行,当前没有用户可见的界面,但还必须将发生的重要事件通知给用户。

6.1.2 解决方案

(API Level 4)

使用 NotificationManager 发送状态栏通知。通知提供了一种温和的方式,可以引起用户的注意。对于接收一条新消息、提醒有可用的更新或者完成了一个长时间运行的工作,通知都可以很完美地完成这些任务。

6.1.3 实现机制

通过所有的系统控件,如 Service、BroadcastReceiver 或 Activity,可以将通知发送到 NotificationManager。在代码清单 6-1 和 6-2 中你会看到一个 Activity,它在用户离开时发送一系列不同的通知类型,并转入主屏幕。

代码清单 6-1 res/layout/activity_notification.xml

```xml
<?xml version="1.0" encoding="utf-8"?>
<LinearLayout xmlns:android="http://schemas.android.com/apk/res/android"
    android:orientation="vertical"
    android:layout_width="match_parent"
    android:layout_height="match_parent">

    <RadioGroup
        android:id="@+id/options_group"
        android:layout_width="match_parent"
        android:layout_height="0dp"
        android:layout_weight="1">
        <TextView
            android:layout_width="wrap_content"
            android:layout_height="wrap_content"
            android:textAppearance="?android:attr/textAppearanceLarge"
            android:text="Rich Styles"/>
        <RadioButton
            android:id="@+id/option_basic"
            android:layout_width="wrap_content"
            android:layout_height="wrap_content"
            android:text="Basic Notification"
            android:checked="true"/>
        <RadioButton
            android:id="@+id/option_bigtext"
            android:layout_width="wrap_content"
            android:layout_height="wrap_content"
            android:text="BigText Style"/>
        <RadioButton
            android:id="@+id/option_bigpicture"
            android:layout_width="wrap_content"
            android:layout_height="wrap_content"
            android:text="BigPicture Style"/>
        <RadioButton
            android:id="@+id/option_inbox"
            android:layout_width="wrap_content"
            android:layout_height="wrap_content"
            android:text="Inbox Style"/>

        <TextView
            android:layout_width="wrap_content"
            android:layout_height="wrap_content"
            android:layout_marginTop="8dp"
            android:textAppearance="?android:attr/textAppearanceLarge"
            android:text="Secured Styles"/>
        <RadioButton
            android:id="@+id/option_private"
```

```xml
        android:layout_width="wrap_content"
        android:layout_height="wrap_content"
        android:text="Public Version Lockscreen"/>
    <RadioButton
        android:id="@+id/option_secret"
        android:layout_width="wrap_content"
        android:layout_height="wrap_content"
        android:text="Secret Lockscreen"/>
    <RadioButton
        android:id="@+id/option_headsup"
        android:layout_width="wrap_content"
        android:layout_height="wrap_content"
        android:text="Heads-Up Notification"/>
</RadioGroup>

<Button
    android:layout_width="match_parent"
    android:layout_height="wrap_content"
    android:text="Post a Notification"
    android:onClick="onPostClick"/>

</LinearLayout>
```

代码清单 6-2　发送通知的 Activity

```java
public class NotificationActivity extends Activity {

    private RadioGroup mOptionsGroup;

    @Override
    public void onCreate(Bundle savedInstanceState) {
        super.onCreate(savedInstanceState);
        setContentView(tv);

        mOptionsGroup = (RadioGroup) findViewById(R.id.options_group);
    }

    public void onPostClick(View v) {
        final int noteId = mOptionsGroup.getCheckedRadioButtonId();
        final Notification note;
        switch(noteId) {
           case R.id.option_basic:
           case R.id.option_bigtext:
           case R.id.option_bigpicture:
           case R.id.option_inbox:
              note = buildStyledNotification(noteId);
              break;
           case R.id.option_private:
           case R.id.option_secret:
```

```java
        case R.id.option_headsup:
            note = buildSecuredNotification(noteId);
            break;
        default:
            throw new IllegalArgumentException("Unknown Type");
    }

    NotificationManager manager =
     (NotificationManager)getSystemService(Context.NOTIFICATION_SERVICE);
     manager.notify(noteId, note);
}

private Notification buildNotification(int type) {
    Intent launchIntent =
            new Intent(this, NotificationActivity.class);
    PendingIntent contentIntent =
            PendingIntent.getActivity(this, 0, launchIntent, 0);

    // 使用发送的时间创建通知
    NotificationCompat.Builder builder = new NotificationCompat.Builder(
            NotificationActivity.this);

    builder.setSmallIcon(R.drawable.ic_launcher)
            .setTicker("Something Happened")
            .setWhen(System.currentTimeMillis())
            .setAutoCancel(true)
            .setDefaults(Notification.DEFAULT_SOUND)
            .setContentTitle("We're Finished!")
            .setContentText("Click Here!")
            .setContentIntent(contentIntent);

    switch(type) {
        case R.id.option_basic:
            // 返回简单的通知
            return builder.build();
        case R.id.option_bigtext:
            // 包括两个动作
            builder.addAction(android.R.drawable.ic_menu_call,
                    "Call", contentIntent);
            builder.addAction(android.R.drawable.ic_menu_recent_history,
                    "History", contentIntent);
            // 在展开时使用BigTextStyle
            NotificationCompat.BigTextStyle textStyle =
                    new NotificationCompat.BigTextStyle(builder);
            textStyle.bigText(
                    "Here is some additional text to be displayed when
                    the notification is "+"in expanded mode. I can fit so
```

```java
                    much more content into this giant view!");

            return textStyle.build();
        case R.id.option_bigpicture:
            // 添加一个额外的动作
            builder.addAction(android.R.drawable.ic_menu_compass,
                    "View Location", contentIntent);
            // 在展开时使用BigPictureStyle
            NotificationCompat.BigPictureStyle pictureStyle =
                    new NotificationCompat.BigPictureStyle(builder);
            pictureStyle.bigPicture(BitmapFactory.decodeResource(get
                    Resources(), R.drawable.dog));

            return pictureStyle.build();
        case R.id.option_inbox:
            // 在展开时使用InboxStyle
            NotificationCompat.InboxStyle inboxStyle =
                    new NotificationCompat.InboxStyle(builder);
            inboxStyle.setSummaryText("4 New Tasks");
            inboxStyle.addLine("Make Dinner");
            inboxStyle.addLine("Call Mom");
            inboxStyle.addLine("Call Wife First");
            inboxStyle.addLine("Pick up Kids");
            return inboxStyle.build();
        default:
            throw new IllegalArgumentException("Unknown Type");
    }
}
//这些属性可以由用户的通知设置重写
private Notification buildSecuredNotification(int type) {
    Intent launchIntent =
        new Intent(this, NotificationActivity.class);
    PendingIntent contentIntent =
        PendingIntent.getActivity(this, 0, launchIntent, 0);
//构造基础通知
NotificationCompat.Builder builder = new
    NotificationCompat.Builder(this)
      .setSmallIcon(R.drawable.ic_launcher)
      .setContentTitle("Account Balance Update")
      .setContentText("Your account balance is -$250.00")
      .setStyle(new NotificationCompat.BigTextStyle()
        .bigText("Your account balance is -$250.00; pay us please"
        + "or we will be forced to take legal action!"))
      .setContentIntent(contentIntent);
```

```java
switch(type) {
    case R.id.option_private:
        //为安全的锁定屏幕提供独特的通知版本
        Notification publicNote = new Notification.Builder(this)
            .setSmallIcon(R.drawable.ic_launcher)
            .setContentTitle("Account Notification")
            .setContentText("An important message has arrived.")
            .setContentIntent(contentIntent)
            .build();
        return builder.setPublicVersion(publicNote)
            .build();
    case R.id.option_secret:
        //从安全的锁定屏幕完全隐藏通知
        return builder.setVisibility(Notification.VISIBILITY_SECRET)
            .build();
    case R.id.option_headsup:
        //在发布时显示警告通知
        return builder.setDefaults(Notification.DEFAULT_SOUND)
            .setPriority(Notification.PRIORITY_HIGH)
            .build();
    default:
        throw new IllegalArgumentException("Unknown Type");
    }
}
```

这个示例中使用 Notification.Builder，在用户离开 Activity 时创建一系列新的通知元素。我们稍后将讨论扩展类型，目前仅关注基本类型。可以提供图标资源和标题字符串，在进行通知时，这些条目将显示在状态栏中。此外，传递一个时间值(以毫秒为单位)，它会作为事件时间显示在通知列表中。这里我们将这个值设为发送通知的时间，但在你自己的应用程序中这个时间可能会有不同的含义。

要点：

在这个示例中我们使用了 NotificationCompat.Builder，它是由支持库提供的，用于在 Android 1.6 中调用新的 API，而无须编写分支代码。如果你的目标平台是原生支持所需通知功能的 Android 版本，则可以在代码中将 NotificationCompat.Builder 替换为 Notification.Builder。

在创建通知前，可以添加其他一些有用参数，比如当用户在状态栏中展开通知时显示在通知列表中的详细文本信息。

我们向 Builder 传入了一个 PendingIntent 参数，它指向我们的 Activity。这个 Intent 使得通知具有可交互功能，即用户在通知列表中点击通知时可以启动一个 Activity。

注意：
对于每个通知事件，这个 Intent 都会启动一个新的 Activity。如果更愿意用一个已经存在的 Activity 实例来响应启动(在栈中就有一个)，则要确定加上 Intent 标志及合适的清单参数，例如 Intent.FLAG_ACTIVITY_CLEAR_TOP 和 android:launchMode="singleTop"。

为了让通知不仅仅以动画的形式出现在状态栏中，我们在 Notification.defaults 中加上系统默认的通知声音，当通知发出时就会播放这个声音。另外还可以加入 Notification.DEFAULT_VIBRATION 和 Notification.DEFAULT_LIGHTS 等各种参数。

提示：
如果想要通知播放自定义的声音，可以将 Notification.sound 参数设置为指向某个文件或 ContentProvider 的 URI 以从中读取声音。

最后，给 Notification 加上各种标志以进一步自定义通知。本例使用 Builder 中的 setAutoCancel()来启用 Notification.FLAG_AUTO_CANCEL，意思是当用户在列表中点击通知后，通知应尽快取消或从列表中移除。如果没有这个标志，通知就会一直在列表中，直到手动调用 NotificationManager.cancel()或 NotificationManager.cancelAll()才会移除或取消。在 Builder 中设置的另一个有用标志是 setOngoing()，它会禁用用户移除通知的能力，且只可以通过编程方式取消。该标志用于通知用户后台操作正在运行，这些操作包括播放音乐或进行中的位置跟踪。

此外，还有其他一些有用的标志，它们在 Builder 内没有对应的方法。完成构造之后，可以直接在通知中设置这些标志：

- FLAG_INSISTENT：重复播放通知声音，直到用户进行响应。
- FLAG_NO_CLEAR：用户点击"Clear Notifications"不会清除通知，只能通过调用 cancel()移除通知。通知准备好后，通过 NotificationManager.notify()方法就可以发送给用户，该方法也会使用 ID 参数。应用程序中每种类型的通知都应该拥有唯一的 ID。同一时刻，管理器只允许相同 ID 的一个通知显示在列表中，新的通知会覆盖具有相同 ID 的旧通知。此外，在手动取消某个特定通知时，也需要传入 ID。

运行这个示例，Activity 会显示，允许用户选择要发布的通知类型。按下按钮后，可以看到在状态栏中发布的所选通知，即使你离开 Activity 且它不再可见，也仍然可以看到此通知(见图 6-1)。

下面按照引入功能的平台版本，仔细剖析此例中丰富的通知功能。

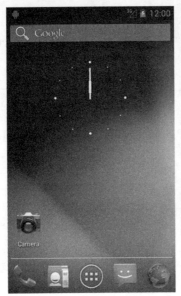 

图 6-1　通知出现(左图)和通知显示在列表中(右图)

1. 扩展通知样式

(API Level 16)

从 Android 4.1 开始，可以在通知视图中直接显示更加丰富的、具有交互性的信息，这就是通知的样式。当前处在屏幕顶端的百叶窗(window shade)中的所有通知默认都是展开的，但用户可以使用两根手指的手势操作展开任意其他的通知。因此，扩展的视图并不会取代传统的视图，相反，在某些情况下会提升用户体验。

Android 默认提供了三种样式(它们都实现了 Notification.Style)：

- BigTextStyle：显示更多的文本信息，例如显示一条消息或公告的全部内容。
- BigPictureStyle：显示全彩的大图片。
- InboxStyle：提供了一个条目列表，样子有点像 Gmail 等应用程序中的收件箱视图。

但是，你不仅限于使用这些样式。Notification.Style 就是一个接口，应用程序可以实现它来显示任意更加适合你的需求的自定义扩展布局。

除了样式，Android 4.1 还为扩展通知添加了内联动作。也就是说，可以直接在百叶窗视图中为用户添加多个动作选项，而不是只有在用户点击整个通知项时才返回一个回调 Intent。这些项会在扩展视图的顶部到底部依次排列。代码清单 6-3 演示了如何修改之前的示例来添加一个 BigTextStyle 样式的扩展通知，结果如图 6-2 所示。

代码清单 6-3　BigTextStyle 样式的通知

```
//使用发送的时间创建通知
NotificationCompat.Builder builder =
        new NotificationCompat.Builder(NotificationActivity.this);

builder.setSmallIcon(R.drawable.icon)
       .setTicker("Something Happened")
```

```
        .setWhen(System.currentTimeMillis())
        .setAutoCancel(true)
        .setDefaults(Notification.DEFAULT_SOUND)
        .setContentTitle("We're Finished!")
        .setContentText("Click Here!")
        .setContentIntent(contentIntent);

//添加一些自定义动作
builder.addAction(android.R.id.drawable.ic_menu_call, "Call Back",
        contentIntent);
builder.addAction(android.R.id.drawable.ic_menu_recent_history,
        "Call History", contentIntent);

//应用一个扩展样式
NotificationCompat.BigTextStyle expandedStyle =
        new NotificationCompat.BigTextStyle(builder);
expandedStyle.bigText("Here is some additional text to be displayed when"
    + " the notification is in expanded mode. "
    + " I can fit so much more content into this giant view!");

Notification note = expandedStyle.build();
```

图 6-2　百叶窗中的 BigTextStyle 样式

可以通过 Builder 的 addAction()方法关联自定义动作。这里演示了如何将动作布局加入整个视图中。本例中，每个动作的响应结果是一样的，但也可以为每个动作都关联一个 PendingIntent，这样每个动作就可以在应用程序中有不同的响应结果了。

之前示例唯一修改的地方就是将已经创建的 Builder 对象封装到 BigTextStyle 对象中并

做了自定义。本例中，另外还设置了 bigText()以在扩展模式下显示文本。然后，通过样式而非 Builder 的 build()方法创建通知。

下面查看一下代码清单 6-4 和图 6-3 所示的 BigPictureStyle。

代码清单 6-4　BigPictureStyle 样式的通知

```
//使用发送的时间创建通知
NotificationCompat.Builder builder =
        new NotificationCompat.Builder(NotificationActivity.this);

builder.setSmallIcon(R.drawable.icon)
        .setTicker("Something Happened")
        .setWhen(System.currentTimeMillis())
        .setAutoCancel(true)
        .setDefaults(Notification.DEFAULT_SOUND)
        .setContentTitle("We're Finished!")
        .setContentText("Click Here!")
        .setContentIntent(contentIntent);

//添加一些自定义动作
builder.addAction(android.R.id.drawable.ic_menu_compass,
        "View Location", contentIntent);

//应用一个扩展样式
NotificationCompat.BigPictureStyle expandedStyle =
        new NotificationCompat.BigPictureStyle(builder);
expandedStyle.bigPicture(
        BitmapFactory.decodeResource(getResources(), R.drawable.icon) );

Notification note = expandedStyle.build();
```

图 6-3　百叶窗中的 BigPictureStyle 样式

这段代码除了使用 bigPicture()方法传递了一张要显示的全彩位图外，剩下的内容几乎和 BigTextStyle 一样。最后，看一下代码清单 6-5 和图 6-4 所示的 InboxStyle。

代码清单 6-5　InboxStyle 样式的通知

```
//使用发送的时间创建通知
NotificationCompat.Builder builder =
        new NotificationCompat.Builder(NotificationActivity.this);

builder.setSmallIcon(R.drawable.icon)
      .setTicker("Something Happened")
      .setWhen(System.currentTimeMillis())
      .setAutoCancel(true)
      .setDefaults(Notification.DEFAULT_SOUND)
      .setContentTitle("We're Finished!")
      .setContentText("Click Here!")
      .setContentIntent(contentIntent);

//应用一个扩展样式
NotificationCompat.InboxStyle expandedStyle =
        new NotificationCompat.InboxStyle(builder);
expandedStyle.setSummaryText("4 New Tasks");
expandedStyle.addLine("Make Dinner");
expandedStyle.addLine("Call Mom");
expandedStyle.addLine("Call Wife First");
expandedStyle.addLine("Pick up Kids");

Notification note = expandedStyle.build();
```

在 Notification.InboxStyle 样式下，可以通过 addLine()方法向列表中添加多个条目。我们仍使用 setSummaryText()方法对所有的条目进行了分类总结，在之前的样式中该方法也是可以使用的。

与以前一样，我们使用了支持库中的 NotificationCompat 类，可以在运行 API Level 4 的设备上的应用程序中调用这些方法。如果应用程序运行的最低目标平台是 Android 4.1，直接使用原生的 Notification.Builder 即可。

本例中，我们看到了支持库的强大之处。调用了 API Level 16 才有的方法，实际上是支持库在底层进行了版本检查，对于某一平台不支持的方法会忽略；不需要建立判断分支代码来使用新 API。

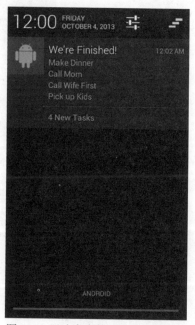

图 6-4　百叶窗中的 InboxStyle 样式

因此，当在 Android 4.0 或更早版本的设备上运行这段同样的代码时，由于我们没有用到新特性，因此只会出现简单的传统通知。

2. 通知的可见性和隐私性

(API Level 21)

从 Android 5.0 开始，无须下拉状态栏就可以使通知在锁定屏幕上完全显示出来，即使设备受到密码保护也是如此。为保护通知中可能存在的隐私信息，可以提供附加的参数来定义在锁定屏幕上可看到的通知。

通知的可见性设置控制其在安全锁定屏幕上的默认行为，这意味着对锁定屏幕启用密码。对于需要简单滑动手势解锁的锁定屏幕，这些功能不会执行任何操作。

图 6-5 给出了在应用任何自定义的可见性值之前默认通知在安全锁定屏幕上的显示效果。

图 6-5 受保护锁定屏幕上的默认通知

系统提供了 3 个通知可见性选项：
- VISIBILITY_PRIVATE：通知的编辑版本，仅包含应用程序标题和图标，在设备解锁时可见。这是默认的可见性设置。
  - ➢ 在本例中，可以提供公共版本的通知，在设备解锁时显示。
- VISIBILITY_PUBLIC：无论设备是解锁还是锁定，都会显示完整的通知。
- VISIBILITY_SECRET：此通知将在锁定屏幕上完全隐藏。用户必须解锁设备才能看到此通知的存在。

警告：

用户可以在设备的通知设置中重写所有这些可见性选项。例如，如果用户选择让所有通知公开可见，那么甚至是设置为保密的通知也会显示在锁定屏幕上。

在 Android 4.1 和以后的版本中，通知支持优先级设置。在 Andorid 5.0 中，该功能使我们可以在浮动显示模式下显示具有较高优先级的通知(如来电)。此模式会将通知与应用程序内容重叠，而不需要等待用户下拉状态栏，从而强迫用户立刻回应此通知。优先级设置为 PRIORITY_HIGH 或 PRIORITY_MAX 的通知将在浮动显示模式下尽可能显示出来。图 6-6 显示了浮动通知的一个示例。

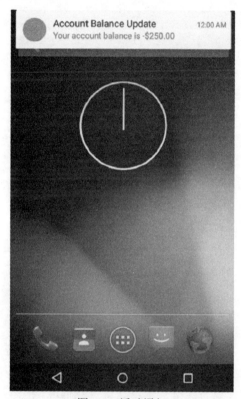

图 6-6　浮动通知

代码清单 6-6 再次给出了示例中的安全生成器方法，从而我们可以更详细地研究这些部分。

代码清单6-6　安全通知功能

```
private Notification buildSecuredNotification(int type) {
    Intent launchIntent =
            new Intent(this, NotificationActivity.class);
    PendingIntent contentIntent =
            PendingIntent.getActivity(this, 0, launchIntent, 0);

    //构造基础通知
    NotificationCompat.Builder builder = new NotificationCompat.Builder(this)
```

```
            .setSmallIcon(R.drawable.ic_launcher)
            .setContentTitle("Account Balance Update")
            .setContentText("Your account balance is -$250.00")
            .setStyle(new NotificationCompat.BigTextStyle()
                .bigText("Your account balance is -$250.00; pay us
                    please " + "or we will be forced to take legal action!"))
            .setContentIntent(contentIntent);

    switch(type) {
        case R.id.option_private:
            //为安全锁定屏幕提供独特的版本
            Notification publicNote = new Notification.Builder(this)
                    .setSmallIcon(R.drawable.ic_launcher)
                    .setContentTitle("Account Notification")
                    .setContentText("An important message has arrived.")
                    .setContentIntent(contentIntent)
                    .build();

            return builder.setPublicVersion(publicNote)
                    .build();
        case R.id.option_secret:
            //在安全锁定屏幕中完全隐藏通知
            return builder.setVisibility(Notification.VISIBILITY_SECRET)
                    .build();
        case R.id.option_headsup:
            //在发布时显示浮动通知
            return builder.setDefaults(Notification.DEFAULT_SOUND)
            .setPriority(Notification.PRIORITY_HIGH)
            .build();
        default:
            throw new IllegalArgumentException("Unknown Type");
    }
}
```

对于此选项块，我们创建了一个基础通知，计划将其用作从包含往来账余额的用户银行发出的警告。如果用户使设备保持安全，这就是我们应该保护的敏感信息。

Android 框架的默认行为将整个通知隐藏在已编辑视图的后面，如图 6-5 所示，但或许我们可以在此方面采取更智能的方式。第一个用户选项被选中时，我们构造第二个 Notification 实例并将其传递给 setPublicVersion()。该版本采用更加友善的消息，在设备解锁时安全地显示此消息。图 6-7 给出了目前显示在安全锁定屏幕上的公共通知。

注意，在前面的选项中并没有修改通知的可见性设置。这是因为我们需要提供给此行为的值 VISIBILITY_PRIVATE 是默认值。然而在第二个用户选项中，我们要在锁定屏幕上完全隐藏通知，直至设备解锁才显示。在此情况下，我们只需要将通知的可见性改为 VISIBILITY_SECRET。

最后，如果通知非常重要，用户应立刻看到，则可以将其优先级升级为 PRIORITY_HIGH。这将使通知立刻以浮动模式发布，如图 6-6 所示。

图 6-7 私有通知的公开版本

提醒：
在不支持这些 API 的旧平台上，Android 框架将会直接忽略此部分中设置的值。

3. NotificationListenerService

(API Level 18)

从 Android 4.3 开始，对于希望监控设备上所有通知状态的应用程序，可以使用一个新的服务。应用程序可以扩展 NotificationListenerService，在应用程序发布新的通知或用户清除现有通知时接收更新。此外，应用程序可以编程取消特定通知或一次性清除所有通知。

启用通知访问

因为此服务为应用程序提供了对活动通知列表的全局访问，所以必须首先授予权限。然而，此例中的应用程序无法将其声明为它想获得的标准权限。相反，用户必须从设备的 Settings 应用程序的 Security 部分明确授予此访问权限。在此部分中提供了 Notification Access 部分，列出了具有已导出 NotificationListenerService 以便用户启用或禁用此功能的所有已安装应用程序。

图 6-8 显示了 Settings 的 Notification Access 部分在 Nexus 设备上的外观，并且显示了一个提示框，用户在点击应用程序条目以允许访问此信息后会看到此提示。

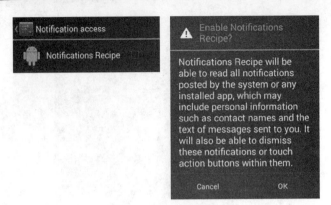

图6-8 通知设置警告

应用程序负责指导用户进行设置，启用此服务以接收事件；框架不会自动完成该工作。在 Settings 中可以直接将用户带到适当的界面，方法是激活通过 android.settings.ACTION_NOTIFICATION_LISTENER_SETTINGS 动作字符串启动 Activity 的 Intent。从 Android 4.3 开始，SDK 中不公开该字符串，因此请注意，它可能在未来的版本中发生改变。

代码清单 6-7 中显示了 NotificationListenerService 的简单扩展。

代码清单 6-7 NotificationListenerService 示例

```java
public class MonitorService extends NotificationListenerService {
    private static final String TAG = "RecipesMonitorService";

    @Override
    public void onNotificationPosted(StatusBarNotification sbn) {
        // 确认通知来自此应用程序
        if(!TextUtils.equals(sbn.getPackageName(), getPackageName())) {
            return;
        }

        Log.i(TAG, "Notification "+sbn.getId()+" Posted");
    }

    @Override
    public void onNotificationRemoved(StatusBarNotification sbn) {
        // 确认通知来自此应用程序
        if(!TextUtils.equals(sbn.getPackageName(), getPackageName())) {
            return;
        }
        // 我们寻求的是基本通知
        if(R.id.option_basic != sbn.getId()) {
            return;
        }

        // 如果基本通知取消，则取消所有的通知
        for(StatusBarNotification note : getActiveNotifications()) {
            if(TextUtils.equals(note.getPackageName(), getPackageName())) {
```

```
                cancelNotification(note.getPackageName(),
                    note.getTag(),
                    note.getId());
            }else {
            cancelNotification(note.getKey());
            }
        }
    }
}
```

必须实现两个抽象方法：onNotificationPosted()和 onNotificationRemoved()。当新通知到来或另一个通知被取消时，框架将分别调用这两个方法。传入的内容是 StatusBar-Notification 实例，该实例只是关于原始通知的基本封装类，它带有一些附加元数据(例如，发布此通知的应用程序的包名以及所应用的 ID 或标记)。原始通知仍然可以作为参数访问。

在此例中，当添加通知时，如果通知来自应用程序，我们只是记录此事件。如果移除通知，我们检查它是否是基本样式通知元素。如果是，则取消应用程序发布的所有仍然活跃的通知。getActiveNotifications()方法有助于获得用户当前可见的所有通知。通过比较每个通知的包名，我们可以确认哪些通知是由我们发布的。当包匹配时，我们使用来自通知元素的元数据调用 cancelNotification()，以编程方式移除该通知。还可以调用 cancelAllNotifications()以清除百叶窗中的所有通知，而不需要知道活跃的通知来自何处。

代码清单 6-8 显示了需要添加的 AndroidManifest.xml 代码片段。

代码清单 6-8　NotificationListenerService 清单元素

```
<service android:name=".MonitorService"
    android:permission="android.permission.BIND_NOTIFICATION_LISTENER_SERVICE">
    <intent-filter>
        <action android:name="android.service.notification.
            NotificationListenerService" />
    </intent-filter>
</service>
```

此处需要的两个元素是<intent-filter>的动作字符串和声明的权限。在确定可以绑定哪个 NotificationListenerService 元素时，框架将查找这两个元素。

6.2　创建定时和周期任务

6.2.1　问题

应用程序想要定时执行某个操作，例如按计划定时更新 UI。

6.2.2 解决方案

(API Level 1)

使用 Handler 进行定时操作。Handler 可以有效地在某一确定时间点或延迟特定时间后执行某个操作。

6.2.3 实现机制

让我们看一个示例 Activity，它通过一个 TextView 显示了当前的时间。参见代码清单 6-9。

代码清单 6-9　使用 Handler 进行更新的 Activity

```java
public class TimingActivity extends Activity {

    TextView mClock;

    @Override
    public void onCreate(Bundle savedInstanceState) {
        super.onCreate(savedInstanceState);
        mClock = new TextView(this);
        setContentView(mClock);
    }

    private Handler mHandler = new Handler();
    private Runnable timerTask = new Runnable() {
        @Override
        public void run() {
            Calendar now = Calendar.getInstance();
            mClock.setText(String.format("%02d:%02d:%02d",
                    now.get(Calendar.HOUR),
                    now.get(Calendar.MINUTE),
                    now.get(Calendar.SECOND)) );
            //1 秒后进行下一次更新
            mHandler.postDelayed(timerTask,1000);
        }
    };

    @Override
    public void onResume() {
        super.onResume();
        mHandler.post(timerTask);
    }

    @Override
    public void onPause() {
        super.onPause();
        mHandler.removeCallbacks(timerTask);
```

        }
    }
```

这里，将读取当前时间并将更新 UI 的操作封装到一个名为 timerTask 的 Runnable 中，该 Runnable 会通过一个已经创建的 Handler 触发。当 Activity 可见时，调用 Handler.post() 会立即执行 timerTask 中的操作。TextView 更新后，timerTask 最后会调用 Handler.postDelayed()，让 Handler 可以在 1 秒(1000 毫秒)后进行再一次的更新操作。

只要 Activity 不被打扰，这个循环会一直进行，即每秒更新一次 UI。而当 Activity 处于暂停状态时(用户关闭了 Activity 或因其他事情改变 Activity 的状态)，Handler.removeCallbacks() 方法会移除所有待处理的操作，这样 timerTask 在 Activity 再次可见之前就不会被调用了。

提示：

在这个示例中，由于 Handler 是在主线程上创建的，因此我们更新 UI 的操作是安全的。Handler 发出的操作都是在创建它的线程中执行的，除非将来自另一个线程的 Looper 明确传递给它的构造函数。我们在后面的范例中会介绍如何将 Looper 用于后台队列，但在此处还值得注意的是，可以从后台线程创建一个 Handler，通过传递 Looper.getMainLooper() 的结果将这个 Handler 发布到主线程，Looper.getMainLooper() 的结果是指向主 UI 线程的 Looper 的静态引用。

## 6.3 定时执行周期任务

### 6.3.1 问题

应用程序需要注册以执行周期任务，例如检查服务器更新或提醒用户执行操作。

### 6.3.2 解决方案

**(API Level 1)**

使用 AlarmManager 管理和执行任务。AlarmManager 可用于计划未来的单次或重复操作，甚至在应用程序没有运行时也可以执行任务。AlarmManager 通过发出 PendingIntent 来触发计划的警告。这个 Intent 可以在提醒触发时执行各种系统组件，例如 BroadcastReceiver 或 Service。

AlarmManager 支持通过类型参数来控制如何调度提醒的条件：
- ELAPSED_REALTIME：根据上次设备启动以来的时间值(以毫秒为单位)触发提醒。
- ELAPSED_REALTIME_WAKEUP：根据经过的时间触发提醒，如果设备处于休眠状态，会将设备激活。
- RTC：根据 UTC 时间触发警告。
- RTC_WAKEUP：根据 UTC 时间触发警告，如果设备处于休眠状态，会将设备激活。
```

> 注意：
> 如果选择唤醒提醒类型，Android 会将设备从睡眠状态唤醒，但不会保持设备的唤醒状态。在执行唤醒事件的后台工作时，必须从 PowerManager 获得 WakeLock。否则，Android 可能快速将设备转入睡眠状态，这会停止你正在执行的操作。

注意，这种方式更加适用于应用程序代码没有运行但依然需要执行某些操作的场景。与应用程序运行时使用的简单定时操作相比，AlarmManager 需要消耗更多的资源。后一种情况最好是使用 Handler 的 postAtTime() 和 postDelayed() 方法进行处理。

(API Level 21)

Android 5.0 和以后版本中的 JobScheduler 系统服务是计划后台工作的更高效解决方案。可以使用该系统服务计划一次性或周期任务以便以后执行。这是为许多任务提供直接提醒的首选方法，因为 Android 框架会尝试尽可能成批执行操作，从而最大限度减少对设备电池和网络使用率的影响。然而，这也意味着默认操作的计时是不准确的。如果需要满足非常严格的计时要求，提醒可能就是更好的选择。

我们计划任务的方式是为每个任务构造独特的 JobInfo 对象，该对象包含应用程序完成任务所需的额外元数据。JobInfo 还支持为任务执行条件设定标准。JobInfo.Builder 包括如下方法，用于对任务请求应用必要的标准：

- setRequiredNetworkType()：描述作业运行必须具备的网络条件。例如，除非设备连接到 Wi-Fi，否则系统不会触发设置了 NETWORK_TYPE_UNMETERED 的作业。默认值表明此作业不需要网络访问。
- setRequiresCharging()：为使作业运行，此设备必须进行充电。对于较少执行且比较耗电的作业，这是较为有用的方法。
- setRequiresDeviceIdle()：为使作业运行，此设备必须处于空闲模式。也就是说，设备必须未激活或处于睡眠状态。
- setPersisted()：控制是否应在设备重启时自动计划作业。默认值是 false，表示应用程序负责在重启时手动计划作业。
- setBackoffCriteria()：控制如何以及何时重新计划失败的作业以使其再次运行。可以使用该方法，在资源(例如网络访问)临时不可用时尽可能减少不必要的重试次数。
- setPeriodic()：表示应在给定时间间隔内定期运行作业，直到其被明确地取消。

为执行工作，应用程序必须提供 Android 框架可关联的 JobService 子集。框架将在预定时间的任务发生时调用此服务。类似于 AlarmManager，即使应用程序进程当前未在运行，计划的作业也会正常运行。这种方法的优点之一是框架自动处理为计划作业获得唤醒锁的工作，从而甚至在设备空闲时，工作也会继续进行。

6.3.3 实现机制

让我们看一下在常规情况下，如何使用 AlarmManager 触发 Service。参见代码清单 6-10 到代码清单 6-12。

代码清单 6-10　要触发的 Service

```java
public class AlarmReceiver extends BroadcastReceiver {
    @Override
    public void onReceive(Context context, Intent intent) {
        //可以执行一个有趣的操作，我们这里只是显示当前时间
        Calendar now = Calendar.getInstance();
        DateFormat formatter = SimpleDateFormat.getTimeInstance();
        Toast.makeText(context, formatter.format(now.getTime()),
            Toast.LENGTH_SHORT).show();
        return START_NOT_STICKY;
    }

    @Override
    public IBinder onBind(Intent intent) {
        return null;
    }
}
```

代码清单 6-11　res/layout/main.xml

```xml
<?xml version="1.0" encoding="utf-8"?>
<LinearLayout xmlns:android="http://schemas.android.com/apk/res/android"
    android:orientation="vertical"
    android:layout_width="match_parent"
    android:layout_height="match_parent">
    <Button
        android:id="@+id/start"
        android:layout_width="match_parent"
        android:layout_height="wrap_content"
        android:text="Start Periodic Task" />
    <Button
        android:id="@+id/stop"
        android:layout_width="match_parent"
        android:layout_height="wrap_content"
        android:text="Cancel Periodic Task" />
</LinearLayout>
```

代码清单 6-12　注册/解除注册提醒的 Activity

```java
public class AlarmActivity extends Activity implements View.OnClickListener {

    private PendingIntent mAlarmIntent;

    @Override
    public void onCreate(Bundle savedInstanceState) {
        super.onCreate(savedInstanceState);
        setContentView(R.layout.main);
        //为两个按钮关联监听器
        findViewById(R.id.start).setOnClickListener(this);
        findViewById(R.id.stop).setOnClickListener(this);
```

```
        //创建启动触发器
        Intent launchIntent = new Intent(this, AlarmReceiver.class);
        mAlarmIntent = PendingIntent.getBroadcast(this, 0, launchIntent, 0);
    }

    @Override
    public void onClick(View v) {
        AlarmManager manager =
            (AlarmManager)getSystemService(Context.ALARM_SERVICE);
        long interval = 5*1000; //5秒

        switch(v.getId()) {
        case R.id.start:
            Toast.makeText(this, "Scheduled", Toast.LENGTH_SHORT).show();
            manager.setRepeating(AlarmManager.ELAPSED_REALTIME,
                    SystemClock.elapsedRealtime()+interval,
                    interval,
                    mAlarmIntent);
            break;
        case R.id.stop:
            Toast.makeText(this, "Canceled", Toast.LENGTH_SHORT).show();
            manager.cancel(mAlarmIntent);
            break;
        default:
            break;
        }
    }
}
```

在这个示例中，我们提供了一个非常简单的 Service，每次触发它时会通过 Toast 简单地显示当前时间。这个 Service 必须在应用程序的清单中通过<service>标签进行注册。否则，应用程序外部的 AlarmManager 将无法知道该如何触发它。这个示例的 Activity 有两个按钮：一个用于启动定时提醒，另一个则是取消所设置的提醒。

触发器的操作由 PendingIntent 引用，PendingIntent 用于设置和取消提醒。我们创建一个直接指向此 Service 的 Intent，然后在使用 getService()获得的 PendingIntent 中封装此 Intent。

提醒：

PendingIntent 也有创造器方法——getActivity()和 getBroadcast()，但在创建 PendingIntent 时，要确保它能正确地指向要触发的应用程序组件。

按下 Start 按钮，Activity 会用 AlarmManager.setRepeating()注册一个会反复发出的提醒。除了 PendingIntent，该方法还有一些参数用于设置触发提醒的时间。第一个参数定义了提醒的类型，包括使用的时间单位和在设备休眠时是否发出提醒。在这个示例中，我们选择了 ELAPSED_REALTIME。

接下来的两个参数分别是首次触发提醒的时间和重复发出提醒的时间间隔。因为选择的提醒类型是 ELAPSED_REALTIME，所以首次触发提醒的时间必须是相对某个时间经过的时间；SystemClock.elapsedRealtime()可以设置这种格式的当前时间。

这个示例中的触发器会在按下按钮 5 秒钟后触发提醒，然后每 5 秒都会触发一次。每 5 秒钟，屏幕上都会通过 Toast 形式显示当前时间，即使应用程序不在前台运行也是如此。当用户打开 Activity 并按下 Stop 按钮时，所有与 PendingIntent 匹配的提醒都会被取消，Toast 流程也会停止。

要点：
在设备重启后，提醒不会得以保存。如果关闭设备电源，然后重新开启，则必须重新计划任何之前已注册的提醒。

1. 更精确的示例

如果要在指定的时间发出提醒，怎么办(如每天上午 9 时)？还是要用 AlarmManager，再加上一些不同的参数即可，参见代码清单 6-13。

代码清单 6-13　定时提醒

```java
long oneDay = 24*3600*1000; //24 hours
long firstTime;

//创建一个日历(默认为当前日期)
//设置提醒时间为 09:00:00
Calendar startTime = Calendar.getInstance();
startTime.set(Calendar.HOUR_OF_DAY, 9);
startTime.set(Calendar.MINUTE, 0);
startTime.set(Calendar.SECOND, 0);

//获取当前时间
Calendar now = Calendar.getInstance();

if(now.before(startTime)) {
    //现在还没有到上午 9 时，从今天算起
    firstTime = startTime.getTimeInMillis();
} else {
    //从明天上午 9 时开始
    startTime.add(Calendar.DATE, 1);
    firstTime = startTime.getTimeInMillis();
}

//设置提醒
manager.setRepeating(AlarmManager.RTC_WAKEUP,
            firstTime,
            oneDay,
            mAlarmIntent);
}
```

这个示例根据真实时间触发提醒。首先判断接下来的上午 9 时是今天还是要等到明天，然后将返回的值作为第一次触发提醒的时间。从 Android 4.4 开始，AlarmManager 默认所

有提醒都是不精确的,这意味着将通过一个小窗口触发它们。连同这个新行为一起还添加了 API 方法 setExact(),使开发人员可以声明下面的提醒不能放入不精确窗口中。在 Android 4.4 版本之前,只需要通过适当的启动时间调用 set()即可。

提示:
在 Android 4.4 之后的版本中,不能计划在准确的时间间隔内重复提醒;始终使用不精确窗口解释这些提醒。必须以完全相同的时间间隔重复的提醒需要由应用程序在每个触发器事件之后重新计划。

2. 使用计划作业

(API Level 21)

接下来查看一个类似的示例,其中改为利用 JobScheduler。代码清单 6-14 给出了修改后的服务,该服务用于将显示时间的工作按照作业进行处理。

代码清单 6-14　作为 JobService 的工作者类

```java
public class WorkerService extends JobService {

    private static final int MSG_JOB = 1;

    //用于处理计划作业的简单队列处理程序
    private Handler mJobProcessor = new Handler(new Handler.Callback() {
        @Override
        public boolean handleMessage(Message msg) {
            JobParameters params = (JobParameters) msg.obj;
            Log.i("WorkerService", "Executing Job "+params.getJobId());
            //完成异步工作之后,必须触发 jobFinished()以使下一个计划任务运行
            doWork();
            jobFinished(params, false);

            return true;
        }
    });

    @Override
    public boolean onStartJob(JobParameters jobParameters) {
        Log.d("WorkerService", "Start Job "+jobParameters.getJobId());
        //为模拟长时间任务,我们延迟执行 7.5 秒
        mJobProcessor.sendMessageDelayed(
                Message.obtain(mJobProcessor, MSG_JOB, jobParameters),
                7500
        );

        /*
         * 如果此处异步完成作业,则返回 false;如果需要执行更多后台工作,
         * 则返回 true。对于后一种情况,必须调用 jobFinished()以通知系统作业完成
         */
```

```
        return true;
    }

    @Override
    public boolean onStopJob(JobParameters jobParameters) {
        Log.w("WorkerService", "Stop Job "+jobParameters.getJobId());
        //当停止请求到来时,我们必须取消挂起的作业
        mJobProcessor.removeMessages(MSG_JOB);

        /*
         * 返回true以计划作业,返回false则删除作业
         */
        return false;
    }

    private void doWork() {
        //执行有趣的操作,在此是显示当前时间
        Calendar now = Calendar.getInstance();
        DateFormat formatter = SimpleDateFormat.getTimeInstance();
        Toast.makeText(this, formatter.format(now.getTime()),
Toast.LENGTH_SHORT).show();
    }
}
```

注意,必须将此服务作为 JobService 的实现提供。这会提供与 JobScheduler 交互所需的回调。我们在 doWork() 方法中封装了显示日期的操作。为了模拟更加复杂的任务,我们通过将其作为 Message 发布到 Handler 来延迟执行 7.5 秒。

系统将通过 onStartJob() 触发之前计划的作业,必须使用此方法开始你的作业任务。onStartJob() 的返回值告诉 Android 框架你的作业在何处可以异步完成(即在方法返回时作业已完成)或是否开始执行异步任务。如果在 onStartJob() 中返回 false,框架就认为任务已完成,并且直到框架触发新的作业时,你的工作才完成。

然而,在此返回 true 以表明任务需要花费较长的时间才能完成。这意味着我们负责在任务完成时通知框架。在当前作业完成之前,JobScheduler 不会触发更多相同的作业。

经过请求的延迟时间之后,handleMessage() 触发以处理 doWork() 方法。在此,我们必须调用 jobFinished() 以表明任务最终完成。该方法需要通过原始的 JobParameters 对象确认作业,因此我们在排队等待的 Message 中传递此对象。如果任务中出现失败情况,jobFinished() 就接受一个布尔型参数,告诉 Android 框架根据在原始 JobInfo 中设置的标准重新计划作业。

JobService 还必须支持取消挂起的作业。如果 JobScheduler 在等待作业完成时收到一个取消请求,则会触发 JobService 的 onStopJob()。该服务负责在发生这种情况时立刻终止作业。在此例中,我们只需要清除 Handler 队列中任何挂起的任务,这样它们就不会激活。

提示:

如果任务是同步的,即从 onStartJob() 返回 false,则 Android 框架假设没有任何挂起的工作,因此取消请求没有任何意义,并且永远不会调用 onStopJob()。

在代码清单 6-15 中可以看到，必须在清单中定义此服务。此外，我们需要将此服务暴露给框架，因此 Android 要求通过 BIND_JOB_SERVICE 权限保护我们的服务。只有框架可以拥有该权限，它保护我们的服务不被其他应用程序访问。未提供此权限会在尝试计划作业时产生异常。

代码清单 6-15　AndroidManifest.xml 的部分代码

```xml
<manifest xmlns:android="http://schemas.android.com/apk/res/android" ...>

    <application ...>

        <service android:name=".WorkerService"
            android:permission="android.permission.BIND_JOB_SERVICE" />

    </application>
</manifest>
```

现在我们的服务已完成，接下来查看代码清单 6-16，其中将新的服务与类似的 Activity 关联起来，可以使用该 Activity 以作业方式管理计划的工作。

代码清单 6-16　计划后台作业的 Activity

```java
public class JobSchedulerActivity extends Activity implements View.OnClickListener {
    private static final int JOB_ID = 1;

    @Override
    protected void onCreate(Bundle savedInstanceState) {
        super.onCreate(savedInstanceState);
        setContentView(R.layout.main);

        //将侦听器附加到两个按钮
        findViewById(R.id.start).setOnClickListener(this);
        findViewById(R.id.stop).setOnClickListener(this);
    }

    @Override
    public void onClick(View view) {
        JobScheduler scheduler = (JobScheduler) getSystemService(JOB_SCHEDULER_SERVICE);
        long interval = 5*1000; //5 秒

        JobInfo info = new JobInfo.Builder(JOB_ID,
                new ComponentName(getPackageName(),
            WorkerService.class.getName()))
                .setPeriodic(interval)
                .build();

        switch(view.getId()) {
```

```
            case R.id.start:
                //Android在将相同的信息传递给Schedule()时返回同一个作业ID,
                //不会重复作业
                int result = scheduler.schedule(info);
                if (result <= 0) {
                    Toast.makeText(this, "Error Scheduling Job",
                        Toast.LENGTH_SHORT).show();
                }
                break;
            case R.id.stop:
                //作业ID必须匹配传递给Schedule()的ID,因此保存该ID
                scheduler.cancel(JOB_ID);
                break;
            default:
                break;
        }
    }
}
```

在此例中,我们将计划的工作表示为从 JobInfo.Builder 中获取的 JobInfo 实例。该信息至少必须包含作业 ID 和将执行工作的服务的名称。我们也提供定期时间间隔,告诉系统每隔 5 秒钟计划此作业。

Start 按钮被按下时,将信息对象传递到 JobScheduler.schedule()以开始触发定期任务。这将在此事件后大约 5 秒启动 WorkerService 上的第一个作业。如果要停止定期任务,Stop 按钮会使用匹配的作业 ID 触发 JobScheduler.cancel()。

注意:

在此例中可以观察到的行为是,大约每隔 12.5 秒显示一次时间。这是因为 JobScheduler 计划下一个定期任务在前一个任务完成后执行。在通知任务完成之前,我们在服务中将任务延迟 7.5 秒,这就延长了整个周期。

6.4 创建粘性操作

6.4.1 问题

应用程序即使在其被用户终止时也可以执行一个或多个后台操作,并且这些后台操作将持续运行直至完成。

6.4.2 解决方案

(API Level 3)

创建一个 IntentService 来处理这项工作。IntentService 是 Android 标准 Service 实现的封装类,是在与用户无交互的后台中执行任务的关键组件。IntentService 会将要执行的任务(用 Intent 表示)放到队列中,然后逐个处理请求,全部处理完之后会终止自己。

IntentService 还会在后台创建一个执行任务所需的工作者线程，所以不需要使用 AsyncTask 或 Java 线程来确保后台操作的正确执行。

这个范例会演示使用 IntentService 创建一个后台操作的中央管理器。本例中，外部应用程序可以调用 Context.startService() 来引用这个管理器。管理器会将所有收到的请求添加到队列中，然后通过调用 onHandleIntent() 逐个处理。

6.4.3 实现机制

下面来看看如何构建一个简单的 IntentService 实现来处理一系列的后台操作。参见代码清单 6-17。

代码清单 6-17　用于处理后台操作的 IntentService

```java
public class OperationsManager extends IntentService {

    public static final String ACTION_EVENT = "ACTION_EVENT";
    public static final String ACTION_WARNING = "ACTION_WARNING";
    public static final String ACTION_ERROR = "ACTION_ERROR";
    public static final String EXTRA_NAME = "eventName";

    private static final String LOGTAG = "EventLogger";

    private IntentFilter matcher;

    public OperationsManager() {
        super("OperationsManager");
        //创建一个过滤器来匹配收到的请求
        matcher = new IntentFilter();
        matcher.addAction(ACTION_EVENT);
        matcher.addAction(ACTION_WARNING);
        matcher.addAction(ACTION_ERROR);
    }

    @Override
    protected void onHandleIntent(Intent intent) {
        //检查是否是有效的请求
        if(!matcher.matchAction(intent.getAction())) {
            Toast.makeText(this, "OperationsManager: Invalid Request",
                    Toast.LENGTH_SHORT).show();
            return;
        }
        //直接在这个方法中处理每个请求，不需要创建其他线程
        if(TextUtils.equals(intent.getAction(), ACTION_EVENT)) {
            logEvent(intent.getStringExtra(EXTRA_NAME));
        }
        if(TextUtils.equals(intent.getAction(), ACTION_WARNING)) {
```

```java
                logWarning(intent.getStringExtra(EXTRA_NAME));
            }
            if(TextUtils.equals(intent.getAction(), ACTION_ERROR)) {
                logError(intent.getStringExtra(EXTRA_NAME));
            }
        }

        private void logEvent(String name) {
            try {
                //通过休眠来模拟长时间的网络操作
                Thread.sleep(5000);
                Log.i(LOGTAG, name);
            } catch(InterruptedException e) {
                e.printStackTrace();
            }
        }

        private void logWarning(String name) {
            try {
                //通过休眠来模拟长时间的网络操作
                Thread.sleep(5000);
                Log.w(LOGTAG, name);
            } catch(InterruptedException e) {
                e.printStackTrace();
            }
        }

        private void logError(String name) {
            try {
                //通过休眠来模拟长时间的网络操作
                Thread.sleep(5000);
                Log.e(LOGTAG, name);
            } catch(InterruptedException e) {
                e.printStackTrace();
            }
        }
    }
```

IntentService 并没有默认的构造函数(除了无参的构造函数)，因此在自定义 IntentService 实现时必须实现一个构造函数，该构造函数会调用父类的构造函数并传入一个服务名。这个名称在技术层面并不重要，只是用作调试而已；Android 会用这个名称命名创建的新工作线程。

所有的请求都是由服务通过 onHandleIntent()方法处理的。这个方法是由提供的工作者线程调用的，所以可以在这里进行所有的工作；不必创建新线程或是执行其他操作。当 onHandleIntent()返回时，就是在通知 IntentService 开始处理队列中的下一个请求。

这个示例还提供了 3 个日志操作，可以在请求 Intent 中用不同的动作字符串调用。为了演示这些操作，每个操作都会用相应的日志级别(INFO、WARNING 或 ERROR)将消息写到设备上。注意，消息的内容是以请求 Intent 的附加信息的形式传递的。用每个 Intent 的 Data 和附加字段保存日志操作的参数，留下 Action 字段用来定义操作的类型。

这个示例服务中还有一个 IntentFilter，用来方便地判断请求是否有效。在创建服务时，所有的有效操作都被添加到这个过滤器中，这样就能通过调用 IntentFilter.matchAction()判断收到的请求是否包含可处理的操作。

代码清单 6-18 和 6-19 是一个调用该服务以执行工作的示例 Activity。

代码清单 6-18　AndroidManifest.xml

```xml
<?xml version="1.0" encoding="utf-8"?>
<manifest xmlns:android="http://schemas.android.com/apk/res/android"
    package="com.examples.sticky"

    <application android:icon="@drawable/icon"
                 android:label="@string/app_name">
        <activity android:name=".ReportActivity"
                  android:label="@string/app_name">
            <intent-filter>
                <action android:name="android.intent.action.MAIN" />
                <category android:name="android.intent.category.LAUNCHER" />
            </intent-filter>
        </activity>
        <service android:name=".OperationsManager"></service>
    </application>
</manifest>
```

注意：
因为 IntentService 是作为 Service 被调用的，所以必须通过<service>标签在应用程序的清单文件中进行声明。

代码清单 6-19　调用 IntentService 的 Activity

```java
public class ReportActivity extends Activity {

    @Override
    public void onCreate(Bundle savedInstanceState) {
        super.onCreate(savedInstanceState);
        logEvent("CREATE");
    }

    @Override
    public void onStart() {
        super.onStart();
        logEvent("START");
    }
```

```java
    @Override
    public void onResume() {
        super.onResume();
        logEvent("RESUME");
    }

    @Override
    public void onPause() {
        super.onPause();
        logWarning("PAUSE");
    }

    @Override
    public void onStop() {
        super.onStop();
        logWarning("STOP");
    }

    @Override
    public void onDestroy() {
        super.onDestroy();
        logWarning("DESTROY");
    }

    private void logEvent(String event) {
        Intent intent = new Intent(this, OperationsManager.class);
        intent.setAction(OperationsManager.ACTION_EVENT);
        intent.putExtra(OperationsManager.EXTRA_NAME, event);
        startService(intent);
    }

    private void logWarning(String event) {
        Intent intent = new Intent(this, OperationsManager.class);
        intent.setAction(OperationsManager.ACTION_WARNING);
        intent.putExtra(OperationsManager.EXTRA_NAME, event);

        startService(intent);
    }
}
```

这个 Activity 中需要值得注意的内容不多,只是通过设备的日志而不是用户界面发送事件。但是,它帮助演示了之前所创建的服务中的队列处理行为。当 Activity 可见时,它会调用其所有的普通生命周期方法,对日志服务发出 3 个请求。每处理完一个请求,就会向日志输出一行内容,然后该服务转向处理下一个请求。

提示:
可以通过 SDK 的 logcat 工具看到这些日志语句。设备或模拟器的 logcat 输出可以通过大多数的开发环境(包括 Eclipse)或命令行(输入 adb logcat)查看。

还要注意在服务处理完所有 3 个请求后，系统会向日志发出通知：服务已停止。只有在需要 IntentService 完成任务时，它才会处于内存中。这个服务是一个非常有用的特性，使其成为整个系统中的好"公民"。

按下 HOME 或 BACK 键会触发更多的生命周期方法，向服务发送更多请求，Pause/Stop/Destroy 部分会调用服务中的一个单独操作，它们的消息会作为警告记录到日志中；只要将请求 Intent 的动作字符串改成其他值就可以设置日志的类型。

注意，即使应用程序已经不可见(抑或打开了另一个应用程序)，消息也会继续发送到日志中。这就是 Android 系统中 Service 组件的强大之处。这类操作受系统保护，无论用户做什么，这些操作都会执行到底。

潜在的不足

在每个操作方法中都有 5 秒的延迟，用来模拟实际请求访问远程 API 或一些类似操作所需的时间。在运行这个示例时，还演示了 IntentService 如何在一个工作者线程上逐个处理队列中的多个请求。这个示例中排队的请求来自应用程序的各个生命周期方法，但在日志中还是每 5 秒记录一条消息。这是因为在处理完当前的请求之前(也就是 onHandleIntent()返回之前)，IntentService 是不会开始处理下一个请求的。

如果应用程序实现粘性后台任务的并发操作，就需要用执行工作的线程池来创建自定义的 Service 实现。Android 作为开源系统的一大魅力就是可以直接看到 IntentService 的源代码，然后以此为基础开发自己的实现，节约开发时间，减少代码量。

6.5 长时间运行的后台操作

6.5.1 问题

应用程序中需要有一个组件一直运行在后台，用于执行一些操作或监控特定事件。

6.5.2 解决方案

(API Level 1)

可以将这个组件放到一个服务中。服务就是后台运行的组件，应用程序可以启动服务，然后让其在后台长期运行。跟后台的其他进程相比，服务的优先级更高，不会因为内存少而被终止。

可以为了某些操作而明确地启动或停止服务，而并不一定要将其与其他的组件(如 Activity)连接起来。然而，如果应用程序必须直接与服务进行交互，服务也提供了相应的绑定接口来传递数据。在这里的实例中，服务的启动和停止都是由系统根据所请求的绑定来明确进行的。

实现服务时要注意的关键之处是确保它的用户友好性。无期限的操作应该由用户主动请求启动。整个应用程序中应该有地方(界面或设置)能使用户控制这类服务的启动和禁用。

6.5.3 实现机制

代码清单 6-20 是一个持久运行的示例服务，用来跟踪和输出用户一定时间段内的位置。

代码清单 6-20　持久运行的跟踪服务

```java
public class TrackerService extends Service implements LocationListener {

    private static final String LOGTAG = "TrackerService";

    private LocationManager manager;
    private ArrayList<Location> storedLocations;

    private boolean isTracking = false;

    /* 服务的创建方法 */
    @Override
    public void onCreate() {
        manager = (LocationManager)getSystemService(LOCATION_SERVICE);
        storedLocations = new ArrayList<Location>();
        Log.i(LOGTAG, "Tracking Service Running...");
    }

    @Override
    public void onDestroy() {
        manager.removeUpdates(this);
        Log.i(LOGTAG, "Tracking Service Stopped...");
    }

    public void startTracking() {
        if(!manager.isProviderEnabled(LocationManager.GPS_PROVIDER)) {
            return;
        }
        Toast.makeText(this, "Starting Tracker", Toast.LENGTH_SHORT).show();
        manager.requestLocationUpdates(LocationManager.GPS_PROVIDER,
            30000, 0, this);

        isTracking = true;
    }

    public void stopTracking() {
        Toast.makeText(this, "Stopping Tracker", Toast.LENGTH_SHORT).show();
        manager.removeUpdates(this);
        isTracking = false;
    }

    public boolean isTracking() {
        return isTracking;
    }
```

```java
/* 服务访问方法 */
public class TrackerBinder extends Binder {
    TrackerService getService() {
        return TrackerService.this;
    }
}

private final IBinder binder = new TrackerBinder();

@Override
public IBinder onBind(Intent intent) {
    return binder;
}

public int getLocationsCount() {
    return storedLocations.size();
}

public ArrayList<Location> getLocations() {
    return storedLocations;
}

/* LocationListener 方法 */
@Override
public void onLocationChanged(Location location) {
    Log.i("TrackerService", "Adding new location");
    storedLocations.add(location);
}

@Override
public void onProviderDisabled(String provider) { }

@Override
public void onProviderEnabled(String provider) { }

@Override
public void onStatusChanged(String provider, int status, Bundle extras) { }
}
```

这个服务会监控和跟踪从 LocationManager 收到的更新信息。当服务被创建时,它建立了一个空白的 Location 项的列表并等待开始跟踪。对于外部组件,如 Activity,则可以调用 startTracking() 和 stopTracking() 来启用和禁用服务的位置更新过程。此外,这个服务还提供了能访问该服务所记录的位置列表的方法。

因为这个服务需要直接与其他 Activity 或组件进行交互,所以需要提供一个 Binder 接口。在服务需要跨进程边界通信时,Binder 的概念会变得很复杂。但在这个示例中,所有的内容都在同一个进程中,所以创建的 Binder 也很简单。这里的 Binder 是在 getService() 方法中创建的,它向调用者返回服务实例本身。稍后我们从 Activity 的角度仔细看看这个服务。

当服务启动跟踪时，它会注册 LocationManager 的更新信息，并将其收到的每个更新都保存到位置列表中。注意，调用 requestLocationUpdates()方法的时间间隔最少为 30 秒。因为该服务会长时间运行，较大的更新时间间隔可以让 GPS(和电池)"稍作休息"。

现在来看一个让用户访问该服务的 Activity，参见代码清单 6-21 到代码清单 6-23。

代码清单 6-21　AndroidManifest.xml

```xml
<?xml version="1.0" encoding="utf-8"?>
<manifest xmlns:android="http://schemas.android.com/apk/res/android"
    package="com.examples.service"

    <application android:icon="@drawable/icon"
                 android:label="@string/app_name">
        <activity android:name=".ServiceActivity"
                  android:label="@string/app_name">
            <intent-filter>
                <action android:name="android.intent.action.MAIN" />
                <category android:name="android.intent.category.LAUNCHER" />
            </intent-filter>
        </activity>

        <service android:name=".TrackerService"></service>
    </application>
    <uses-permission android:name="android.permission.ACCESS_FINE_LOCATION"/>
</manifest>
```

提醒：

服务必须在应用程序清单中通过<service>标签进行声明，这样 Android 系统才能知道如何以及到哪里调用它。同样，因为需要用到 GPS，所以这个示例中还必须添加 android.permission.ACCESS_FINE_LOCATION 权限。

代码清单 6-22　res/layout/main.xml

```xml
<?xml version="1.0" encoding="utf-8"?>
<LinearLayout xmlns:android="http://schemas.android.com/apk/res/android"
    android:orientation="vertical"
    android:layout_width="match_parent"
    android:layout_height="match_parent">
    <Button
        android:id="@+id/enable"
        android:layout_width="match_parent"
        android:layout_height="wrap_content"
        android:text="Start Tracking" />
    <Button
        android:id="@+id/disable"
        android:layout_width="match_parent"
        android:layout_height="wrap_content"
        android:text="Stop Tracking" />
    <TextView
```

```xml
            android:id="@+id/status"
            android:layout_width="match_parent"
            android:layout_height="wrap_content" />
</LinearLayout>
```

代码清单 6-23   与服务进行交互的 Activity

```java
public class ServiceActivity extends Activity implements View.OnClickListener {

    Button enableButton, disableButton;
    TextView statusView;

    TrackerService trackerService;
    Intent serviceIntent;

    @Override
    public void onCreate(Bundle savedInstanceState) {
        super.onCreate(savedInstanceState);
        setContentView(R.layout.main);
        enableButton = (Button)findViewById(R.id.enable);
        enableButton.setOnClickListener(this);
        disableButton = (Button)findViewById(R.id.disable);
        disableButton.setOnClickListener(this);
        statusView = (TextView)findViewById(R.id.status);

        serviceIntent = new Intent(this, TrackerService.class);
    }

    @Override
    public void onResume() {
        super.onResume();
        //启动服务使其持久化，无视绑定
        startService(serviceIntent);
        //绑定服务
        bindService(serviceIntent, serviceConnection, Context.BIND_AUTO_CREATE);
    }

    @Override
    public void onPause() {
        super.onPause();
        if(!trackerService.isTracking()) {
            //停止服务，解除绑定后永久移除
            stopService(serviceIntent);
        }
        //解除服务绑定
        unbindService(serviceConnection);
    }

    @Override
    public void onClick(View v) {
        switch(v.getId()) {
```

```java
        case R.id.enable:
            trackerService.startTracking();
            break;
        case R.id.disable:
            trackerService.stopTracking();
            break;
        default:
            break;
    }
    updateStatus();
}

private void updateStatus() {
    if(trackerService.isTracking()) {
        statusView.setText(
            String.format("Tracking enabled. %d locations
                logged.",trackerService.getLocationsCount()));
    } else {
        statusView.setText("Tracking not currently enabled.");
    }
}

private ServiceConnection serviceConnection = new ServiceConnection() {
    public void onServiceConnected(ComponentName className, IBinder
        service) {
        trackerService = ((TrackerService.TrackerBinder)service).getService();
        updateStatus();
    }

    public void onServiceDisconnected(ComponentName className) {
        trackerService = null;
    }
};
}
```

图 6-9 显示了一个有两个按钮的 Activity,允许用户启用和禁用位置跟踪行为,并将当前服务的状态通过文本的形式显示出来。

当 Activity 可见时,它会被绑定到 TrackerService 上。这是通过 ServiceConnection 接口实现的,该接口在绑定和解绑操作完成后都会提供相应的回调方法。在服务与 Activity 绑定后,就可以直接调用服务所提供的所有公共方法。

然而,只使用绑定还不能让服务长时间运行;仅仅通过 Binder 接口访问服务会使服务随着 Activity 的生命周期而自动创建或销毁。在这个示例中,我们希望即使 Activity 没有运行,服务也能继续运行。为了实现这个目的,在将服务与 Activity 绑定之前,先用 startService()明确启动服务。给正在运行中的服务发送启动命令不会造成任何问题,所以在 onResume()方法中这么做也完全可行。

现在,即使在 Activity 取消绑定之后,服务也还会在内存中运行。这个示例的 onPause() 方法会检查用户是否正在启动跟踪,如果没有就会先停止服务。这样在不需要跟踪时,服务也会停止。在没有任务需要完成时，服务就可从内存中永久移除。

图 6-9　ServiceActivity 布局

运行这个示例，按下 Start Tracking 按钮就会启动持久运行的服务和 LocationManager。这时用户也许会离开应用程序，但服务会一直运行，而且会记录从 GPS 收到的位置更新信息。当用户再次回到应用程序时，用户会看到一直运行着的服务和当前保存的位置数量。按下 Stop Tracking 按钮会终止该进程，然后在用户退出 Activity 后立即结束服务。

6.6　启动其他应用程序

6.6.1　问题

设备上的其他应用程序已经具备应用程序需要的功能。为了避免功能的重复开发，需要能够启动其他应用程序来执行该功能。

6.6.2　解决方案

(API Level 1)

使用隐式的 Intent 告知系统你想要做的事情，系统会判断目前是否有满足需要的应用程序。通常，开发人员会使用显式的 Intent 启动其他的 Activity 或服务，比如：

```
Intent intent = new Intent(this, NewActivity.class);
```

```
startActivity(intent);
```

声明要启动的特定组件后，Intent 的目的也就很明确了。同样，我们还可以使用动作、类别、数据和类型定义一个 Intent 来实现所要完成任务的更加隐性的需求。

当应用程序通过这种方式启动外部应用程序时，通常它们会处在同一个 Android 任务栈中，因此在操作完成(或者用户退出外部应用程序)后，用户还会回到应用程序中。这就确保了用户体验的无缝衔接，从用户的角度来看，这些应用程序是浑然一体的。

6.6.3 实现机制

以此方式定义 Intent 时，需要设定哪些信息并不明确，因为没有统一发布的标准，有可能两个应用程序所能提供的服务(例如读取 PDF 文件)是一样的，但监听所收到的 Intent 的过滤器却不尽相同。你需要向系统(或用户)提供足够的信息来选择最合适的应用程序以完成所需的任务。

定义隐式 Intent 的核心信息是动作，这是一个字符串值，可以通过构造函数传递，也可以用 Intent.setAction()设置。这个值告诉 Android 你要做什么，比如查看某些内容、发送消息、做选择或其他目的。这里要根据具体情况设置，而且通常不同的组合可以得到相同的结果。让我们看几个实用的示例。

1. 读取 PDF 文件

几乎每台 Android 设备上都预装了 PDF 阅读器，在 Android 电子市场里面也有无数的 PDF 阅读器，但在核心 SDK 中并没有提供显示 PDF 文档的组件。所以，在应用程序中内嵌显示 PDF 的功能是比较麻烦的，而且没有必要。

相反，代码清单 6-24 演示了如何找到并启动其他应用程序来查看 PDF 文件。

代码清单 6-24　查看 PDF 的方法

```
private void viewPdf(Uri file) {
    Intent intent;
    intent = new Intent(Intent.ACTION_VIEW);
    intent.setDataAndType(file, "application/pdf");
    try {
        startActivity(intent);
    } catch(ActivityNotFoundException e) {
        //如果没有可用的应用程序，就提示下载一个
        AlertDialog.Builder builder = new AlertDialog.Builder(this);
        builder.setTitle("No Application Found");
        builder.setMessage("We could not find an application to view PDFs."
            +" Would you like to download one from Android Market?");
        builder.setPositiveButton("Yes, Please",
            new DialogInterface.OnClickListener() {
            @Override
            public void onClick(DialogInterface dialog, int which) {
                Intent marketIntent = new Intent(Intent.ACTION_VIEW);
                marketIntent.setData(
```

```
                    Uri.parse("market://details?id=com.adobe.reader"));
            startActivity(marketIntent);
        }
    });
    builder.setNegativeButton("No, Thanks", null);
    builder.create().show();
}
```

这个示例会找到最合适的应用程序并打开设备上任意的本地 PDF 文件(内部或外部存储器)。如果在设备上没有找到可以查看 PDF 的应用程序，会提示用户到 Google Play 上下载一个。

创建并构造 Intent 时使用的是通用的 Intent.ACTION_VIEW 动作字符串，它会告诉系统我们想要查看该 Intent 中提供的数据文件，而数据文件本身及其 MIME 类型则可以让系统了解要查看的数据类型。

提示：

使用 Intent.setData()和 Intent.setType()会相互清除对方设置的值。如果想要同时设置两者，就跟示例中一样使用 Intent.setDataAndType()即可。

如果 startActivity()执行失败得到 ActivityNotFoundException 异常，则意味着在用户的设备上并没有安装可以查看 PDF 的应用程序。出现这种情况时，为了可以让用户更好地体验，我们将问题告知用户并提示用户可以去 Market 上下载一个阅读器。如果用户按下 Yes，就会发送另一个隐式 Intent 请求，直接打开 Google Play 上 Adobe Reader 的应用程序下载页，用户可以免费下载此应用程序以查看 PDF 文件。在下一个范例中我们会讨论在这个 Intent 中使用的 Uri 策略。

注意，向这个示例方法传入了一个指向本地文件的 Uri 参数。下面的示例演示了如何获取内部存储器文件的 Uri：

```
String filename = NAME_OF_YOUR_FILE;
File internalFile = getFileStreamPath(filename);
Uri internal = Uri.fromFile(internalFile);
```

getFileStreamPath()是通过 Context 调用的，因此如果这段代码不在 Activity 中，你还必须引用一个 Context 对象才能调用它。下面给出如何构建指向外部储存器文件的 Uri：

```
String filename = NAME_OF_YOUR_FILE;
File externalFile = new File(Environment.getExternalStorageDirectory(),
    filename);
Uri external = Uri.fromFile(externalFile);
```

这个示例也适用于其他类型的文档，只需要修改附加到 Intent 的 MIME 类型即可。

2. 与好友分享内容

在开发人员中很流行的另一个功能就是让用户可以通过电子邮件、短信或主流的社交网络与朋友分享应用程序中的内容。所有的 Android 设备都有电子邮件和短信应用程序，大部分想通过社交网络(例如 Facebook 或 Twitter)分享内容的用户也会在自己的设备上安装相应的移动应用。

恰好，这个任务也可以用隐式 Intent 来完成，因为大多数此类应用程序都会以某种方式响应 Intent.ACTION_SEND 动作字符串。代码清单 6-25 演示了如何通过单个 Intent 请求发送各种内容到媒体。

代码清单 6-25　分享 Intent

```
private void shareContent(String update) {
    Intent intent = new Intent(Intent.ACTION_SEND);
    intent.setType("text/plain");
    intent.putExtra(Intent.EXTRA_TEXT, update);
    startActivity(Intent.createChooser(intent, "Share..."));
}
```

这里，我们告诉系统要以 Intent 的附加信息的形式发送一段文本。这是一种常见的请求，我们希望能有多个应用程序处理这种请求。默认情况下，Android 会弹出一个应用程序列表供用户选择打开。此外，部分设备还提供了一个选择框，让用户选择完成这种操作的默认应用程序，以后就不会再弹出这个列表了。

我们可以对这个过程施加更多的控制，因为我们也希望每次能给用户提供多种选择。要实现这个目的，就不要直接把 Intent 传递给 startActivity()，而是先传递给 Intent.createChooser()，这样就能自定义标题，确保始终向用户显示可供选择的列表。

用户做出选择后，指定的应用程序就会启动，在消息输入框中预先填充 EXTRA_TEXT 中的信息，做好分享内容的准备！

3. 使用 ShareActionProvider

(API Level 14)

从 Android 4.0 开始，引入了一个新的 ShareActionProvider 小部件，使用一种更加通用的机制来帮助应用程序分享内容。该小部件被添加到选项菜单中，从而在 Action Bar 或溢出菜单上显示。它还有一个附加的功能，默认情况下，它会根据用户的使用习惯对分享选项排序。也就是说，用户频繁点击的选项会排到列表的最上方。

在菜单中实现 ShareActionProvider 非常简单，与创建分享 Intent 比起来，只需要很少的几行代码即可。代码清单 6-26 显示了如何将 ShareActionProvider 关联到一个菜单项上。

代码清单 6-26　res/menu/options.xml

```
<menu xmlns:android="http://schemas.android.com/apk/res/android">
    <item android:id="@+id/menu_share"
        android:showAsAction="ifRoom"
        android:title="Share"
        android:actionProviderClass="android.widget.ShareActionProvider"/>
</menu>
```

注意：
如果你的 Menu 不是采用 XML 定义的，可以在 Java 代码中调用 setActionProvider() 来关联 ShareActionProvider。

代码清单 6-27 显示了如何在 Activity 中将分享 Intent 关联到 ShareActionProvider 小部件上。

代码清单 6-27　提供分享 Intent

```
@Override
public boolean onCreateOptionsMenu(Menu menu) {
    //填充菜单
    getMenuInflater().inflate(R.menu.options, menu);

    //找到项并设置分享 Intent
    MenuItem item = menu.findItem(R.id.menu_share);
    ShareActionProvider provider = (ShareActionProvider) item.getActionProvider();

    Intent intent = new Intent(Intent.ACTION_SEND);
    intent.setType("text/plain");
    intent.putExtra(Intent.EXTRA_TEXT, update);
    provider.setShareIntent(intent);

    return true;
}
```

好了！ShareActionProvider 处理了所有的用户交互，因此你的应用程序甚至不需要处理用户的 MenuItem 选择事件。

6.7　启动系统应用程序

6.7.1　问题

对于应用程序需要的功能，设备上的某个系统应用程序已经具备。为了避免功能的重复开发，需要能够启动系统应用程序来执行该功能。

6.7.2　解决方案

(API Level 1)

用隐式 Intent 告诉系统你所需要的应用程序。每个系统应用程序都采用自定义的 Uri 方案，可以将其以数据的形式插入隐式 Intent 中来启动你所需要的应用程序。

通过这种方式启动时，外部应用程序通常会处在与你的应用程序相同的 Android 任务栈中，因此在任务完成(或用户退出系统应用程序)后，用户还会回到你的应用程序中。这就确保了用户体验的无缝衔接，从用户的角度来看，这些应用程序是浑然一体的。

6.7.3 实现机制

下面几个示例都构建了用于在不同状态下启动系统应用程序的 Intent。在创建 Intent 之后，就可以将 Intent 发送给 startActivity()来启动相应的应用程序了。

1. 浏览器

浏览器应用程序可以用来显示网页或执行 Web 搜索。
要显示网页，可以构建并启动下面的 Intent：

```
Intent pageIntent = new Intent();
pageIntent.setAction(Intent.ACTION_VIEW);
pageIntent.setData(Uri.parse("http://WEB_ADDRESS_TO_VIEW"));

startActivity(pageIntent);
```

用自己想要查看的网页替换 data 字段中的 Uri。要在浏览器中加载 Web 搜索，需要构建和启动下面的 Intent：

```
Intent searchIntent = new Intent();
searchIntent.setAction(Intent.ACTION_WEB_SEARCH);
searchIntent.putExtra(SearchManager.QUERY, STRING_TO_SEARCH);

startActivity(searchIntent);
```

用自己想要执行的搜索词替换掉 Intent 中的附加信息数据。

2. 拨号器

拨号器应用程序用于拨打特定号码，加载拨号器的 Intent 如下：

```
Intent dialIntent = new Intent();
dialIntent.setAction(Intent.ACTION_DIAL);
dialIntent.setData(Uri.Parse("tel:8885551234"));

startActivity(dialIntent);
```

将数据中的 Uri 直接替换成要拨打的电话号码。

注意：
这个动作只显示拨号器，并不会真正将这个号码拨出去。如果要直接拨打电话，需要用 Intent.ACTION_CALL。但在绝大多数情况下，Google 并不鼓励这么做。使用 Intent.ACTION_CALL 还需要在应用程序的清单文件中声明 android.permission.CALL_PHONE 权限。

3. 地图

设备上的地图应用程序可以用于显示所处位置或是提供两点间的路径。如果知道要显示位置的经纬度，可以使用下面的 Intent：

```
Intent mapIntent = new Intent();
mapIntent.setAction(Intent.ACTION_VIEW);
mapIntent.setData(Uri.parse("geo:latitude,longitude"));

startActivity(mapIntent);
```

设置当前位置的经纬度坐标。例如，Uri "geo:37.422,-122.084" 会在地图上显示 Google 总部的位置。如果知道要显示的位置地址，可以使用下面的 Intent：

```
Intent mapIntent = new Intent();
mapIntent.setAction(Intent.ACTION_VIEW);
mapIntent.setData(Uri.parse("geo:0,0?q=ADDRESS"));

startActivity(mapIntent);
```

这会在地图上加入要显示位置的地址。例如 Uri "geo:0,0?q=1600 Amphitheatre Parkway, Mountain View, CA 94043" 会在地图上显示 Google 总部的地址。

提示：
地图应用程序中的 Uri 可以用 "+" 号代替地址查询中的空格。如果在编码带空格的字符串时遇到问题，可以尝试用 "+" 代替。

如果要显示两个位置间的路径，可以使用下面的 Intent：

```
Intent mapIntent = new Intent();
mapIntent.setAction(Intent.ACTION_VIEW);
mapIntent.setData(Uri.parse("http://maps.google.com/maps?saddr=lat,
    lng&daddr=lat,lng"));

startActivity(mapIntent);
```

这会在地图上加入起点和终点的位置。

也可以仅用一个地址参数(既是起点地址，也是终点地址)打开地图应用程序，例如 "Uri：http://maps.google.com/maps?&daddr=37.422,-122.084" 会打开地图应用程序，显示终点位置，让用户自行输入起点地址。

4. 电子邮件

用下面的 Intent 可以按照撰写模式启动设备上的各种电子邮件应用程序：

```
Intent mailIntent = new Intent();
mailIntent.setAction(Intent.ACTION_SEND);
mailIntent.setType("message/rfc822");
mailIntent.putExtra(Intent.EXTRA_EMAIL, new String[] {"recipient@gmail.com"});
mailIntent.putExtra(Intent.EXTRA_CC, new String[] {"carbon@gmail.com"});
mailIntent.putExtra(Intent.EXTRA_BCC, new String[] {"blind@gmail.com"});
mailIntent.putExtra(Intent.EXTRA_SUBJECT, "Email Subject");
mailIntent.putExtra(Intent.EXTRA_TEXT, "Body Text");
```

```
mailIntent.putExtra(Intent.EXTRA_STREAM, URI_TO_FILE);

startActivity(mailIntent);
```

在此场景中,如果是空白电子邮件,就只需要设置 action 和 type 字段。其他信息就是预填充电子邮件的各个组成部分了。注意,就算此处只有一个收件人,EXTRA_EMAIL(收件人)、EXTRA_CC(抄送)和 EXTRA_BCC(密件抄送)也都必须是字符串数组。文件附件则是用 Intent 的 EXTRA_STREAM 设置的。传递给这个参数的值应该是 Uri,指向要发送的本地文件。

如果需要给电子邮件添加多个附件,就需要稍作改动,如下所示:

```
Intent mailIntent = new Intent();
mailIntent.setAction(Intent.ACTION_SEND_MULTIPLE);
mailIntent.setType("message/rfc822");
mailIntent.putExtra(Intent.EXTRA_EMAIL, new String[] {"recipient@gmail.com"});
mailIntent.putExtra(Intent.EXTRA_CC, new String[] {"carbon@gmail.com"});
mailIntent.putExtra(Intent.EXTRA_BCC, new String[] {"blind@gmail.com"});
mailIntent.putExtra(Intent.EXTRA_SUBJECT, "Email Subject");
mailIntent.putExtra(Intent.EXTRA_TEXT, "Body Text");

ArrayList<Uri> files = new ArrayList<Uri>();
files.add(URI_TO_FIRST_FILE);
files.add(URI_TO_SECOND_FILE);
//……根据具体情况重复执行 add()以便添加需要的所有附件
mailIntent.putParcelableArrayListExtra(Intent.EXTRA_STREAM, files);

startActivity(mailIntent);
```

注意,现在 Intent 的动作字符串是 ACTION_SEND_MULTIPLE。其中所有主要的字段跟前面的代码都是一样的,只有附件是以 EXTRA_STREAM 添加的。这个示例会创建一个指向要添加文件的 Uri 元素列表,然后用 putParcelableArrayListExtra()将其添加到 Intent 中。

用户的设备上经常会有多个应用程序可以处理这种内容,所以通常在将上述 Intent 传递给 startActivity()之前会用 Intent.createChooser()处理。

5. SMS(短信)

应用程序可以调用短信应用程序来编写新的 SMS(短信),相关的 Intent 如下所示:

```
Intent smsIntent = new Intent();
smsIntent.setAction(Intent.ACTION_VIEW);
smsIntent.setType("vnd.android-dir/mms-sms");
smsIntent.putExtra("address", "8885551234");
smsIntent.putExtra("sms_body", "Body Text");

startActivity(smsIntent);
```

与编写电子邮件一样，至少要设置 action 和 type 字段才能启动带有空白消息的短信应用程序。address 和 sms_body 附加信息则用于允许应用程序设置短信的接收人(address)和短信正文(sms_body)。

注意，在 Android 框架中，这些关键字并没有完全确定。这意味着在将来的 Android 框架中，这些关键字可能会发生变化。不过在撰写本书时，这些关键字在所有版本的 Android 中都能正常使用。

6. 联系人选择器

应用程序可以启动默认的联系人选择器，让用户从联系人数据库中选择联系人，相关的 Intent 如下所示：

```
static final int REQUEST_PICK = 100;

Intent pickIntent = new Intent();
pickIntent.setAction(Intent.ACTION_PICK);
pickIntent.setData(URI_TO_CONTACT_TABLE);

startActivityForResult(pickIntent, REQUEST_PICK);
```

这个 Activity 用于返回代表用户选项的 Uri，因此需要使用 startActivityForResult()启动它。

6. Google Play

可以在应用程序中启动 Google Play 以显示某个应用程序的具体信息，或是搜索某个关键字。加载某个应用程序的页面的 Intent 如下所示：

```
Intent marketIntent = new Intent();
marketIntent.setAction(Intent.ACTION_VIEW);
marketIntent.setData(Uri.parse("market://details?id=PACKAGE_NAME_HERE"));

startActivity(marketIntent);
```

将要显示的应用程序的包名(例如"com.adobe.reader")插入即可。如果要打开 Google Play 进行搜索，使用下面的 Intent：

```
Intent marketIntent = new Intent();
marketIntent.setAction(Intent.ACTION_VIEW);
marketIntent.setData(Uri.parse("market://search?q=SEARCH_QUERY"));

startActivity(marketIntent);
```

这会插入要搜索的查询字符串。搜索查询有以下 3 种形式：

- q=<简单的文本字符串>：在市场中做关键字搜索。
- q=pname:<包名>：搜索包的名称，只会返回完全匹配的结果。
- q=pub:<开发人员姓名>：搜索开发人员的姓名字段，只会返回完全匹配的结果。

6.8 让其他应用程序启动你的应用程序

6.8.1 问题

你的应用程序能够很好地完成某项工作,所以想提供一个接口,让设备上的其他应用程序也可以运行你的应用程序。

6.8.2 解决方案

(API Level 1)

在应用程序的 Activity 或服务上创建一个 IntentFilter,然后在文档中说明访问该 Activity 或服务所需的 Intent 中的动作、数据类型和附加信息。前面提及,Intent 的 action、category、data/type 用来判断是否匹配应用程序的需求。其他必需的信息或是可选参数都应该作为 Intent 的附加信息传递。

6.8.3 实现机制

假设你已经创建了一个应用程序,该应用程序包含一个 Activity,它会在播放视频的同时在屏幕的顶端显示视频标题的跑马灯。你想让其他的应用程序通过传递必要的数据调用这个应用程序来播放视频,这就需要为应用程序传入所需的数据以定义一个有用的 Intent 结构,然后在应用程序的清单文件中给这个 Activity 加上一个 IntentFilter 来判断是否匹配。

这个假想的 Activity 需要如下两部分数据来完成工作:
- 视频文件的 Uri,可以是本地视频或远程视频。
- 视频标题的字符串。

如果应用程序只能播放某种视频类型,我们可以定义一个通用的动作(如 ACTION_VIEW),然后在 Data 类型中限定为应用程序所能处理的视频类型。代码清单 6-28 演示了如何在清单文件中定义 Activity,以便以这种方式过滤 Intent。

代码清单 6-28　AndroidManifest.xml 中带数据类型过滤器的<activity>元素

```
<activity android:name=".PlayerActivity">
    <intent-filter>
        <action android:name="android.intent.action.VIEW" />
        <category android:name="android.intent.category.DEFAULT" />
        <data android:mimeType="video/h264" />
    </intent-filter>
</activity>
```

这个过滤器会匹配所有声明为 H.264 格式或根据 Uri 文件判断为 H.264 格式的视频片段。外部应用程序调用该 Activity 播放视频的代码如下:

```
Uri videoFile = A_URI_OF_VIDEO_CONTENT;
Intent playIntent = new Intent(Intent.ACTION_VIEW);
```

```
playIntent.setDataAndType(videoFile, "video/h264");
playIntent.putExtra(Intent.EXTRA_TITLE, "My Video");
startActivity(playIntent);
```

在某些情况下，无论要传入的视频是什么格式，外部应用程序都可以直接调用该播放器，这样用起来会更方便些。在此，我们可以给 Intent 定义一个自己的动作字符串，让清单文件中关联 Activity 的过滤器只接受这个自定义的动作字符串。参见代码清单 6-29。

代码清单 6-29　AndroidManifest.xml 中带自定义动作过滤器的 \<activity\>元素

```xml
<activity android:name=".PlayerActivity">
    <intent-filter>
        <action android:name="com.examples.myplayer.PLAY" />
        <category android:name="android.intent.category.DEFAULT" />
    </intent-filter>
</activity>
```

外部应用程序可以通过下面的代码调用这个 Activity 来播放视频：

```java
Uri videoFile = A_URI_OF_VIDEO_CONTENT;
Intent playIntent = new Intent("com.examples.myplayer.PLAY");
playIntent.setData(videoFile);
playIntent.putExtra(Intent.EXTRA_TITLE, "My Video");
startActivity(playIntent);
```

处理一次成功的启动

无论 Intent 是如何匹配 Activity 的，只要 Activity 启动，都需要检查收到的 Intent 中的 Activity 完成预期工作所需的两份数据，参见代码清单 6-30。

代码清单 6-30　检查 Intent 的 Activity

```java
public class PlayerActivity extends Activity {

    public static final String ACTION_PLAY = "com.examples.myplayer.PLAY";

    @Override
    public void onCreate(Bundle savedInstanceState) {
        super.onCreate(savedInstanceState);
        setContentView(R.layout.main);

        //检查启动 Activity 的 Intent
        Intent incoming = getIntent();
        //从 data 字段中获取视频 URI
        Uri videoUri = incoming.getData();
        //获取可选的视频标题字符串(如果存在的话)
        String title;
        if(incoming.hasExtra(Intent.EXTRA_TITLE)) {
            title = incoming.getStringExtra(Intent.EXTRA_TITLE);
        } else {
            title = "";
```

```
                }
                /* 开始播放视频和显示标题*/
        }
        /* 其他 Activity 代码 */
}
```

Activity 启动后,调用 Activity.getIntent()可以得到主调 Intent。视频内容的 Uri 是通过 Intent 的 data 字段传递的,可以通过 Intent.getData()获得该 Uri。主调 Intent 中的视频标题是可选字段,我们会检查附加信息包,看看是否传入了这个字段;如果传入,就从 Intent 中得到这个值。

注意,这个示例中的 PlayerActivity 自定义了一个动作字符串常量,但上面使用 Intent 启动 Activity 的过程中并没有引用这个常量。这是因为外部应用程序无法访问我们在应用程序中定义的公共常量。

正因为如此,要尽量重用 SDK 中已有的 Intent 附加信息键,而不要定义新的常量。在这个示例中,我们选择标准的 Intent.EXTRA_TITLE 来定义可选的附加信息,而非自己为这个值新建一个键。

6.9 与联系人交互

6.9.1 问题

应用程序需要使用 Android 提供的 ContentProvider,对数据库中的用户联系人信息直接进行添加、查看、修改或删除操作。

6.9.2 解决方案

(API Level 5)

用 ContactsContract 提供的接口来访问数据。ContactsContract 包含一系列的 ContentProvider API,用于将系统中多个用户账户的联系人信息整合到一个数据库中,生成大量可以访问和修改其中数据的 Uri、表和列。

Contact 采用了分层结构,分为三层:Contacts、RawContacts 和 Data:

- 一个 Contact 在概念上代表一个人,是 Android 认定为同一个人的所有 RawContacts 的集合。
- RawContacts 表示设备中某个类型的账号数据集。这个账号可以是用户的电子邮件地址簿或 Facebook 账号。
- Data 元素是 RawContacts 中的一条信息,例如电子邮件地址、电话号码、邮寄地址等。

完整的 API 有太多的组合和选项,受篇幅限制,这里不可能一一详述,请参阅 SDK 文档。接下来介绍如何实现联系人数据集的基本查询和简单修改。

6.9.3 实现机制

Android Contacts API 面对的是一个由多个表和各种连接(join)组成的复杂数据库,但在应用程序中访问这些数据的方法与访问其他 SQLite 数据库的方法并没有本质的区别。

1. 列举/查看联系人信息

让我们看一个示例,其中的 Activity 会列出通讯录数据库中所有的联系人,在选中某个联系人后会显示其详细信息,参见代码清单 6-31。

要点:

要在应用程序中显示来自 Contacts API 的信息,需要在应用程序的清单文件中声明 android.permission.READ_CONTACTS 权限。

代码清单 6-31 显示联系人信息的 Activity

```java
public class ContactsActivity extends FragmentActivity {

    private static final int ROOT_ID = 100;

    @Override
    protected void onCreate(Bundle savedInstanceState) {
        super.onCreate(savedInstanceState);
        FrameLayout rootView = new FrameLayout(this);
        rootView.setId(ROOT_ID);
        setContentView(rootView);

        // 创建并添加新的列表 Fragment
        getSupportFragmentManager().beginTransaction()
            .add(ROOT_ID, ContactsFragment.newInstance())
            .commit();
    }

    public static class ContactsFragment extends ListFragment
            implements AdapterView.OnItemClickListener,
            LoaderManager.LoaderCallbacks<Cursor> {

        public static ContactsFragment newInstance() {
            return new ContactsFragment();
        }

        private SimpleCursorAdapter mAdapter;

        @Override
        public void onActivityCreated(Bundle savedInstanceState) {
            super.onActivityCreated(savedInstanceState);
```

```java
        // 在ListView中显示所有联系人
        mAdapter = new SimpleCursorAdapter(getActivity(),
                android.R.layout.simple_list_item_1, null,
                new String[] { ContactsContract.Contacts.DISPLAY_NAME },
                new int[] { android.R.id.text1 },
                0);
        setListAdapter(mAdapter);
        // 侦听选择的项
        getListView().setOnItemClickListener(this);

        getLoaderManager().initLoader(0, null, this);
    }

    @Override
    public Loader<Cursor> onCreateLoader(int id, Bundle args) {
        // 返回按姓名排序的所有联系人
        String[] projection = new String[] {
                ContactsContract.Contacts._ID,
                ContactsContract.Contacts.DISPLAY_NAME
        };

        return new CursorLoader(getActivity(),
                ContactsContract.Contacts.CONTENT_URI,
                projection, null, null,
                ContactsContract.Contacts.DISPLAY_NAME);
    }

    @Override
    public void onLoadFinished(Loader<Cursor> loader, Cursor data) {
        mAdapter.swapCursor(data);
    }

    @Override
    public void onLoaderReset(Loader<Cursor> loader) {
        mAdapter.swapCursor(null);
    }

    @Override
    public void onItemClick(AdapterView<?> parent, View v,
            int position, long id) {
        final Cursor contacts = mAdapter.getCursor();
        if(contacts.moveToPosition(position)) {
            int selectedId = contacts.getInt(0); // _ID column
            // 从email表中获取电子邮件数据
            Cursor email = getActivity().getContentResolver()
                .query(ContactsContract.CommonDataKinds.Email.CONTENT_URI,
                        new String[] {ContactsContract.
                    CommonDataKinds.Email.DATA},
                        ContactsContract.Data.CONTACT_ID
                            + " = " + selectedId, null, null);
```

```java
    // 从phone表中获取手机号
    Cursor phone = getActivity().getContentResolver()
        .query(ContactsContract.CommonDataKinds.Phone.CONTENT_URI,
                new String[] {ContactsContract.
            CommonDataKinds.Phone.NUMBER},
                    ContactsContract.Data.CONTACT_ID
                            + " = " + selectedId, null, null);
    // 从address表中获取邮寄地址
    Cursor address = getActivity().getContentResolver()
        .query(ContactsContract.CommonDataKinds.StructuredPostal.
                CONTENT_URI,new String[]
    {ContactsContract.CommonDataKinds.StructuredPostal.FORMATTED
_ADDRESS},ContactsContract.Data.CONTACT_ID
                            + " = " + selectedId, null, null);

    // 构建对话框消息
    StringBuilder sb = new StringBuilder();
    sb.append(email.getCount() + " Emails\n");
    if(email.moveToFirst()) {
        do {
            sb.append("Email: " + email.getString(0));
            sb.append('\n');
        } while(email.moveToNext());
        sb.append('\n');
    }
    sb.append(phone.getCount() + " Phone Numbers\n");
    if(phone.moveToFirst()) {
        do {
            sb.append("Phone: " + phone.getString(0));
            sb.append('\n');
        } while(phone.moveToNext());
        sb.append('\n');
    }
    sb.append(address.getCount() + " Addresses\n");
    if(address.moveToFirst()) {
        do {
            sb.append("Address:\n"
                    + address.getString(0));
        } while(address.moveToNext());
        sb.append('\n');
    }

    AlertDialog.Builder builder = new AlertDialog.Builder(getActivity());
    builder.setTitle(contacts.getString(1)); // 显示姓名
    builder.setMessage(sb.toString());
    builder.setPositiveButton("OK", null);
    builder.create().show();

    // 关闭临时光标
```

```
                email.close();
                phone.close();
                address.close();
            }
        }
    }
}
```

从以上代码可以看出，在这些 API 中引用表和字段的代码非常烦琐。在这个示例中，所有的 Uri 元素、表和字段的引用都是 ContactsContract 的内部类。在与 Contacts API 交互时，一定要注意检查是否使用了正确的类，所有不是来自 ContactsContract 的 Contacts 类都不宜使用且不兼容。

在创建包含 Activity 用户界面的 Fragment 时，我们通过引用 Contacts.CONTENT_URI 的 CursorLoader 简单查询核心的 Contacts 表，只请求将 Cursor 封装到 ListAdapter 中所需的列。得到的 Cursor 以列表的形式显示在用户界面上。这个示例利用 ListFragment 的便利行为，用 ListView 作为内容视图，这样我们就不必管理这些组件了。

此刻用户可能会上下滑动设备上的联系人条目，然后点击其中某个联系人以查看其详细信息。在用户选中某个联系人后，该联系人的 ID 值会被记录下来，应用程序会根据这个值到另一个 ContactsContract.Data 表中获取更详细的信息。要注意的是，一个联系人的数据是分布在多个表中的(电子邮件在 email 表中，电话号码在 phone 表中，等等)，所以需要多次查询才能获得完整信息。

每个 CommonDataKinds 表都有唯一的 CONTENT_URI 用于在查询时引用该表，同时还有一组字段别名用于请求数据。这些数据表中的每一行数据都通过 Data.CONTACT_ID 与一个联系人关联，所以每个 Cursor 只返回与这个值匹配的行。

在收集用户选中的联系人的所有数据后，我们遍历结果以将其显示在对话框中供用户查看。因为这些表中的数据是多个数据来源的集合，所以所有这些查询返回多个结果的情况并不少见。对于每个 Cursor，会显示返回结果的数量，然后附上其所包含的每个值。在所有数据都组合好之后，就会创建一个对话框将这些信息显示给用户。

最后及时关闭所有的临时 Cursor 和非托管 Cursor。

2. 运行应用程序

在设置有多个账户的设备上运行这个应用程序时，你会发现所显示的列表过长，其长度远超过设备上预装的 Contacts 应用程序中联系人列表的长度。Contacts API 可以保存可能被用户隐藏或是内部使用的分组内容。例如，Gmail 经常会将收到的电子邮件的地址保存下来，即使这个电子邮件地址跟任何实际联系人都没有关联也会保存，这是为了以后使用方便。

在下一个示例中，我们将演示如何过滤这个列表，但目前的 Contacts 表中存放的数据实在太多了。

3. 修改/添加联系人

下面查看一个示例 Activity，操作具体联系人的数据，参见代码清单 6-32。

要点：

要在应用程序中使用 Contacts API，需要在应用程序的清单文件中声明 android.permission.READ_CONTACTS 和 android.permission.WRITE_CONTACTS 权限。

代码清单 6-32　写入 Contacts API 的 Activity

```java
public class ContactsEditActivity extends FragmentActivity {

    private static final String TEST_EMAIL = "tester@email.com";
    private static final int ROOT_ID = 100;

    @Override
    protected void onCreate(Bundle savedInstanceState) {
        super.onCreate(savedInstanceState);
        FrameLayout rootView = new FrameLayout(this);
        rootView.setId(ROOT_ID);

        setContentView(rootView);

        // 创建并添加新的列表 Fragment
        getSupportFragmentManager().beginTransaction()
            .add(ROOT_ID, ContactsEditFragment.newInstance())
            .commit();
    }

    public static class ContactsEditFragment extends ListFragment implements
            AdapterView.OnItemClickListener,
            DialogInterface.OnClickListener,
            LoaderManager.LoaderCallbacks<Cursor> {

        public static ContactsEditFragment newInstance() {
            return new ContactsEditFragment();
        }

        private SimpleCursorAdapter mAdapter;
        private Cursor mEmail;
        private int selectedContactId;

        @Override
        public void onActivityCreated(Bundle savedInstanceState) {
            super.onActivityCreated(savedInstanceState);

            // 在 ListView 中显示所有联系人
            mAdapter = new SimpleCursorAdapter(getActivity(),
                    android.R.layout.simple_list_item_1, null,
                    new String[] { ContactsContract.Contacts.DISPLAY_NAME },
                    new int[] { android.R.id.text1 },
                    0);
            setListAdapter(mAdapter);
            // 侦听选择的项
```

```java
        getListView().setOnItemClickListener(this);

        getLoaderManager().initLoader(0, null, this);
    }

    @Override
    public Loader<Cursor> onCreateLoader(int id, Bundle args) {
        // 返回按姓名排序的所有联系人
        String[] projection = new String[] { ContactsContract.Contacts._ID,
                ContactsContract.Contacts.DISPLAY_NAME };
        // 仅列出对用户可见的联系人
        return new CursorLoader(getActivity(),
                ContactsContract.Contacts.CONTENT_URI,
                 projection, ContactsContract.Contacts.IN_VISIBLE_
                 GROUP+" = 1",null,
                ContactsContract.Contacts.DISPLAY_NAME);
    }

    @Override
    public void onLoadFinished(Loader<Cursor> loader, Cursor data) {
        mAdapter.swapCursor(data);
    }

    @Override
    public void onLoaderReset(Loader<Cursor> loader) {
        mAdapter.swapCursor(null);
    }

    @Override
    public void onItemClick(AdapterView<?> parent, View v, int position,
       long id) {
        final Cursor contacts = mAdapter.getCursor();
        if(contacts.moveToPosition(position)) {
            selectedContactId = contacts.getInt(0); // _ID 字段
            // 从 email 表中获取电子邮件数据
            String[] projection = new String[] {
                    ContactsContract.Data._ID,
                    ContactsContract.CommonDataKinds.Email.DATA };
            mEmail = getActivity().getContentResolver().query(
                    ContactsContract.CommonDataKinds.Email.CONTENT_URI,
                    projection,
                    ContactsContract.Data.CONTACT_ID + " = " +
                     selectedContactId,
                    null,
                    null);

            AlertDialog.Builder builder = new AlertDialog.Builder(
              getActivity());
            builder.setTitle("Email Addresses");
            builder.setCursor(mEmail, this, ContactsContract.
```

```java
                    CommonDataKinds.Email.DATA);
            builder.setPositiveButton("Add", this);
            builder.setNegativeButton("Cancel", null);
            builder.create().show();
    }
}

@Override
public void onClick(DialogInterface dialog, int which) {
    // 数据只能与一个RAW联系人关联,获取第一个匹配的行id
    Cursor raw = getActivity().getContentResolver().query(
            ContactsContract.RawContacts.CONTENT_URI,
            new String[] { ContactsContract.Contacts._ID },
            ContactsContract.Data.CONTACT_ID + " = " +
        selectedContactId, null, null);
    if(!raw.moveToFirst()) {
        return;
    }

    int rawContactId = raw.getInt(0);
    ContentValues values = new ContentValues();
    switch(which) {
        case DialogInterface.BUTTON_POSITIVE:
            // 用户想要添加一个新的电子邮件地址
            values.put(ContactsContract.CommonDataKinds.Email.RAW_
                CONTACT_ID, rawContactId);
            values.put(ContactsContract.Data.MIMETYPE, ContactsContract
                    .CommonDataKinds.Email.CONTENT_ITEM_TYPE);
            values.put(ContactsContract.CommonDataKinds.Email.DATA,
                    TEST_EMAIL);
            values.put(ContactsContract.CommonDataKinds.Email.TYPE,
                    ContactsContract.CommonDataKinds.Email.TYPE_OTHER);
            getActivity().getContentResolver()
                    .insert(ContactsContract.Data.CONTENT_URI, values);
            break;
        default:
            // 用户想要编辑选择项
            values.put(ContactsContract.CommonDataKinds.Email.DATA,
                TEST_EMAIL);
            values.put(ContactsContract.CommonDataKinds.Email.TYPE,
                    ContactsContract.CommonDataKinds.Email.TYPE_OTHER);
            getActivity().getContentResolver()
                    .update(ContactsContract.Data.CONTENT_URI, values,
                            ContactsContract.Data._ID+" = 
                            "+mEmail.getInt(0), null);
            break;
    }

    // 不再需要电子邮件Cursor
    mEmail.close();
```

            }
        }
    }

　　这个示例中，我们还是和以前一样，会查询 Contacts 数据库中的所有数据。这次，只有一个选择条件：

```
ContactsContract.Contacts.IN_VISIBLE_GROUP+" = 1"
```

　　这个限制条件是：返回用户在联系人用户界面中可以看到的内容。这会缩短(在个别情况下，会大大缩短)在 Activity 中显示的联系人列表的长度，使其中显示的内容更接近 Contacts 应用程序中显示的联系人列表。

　　当用户从此列表中选中某个联系人时，就会显示一个对话框，列出该联系人的所有电子邮件信息。如果选中列表中的某个地址，就可以进行编辑；如果点击 Add 按钮，就可以添加新的电子邮件地址条目。为了保持示例的简洁性，我们没有提供输入新电子邮件地址的界面，而只是插入一个常量作为新记录或是更新选中的记录。

　　诸如电子邮件地址一类的数据元素只能与一个 RawContact 关联。因此，在添加新的电子邮件地址时，必须获取用户所选的高层 Contacts 的一个 RawContact 的 ID。在这个示例中，我们并不关心到底是哪个 RawContact，所以我们就获取第一个匹配的 RawContact 的 ID。只有在进行插入操作时才需要这个值，因为如果是更新已有的电子邮件记录，那么所需的行 ID 就已经保存在表中了。

　　还要注意，CommonDataKinds 中提供的作为别名以读取数据的 Uri 不能用来修改和更新数据。插入和更新必须使用 ContactsContract.Data Uri 直接调用。这意味着(除了在操作方法中引用不同的 Uri)还必须指定元数据，即 MIMETYPE。如果没有设置所插入数据的 MIMETYPE 字段，接下来的查询就不会将其识别为联系人的电子邮件地址。

4. 聚合实例

　　因为这个示例是用同样的数据添加或编辑电子邮件地址，所以从这个示例中可以实时地看到 Android 的聚合操作。在运行这个示例应用程序时，你可能会注意到，在添加同样的电子邮件地址或将已有的记录修改为与其他电子邮件相同时，Android 会思考现在这个联系人和之前的联系人是不是同一个。在这个示例应用程序中，随着对核心 Contacts 表的托管查询的更新，你会发现有些联系人消失了，因为这些联系人会与其他的相关联系人聚合在一起。

　　注意：
　　在 Android 模拟器中还没有完全实现联系人聚合行为。要看到上述效果，需要在真机上运行这段代码。

5. 维护引用

　　Android 的 Contacts API 引入了一个对某些应用程序很重要的新概念。因为这个聚合过程的发生，表示一个联系人的行 ID 的含义就显得不那么明确了；当一个联系人跟另一个联系人聚合在一起之后，他会获得一个新的_ID。

如果应用程序需要一个指向特定联系人的长时间不变的引用，建议你使用 ContactsContract.Contacts.LOOKUP_KEY，而不要使用行 ID。在使用这个键查询联系人时，还需要一个 ContactsContract.Contacts.CONTENT_LOOKUP_URI 形式的特殊 Uri。长期用这些值查询联系人数据库，可以使应用程序避免自动聚合过程所带来的麻烦。

6.10 读取设备媒体和文档

6.10.1 问题

应用程序需要导入用户选择的文档项(文本文件、音频、视频或图片)来显示或播放。

6.10.2 解决方案

(API Level 1)

使用 Intent.ACTION_GET_CONTENT 这个隐式 Intent 从应用程序唤起选择器界面。发送这个 Intent 时加入感兴趣媒体的匹配内容类型(音频、视频或图片最为常见)就会唤起选择器界面，用户可以通过选择器界面选择其中的一个文件，这个 Intent 的结果会包含一个指向用户所选媒体文件的 Uri。

(API Level 19)

使用一个隐式 Intent，目标是通过 Intent.ACTION_OPEN_DOCUMENT 打开系统的文档选择器界面。在此公共界面中，支持所请求内容类型的所有应用程序将列出用户可能选择的条目。填充此界面的内容来自为所请求类型提供 DocumentProvider 的应用程序。这些提供程序元素可以来自系统或其他应用程序。我们将在本章后面介绍如何创建自己的提供程序元素。

提示：

在 API Level 19+版本中，ACTION_GET_CONTENT 仍然可以使用，它将以兼容模式启动标准文档选择器，该选择器包括较新的集成提供程序以及选择单个应用程序的选择器界面的选项。

6.10.3 实现机制

让我们看一下在一个示例 Activity 中使用这项技术的示例。参见代码清单 6-33 和 6-34。

代码清单 6-33　res/layout/main.xml

```
<?xml version="1.0" encoding="utf-8"?>
    <LinearLayout
        xmlns:android="http://schemas.android.com/apk/res/android"
        android:orientation="vertical"
```

```xml
    android:layout_width="match_parent"
    android:layout_height="match_parent">
    <Button
        android:id="@+id/imageButton"
        android:layout_width="match_parent"
        android:layout_height="wrap_content"
        android:text="Images" />
    <Button
        android:id="@+id/videoButton"
        android:layout_width="match_parent"
        android:layout_height="wrap_content"
        android:text="Video" />
    <Button
        android:id="@+id/audioButton"
        android:layout_width="match_parent"
        android:layout_height="wrap_content"
        android:text="Audio" />
</LinearLayout>
```

代码清单 6-34　使用媒体选择器的 Activity

```java
public class MediaActivity extends Activity implements View.OnClickListener {

    private static final int REQUEST_AUDIO = 1;
    private static final int REQUEST_VIDEO = 2;
    private static final int REQUEST_IMAGE = 3;

    @Override
    public void onCreate(Bundle savedInstanceState) {
        super.onCreate(savedInstanceState);
        setContentView(R.layout.main);

        Button images = (Button)findViewById(R.id.imageButton);
        images.setOnClickListener(this);
        Button videos = (Button)findViewById(R.id.videoButton);
        videos.setOnClickListener(this);
        Button audio = (Button)findViewById(R.id.audioButton);
        audio.setOnClickListener(this);
    }

    @Override
    protected void onActivityResult(int requestCode, int resultCode, Intent
      data) {

        if(resultCode == Activity.RESULT_OK) {
            //返回的 Intent 中包含用户所选文件的 Uri
            Uri selectedContent = data.getData();

            if(requestCode == REQUEST_IMAGE) {
                //向 BitmapFactory 传递 InputStream
            }
```

```java
            if(requestCode == REQUEST_VIDEO) {
                //向MediaPlayer传递Uri或FileDescriptor
            }
            if(requestCode == REQUEST_AUDIO) {
                //向MediaPlayer传递Uri或FileDescriptor
            }
        }
    }

    @Override
    public void onClick(View v) {
        Intent intent = new Intent();
        //使用适当的Intent动作
        if(Build.VERSION.SDK_INT >= Build.VERSION_CODES.KITKAT) {
            intent.setAction(Intent.ACTION_OPEN_DOCUMENT);
        } else {
            intent.setAction(Intent.ACTION_GET_CONTENT);
        }
        //仅返回可以打开流的文件
        intent.addCategory(Intent.CATEGORY_OPENABLE);
        //设置正确的MIME类型并启动
        switch(v.getId()) {
            case R.id.imageButton:
                intent.setType("image/*");
                startActivityForResult(intent, REQUEST_IMAGE);
                return;
            case R.id.videoButton:
                intent.setType("video/*");
                startActivityForResult(intent, REQUEST_VIDEO);
                return;
            case R.id.audioButton:
                intent.setType("audio/*");
                startActivityForResult(intent, REQUEST_AUDIO);
                return;
            default:
                return;
        }
    }
}
```

这个示例中有 3 个用户可以按下的按钮，每个按钮对应一种特定媒体类型。当用户按下任意一个按钮时，就会向系统发送一个带有适当平台级动作的 Intent。在运行 Android 4.4 或以后版本的设备上，这将显示系统文档选择器。以前的设备将从必要的应用程序启动适当的拾取器 Activity，如果多个应用程序可以处理此内容类型，则显示一个选择窗口。我们还为此 Intent 包括 CATEGORY_OPENABLE，这向系统表明选择器中只会显示应用程序可以为其打开流的项。

如果用户选择了一个有效的文件，这时会在返回的结果 Intent 中包含一个指向所选文件的内容 Uri，Intent 的状态为 RESULT_OK。如果用户取消或退出媒体选择器，结果 Intent

的状态将为 RESULT_CANCELED，而相应的 data 字段则为 null。

接收媒体文件的 Uri 后，应用程序可以自由地选择播放或不播放所选媒体内容。MediaPlayer 和 VideoView 这样的类可以直接通过 Uri 播放媒体内容，而其他大多数类则采用 InputStream 或 FileDescriptor 引用。两者分别可以通过 ContentResolver.openInputStream() 和 ContentResolver.openFileDescriptor() 从 Uri 获得。

选择整个目录

在 API Level 21 及更高版本的平台中，用户可以使用 Intent.OPEN_DOCUMENT_TREE 动作请求选择目录条目。

这将在 Intent 结果中返回特殊的 Uri，该 Uri 用于列出目录的内容或在此位置创建新的文档。下面的代码片段将获取目录选择的结果并列出目录的内容。

```java
protected void onActivityResult(int requestCode, int resultCode, Intent data) {
    if(data == null || data.getData() == null) return;
    final Uri result = data.getData();

    //构造用于查询所选目录内容的 Uri
    String subDocumentId = DocumentsContract.getTreeDocumentId(result);
    Uri subTree = DocumentsContract.buildChildDocumentsUriUsingTree(result, subDocumentId);

    //查询目录并列出其内容
    Cursor cursor = getContentResolver().query(subTree, null, null, null, null);
    if(cursor != null) {
        if(cursor.getCount() == 0) {
            //目录为空
        } else {
            StringBuilder sb = new StringBuilder();
            sb.append("Contents of Directory:\n");

            while(cursor.moveToNext()) {
                //获得包含文档名称的列
                int index =
                cursor.getColumnIndex(DocumentsContract.Document.COLUMN_
                    DISPLAY_NAME);
                sb.append(cursor.getString(index));
                sb.append("\n");
            }

            //将文件列表输出到 logcat
            Log.d("DirectoryList", sb.toString());
        }

        cursor.close();
```

```
    } else {
        //读取目录内容出错
    }
}
```

使用 API DocumentContract，我们可以将用户所选目录的 Uri 转换为可用于通过 ContentResolver 查询相关 DocumentsProvider 的 Uri。得到的 Cursor 将包含提供程序具有的关于所选目录中每个文档的所有元数据。

6.11 保存设备媒体和文档

6.11.1 问题

应用程序想要创建新的文档或媒体并把它们插入设备的全局提供程序中，这样所有的应用程序都可以看到它们。

6.11.2 解决方案

(API Level 1)
使用 MediaStore 提供的 ContentProvider 接口可以插入新的媒体内容。除了媒体内容本身，通过这个接口还可以插入用来标记媒体内容的各种元数据，例如标题、描述和创建时间等。ContentProvider 插入操作完成后会返回一个 Uri，应用程序可以通过这个 Uri 访问新的媒体。

(API Level 19)
在装配了 Android 4.4+ 的设备上，我们还可以触发一个隐式 Intent，目标是通过 Intent.ACTION_CREATE_DOCUMENT 在任意设备的已注册 DocumentProvider 实例中保存新文档。这可以是任意类型的文档内容，包括(但不限于)媒体文件。然而，它并不旨在取代 MediaStore，后者仍是将内容直接保存到系统的核心 ContentProvider 中的最佳方法。如果需要让用户更直接地涉及保存内容(包括媒体)，文档框架在此就是更好的路径。

6.11.3 实现机制

让我们看一个向 MediaStore 中插入图片或视频片段的示例。参见代码清单 6-35 和 6-36。

代码清单 6-35　res/layout/main.xml

```xml
<?xml version="1.0" encoding="utf-8"?>
<LinearLayout xmlns:android="http://schemas.android.com/apk/res/android"
    android:layout_width="fill_parent"
    android:layout_height="fill_parent"
    android:orientation="vertical" >
    <Button
```

```xml
            android:id="@+id/imageButton"
            android:layout_width="fill_parent"
            android:layout_height="wrap_content"
            android:text="Images" />
    <Button
            android:id="@+id/videoButton"
            android:layout_width="fill_parent"
            android:layout_height="wrap_content"
            android:text="Video" />
    <Button
            android:id="@+id/textButton"
            android:layout_width="fill_parent"
            android:layout_height="wrap_content"
            android:text="Text Document" />

</LinearLayout>
```

代码清单 6-36　保存数据到 MediaStore 中的 Activity

```java
public class StoreActivity extends Activity implements View.OnClickListener {

    private static final int REQUEST_CAPTURE = 100;
    private static final int REQUEST_DOCUMENT = 101;

    @Override
    public void onCreate(Bundle savedInstanceState) {
        super.onCreate(savedInstanceState);
        setContentView(R.layout.save);

        Button images = (Button) findViewById(R.id.imageButton);
        images.setOnClickListener(this);
        Button videos = (Button) findViewById(R.id.videoButton);
        videos.setOnClickListener(this);
        // 我们仅可以在 API Level 19 以上版本中创建新文档
        Button text = (Button) findViewById(R.id.textButton);
        if(Build.VERSION.SDK_INT >= Build.VERSION_CODES.KITKAT) {
            text.setOnClickListener(this);
        } else {
            text.setVisibility(View.GONE);
        }
    }

    @Override
    protected void onActivityResult(int requestCode, int resultCode, Intent data) {
        if(requestCode == REQUEST_CAPTURE && resultCode ==
                Activity.RESULT_OK) {
            Toast.makeText(this, "All Done!", Toast.LENGTH_SHORT).show();
        }
        if(requestCode == REQUEST_DOCUMENT && resultCode ==
                Activity.RESULT_OK) {
```

```java
            // 一旦用户选择在何处保存新文档,
            // 就可以在其中写入内容
            Uri document = data.getData();
            writeDocument(document);
        }
    }

    private void writeDocument(Uri document) {
        try {
            ParcelFileDescriptor pfd =
                getContentResolver().openFileDescriptor(document, "w");
            FileOutputStream out = new
                FileOutputStream(pfd.getFileDescriptor());
            // 为文件构造一些内容
            StringBuilder sb = new StringBuilder();
            sb.append("Android Recipes Log File:");
            sb.append("\n");
            sb.append("Last Written at: ");
            sb.append(DateFormat.getLongDateFormat(this).format(new
                Date()));

            out.write(sb.toString().getBytes());

            // 通过结束流让文档提供程序知道在何处完成
            out.flush();
            out.close();
            // 关闭文件句柄
            pfd.close();
        } catch(FileNotFoundException e) {
            Log.w("AndroidRecipes", e);
        } catch(IOException e) {
            Log.w("AndroidRecipes", e);
        }
    }

    @Override
    public void onClick(View v) {
        ContentValues values;
        Intent intent;
        Uri storeLocation;
        final long nowMillis = System.currentTimeMillis();

        switch(v.getId()) {
        case R.id.imageButton:
            // 创建图片的元数据
            values = new ContentValues(5);
            values.put(MediaStore.Images.ImageColumns.DATE_TAKEN,
                nowMillis);
            values.put(MediaStore.Images.ImageColumns.DATE_ADDED,
                nowMillis / 1000);
```

```java
        values.put(MediaStore.Images.ImageColumns.DATE_MODIFIED,
            nowMillis / 1000);
        values.put(MediaStore.Images.ImageColumns.DISPLAY_NAME,
            "Android Recipes Image Sample");
        values.put(MediaStore.Images.ImageColumns.TITLE, "Android
            Recipes Image Sample");

        // 插入元数据并检索文件的 Uri 位置
        storeLocation = getContentResolver()
            .insert(MediaStore.Images.Media.EXTERNAL_CONTENT_URI, values);
        // 用获取的 Uri 作为媒体的存放目标
        intent = new Intent(MediaStore.ACTION_IMAGE_CAPTURE);
        intent.putExtra(MediaStore.EXTRA_OUTPUT, storeLocation);
        startActivityForResult(intent, REQUEST_CAPTURE);
        return;
case R.id.videoButton:
        // 创建视频的元数据
        values = new ContentValues(7);
        values.put(MediaStore.Video.VideoColumns.DATE_TAKEN,
            nowMillis);
        values.put(MediaStore.Video.VideoColumns.DATE_ADDED,
            nowMillis / 1000);
        values.put(MediaStore.Video.VideoColumns.DATE_MODIFIED,
            nowMillis / 1000);
        values.put(MediaStore.Video.VideoColumns.DISPLAY_NAME,
            "Android Recipes Video Sample");
        values.put(MediaStore.Video.VideoColumns.TITLE, "Android
            Recipes Video Sample");
        values.put(MediaStore.Video.VideoColumns.ARTIST, "Yours
            Truly");
        values.put(MediaStore.Video.VideoColumns.DESCRIPTION,
            "Sample Video Clip");

        // 插入元数据并检索文件的 Uri 位置
        storeLocation = getContentResolver()
            .insert(MediaStore.Video.Media.EXTERNAL_CONTENT_URI, values);
        // 用获取的 Uri 作为媒体的存放目标
        intent = new Intent(MediaStore.ACTION_VIDEO_CAPTURE);
        intent.putExtra(MediaStore.EXTRA_OUTPUT, storeLocation);
        startActivityForResult(intent, REQUEST_CAPTURE);
        return;
case R.id.textButton:
        // 创建新的文档
        intent = new Intent(Intent.ACTION_CREATE_DOCUMENT);
        intent.addCategory(Intent.CATEGORY_OPENABLE);

        // 这是一个文本文档
        intent.setType("text/plain");
```

```
                // 在文档上预设的可选标题
                intent.putExtra(Intent.EXTRA_TITLE, "Android Recipes");
                startActivityForResult(intent, REQUEST_DOCUMENT);
            default:
                return;
            }
        }
    }
```

注意：

因为本例用到了摄像头硬件，所以应该在真实设备上运行这个示例，这样才能看到完全的效果。模拟器将适当执行代码，但如果没有硬件支持的话，这个示例就失去意义了。

在这个示例中，当用户点击图片或视频按钮时，媒体文件相关的元数据都会被插入到一个 ContentValues 实例中。有些元数据字段是图片文件和视频文件所共有的，如下所示：

- TITLE：表示内容标题的字符串。作为内容名称显示在 Gallery 应用程序中。
- DISPLAY_NAME：在大多数选择界面中显示的名称，如系统文档选择器。
- DATE_TAKEN：描述媒体文件创建时间的整数值。注意此值以毫秒为单位。
- DATE_ADDED：描述将媒体文件添加到 MediaStore 的整数值。注意此值以秒而不是毫秒为单位。
- DATE_MODIFIED：描述上次更改媒体文件的整数值。该值用于在系统文档选择器中排序项。注意该值也以秒为单位。

接下来，通过相应的 CONTENT_URI 引用就可以将 ContentValues 插入到 MediaStore 中。注意，要在真正采集媒体文件之前插入元数据。插入成功后会返回一个完全限定的 Uri，然后应用程序就可以将此 Uri 作为媒体内容的存放目标。

在前面的示例中，我们使用了第 4 章中关于录音和摄像的简单方法，即要求系统应用程序来处理这个过程。回忆一下第 4 章的内容，录音和摄像的 Intent 都可以传入标识文件保存位置的附加信息。这里我们传入的就是插入成功后返回的 Uri。

当录音和摄像的 Activity 成功返回后，这个应用程序就结束了。外部应用程序将采集到的图片和视频保存到 MediaStore 引用的位置。现在所有的应用程序都可以看到这些数据，包括系统的 Gallery 应用程序。

创建文档

注意第三个按钮 Text Document，只有在 Android 4.4 或以后版本的设备上运行时，该按钮才可见并启用。如果用户点击此按钮，就使用 ACTION_CREATE_DOCUMENT 构造一个 Intent 请求，以启动系统的文档界面。然而，在此会启动界面，允许用户选择应将新文件保存在何处（即哪个提供程序）以及标题应是什么。连同该请求一起，我们设置 MIME 类型以指示想要创建的文档类型，在此例中是纯文本。最后，我们可以通过随 Intent 一起传入 EXTRA_TITLE 来建议标题，但用户始终有权在后面更改此标题。

一旦用户选择在何处保存新文档，就会在 onActivityResult()中提供一个内容 Uri，我们可以打开流，将文档的数据写入存储空间。该例中的 writeDocument()方法从该 Uri 打开

FileDescriptor，并将一些基本文本内容写入新的文档。关闭该流和描述符就会告诉作为所有者的提供程序，文档更新已完成。

> 提示：
> ACTION_CREATE_DOCUMENT 用于建立想要保存的新文档。对于编辑已经就绪的现有文档，在之前的示例中使用 ACTION_CREATE_DOCUMENT 就会获得指向现有文件的工作 Uri。然而请注意，并不是所有提供程序都支持写入数据。需要检查所提供的 Uri 的权限，之后才可以尝试编辑以此方式收到的文档。

6.12 读取消息数据

6.12.1 问题

需要查询 ContentProvider 或设备上本地保存的信息，查找发送或接收的 SMS/MMS 短信。

6.12.2 解决方案

(API Level 19)

使用通过 Telephony 框架提供的契约接口。Telephony 的内部类定义了用于读取 SMS 短信、MMS 短信和附加元数据的所有 Uri 和数据字段。

要点：
必须在清单文件中请求 android.permission.READ_SMS 才可以获得 Telephony 提供程序的读取权限。

Telephony 提供程序为如下数据块提供了接口：
- Telephony.Sms：包含所有 SMS 短信的消息内容和接收人/投递元数据。
- Telephony.Mms：包含所有 MMS 短信的消息内容和接收人/投递元数据。
- Telephony.MmsSms：包含 SMS 和 MMS 的组合消息。还包括自定义的 Uri，用于请求对话、草稿的列表以及搜索消息。
- Telephony.Threads：提供关于对话的额外元数据，例如对话线程的消息计数和读取状态。

基于文本的 SMS 短信相对直观，它们的整个内容驻留在 Telephony.Sms 表的几个字段内。如果短信有多个接收人，这些短信会分解为包含相同文本内容的多条消息。而 MMS 短信则由多个部分组成，这些部分单独存储在各自的表中：
- Mms.Addr：包含关于每条 MMS 短信所涉及的所有接收人的元数据。每条短信可以有唯一的一组收件人。
- Mms.Part：包含 MMS 短信中每部分的内容。短信文本作为一个部分存储，而图片、视频或其他附件作为附加的部分存储。MIME 字符串指定每个部分的内容类型。

可以通过查询 Telephony.Sms 内容 Uri 来显示单条 SMS 短信。然而，显示单条 MMS 短信需要在 Telephony.Mms 中遍历所有这些子部分以收集我们需要的数据。

> **提示：**
> SMS/MMS 短信数据仅存在于有手机硬件的设备上，因此你可能要在清单文件中添加 <uses-feature android:name="android.hardware.telephony">声明，以便过滤掉没有适当功能的设备。

写入 Telephony 提供程序

将消息数据和元数据写入 Telephony 提供程序有一些相关的特殊规则。虽然任何应用程序都可以请求 WRITE_SMS 权限(并获得授权)，但只有用户在 Settings 部分的 Default Messaging Application 中选择的应用程序才可以将数据写入内容提供程序。

使用第 3 章介绍的描述机制发送 SMS 短信的非默认应用程序会由框架将短信内容自动写入提供程序，而默认应用程序(作为提供程序可以访问的单独应用程序)将负责直接写入它自己的内容。

为了将某个消息应用程序选择为默认应用程序，在应用程序的清单文件中必须存在如下标准：

- 为 android.provider.Telephony.SMS_DELIVER 动作注册的广播接收器负责接收新的 SMS 短信。
- 为 android.provider.Telephony.WAP_PUSH_DELIVER 动作注册的广播接收器负责接收新的 MMS 短信。
- 针对 android.intent.action.SENDTO 动作过滤的 Activity 发送新的 SMS/MMS 短信。
- 针对 android.intent.action.RESPOND_VIA_MESSAGE 动作过滤的服务发送快速回复短信给传入短信的呼叫者。

如果正在编写必须具有提供程序写入权限的消息应用程序，则我们在本节中讨论的 Telephony 契约中定义的相同 Uri 和字段结构也可用于 insert()、update()或 delete()提供程序内容。

6.12.3 实现机制

在此例中，我们将创建一个简单的消息应用程序，该应用程序从 Telephony 提供程序读取对话数据并将其显示在列表中。首先查看查询和分析来自提供程序的数据的代码。在代码清单 6-37 中，我们创建了自定义的 AsyncTaskLoader 实现，该实现用于在后台线程上查询提供程序并将结果轻松返回到 UI。

代码清单 6-37 对话数据的 Loader

```
public class ConversationLoader extends AsyncTaskLoader<List<MessageItem>> {

    public static final String[] PROJECTION = new String[] {
            // 确定消息是 SMS 还是 MMS
```

```java
        MmsSms.TYPE_DISCRIMINATOR_COLUMN,
        // 基础条目 ID
        BaseColumns._ID,
        // 对话(线程) ID
        Conversations.THREAD_ID,
        // 日期值
        Sms.DATE,
        Sms.DATE_SENT,
        // 仅适用于 SMS
        Sms.ADDRESS,
        Sms.BODY,
        Sms.TYPE,
        // 仅适用于 MMS
        Mms.SUBJECT,
        Mms.MESSAGE_BOX
};

// 正在加载的对话的线程 ID
private long mThreadId;
// 该设备的编号
private String mDeviceNumber;

public ConversationLoader(Context context) {
    this(context, -1);
}

public ConversationLoader(Context context, long threadId) {
    super(context);
    mThreadId = threadId;
    // 如果可能的话，获得此设备的电话号码
    TelephonyManager manager =
      (TelephonyManager) context.getSystemService(Context.TELEPHONY_SERVICE);
    mDeviceNumber = manager.getLine1Number();
}

@Override
protected void onStartLoading() {
    // 在每次初始请求时重新加载
    forceLoad();
}

@Override
public List<MessageItem> loadInBackground() {
    Uri uri;
    String[] projection;
    if(mThreadId < 0) {
        // 加载所有对话
        uri = MmsSms.CONTENT_CONVERSATIONS_URI;
        projection = null;
    } else {
```

```
        // 仅加载请求的线程
        uri = ContentUris.withAppendedId(MmsSms.CONTENT_
            CONVERSATIONS_URI, mThreadId);
        projection = PROJECTION;
    }

    Cursor cursor = getContext().getContentResolver().query(
            uri,
            projection,
            null,
            null,
            null);

    return MessageItem.parseMessages(getContext(), cursor, mDeviceNumber);
    }
}
```

AsyncTaskLoader 可以相当简单地自定义，只需要提供 loadInBackground()的实现来完成感兴趣的工作，并在 onStartLoading()中包括一些使进程运行的逻辑。在框架中，通常会缓存结果，仅在内容改变时调用 forceLoad()，但出于简化考虑，我们在每次请求时从提供程序加载数据。

ConversationLoader 将用于获得两个消息列表：目前存在的所有对话的列表以及所选对话(或线程)的所有消息的列表。因此在实例化 ConversationLoader 时，传入或忽略对话线程的 ID 以确定输出结果。我们还从 TelephonyManager 获得了设备的电话号码，供以后使用。该步骤并不是读取提供程序必不可少的部分，但可以帮助我们在后面清理显示内容。

要点：

使用 TelephonyManager 获得设备信息时，应用程序还必须在清单文件中声明 android.permission.READ_PHONE_STATE 权限。

在两种请求模式中，我们在组合消息表中建立对 MmsSms.CONTENT_CONVERSATIONS_URI 的查询。借助此 Uri 可以方便地获得对话概述，因为通过返回每个线程中最新的消息，它将返回所有已知对话线程的列表。这样就可以轻松地向每个用户直接显示结果。

列出所有的对话时，我们不需要提供自定义的投影，这将返回所有的字段。然而，在查看特定线程时，我们传入特定的字段子集以供检查。这样做主要是因为我们可以获得 MmsSms.TYPE_DISCRIMINATOR_COLUMN 值，该值告诉我们每条消息是 SMS 还是 MMS。该字段不可用于主对话列表，并且默认不会对 null 投影返回该值。

排序组合结果

可能你想要按日期排序这些查询的结果。常见的实现方式是在提供程序查询中使用排序子句来排序返回的结果。然而，对于我们前面完成的 SMS/MMS 组合查询，执行排序并不容易。SMS 短信的 DATE 和 DATE_SENT 字段使用从纪元开始经过的毫秒来展示它们的时间戳，而 MMS 短信的这些相同字段使用从纪元开始经过的秒数来展示它们的时间戳。

排序组合结果的最简单方法是在解析到模型(如 MessageItem)中时规范化时间戳，然后使用 Collections 的排序功能对产生的对象列表进行排序。

成功查询提供程序之后，我们想要将内容解析到公共模型对象中，我们可以方便地在列表中显示该对象。为此，我们将结果 Cursor 传递给已创建的 MessageItem 类中的工厂方法，如代码清单 6-38 所示。

代码清单 6-38　MessageItem 模型和解析

```java
public class MessageItem {
    /* 消息类型标识符 */
    private static final String TYPE_SMS = "sms";
    private static final String TYPE_MMS = "mms";

    static final String[] MMS_PROJECTION = new String[] {
        // 基础条目 ID
        BaseColumns._ID,
        // 此部分内容的 MIME 类型
        Mms.Part.CONTENT_TYPE,
        // 文本/纯文本部分的文本内容
        Mms.Part.TEXT,
        // 非文本部分的二进制内容路径
        Mms.Part._DATA
    };

    /* 消息 ID */
    public long id;
    /* 线程(对话)ID */
    public long thread_id;
    /* 消息的地址字符串 */
    public String address;
    /* 消息的正文字符串 */
    public String body;
    /* 在此设备上发送或接收该消息 */
    public boolean incoming;
    /* MMS 图片附件 */
    public Uri attachment;

    /*
     * 根据通过 Loader 查询的 Cursor 数据构造消息列表
     */
    public static List<MessageItem> parseMessages(Context context, Cursor cursor,String myNumber) {

        List<MessageItem> messages = new ArrayList<MessageItem>();
        if(!cursor.moveToFirst()) {
            return messages;
        }
        // 基于类型标识符解析每条消息
        do {
            String type = getMessageType(cursor);
            if(TYPE_SMS.equals(type)) {
                MessageItem item = parseSmsMessage(cursor);
```

```java
                messages.add(item);
            } else if(TYPE_MMS.equals(type)) {
                MessageItem item = parseMmsMessage(context, cursor, myNumber);
                messages.add(item);
            } else {
                Log.w("TelephonyProvider", "Unknown Message Type");
            }
        } while(cursor.moveToNext());
    cursor.close();
    
    return messages;
}

/*
 * 如果Cursor中存在消息类型的话，读取该类型
 * 否则，根据Cursor中存在的字段值推断类型
 */
private static String getMessageType(Cursor cursor) {
    int typeIndex = 
        cursor.getColumnIndex(MmsSms.TYPE_DISCRIMINATOR_COLUMN);
    if(typeIndex < 0) {
        // 类型字段未投影，使用另一个识别字段
        String cType = 
        cursor.getString(cursor.getColumnIndex(Mms.CONTENT_TYPE));
        // 如果存在内容类型，这就是MMS短信
        if(cType != null) {
            return TYPE_MMS;
        } else {
            return TYPE_SMS;
        }
    } else {
        return cursor.getString(typeIndex);
    }
}

/*
 * 从SMS短信中解析出带内容的MessageItem
 */
private static MessageItem parseSmsMessage(Cursor data) {
    MessageItem item = new MessageItem();
    item.id = data.getLong(data.getColumnIndexOrThrow(BaseColumns._ID));
    item.thread_id = data.getLong(data.getColumnIndexOrThrow(
        Conversations.THREAD_ID));
    item.address = data.getString(data.getColumnIndexOrThrow(Sms.ADDRESS));
    item.body = data.getString(data.getColumnIndexOrThrow(Sms.BODY));
    item.incoming = isIncomingMessage(data, true);
    return item;
}

/*
```

```java
 * 从 MMS 短信中解析出带内容的 MessageItem
 */
private static MessageItem parseMmsMessage(Context context, Cursor data,
        String myNumber) {
    MessageItem item = new MessageItem();
    item.id =
        data.getLong(data.getColumnIndexOrThrow(BaseColumns._ID));
    item.thread_id =
        data.getLong(data.getColumnIndexOrThrow(Conversations.THREAD_ID));

    item.incoming = isIncomingMessage(data, false);

    long _id =
        data.getLong(data.getColumnIndexOrThrow(BaseColumns._ID));

    // 查询此消息的地址信息
    Uri addressUri = Uri.withAppendedPath(Mms.CONTENT_URI, _id +
        "/addr");
    Cursor addr = context.getContentResolver().query(
            addressUri,
            null,
            null,
            null,
            null);
    HashSet<String> recipients = new HashSet<String>();
    while(addr.moveToNext()) {
        String address =
    addr.getString(addr.getColumnIndex(Mms.Addr.ADDRESS));
        // 不要将我们自己的号码添加到显示列表中
        if(myNumber == null || !address.contains(myNumber)) {
            recipients.add(address);
        }
    }
    item.address = TextUtils.join(", ", recipients);
    addr.close();

    // 查询与此消息关联的所有 MMS 部分
    Uri messageUri = Uri.withAppendedPath(Mms.CONTENT_URI, _id +
        "/part");
    Cursor inner = context.getContentResolver().query(
            messageUri,
            MMS_PROJECTION,
            Mms.Part.MSG_ID + " = ?",
            new String[] {String.valueOf(data.getLong(
                data.getColumnIndex(Mms._ID)))}, null);

    while(inner.moveToNext()) {
        String contentType = inner.getString(inner.getColumnIndex(
    Mms.Part.CONTENT_TYPE));
```

```java
            if(contentType == null) {
                continue;
            } else if(contentType.matches("image/.*")) {
                // 查找作为图片附件的部分
                long partId = inner.getLong(inner.getColumnIndex(
                    BaseColumns._ID));
                item.attachment = Uri.withAppendedPath(Mms.CONTENT_URI,
                    "part/" + partId);
            } else if(contentType.matches("text/.*")) {
                // 查找作为文本数据的部分
                item.body = inner.getString(inner.getColumnIndex(
                    Mms.Part.TEXT));
            }
        }

        inner.close();
        return item;
    }

    /*
     * 通过列在提供程序中的类型/箱信息验证是传入的消息还是传出的消息
     */
    private static boolean isIncomingMessage(Cursor cursor, boolean isSms) {
        int boxId;
        if(isSms) {
            boxId =
        cursor.getInt(cursor.getColumnIndexOrThrow(Sms.TYPE));
            return (boxId == TextBasedSmsColumns.MESSAGE_TYPE_INBOX ||
                    boxId == TextBasedSmsColumns.MESSAGE_TYPE_ALL) ?
                    true : false;
        } else {
            boxId =
        cursor.getInt(cursor.getColumnIndexOrThrow(Mms.MESSAGE_BOX));
            return (boxId == Mms.MESSAGE_BOX_INBOX || boxId ==
                Mms.MESSAGE_BOX_ALL) ? true : false;
        }
    }
}
```

MessageItem 自身是一个标准的占位符对象，适用于消息标识符、名称、文本内容和图片附件(用于 MMS 短信)。在 parseMessages()内部，我们遍历 Cursor 数据并根据每一行构造一个新的 MessageItem。SMS 和 MMS 的解析方式有所区别，因此我们必须首先确定消息类型。如果存在 TYPE_DISCRIMINATOR_COLUMN，则可以简单地确定消息类型，检查该字段的值即可。在其他情况下，我们可以基于每条消息的字段条目推断类型。

解析 SMS 短信非常简单，因为我们只需要直接从主 Cursor 读取作为字段的消息 ID、线程 ID、地址、正文和传入状态。解析 MMS 短信稍微复杂一些，因为这种消息的内容划分为多个部分。我们可以从主 Cursor 读取消息 ID、线程 ID 和传入状态，但地址和内容信息需要从附加表中检索获得。

首先，我们需要从 Mms.Addr 表获得接收人信息。MMS 短信可以发给多个接收人，每个接收人在 Mms.Addr 表中表示为一行，而该表的 MSG_ID 匹配对应的 MMS 短信。我们遍历这些元素，构造一个逗号分隔的结果列表，将其附加到 MessageItem 对象上。还请注意，我们在此列表中检查是否有自己的号码，避免将其添加到列表中。每条消息还有一个用于本地设备号码的 Addr 条目，我们不希望在 UI 中每次显示该条目，因此将其过滤掉。

接下来我们需要解析消息内容。这些值存储在同样以 MSG_ID 作为关键字段的 Mms.Part 表中。MMS 短信可以关联许多类型的内容(联系人数据、视频、图片等)，但我们只想显示可能存在的文本或图片数据。在遍历这些部分时，我们验证内容类型的 MIME 字符串，找出文本或图片组成部分，然后将其添加到 MessageItem 中。对于图片附件，我们只是存储指向内容的 Uri，而不是解码图片并保存。

注意：

在编写本书时，SDK 为 Mms.Addr 和 Mms.Part 提供了字段常量，但没有提供内容 Uri。这种情况在未来可能得到改观，但目前我们不得不根据基础的 Mms.CONTENT_URI 常量硬编码路径。

对于 SMS 和 MMS 短信，都可以通过查看消息的箱类型来确定消息的状态是传入还是传出。在收件箱中标记的消息或没有箱名的消息都被视为传入的消息，而所有其他消息箱名(发件箱、已发送、草稿等)都表示是传出的消息。

现在我们有了经过解析的消息列表，接下来查看代码清单 6-39 和 6-40，检查在此例中实现的用户界面。

代码清单 6-39　显示 SMS/MMS 短信的 Activity

```
public class SmsActivity extends Activity
        implements OnItemClickListener, LoaderCallbacks<List<MessageItem>> {

    private MessagesAdapter mAdapter;

    @Override
    protected void onCreate(Bundle savedInstanceState) {
        super.onCreate(savedInstanceState);
        ListView list = new ListView(this);
        mAdapter = new MessagesAdapter(this);
        list.setAdapter(mAdapter);

        final Intent intent = getIntent();

        if(!intent.hasExtra("threadId")) {
            // 条目在对话未显示时可点击
            list.setOnItemClickListener(this);
        }
```

```java
        // 加载消息数据
        getLoaderManager().initLoader(0, getIntent().getExtras(), this);

        setContentView(list);
    }

    @Override
    public void onItemClick(AdapterView<?> parent, View view, int position,
long id) {
        final MessageItem item = mAdapter.getItem(position);
        long threadId = item.thread_id;
        // 启动新的实例以显示此对话
        Intent intent = new Intent(this, SmsActivity.class);
        intent.putExtra("threadId", threadId);
        startActivity(intent);
    }

    @Override
    public Loader<List<MessageItem>> onCreateLoader(int id, Bundle args) {
        if(args != null && args.containsKey("threadId")) {
            return new ConversationLoader(this, args.getLong("threadId"));
        } else {
            return new ConversationLoader(this);
        }
    }

    @Override
    public void onLoadFinished(Loader<List<MessageItem>> loader,
List<MessageItem> data) {
        mAdapter.clear();
        mAdapter.addAll(data);
        mAdapter.notifyDataSetChanged();
    }

    @Override
    public void onLoaderReset(Loader<List<MessageItem>> loader) {
        mAdapter.clear();
        mAdapter.notifyDataSetChanged();
    }

    private static class MessagesAdapter extends ArrayAdapter<MessageItem> {

        int cacheSize = 4 * 1024 * 1024; // 4MB
        private LruCache<String, Bitmap> bitmapCache = new LruCache<String,
Bitmap>(cacheSize) {
            protected int sizeOf(String key, Bitmap value) {
                return value.getByteCount();
            }
        };
```

```java
    public MessagesAdapter(Context context) {
        super(context, 0);
    }

    @Override
    public View getView(int position, View convertView, ViewGroup parent) {
        if(convertView == null) {
            convertView = LayoutInflater.from(getContext())
                    .inflate(R.layout.message_item, parent, false);
        }

        MessageItem item = getItem(position);

        TextView text1 = (TextView) convertView.findViewById(R.id.text1);
        TextView text2 = (TextView) convertView.findViewById(R.id.text2);
        ImageView image = (ImageView) convertView.findViewById(R.id.image);
        text1.setText(item.address);
        text2.setText(item.body);
        // 基于传入/传出状态设置文本样式
        Typeface tf = item.incoming ?
          Typeface.defaultFromStyle(Typeface.ITALIC) : Typeface.DEFAULT;
        text2.setTypeface(tf);
        image.setImageBitmap(getAttachment(item));

        return convertView;
    }

    private Bitmap getAttachment(MessageItem item) {
        if(item.attachment == null) return null;

        final Uri imageUri = item.attachment;
        // 从缓存中提取图片缩略图(如果有的话)
        Bitmap cached = bitmapCache.get(imageUri.toString());
        if(cached != null) {
            return cached;
        }

        // 如果缓存中没有缩略图,则从提供程序解码该资源
        try {
            BitmapFactory.Options options = new
              BitmapFactory.Options();
            options.inJustDecodeBounds = true;
            int cellHeight = getContext().getResources()
                    .getDimensionPixelSize(R.dimen.message_height);
            InputStream is = getContext().
              getContentResolver().openInputStream(imageUri);
            BitmapFactory.decodeStream(is, null, options);

            options.inJustDecodeBounds = false;
            options.inSampleSize = options.outHeight/cellHeight;
            is = getContext().getContentResolver().openInputStream(
```

```
                    imageUri);
            Bitmap bitmap = BitmapFactory.decodeStream(is, null, options);

            bitmapCache.put(imageUri.toString(), bitmap);
            return bitmap;
        } catch(Exception e) {
            return null;
        }
    }
}
```

代码清单 6-40　res/layout/message_item.xml

```
<?xml version="1.0" encoding="utf-8"?>
<RelativeLayout
    xmlns:android="http://schemas.android.com/apk/res/android"
    android:layout_width="match_parent"
    android:layout_height="wrap_content"
    android:minHeight="@dimen/message_height">
    <ImageView
        android:id="@+id/image"
        android:layout_width="@dimen/message_height"
        android:layout_height="@dimen/message_height"
        android:layout_alignParentRight="true"
        android:layout_centerVertical="true" />
    <TextView
        android:id="@+id/text1"
        android:layout_width="match_parent"
        android:layout_height="wrap_content"
        android:layout_toLeftOf="@id/image"
        android:layout_marginLeft="6dp"
        android:textStyle="bold" />
    <TextView
        android:id="@+id/text2"
        android:layout_width="match_parent"
        android:layout_height="wrap_content"
        android:layout_below="@id/text1"
        android:layout_toLeftOf="@id/image"
        android:layout_marginLeft="12dp" />

</RelativeLayout>
```

　　我们的 Activity 用于显示两种类型的消息数据。该 Activity 在创建时会调用 initLoader()，构造新的 ConversationLoader 来查询提供程序。传入的参数是 Activity Intent 收到的附加信息。应用程序初次启动时，没有传入 Intent 附加信息，因此系统构造 ConversationLoader 以加载所有的对话线程。后面使用特定线程 ID 启动 Activity 时，ConversationLoader 将查询对话中的所有消息。

　　一旦加载完成，系统就使用自定义的 MessageAdapter 在 ListView 中展示数据。该适配器使用两个文本行和存放图片的空间填充自定义条目布局(见代码清单 6-40)。MessageItem 地址信息加载到顶部的标签中，文本内容则加载到底部的标签中。如果消息是 MMS 且存在图片附件，我们就尝试通过 getAttachment()返回图片并将其插入 ImageView。

每次从磁盘加载这些图片都要浪费大量内存，并且执行起来有点缓慢，因此为改进在列表中进行滚动的性能，我们添加了 LruCache，将最近加载的位图存入内存。缓存大小设置为 4MB，这样就不会随着时间推移而过度填充应用程序的堆空间。此外，从 BitmapFactory 返回(通过 BitmapFactory.Options.inSampleSize)的每张图片都会降低采样率，从而避免加载的图片大于列表行中的可用空间，同时避免浪费内存。

我们现在有了基本的消息应用程序，它在我们的设备上显示一个存放 SMS/MMS 短信的只读窗口。该应用程序在启动时会显示每个线程的最新消息，从而列出所有的对话。点击某个对话时，新的 Activity 将会显示，其中列出了线程内的所有单条消息。按 Back 按钮将返回到主列表，从而用户可以选择另一个对话来查看。

6.13 与日历交互

6.13.1 问题

应用程序需要与 Android 框架提供的 ContentProvider 直接交互，以便在设备上添加、浏览、更改或删除日历事件。

6.13.2 解决方案

(API Level 14)
使用 CalendarContract 接口读取/写入系统的 ContentProvider 中的事件数据。CalendarContract 提供了可以访问设备日历、事件、出席者、提醒等信息的 API。和 ContactsContract 非常类似，这个接口定义了很多执行查询所需的数据。相关方法的使用方式和其他系统 ContentProvider 的使用方式是一样的。

6.13.3 实现机制

使用 CalendarContract 和使用 ContactsContract 非常相似，它们都提供了相应的 Uri 标识符和在使用 ContentResolver 构建查询时所需的字段值。代码清单 6-41 显示了获取和显示设备上日历事件列表的 Activity。

代码清单 6-41　列出设备上日历事件的 Activity

```
public class CalendarListActivity extends ListActivity implements
        LoaderManager.LoaderCallbacks<Cursor>, AdapterView.OnItemClickListener {
    private static final int LOADER_LIST = 100;

    SimpleCursorAdapter mAdapter;
```

```java
@Override
public void onCreate(Bundle savedInstanceState) {
    super.onCreate(savedInstanceState);
    getLoaderManager().initLoader(LOADER_LIST, null, this);

    //在 ListView 中显示所有的日历事件
    mAdapter = new SimpleCursorAdapter(this,
    android.R.layout.simple_list_item_2, null,
    new String[] {
            CalendarContract.Calendars.CALENDAR_DISPLAY_NAME,
            CalendarContract.Calendars.ACCOUNT_NAME },
    new int[] {
             android.R.id.text1, android.R.id.text2 }, 0);
    setListAdapter(mAdapter);
    //监听条目的选择事件
    getListView().setOnItemClickListener(this);
}

@Override
public void onItemClick(AdapterView<?> parent, View view, int position,
        long id) {
    Cursor c = mAdapter.getCursor();
    if(c != null && c.moveToPosition(position)) {
        Intent intent = new Intent(this, CalendarDetailActivity.class);
        //将选择的日历事件的_ID 和 TITLE 信息传递给下一个 Activity
        intent.putExtra(Intent.EXTRA_UID, c.getInt(0));
        intent.putExtra(Intent.EXTRA_TITLE, c.getString(1));
        startActivity(intent);
    }
}

@Override
public Loader<Cursor> onCreateLoader(int id, Bundle args) {
    //返回所有的日历事件,通过名称排序
    String[] projection = new String[] { CalendarContract.Calendars._ID,
            CalendarContract.Calendars.CALENDAR_DISPLAY_NAME,
            CalendarContract.Calendars.ACCOUNT_NAME };

    return new CursorLoader(this, CalendarContract.Calendars.CONTENT_URI,
            projection, null, null,
            CalendarContract.Calendars.CALENDAR_DISPLAY_NAME);
}

@Override
public void onLoadFinished(Loader<Cursor> loader, Cursor data) {
    mAdapter.swapCursor(data);
}

@Override
```

```java
    public void onLoaderReset(Loader<Cursor> loader) {
        mAdapter.swapCursor(null);
    }
}
```

与之前的联系人示例相比，这里我们使用 Android 的 Loader 模式来查询数据并把得到的结果 Cursor 加载到列表中。这种方式相对于 managedCursor() 有很多优点，最大的好处就是所有的查询操作都是自动在后台线程中执行的，这样就可以保证 UI 线程可以及时响应。另外，Loader 模式是可以内部重用的，即请求相同数据的多个客户端实际上访问的是同一个由 LoaderManager 管理的 Loader。

使用 Loader 时，如果有新的日历数据可用，我们的 Activity 会接收很多回调方法。实际上，CursorLoader 也像 ContentObserver 一样注册，因此在它所对应的数据集发生变化时，甚至不需要重新加载就可以得到一个回调，该回调中包含数据集新的 Cursor，但是会返回到日历。

要想得到设备的日历事件列表，需要先使用 Calendars.CONTENT_URI 以及相应的字段名(这里是记录的 ID、日历事件名称和拥有该事件的账户名称)构建一个查询。查询完成后，会调用 onLoadFinished() 方法，这时会返回一个指向查询结果数据的新的 Cursor，然后我们会把这个 Cursor 传递给列表适配器。在用户点击一个日历事件条目后，就会初始化一个新的 Activity 来浏览这个日历事件。下一小节会查看关于这个示例的更多细节信息。

浏览/修改日历事件

代码清单 6-42 展示了本例中另一个 Activity 的内容，该 Activity 显示了选中日历的所有事件列表。

代码清单 6-42　列出并修改日历事件的 Activity

```java
public class CalendarDetailActivity extends ListActivity implements
        LoaderManager.LoaderCallbacks<Cursor>, AdapterView.OnItemClickListener,
        AdapterView.OnItemLongClickListener {
    private static final int LOADER_DETAIL = 101;

    SimpleCursorAdapter mAdapter;

    int mCalendarId;

    @Override
    protected void onCreate(Bundle savedInstanceState) {
        super.onCreate(savedInstanceState);

        mCalendarId = getIntent().getIntExtra(Intent.EXTRA_UID, -1);

        String title = getIntent().getStringExtra(Intent.EXTRA_TITLE);
        setTitle(title);

        getLoaderManager().initLoader(LOADER_DETAIL, null, this);
```

```java
        //在 ListView 中显示所有事件
        mAdapter = new SimpleCursorAdapter(this,
            android.R.layout.simple_list_item_2, null,
            new String[] {
                CalendarContract.Events.TITLE,
                CalendarContract.Events.EVENT_LOCATION },
            new int[] {
                android.R.id.text1, android.R.id.text2 }, 0);
        setListAdapter(mAdapter);
        //监听条目的选择事件
        getListView().setOnItemClickListener(this);
        getListView().setOnItemLongClickListener(this);
    }

    @Override
    public boolean onCreateOptionsMenu(Menu menu) {
        menu.add("Add Event")
            .setIcon(android.R.drawable.ic_menu_add)
            .setShowAsAction(MenuItem.SHOW_AS_ACTION_ALWAYS);

        return true;
    }

    @Override
    public boolean onOptionsItemSelected(MenuItem item) {
        showAddEventDialog();
        return true;
    }

    //显示一个对话框来添加新事件
    private void showAddEventDialog() {
        final EditText nameText = new EditText(this);
        AlertDialog.Builder builder = new AlertDialog.Builder(this);
        builder.setTitle("New Event");
        builder.setView(nameText);
        builder.setNegativeButton("Cancel", null);
        builder.setPositiveButton("Add Event",
                new DialogInterface.OnClickListener() {
                    @Override
                    public void onClick(DialogInterface dialog, int which) {
                        addEvent(nameText.getText().toString());
                    }
                });
        builder.show();
    }

    //使用特定的名称和当前时间(作为事件的开始日期)向日历中添加一个事件
    private void addEvent(String eventName) {
        long start = System.currentTimeMillis();
        //当前事件 1 小时后结束
```

```java
        long end = start + (3600 * 1000);

        ContentValues cv = new ContentValues(5);
        cv.put(CalendarContract.Events.CALENDAR_ID, mCalendarId);
        cv.put(CalendarContract.Events.TITLE, eventName);
        cv.put(CalendarContract.Events.DESCRIPTION,
                "Event created by Android Recipes");
        cv.put(CalendarContract.Events.EVENT_TIMEZONE,
                Time.getCurrentTimezone());
        cv.put(CalendarContract.Events.DTSTART, start);
        cv.put(CalendarContract.Events.DTEND, end);

        getContentResolver().insert(CalendarContract.Events.CONTENT_URI, cv);
    }

    //在日历中将选中的事件删除
    private void deleteEvent(int eventId) {
        String selection = CalendarContract.Events._ID + " = ?";
        String[] selectionArgs = { String.valueOf(eventId) };
        getContentResolver().delete(CalendarContract.Events.CONTENT_URI,
                selection, selectionArgs);
    }

    @Override
    public void onItemClick(AdapterView<?> parent, View view, int position,
            long id) {
        Cursor c = mAdapter.getCursor();
        if(c != null && c.moveToPosition(position)) {
            //单击时会显示一个包含日历事件详细信息的对话框
            SimpleDateFormat sdf = new SimpleDateFormat("yyyy-MM-dd HH:mm:ss");
            StringBuilder sb = new StringBuilder();

            sb.append("Location: " + c.getString(c.getColumnIndex(
                    CalendarContract.Events.EVENT_LOCATION))+ "\n\n");
            int startDateIndex = c.getColumnIndex(CalendarContract.Events.DTSTART);
            Date startDate = c.isNull(startDateIndex) ? null
                    : new Date( Long.parseLong(c.getString(startDateIndex)) );
            if(startDate != null) {
                sb.append("Starts At: " + sdf.format(startDate) + "\n\n");
            }
            int endDateIndex = c.getColumnIndex(CalendarContract.Events.DTEND);
            Date endDate = c.isNull(endDateIndex) ? null
                    : new Date( Long.parseLong(c.getString(endDateIndex)) );
            if(endDate != null) {
                sb.append("Ends At: " + sdf.format(endDate) + "\n\n");
            }
            AlertDialog.Builder builder = new AlertDialog.Builder(this);
            builder.setTitle(
                    c.getString(c.getColumnIndex(CalendarContract.
```

```java
                    Events.TITLE)) );
            builder.setMessage(sb.toString());
            builder.setPositiveButton("OK", null);
            builder.show();
        }
    }

    @Override
    public boolean onItemLongClick(AdapterView<?> parent, View view,
            int position, long id) {
        Cursor c = mAdapter.getCursor();
        if(c != null && c.moveToPosition(position)) {
            //用户长按时,会删除选中的日历事件
            final int eventId = c.getInt(
                    c.getColumnIndex(CalendarContract.Events._ID));
            String eventName = c.getString(
                    c.getColumnIndex(CalendarContract.Events.TITLE));
            AlertDialog.Builder builder = new AlertDialog.Builder(this);
            builder.setTitle("Delete Event");
            builder.setMessage(String.format(
                    "Are you sure you want to delete %s?",
                    TextUtils.isEmpty(eventName) ? "this event" : eventName));
            builder.setNegativeButton("Cancel", null);
            builder.setPositiveButton("Delete Event",
                    new DialogInterface.OnClickListener() {
                        @Override
                        public void onClick(DialogInterface dialog, int which) {
                            deleteEvent(eventId);
                        }
                    });
            builder.show();
        }

        return true;
    }

    @Override
    public Loader<Cursor> onCreateLoader(int id, Bundle args) {
        //返回所有的日历,按照名称排序
        String[] projection = new String[] { CalendarContract.Events._ID,
                CalendarContract.Events.TITLE,
                CalendarContract.Events.DTSTART,
                CalendarContract.Events.DTEND,
                CalendarContract.Events.EVENT_LOCATION };
        String selection = CalendarContract.Events.CALENDAR_ID + " = ?";
        String[] selectionArgs = { String.valueOf(mCalendarId) };
        return new CursorLoader(this, CalendarContract.Events.CONTENT_URI,
                projection, selection, selectionArgs,
                CalendarContract.Events.DTSTART + " DESC");
    }
```

```
        @Override
        public void onLoadFinished(Loader<Cursor> loader, Cursor data) {
            mAdapter.swapCursor(data);
        }

        @Override
        public void onLoaderReset(Loader<Cursor> loader) {
            mAdapter.swapCursor(null);
        }
    }
```

可以看到，查询日历事件列表的代码和显示这些事件的代码非常相似，本例中是通过选择的日历 ID 作为 selection 参数来查询 Events.CONTENT_URI 的。这里，在用户点击一个日历事件后，系统就会向用户显示一个简单的对话框，对话框中显示的则是日历事件的详细信息。此外，这个 Activity 还包含一些创建和删除日历事件的方法。

要想添加新的日历事件，可以在选项菜单中添加一个新的条目，如果设备只能看到一个条目，该条目会显示在 Action Bar 的最前面。按下该条目后，会弹出一个对话框让用户输入事件的名称。如果用户选择继续，就会创建一个含有建立日历事件所需信息的 ContentValues 对象。因为这个事件是一次性发生的，所以必须指定开始时间和结束时间以及有效的时区。我们还必须指定所查看日历的 ID，这样事件才可以和日历有效地关联起来。然后，通过 ContentResolver 将数据插入 Event 表中。

要想删除一个事件，用户只需要在列表中特定的条目上长按，然后在显示的对话框中确认删除即可。这种情况下，只需要获得选中事件的唯一记录 ID 并把它传入 ContentResolver 的 selection 字符串中。

你是否发现在这两个示例中，在添加/删除事件之后我们并没有编写任何代码来更新 Cursor 或 CursorAdapter？这就是 Loader 模式的强大之处！CursorLoader 会监控数据集，当数据集变化时，它会自动更新并赋予适配器一个新的 Cursor，该适配器则会刷新界面的显示。

注意：

很多 Loader 都是从 Android 3.0(API Level 11)开始引入的，不过在 Android 支持库中也包含它们。因此，可以在 Android 1.6 中通过支持库使用它们。

6.14 执行日志代码

6.14.1 问题

为了调试或测试，需要在代码中加入日志相关代码，并且需要在代码发布前移除这些日志代码。

6.14.2 解决方案

(API Level 1)

在 Log 类中使用 BuildConfig.DEBUG 标识确保一些语句只在应用程序的调试阶段才会打印。出于对将来测试和开发的考虑,即使在应用程序已经发布的情况下,在代码中保留一些日志语句还是极为方便的。但如果这些语句没有得到相应的保护,可能会在用户设备的控制台上打印出应用程序的一些隐私信息。通过对 Log 创建简单的封装类来监控 BuildConfig.DEBUG,就可以放心地保留日志语句,而不必担心它们会被错误显示。

6.14.3 实现机制

代码清单 6-43 显示了一个对于默认 Android 日志功能的简单封装类。

代码清单 6-43　日志封装类

```java
public class Logger {
    private static final String LOGTAG = "AndroidRecipes";

    private static String getLogString(String format, Object... args) {
        //次要优化,如果需要的话,只调用 String.format
        if(args.length == 0) {
            return format;
        }

        return String.format(format, args);
    }

    /*打印常用的 INFO、WARNING、ERROR 级别的日志*/

    public static void e(String format, Object... args) {
        Log.e(LOGTAG, getLogString(format, args));
    }

    public static void w(String format, Object... args) {
        Log.w(LOGTAG, getLogString(format, args));
    }

    public static void w(Throwable throwable) {
        Log.w(LOGTAG, throwable);
    }

    public static void i(String format, Object... args) {
        Log.i(LOGTAG, getLogString(format, args));
    }

    /*用 DEBUG 标识来保护 DEBUG 和 VERBOSE 日志级别*/
```

```java
    public static void d(String format, Object... args) {
        if(!BuildConfig.DEBUG) return;

        Log.d(LOGTAG, getLogString(format, args));
    }

    public static void v(String format, Object... args) {
        if(!BuildConfig.DEBUG) return;

        Log.v(LOGTAG, getLogString(format, args));
    }
}
```

这个类对 Android 框架版本提供的日志做了一些简单的优化，使之更加实用。首先，它整合了整个应用程序的日志标记，这样 logcat 中的日志就会有统一的标题。其次，它的输入是一个格式化的字符串，这样变量不必分隔日志字符串就可以打印。另一个优化就是，String.format()的执行可能会很慢，所以我们只在有参数特意指定格式化时才会调用它。否则，直接返回原始字符串。

最后，它通过 BuildConfig.DEBUG 标识保护了 5 个主要日志级别中的两个，这两个级别的日志语句只有在应用程序调试时才会打印。在应用程序发布后也有一些情况(如应用程序出现错误)需要打印日志语句，所以最好不要隐藏 debug 标识后面的日志级别的日志。代码清单 6-44 快速展示了这个封装类是如何取代传统的日志功能的。

代码清单6-44 使用了 Logger 的 Activity

```java
public class LoggerActivity extends Activity {

    @Override
    public void onCreate(Bundle savedInstanceState) {
        super.onCreate(savedInstanceState);
        setContentView(R.layout.main);

        //只有在调试时才会打印这条语句
        Logger.d("Activity Created");
    }

    @Override
    protected void onResume() {
        super.onResume();

        //只有在调试时才会打印这条语句
        Logger.d("Activity Resume at %d", System.currentTimeMillis());
        //总会打印这条语句
        Logger.i("It is now %d", System.currentTimeMillis());
    }

    @Override
```

```java
protected void onPause() {
    super.onPause();

    //只有在调试时才会打印这条语句
    Logger.d("Activity Pause at %d", System.currentTimeMillis());
    //总会打印这条语句
    Logger.w("No, don't leave!");
}
```

6.15 创建后台工作线程

6.15.1 问题

需要创建一个长时间运行的后台线程来等待执行某项任务,并且该线程在不用时可以很容易地终止。

6.15.2 解决方案

(API Level 1)

HandlerThread 可以帮助创建一个拥有有效 Looper 的后台线程,该 Looper 会关联一个 Handler,而 Handler 中的 MessageQueue 会处理所有的任务。Android 中最常用的后台技术之一就是 AsyncTask,这个类非常好用,应该在应用程序中使用它。但是,它也有一些缺点,这些缺点在某些情况下会使其他实现显得更加高效。一个缺点就是 AsyncTask 只能执行一次且受限,如果想要在 Activity 或服务这样的组件的生命周期中执行重复或无限期的任务,AsyncTask 显示有些工作繁重。通常,需要创建多个 AsyncTask 实例才能完成这些任务。

在这种情况下,HandlerThread 的优势就是只需要创建一个工作对象,它可以在后台接受多个任务,并且通过 Looper 所维护的内部队列逐个处理这些任务。

6.15.3 实现机制

代码清单 6-45 展示了 HandlerThread 的一个扩展类,用来操作一些简单的图片数据。修改图片需要一些时间,所以我们希望在一个后台线程中执行此任务,这样就可以保证应用程序的用户界面不会被阻塞。

代码清单 6-45　后台工作线程

```java
public class ImageProcessor extends HandlerThread implements Handler.Callback {
    public static final int MSG_SCALE = 100;
    public static final int MSG_CROP = 101;

    private Context mContext;
```

```java
    private Handler mReceiver, mCallback;

    public ImageProcessor(Context context) {
        this(context, null);
    }

    public ImageProcessor(Context context, Handler callback) {
        super("AndroidRecipesWorker");
        mCallback = callback;
        mContext = context.getApplicationContext();
    }

    @Override
    protected void onLooperPrepared() {
        mReceiver = new Handler(getLooper(), this);
    }

    @Override
    public boolean handleMessage(Message msg) {
        Bitmap source, result;
        //从传入的消息中解析参数
        int scale = msg.arg1;
        switch(msg.what) {
        case MSG_SCALE:
            source = BitmapFactory.decodeResource(mContext.getResources(),
                    R.drawable.ic_launcher);
            //创建一张新的、缩放的图片
            result = Bitmap.createScaledBitmap(source,
                    source.getWidth() * scale, source.getHeight() * scale, true);
            break;
        case MSG_CROP:
            source = BitmapFactory.decodeResource(mContext.getResources(),
                    R.drawable.ic_launcher);
            int newWidth = source.getWidth() / scale;
            //创建一张新的、横向裁剪的图片
            result = Bitmap.createBitmap(source,
                    (source.getWidth() - newWidth) / 2, 0,
                    newWidth, source.getHeight());
            break;
        default:
            throw new IllegalArgumentException("Unknown Worker Request");
        }

        //将图片返回给主线程
        if(mCallback != null) {
            mCallback.sendMessage(Message.obtain(null, 0, result));
        }
        return true;
    }
```

```java
//添加/删除回调 Handler
public void setCallback(Handler callback) {
    mCallback = callback;
}

/*队列操作相关方法*/

//缩放图标为特定的值
public void scaleIcon(int scale) {
    Message msg = Message.obtain(null, MSG_SCALE, scale, 0, null);
    mReceiver.sendMessage(msg);
}

//居中裁剪图标，然后缩放为特定的值
public void cropIcon(int scale) {
    Message msg = Message.obtain(null, MSG_CROP, scale, 0, null);
    mReceiver.sendMessage(msg);
}
}
```

HandlerThread 这个名称可能不是特别恰当，这是因为它实际上并没有可以用来处理输入的 Handler。相反，它是一个线程，即和外部 Handler 一起创建的一个后台线程。正因为如此，我们必须自己实现一个 Handler 来真正处理我们想要执行的工作。本例中，我们自定义的处理器实现了 Handler.Callback 接口并传入了线程拥有的新 Handler。这样做也可以，而且可以避免使用 Handler 的子类。在 onLooperPrepared()回调之后接收器 Handler 才会被创建，这是因为我们需要用 HandlerThread 所创建的 Looper 对象将要执行的任务发送到后台线程。

我们创建的允许其他对象进入队列工作的外部 API 都创建了一个 Message 并将它发送到接收器 Handler 的 handleMessage()方法中进行处理，该方法会检查 Message 的内容并创建适当修改过的图片。handleMessage()中的所有代码都会运行在我们的后台线程中。

在工作完成后，就需要另外使用一个附加到主线程的 Handler，从而发送相关处理结果并更新 UI。

提醒：
任何和 UI 元素操作相关的代码必须在主线程中执行。请记住这一点。

该回调 Handler 会接收另一个 Message，该 Message 包含通过图片代码产生的 Bitmap。这也是使用 Message 接口在不同的线程之间传递数据的好处之一；每个实例都可以带两个整型参数以及一个任意对象，所以在传递参数或访问结果时也不需要额外的代码。在这个示例中，传入的一个整型参数是转换图片的缩放值，Object 字段则用来返回一张位图。要想了解这个示例的实际结果，请参见代码清单 6-46 和 6-47 中的示例应用程序。

代码清单 6-46   res/layout/main.xml

```xml
<LinearLayout xmlns:android="http://schemas.android.com/apk/res/android"
    android:layout_width="match_parent"
```

```xml
        android:layout_height="match_parent"
        android:orientation="vertical" >

    <Button
        android:layout_width="match_parent"
        android:layout_height="wrap_content"
        android:text="Scale Icon"
        android:onClick="onScaleClick" />
    <Button
        android:layout_width="match_parent"
        android:layout_height="wrap_content"
        android:text="Crop Icon"
        android:onClick="onCropClick" />
    <ImageView
        android:id="@+id/image_result"
        android:layout_width="match_parent"
        android:layout_height="match_parent"
        android:scaleType="center" />
</LinearLayout>
```

代码清单 6-47　和工作线程进行交互的 Activity

```java
public class WorkerActivity extends Activity implements Handler.Callback {

    private ImageProcessor mWorker;
    private Handler mResponseHandler;

    private ImageView mResultView;

    @Override
    public void onCreate(Bundle savedInstanceState) {
        super.onCreate(savedInstanceState);
        setContentView(R.layout.main);

        mResultView = (ImageView) findViewById(R.id.image_result);
        //该 Activity 关联的后台回调 Handler
        mResponseHandler = new Handler(this);
    }

    @Override
    protected void onResume() {
        super.onResume();
        //启动一个新的工作线程
        mWorker = new ImageProcessor(this, mResponseHandler);
        mWorker.start();
    }

    @Override
    protected void onPause() {
        super.onPause();
        //终止工作线程
```

```java
        mWorker.setCallback(null);
        mWorker.quit();
        mWorker = null;
    }

    /*
     *后台执行结果的回调方法
     *运行在 UI 线程上
     */
    @Override
    public boolean handleMessage(Message msg) {
        Bitmap result = (Bitmap) msg.obj;
        mResultView.setImageBitmap(result);
        return true;
    }

    /*发送后台工作线程的动作方法 */

    public void onScaleClick(View v) {
        for(int i=1; i < 10; i++) {
            mWorker.scaleIcon(i);
        }
    }

    public void onCropClick(View v) {
        for(int i=1; i < 10; i++) {
            mWorker.cropIcon(i);
        }
    }
}
```

这个示例利用工作线程的方式是在 Activity 处于前台时创建一个单独的后台线程运行实例，并且在用户点击按钮时将图片操作的请求发送给该线程。为了进一步演示此模式的缩放效果，在每次点击按钮时都会发送很多的操作请求。这个 Activity 也实现了 Handler.Callback，用一个简单的 Handler(运行在主线程上)接收后台工作线程发送过来的操作结果。

要想启动后台线程，只需要调用 HandlerThread 的 start()方法，这时会设置 Looper 和 Handler，然后等待任务的输入。终止后台线程就更简单了，只需要调用 quit()就可以停止 Looper 并立即移除队列中未处理的消息。我们还将回调设为 null，这样此时可能正在运行的任务就不会再通知 Activity 了。

运行这个应用程序后，你会发现不管按钮按下的速度和频率有多快，后台工作线程都不会拖慢 UI 线程。所有的请求都会被添加到消息队列中，在用户关闭 Activity 之前，这些请求会被尽可能处理。程序运行的可见结果就是在每个请求处理完成后，在按钮的下方显示刚刚创建的图片。

6.16 自定义任务栈

6.16.1 问题

应用程序允许外部应用程序直接启动它的某些 Activity，你需要实现适当的 BACK 和 UP 导航行为。

6.16.2 解决方案

(API Level 4)

Android 支持库中的 NavUtils 和 TaskStackBuilder 类可以很容易地在应用程序中构建和生成合适的导航栈。实际上，这两个类的功能是 Android 4.1 及以后版本原生的功能，但如果要在稍早版本的应用程序中也支持该功能，可以使用 Android 支持库提供的兼容 API，实际调用的也是原生方法。

BACK 与 UP

Android 的界面导航有两类用户动作行为：一类就是用户按下 BACK 按钮后的行为，另一类则是在 Action Bar 上按下 Home 图标(称为 UP 动作)后的行为。对于那些对平台还不太熟悉的开发人员来说，经常会混淆它们，特别是很多情况下这两类动作都会执行相同的功能。

从概念上讲，BACK 应该让用户返回到上一个浏览的界面(相对于当前界面)，而 UP 动作应该返回当前界面的父界面。对于大多数应用程序来说，用户都是从主界面进入到带有特定内容的子界面，因此 BACK 和 UP 都会回到相同的地方，这也会让开发人员对它们的用法产生质疑。

然而，考虑一下对于应用程序的一个或多个 Activity 可以由外部应用程序直接启动的情况。例如，某个 Activity 的作用是浏览图片文件。或者发送通知消息的应用程序在事件发生时会允许用户直接进入底层的 Activity。在这些情况下，BACK 动作会将用户返回到调用应用程序之前的应用程序。而对于 UP 动作，如果用户希望继续使用应用程序而不是回到之前的应用程序，UP 动作可以让用户返回到应用程序的界面栈。这种情况下，应用程序的整个 Activity 元素栈通常还没有构建，这时就需要 TaskStackBuilder 和应用代码清单中一些关键属性的帮助了。

6.16.3 实现机制

让我们定义两个应用程序来演示这个范例是如何工作的。首先查看一下代码清单 6-48，它显示了清单的 `<application>` 元素中的内容。

代码清单 6-48　AndroidManifest.xml 中的 application 标签

```
<application
    android:icon="@drawable/ic_launcher"
```

```xml
            android:label="TaskStack"
            android:theme="@style/AppTheme" >
        <activity
            android:name=".RootActivity"
            android:label="@string/title_activity_root" >
            <intent-filter>
                <action android:name="android.intent.action.MAIN" />
                <category android:name="android.intent.category.LAUNCHER" />
            </intent-filter>
        </activity>
        <activity android:name=".ItemsListActivity"
            android:parentActivityName=".RootActivity">
            <!--为支持库定义的父界面-->
            <meta-data android:name="android.support.PARENT_ACTIVITY"
                android:value=".RootActivity" />
        </activity>
        <activity android:name=".DetailsActivity"
                android:parentActivityName=".ItemsListActivity">
            <!--为支持库定义的父界面-->
            <meta-data android:name="android.support.PARENT_ACTIVITY"
                android:value=".ItemsListActivity" />
            <!--提供一个过滤器，允许外部应用程序启动-->
            <intent-filter>
                <action android:name="com.examples.taskstack.ACTION_NEW_ARRIVAL" />
                <category android:name="android.intent.category.DEFAULT" />
            </intent-filter>
        </activity>
    </application>
```

定义这种导航的第一步就是确定每个 Activity 之间的亲子关系。在 Android 4.1 中，引入了 android:parentActivityName 属性来创建这种关系。想要在老版本中也实现相同功能，需要使用支持库定义的<meta-data>值来为每个 Activity 定义父界面。我们的示例则采用两种方式为每个底层 Activity 定义了亲子关系，既可以运行在本地 API 上，也可以运行在支持库上。

在 DetailsActivity 中还有一个自定义的<intent-filter>，它允许外部应用程序直接启动 DetailsActivity。

注意：
如果你的应用程序只支持 Android 4.1 及以后的版本，就可以到此为止了。因为构建栈和导航的其余功能在这些版本中已经内置到 Activity 的默认行为中，不必额外编写代码。这种情况下，如果想在一些特殊的情形下自定义任务栈，只需要实现 TaskStackBuilder 即可。

定义好亲子关系后，开始编写每个 Activity 的代码。参见代码清单 6-49 到 6-51。

代码清单 6-49　根 Activity

```java
public class RootActivity extends Activity implements View.OnClickListener {

    @Override
    public void onCreate(Bundle savedInstanceState) {
        super.onCreate(savedInstanceState);
        Button listButton = new Button(this);
        listButton.setText("Show Family Members");
        listButton.setOnClickListener(this);

        setContentView(listButton,
                new ViewGroup.LayoutParams(LayoutParams.MATCH_PARENT,
                        LayoutParams.WRAP_CONTENT));
    }

    public void onClick(View v) {
        //启动下一个 Activity
        Intent intent = new Intent(this, ItemsListActivity.class);
        startActivity(intent);
    }
}
```

代码清单 6-50　第二级 Activity

```java
public class ItemsListActivity extends Activity implements OnItemClickListener {

    private static final String[] ITEMS = {"Mom", "Dad", "Sister", "Brother",
      "Cousin"};

    @Override
    protected void onCreate(Bundle savedInstanceState) {
        super.onCreate(savedInstanceState);
        //启用带向上箭头的 Action Bar Home 按钮
        getActionBar().setDisplayHomeAsUpEnabled(true);
        //创建并显示家庭成员的列表
        ListView list = new ListView(this);
        ArrayAdapter<String> adapter = new ArrayAdapter<String>(this,
                android.R.layout.simple_list_item_1, ITEMS);
        list.setAdapter(adapter);
        list.setOnItemClickListener(this);

        setContentView(list);
    }

    @Override
    public boolean onOptionsItemSelected(MenuItem item) {
        switch (item.getItemId()) {
        case android.R.id.home:
```

```java
        //创建父 Activity 的 Intent
        Intent upIntent = NavUtils.getParentActivityIntent(this);
        //检查是否需要创建整个栈
        if(NavUtils.shouldUpRecreateTask(this, upIntent)) {
            //若该栈不存在,必须生成一个
            TaskStackBuilder.create(this)
                    .addParentStack(this)
                    .startActivities();
        } else {
            //如果栈已经存在,执行 UP 的导航动作
            NavUtils.navigateUpFromSameTask(this);
        }
        return true;
    default:
        return super.onOptionsItemSelected(item);
    }
}


@Override
public void onItemClick(AdapterView<?> parent, View v, int position,
long id) {
    //启动最终的 Activity,传入选择的条目的名称
    Intent intent = new Intent(this, DetailsActivity.class);
    intent.putExtra(Intent.EXTRA_TEXT, ITEMS[position]);
    startActivity(intent);
}
}
```

代码清单 6-51　第三级 Activity

```java
public class DetailsActivity extends Activity {
    //自定义 Action 字符串,用于外部 Activity 加载
    public static final String ACTION_NEW_ARRIVAL =
            "com.examples.taskstack.ACTION_NEW_ARRIVAL";

    @Override
    protected void onCreate(Bundle savedInstanceState) {
        super.onCreate(savedInstanceState);
        //启用带向上箭头的 Action Bar Home 按钮
        getActionBar().setDisplayHomeAsUpEnabled(true);

        TextView text = new TextView(this);
        text.setGravity(Gravity.CENTER);
        String item = getIntent().getStringExtra(Intent.EXTRA_TEXT);
        text.setText(item);

        setContentView(text);
```

```java
    }

    @Override
    public boolean onOptionsItemSelected(MenuItem item) {
        switch(item.getItemId()) {
        case android.R.id.home:
            //创建父 Activity 的 Intent
            Intent upIntent = NavUtils.getParentActivityIntent(this);
            //检查是否需要创建整个栈
            if(NavUtils.shouldUpRecreateTask(this, upIntent)) {
                //若该栈不存在，必须生成一个
                TaskStackBuilder.create(this)
                    .addParentStack(this)
                    .startActivities();
            } else {
                //如果栈已经存在，执行 UP 的导航动作
                NavUtils.navigateUpFromSameTask(this);
            }
            return true;
        default:
            return super.onOptionsItemSelected(item);
        }
    }
}
```

这个示例应用程序由三个界面组成，根界面上只有一个按钮，可以启动第二级 Activity。第二级 Activity 中是一个含有一些选项的 ListView。选中一个选项后，就会启动第三级 Activity，该 Activity 会在视图中间显示刚刚选择的选项。和预期一样，用户可以使用 BACK 按钮在这个界面栈中进行回退动作。然而，本例中还启用了 UP 动作来实现相同的界面导航效果。

在两个底层 Activity 中启用 UP 导航时会有一些相同的代码。首先会调用 Action Bar 的 setDisplayHomeAsUpEnabled()。此时 Action Bar 上的 Home 图标可以被点击并显示默认的 BACK 箭头，表明 UP 动作是可执行的。当用户点击这个箭头时，就会触发 onOptionsItemSelected()，该项目的 ID 为 android.R.id.home，我们正是通过这个 ID 判断用户是否执行了 UP 导航请求。

在处理 UP 导航请求时，则要判断我们需要的 Activity 栈是否已经存在，是否需要创建，shouldUpRecreateTask()方法起此作用。在 Android 4.1 版本之前的平台上，采用的方法是判断目标 Intent 中是否包含非 Intent.ACTION_MAIN 的有效动作字符串。而在 Android 4.1 及以后版本中，则是检查目标 Intent 的 taskAffinity 属性，从而得知与应用程序的其他界面是否存在亲子关系。

Activity 栈不存在，主要是因为 Activity 是直接启动的，而不是在应用程序中通过界面跳转而启动的，必须重新创建。TaskStackBuilder 中包含了很多方法用来创建界面栈，从而满足应用程序不同的需求。这里使用了便捷的 addParentStack()方法，它会遍历所有的

parentActivityName 属性(或 PARENT_ACTIVITY，如果平台支持的话)和必要的 Intent 来重新创建这个 Activity 到根 Activity 之间的界面栈。之后，只需要调用 startActivities()构建栈并跳转到下一级界面。

如果栈已经存在，调用 NavUtils 的 navigateUpFromSameTask()方法就可以回到上级界面。这种方法非常方便，不必像 navigateUpTo()一样需要调用 getParentActivityIntent()来构建目标 Intent。

现在，应用程序已经可以合理地响应 BACK/UP 导航模式，那么该如何测试呢？直接运行这个应用程序会发现 BACK 和 UP 动作得到的是相同的结果。

让我们创建另一个简单的应用程序以启动 DetailsActivity，这样就可以很好地演示 BACK 和 UP 的导航效果。参见代码清单 6-52 到 6-54。

代码清单 6-52   AndroidManifest.xml

```xml
<manifest xmlns:android="http://schemas.android.com/apk/res/android"
    package="com.examples.taskstacklaunch"
    android:versionCode="1"
    android:versionName="1.0">

    <application android:label="TaskStackLaunch"
        android:icon="@drawable/ic_launcher"
        android:theme="@style/AppTheme">
        <activity
            android:name=".MainActivity">
            <intent-filter>
                <action android:name="android.intent.action.MAIN" />
                <category android:name="android.intent.category.LAUNCHER" />
            </intent-filter>
        </activity>
    </application>

</manifest>
```

代码清单 6-53   res/layout.main.xml

```xml
<?xml version="1.0" encoding="utf-8"?>
<LinearLayout xmlns:android="http://schemas.android.com/apk/res/android"
    android:layout_width="match_parent"
    android:layout_height="match_parent"
    android:orientation="vertical" >
    <Button
        android:id="@+id/button_nephew"
        android:layout_width="match_parent"
        android:layout_height="wrap_content"
        android:text="Add a New Nephew" />
    <Button
        android:id="@+id/button_niece"
        android:layout_width="match_parent"
```

```xml
        android:layout_height="wrap_content"
        android:text="Add a New Niece" />
    <Button
        android:id="@+id/button_twins"
        android:layout_width="match_parent"
        android:layout_height="wrap_content"
        android:text="Add Twin Nieces!" />
</LinearLayout>
```

代码清单 6-54 在 Task 栈中启动的 Activity

```java
public class MainActivity extends Activity implements View.OnClickListener {
    //自定义 Action 字符串，用于外部程序启动该 Activity
    public static final String ACTION_NEW_ARRIVAL =
            "com.examples.taskstack.ACTION_NEW_ARRIVAL";

    @Override
    protected void onCreate(Bundle savedInstanceState) {
        super.onCreate(savedInstanceState);
        setContentView(R.layout.main);
        //添加按钮监听器
        findViewById(R.id.button_nephew).setOnClickListener(this);
        findViewById(R.id.button_niece).setOnClickListener(this);
        findViewById(R.id.button_twins).setOnClickListener(this);
    }

    @Override
    public void onClick(View v) {
        String newArrival;
        switch (v.getId()) {
        case R.id.button_nephew:
            newArrival = "Baby Nephew";
            break;
        case R.id.button_niece:
            newArrival = "Baby Niece";
            break;
        case R.id.button_twins:
            newArrival = "Twin Nieces!";
            break;
        default:
            return;
        }

        Intent intent = new Intent(ACTION_NEW_ARRIVAL);
        intent.putExtra(Intent.EXTRA_TEXT, newArrival);
        startActivity(intent);
    }
}
```

这个应用程序提供了一些选项供传入名称，然后会直接启动我们之前应用程序的 DetailActivity。本例中，我们看到了 BACK 和 UP 表现出来的不同行为。按下 BACK 动作按钮会将用户带回选项选择界面，这是因为正是这个 Activity 启动了 DetailActivity。而按下 UP 动作按钮则会让用户返回原始应用程序的界面栈，因此显示项目的 ListView。从这以后，用户的界面栈就发生了变化，此时按下 BACK 按钮同样会在原始应用程序的界面栈中进行导航，从而匹配后续的 UP 动作。图 6-10 展示了此用例。

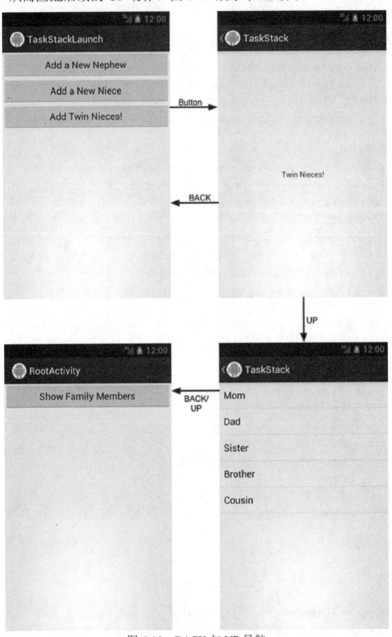

图 6-10　BACK 与 UP 导航

6.17 实现 AppWidget

6.17.1 问题

应用程序提供的信息需要用户快速频繁地访问。你需要在用户的主界面中添加应用程序的可交互组件。

6.17.2 解决方案

(API Level 3)

用户可以选择将应用程序的部分安装到主界面上来构建 AppWidget。AppWidget 也是 Android 能够胜过其他移动操作系统的核心功能。AppWidget 的最吸引人之处就是用户可以在手机的主页面上定制对应用程序的快速访问。

AppWidget 是视图元素,运行在 Launcher 应用程序的进程中,但在应用程序的进程中进行控制。正因为如此,必须使用一种特殊的可以支持远程进程连接的框架。特别地,需要将此小部件的视图结构封装到一个 RemoteViews 对象中,该对象可以通过 ID 更新视图元素而不必直接访问它们。RemoteViews 只支持框架中的一部分布局和小部件。下面的列表显示了 RemoteViews 当前支持的布局和小部件:

- 布局
  - FrameLayout
  - GridLayout
  - LinearLayout
  - RelativeLayout
- 小部件
  - AdapterViewFlipper
  - AnalogClock
  - Button
  - Chronometer
  - GridView
  - ImageButton
  - ImageView
  - ListView
  - ProgressBar
  - StackView
  - TextView
  - ViewFlipper

AppWidget 中的视图只能由以上这些元素组成,否则可能无法正确显示。

在远程进程中工作意味着大多数的用户交互都要通过 PendingIntent 实例来传递,而不是通过传统的监听器接口。PendingIntent 可以让应用程序拦截 Intent 动作以及有权限执行该 Intent 的 Context,这样这个动作就可以自由地传递给其他进程并在某个特定的时间执行,整个过程就好像这个动作是从原始应用程序的 Context 直接传过来一样。

调整大小

手机上的 Android 启动器界面通常都是 4×4 的网格界面空间,在这里可以调整 AppWidget 的大小。而平板电脑的空间会更大一些,在确定小部件的最小高度和宽度时需要考虑一下这种度量情况。Android 3.1 在 AppWidget 放置好以后还允许用户改变它的大小,但在之前的平台版本中,AppWidget 的大小是固定的。摘自 Android 文档,表 6-1 定义了一条比较好的经验规则,即给定的最小尺寸会占用多少网格。

表 6-1 主界面网格的大小

网格数	可用的空间
1	40dp
2	110dp
3	180dp
4	250dp
n	(70×n) − 30

例如,如果小部件最小为 200dp×48dp,那么它在启动器界面上显示时需要三列一行。

6.17.3 实现机制

首先,我们看一下如何构建一个简单的 AppWidget,该 AppWidget 既可以通过小部件本身更新,也可以使用它所关联的 Activity 更新。这个示例会构造一个随机数生成器(可以肯定大家都希望把它放置在启动器界面上)的 AppWidget。首先看一下代码清单 6-55 中的应用程序清单文件。

代码清单 6-55 AndroidManifest.xml

```xml
<application android:label="@string/app_name"
        android:icon="@drawable/ic_launcher">
    <!--简单的 AppWidget 组件 -->
    <activity android:name=".MainActivity">
        <intent-filter>
            <action android:name="android.intent.action.MAIN" />
            <category android:name="android.intent.category.LAUNCHER" />
        </intent-filter>
    </activity>

    <receiver android:name=".SimpleAppWidget">
```

```xml
        <intent-filter>
          <action android:name="android.appwidget.action.APPWIDGET_UPDATE" />
        </intent-filter>
        <!--配置AppWidget 所需的数据 -->
        <meta-data android:name="android.appwidget.provider"
                   android:resource="@xml/simple_appwidget" />
      </receiver>

      <service android:name=".RandomService" />
</application>
```

这里创建 AppWidget 唯一需要做的就是添加名为 SimpleAppWidget 的<receiver>元素。这个元素必须指向 AppWidgetProvider 的子类，可能你已经想到了，该子类是一个自定义的 BroadcastReceiver。它必须在清单中注册 APPWIDGET_UPDATE 广播动作。它还会处理很多其他的广播，但 APPWIDGET_UPDATE 广播动作必须在清单中声明。然后必须附加一个指向<appwidget-provider>的<meta-data>元素，它最终会被添加到 AppWidgetProviderInfo 中。代码清单 6-56 展示了 simple_appwidget.xml。

代码清单 6-56　res/xml/simple_appwidget.xml

```xml
<?xml version="1.0" encoding="utf-8"?>
<appwidget-provider xmlns:android="http://schemas.android.com/apk
                                   /res/android"
    android:minWidth="180dp"
    android:minHeight="40dp"
    android:updatePeriodMillis="86400000"
    android:initialLayout="@layout/simple_widget_layout"/>
```

这些属性定义了 AppWidget 的配置。除了定义大小外，updatePeriodMillis 还定义 Android 应自动更新该小部件以刷新的时间间隔。这个值最好不要设置得比你需要的时间长。很多情况下，使用其他的服务或观察者进行 AppWidget 的更新通知会更加高效。事实上，如果更新频率低于 30 秒，Android 系统是不会进行 AppWidget 更新的。我们将 AppWidget 设置为一天更新一次。这个示例还定义了 initialLayout 属性，它指定 AppWidget 所使用的布局。

另外这里还可以指定其他很多有用的属性：

- android:configure 指定一个配置 AppWidget 的 Activity，该 Activity 是在 AppWidget 添加到启动器之前启动的。
- android:icon 属性通过引用资源指定系统选择用户界面中 AppWidget 的小部件图标的样子。
- android:previewImage 属性通过引用资源指定系统选择用户界面(API Level 11)中 AppWidget 的全尺寸预览。
- android:resizeMode 属性定义了在不同平台上如何重新调整大小：水平、垂直或二者兼有(API Level 12)。

代码清单 6-57 和 6-58 展示了 AppWidget 的布局。

代码清单 6-57　res/layout/simple_widget_layout.xml

```xml
<?xml version="1.0" encoding="utf-8"?>
<LinearLayout xmlns:android="http://schemas.android.com/apk/res/android"
    android:layout_width="match_parent"
    android:layout_height="match_parent"
    android:background="@drawable/widget_background"
    android:orientation="horizontal"
    android:padding="10dp" >
    <LinearLayout
        android:id="@+id/container"
        android:layout_width="0dp"
        android:layout_height="wrap_content"
        android:layout_weight="1"
        android:layout_gravity="center_vertical"
        android:orientation="vertical">
        <TextView
            android:id="@+id/text_title"
            android:layout_width="wrap_content"
            android:layout_height="wrap_content"
            android:layout_gravity="center_horizontal"
            android:textAppearance="?android:attr/textAppearanceMedium"
            android:text="Random Number" />
        <TextView
            android:id="@+id/text_number"
            android:layout_width="wrap_content"
            android:layout_height="wrap_content"
            android:layout_gravity="center_horizontal"
            android:textStyle="bold"
            android:textAppearance="?android:attr/textAppearanceLarge"/>
    </LinearLayout>

    <ImageButton
        android:id="@+id/button_refresh"
        android:layout_width="55dp"
        android:layout_height="55dp"
        android:layout_gravity="center_vertical"
        android:background="@null"
        android:src="@android:drawable/ic_menu_rotate" />

</LinearLayout>
```

代码清单 6-58　res/drawable/widget_background.xml

```xml
<?xml version="1.0" encoding="utf-8"?>
<shape xmlns:android="http://schemas.android.com/apk/res/android"
    android:shape="rectangle">
    <corners
        android:radius="10dp" />
    <solid
        android:color="#A333" />
```

```xml
    <stroke
        android:width="2dp"
        android:color="#333" />
</shape>
```

通常最好为 AppWidget 创建在尺寸变化的容器中可以很容易拉伸和适配的布局(特别是在 AppWidget 可以调整大小的之后版本平台中)。在这个示例中，我们通过 XML 为小部件定义了半透明的圆角矩形背景，该背景可以适应任何尺寸。布局的子视图也使用 weight 属性定义,这样就可以填充多余的空间。这个布局有两个 TextView 元素和一个 ImageButton。所有这些视图需要设置相应的 android:id 属性，这是因为它们后面一旦被封装为 RemoteViews 实例，将无法使用其他方式访问它们。代码清单 6-59 显示了之前提到的 AppWidgetProvider。

代码清单 6-59　AppWidgetProvider 实例

```java
public class SimpleAppWidget extends AppWidgetProvider {
    /*
     * 更新由这个提供程序创建的小部件时，这个方法会被调用
     * 通常情况下，以下情况会调用该方法:
     * 1. 开始创建小部件
     * 2. 达到 AppWidgetProviderInfo 中定义的 updatePeriodMillis 时间间隔
     * 3. AppWidgetManager 中的 updateAppWidget()方法被手动调用
     */
    @Override
    public void onUpdate(Context context, AppWidgetManager appWidgetManager,
            int[] appWidgetIds) {
        //启动一个后台服务来更新此小部件
        context.startService(new Intent(context, RandomService.class));
    }
}
```

这里只需要实现 onUpdate()方法即可，该方法在用户添加小部件时会被调用。另外，之后 Android 框架或应用程序需要更新小部件时也会调用该方法。很多情况下，可以直接在该方法中创建视图和更新 AppWidget。但由于 AppWidgetProvider 是一个 BroadcastReceiver，最好不要在它里面执行时间过长的操作。如果构建 AppWidget 需要做很多工作，应该启动一个服务或后台线程来完成，我们在这里就是这样做的。

为了方便起见，这个方法会传入一个 AppWidgetManager 实例，如果想要在这个方法中更新 AppWidget，必须用到 AppWidgetManager。在同一个启动器界面中可以加载多个相同的 AppWidget。ID 数组就是每个 AppWidget 的引用，因此可以全部更新。下面看一下代码清单 6-60 所示的服务代码。

代码清单 6-60　AppWidget 服务

```java
public class RandomService extends Service {
    /*更新完成后的广播动作 */
    public static final String ACTION_RANDOM_NUMBER =
            "com.examples.appwidget.ACTION_RANDOM_NUMBER";
```

```java
/* 当前数据,通过静态值保存*/
private static int sRandomNumber;
public static int getRandomNumber() {
    return sRandomNumber;
}

@Override
public int onStartCommand(Intent intent, int flags, int startId) {
    //更新随机数数据
    sRandomNumber = (int)(Math.random() * 100);

    //创建 AppWidget 视图
    RemoteViews views = new RemoteViews(getPackageName(),
            R.layout.simple_widget_layout);
    views.setTextViewText(R.id.text_number, String.valueOf(sRandomNumber));

    //为刷新按钮设置一个 Intent,该 Intent 会再次启动这个服务
    PendingIntent refreshIntent = PendingIntent.getService(this, 0,
            new Intent(this, RandomService.class), 0);
    views.setOnClickPendingIntent(R.id.button_refresh, refreshIntent);

    //设置一个 Intent,在单击小部件文本时会打开一个 Activity
    PendingIntent appIntent = PendingIntent.getActivity(this, 0,
            new Intent(this, MainActivity.class), 0);
    views.setOnClickPendingIntent(R.id.container, appIntent);

    //更新小部件
    AppWidgetManager manager = AppWidgetManager.getInstance(this);
    ComponentName widget = new ComponentName(this, SimpleAppWidget.class);
    manager.updateAppWidget(widget, views);

    //发送一个广播,通知所有的监听者
    Intent broadcast = new Intent(ACTION_RANDOM_NUMBER);
    sendBroadcast(broadcast);

    //这个服务不应继续运行下去
    stopSelf();
    return START_NOT_STICKY;
}

/*
 * 这里我们并没有绑定这个服务,所以这个方法返回 null 即可
 */
@Override
public IBinder onBind(Intent intent) {
    return null;
}
}
```

这个 RandomService 在启动时会执行两个操作。首先，重新生成一个随机数数据并保存到一个静态变量中。其次，它为我们的 AppWidget 构建了一个新的视图。通过这种方式，我们就可以在需要时使用该服务刷新 AppWidget 了。首先必须创建一个 RemoteViews 实例，传入我们的小部件布局，然后通过 setTextViewText() 方法更新布局中 TextView 的数字，通过 setOnClickPendingIntent() 方法关联控件点击时的监听器。第一个 PendingIntent 关联的是 AppWidget 的刷新按钮，该 Intent 会设置为激活以重新启动这个服务。第二个 PendingIntent 关联的是小部件的主布局，点击主布局的任何地方都会启动应用程序的主 Activity。

初始化 RemoteViews 的最后一步就是更新 AppWidget。这里是通过先获得 AppWidget-Manager 的实例，然后调用 updateAppWidget() 实现的。这里我们并没有提供程序关联的每个 AppWidget 的 ID，如果有的话，也可以使用 ID 更新每个 AppWidget。相反，我们可以传入一个引用 AppWidgetProvider 的 ComponentName，这时 AppWidgetProvider 关联的所有 AppWidget 都会被更新。

最后，我们发送了一个广播，通知所有的监听器已经重新生成了一个随机数，然后停止这个服务。至此，我们完成了 AppWidget 工作于此设备上需要的所有代码。最后我们需要额外添加一些组件和一个用于与相同数据交互的 Activity。参见代码清单 6-61 和 6-62。

代码清单 6-61　res/layout/main.xml

```xml
<?xml version="1.0" encoding="utf-8"?>
<LinearLayout xmlns:android="http://schemas.android.com/apk/res/android"
    android:layout_width="match_parent"
    android:layout_height="match_parent"
    android:orientation="vertical" >
    <Button
        android:layout_width="match_parent"
        android:layout_height="wrap_content"
        android:text="Generate New Number"
        android:onClick="onRandomClick" />
    <TextView
        android:layout_width="wrap_content"
        android:layout_height="wrap_content"
        android:layout_gravity="center_horizontal"
        android:textAppearance="?android:attr/textAppearanceLarge"
        android:text="Current Random Number" />
    <TextView
        android:id="@+id/text_number"
        android:layout_width="wrap_content"
        android:layout_height="wrap_content"
        android:layout_gravity="center_horizontal"
        android:textSize="55dp"
        android:textStyle="bold" />

</LinearLayout>
```

代码清单 6-62　应用程序的主 Activity

```java
public class MainActivity extends Activity {

    private TextView mCurrentNumber;

    @Override
    protected void onCreate(Bundle savedInstanceState) {
        super.onCreate(savedInstanceState);
        setContentView(R.layout.main);

        mCurrentNumber = (TextView) findViewById(R.id.text_number);
    }

    @Override
    protected void onResume() {
        super.onResume();
        updateNumberView();
        //注册一个接收器，在服务结束时接收更新
        IntentFilter filter = new IntentFilter(RandomService.ACTION_RANDOM_NUMBER);
        registerReceiver(mReceiver, filter);
    }

    @Override
    protected void onPause() {
        super.onPause();
        //解除接收器的注册
        unregisterReceiver(mReceiver);
    }

    public void onRandomClick(View v) {
        //调用服务，更新随机数
        startService(new Intent(this, RandomService.class));
    }

    private void updateNumberView() {
        //用最新的数字更新视图
        mCurrentNumber.setText(String.valueOf(RandomService.getRandomNumber()));
    }

    private BroadcastReceiver mReceiver = new BroadcastReceiver() {
        @Override
        public void onReceive(Context context, Intent intent) {
            //使用新的数字更新视图
            updateNumberView();
        }
```

```
        };
    }
```

这个 Activity 显示了 RandomService 所提供的当前随机数的值。点击按钮时还会启动服务以生成一个新的数字。它最大的好处就是还会更新我们的 AppWidget，从而保证两者同步。我们还注册了一个 BroadcastReceiver 来监听服务生成新数字的结束事件，这样就可以及时更新当前的用户界面。图 6-11 展示了应用程序的 Activity 以及添加到手机主界面上的对应 AppWidget。

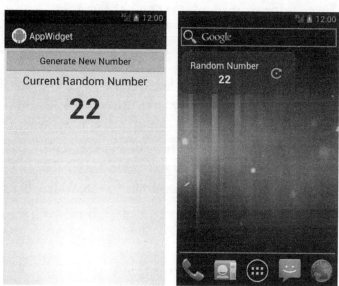

图 6-11　应用程序的随机数 Activity(左图)和 AppWidget(右图)

基于集合的 AppWidget

(API Level 12)

从 Android 3.0 开始，在 AppWidget 框架中加入了集合视图后，AppWidget 上可以显示的内容也更多了。这允许应用程序通过列表、网格或栈显示信息。在 Android 3.1 中，AppWidget 在放置好后还可以调整大小。让我们查看一个 AppWidget 示例，该 AppWidget 允许用户查看自己的媒体集合信息。同样，首先看一下代码清单 6-63。

代码清单 6-63　AndroidManifest.xml

```
<application android:label="@string/app_name"
    android:icon="@drawable/ic_launcher">
    <!--集合 AppWidget 组件 -->
    <activity android:name=".ListWidgetConfigureActivity">
        <intent-filter>
            <action android:name="android.
                appwidget.action.APPWIDGET_CONFIGURE"/>
        </intent-filter>
    </activity>
```

```xml
<receiver android:name=".ListAppWidget">
    <intent-filter>
        <action android:name="android.
            appwidget.action.APPWIDGET_UPDATE" />
    </intent-filter>
    <meta-data android:name="android.appwidget.provider"
        android:resource="@xml/list_appwidget" />
</receiver>

<service android:name=".ListWidgetService"
    android:permission="android.permission.BIND_REMOTEVIEWS" />
<service android:name=".MediaService" />
</application>
```

这个示例中的 AppWidgetProvider 定义和之前类似,只是名称变为 ListAppWidget。我们还定义了一个带有特殊权限(BIND_REMOTEVIEWS)的服务。稍后会看到它实际上是一个 RemoteViewsService,Android 框架会使用这个服务为 AppWidget 的列表提供数据,这与 ListAdapter 使用 ListView 的工作原理相似。最后,我们定义了一个 Activity,它会在用户添加 AppWidget 前对 AppWidget 进行设置。要想实现这个功能,这个 Activity 必须为 APPWIDGET_CONFIGURE 动作包含一个<intent-filter>。AppWidget 相关的 AppWidgetProviderInfo 定义可以参见代码清单 6-64。

代码清单 6-64 res/xml/list_appwidget.xml

```xml
<?xml version="1.0" encoding="utf-8"?>
<appwidget-provider
    xmlns:android="http://schemas.android.com/apk/res/android"
    android:minWidth="110dp"
    android:minHeight="110dp"
    android:updatePeriodMillis="86400000"
    android:initialLayout="@layout/list_widget_layout"
    android:configure="com.examples.appwidget.ListWidgetConfigureActivity"
    android:resizeMode="horizontal|vertical"/>
```

除了之前示例中介绍的一些标准属性,这里还添加 android:configure 属性(指向我们的配置 Activity)和 android:resizeMode 属性(允许 AppWidget 在两个方向调整大小)。代码清单 6-65 到 6-67 显示了 AppWidget 自身和 ListView 每个条目的布局文件代码。

代码清单 6-65 res/layout/list_widget_layout.xml

```xml
<?xml version="1.0" encoding="utf-8"?>
<LinearLayout xmlns:android="http://schemas.android.com/apk/res/android"
    android:layout_width="match_parent"
    android:layout_height="match_parent"
    android:orientation="vertical"
    android:background="@drawable/list_widget_background">
    <TextView
        android:id="@+id/text_title"
        android:layout_width="match_parent"
```

```xml
            android:layout_height="45dp"
            android:gravity="center"
            android:textAppearance="?android:attr/textAppearanceMedium" />
    <FrameLayout
        android:layout_width="match_parent"
        android:layout_height="match_parent" >
        <ListView
            android:id="@+id/list"
            android:layout_width="match_parent"
            android:layout_height="match_parent" />
        <TextView
            android:id="@+id/list_empty"
            android:layout_width="wrap_content"
            android:layout_height="wrap_content"
            android:layout_gravity="center"
            android:text="No Items Available" />
    </FrameLayout>
</LinearLayout>
```

代码清单6-66   res/drawable/list_widget_background.xml

```xml
<?xml version="1.0" encoding="utf-8"?>
<shape xmlns:android="http://schemas.android.com/apk/res/android"
       android:shape="rectangle">
    <solid
        android:color="#A333" />
</shape>
```

代码清单6-67   res/layout/list_widget_item.xml

```xml
<?xml version="1.0" encoding="utf-8"?>
<LinearLayout xmlns:android="http://schemas.android.com/apk/res/android"
    android:id="@+id/list_widget_item"
    android:layout_width="match_parent"
    android:layout_height="?android:attr/listPreferredItemHeight"
    android:paddingLeft="10dp"
    android:gravity="center_vertical"
    android:orientation="vertical" >

    <TextView
        android:id="@+id/line1"
        android:layout_width="wrap_content"
        android:layout_height="wrap_content" />

    <TextView
        android:id="@+id/line2"
        android:layout_width="wrap_content"
        android:layout_height="wrap_content" />
```

```
</LinearLayout>
```

AppWidget 的布局就是一个简单的 ListView，ListView 的上面是一个 TextView 标题。我们将 ListView 封装到一个 FrameLayout 中，这样就可以提供一个完全空白的视图。

提示：

尝试在 AppWidget 中使用大多数 Android 标准的 ListView 行布局(例如 android.R.id.simple_list_item_1)将不会成功。这是因为这些行布局中通常会含有 CheckedTextView 这种 RemoteViews 不支持的视图。必须自己创建每行的布局。

在查看 AppWidgetProvider 之前，首先让我们看一下配置 Activity。这个界面是用户把 AppWidget 添加到手机主屏幕后(安装之前)看到的第一个界面。这个 Activity 的结果实际会控制 AppWidgetProvider 是否已经被调用。参见代码清单 6-68 和 6-69。

代码清单 6-68　res/layout/configure.xml

```
<?xml version="1.0" encoding="utf-8"?>
<RelativeLayout xmlns:android="http://schemas.android.com/apk/res/android"
    android:layout_width="match_parent"
    android:layout_height="match_parent">
    <TextView
        android:id="@+id/text_title"
        android:layout_width="wrap_content"
        android:layout_height="wrap_content"
        android:textAppearance="?android:attr/textAppearanceLarge"
        android:text="Select Media Type:" />
    <RadioGroup
        android:id="@+id/group_mode"
        android:layout_width="wrap_content"
        android:layout_height="wrap_content"
        android:layout_below="@id/text_title"
        android:orientation="vertical">
        <RadioButton
            android:id="@+id/mode_image"
            android:layout_width="wrap_content"
            android:layout_height="wrap_content"
            android:text="Images"/>
        <RadioButton
            android:id="@+id/mode_video"
            android:layout_width="wrap_content"
            android:layout_height="wrap_content"
            android:text="Videos"/>
    </RadioGroup>

    <Button
        android:layout_width="match_parent"
        android:layout_height="wrap_content"
        android:layout_alignParentBottom="true"
        android:text="Add Widget"
        android:onClick="onAddClick" />
```

```
</RelativeLayout>
```

代码清单 6-69　配置 Activity

```java
public class ListWidgetConfigureActivity extends Activity {

    private int mAppWidgetId;
    private RadioGroup mModeGroup;

    @Override
    protected void onCreate(Bundle savedInstanceState) {
        super.onCreate(savedInstanceState);
        setContentView(R.layout.configure);

        mModeGroup = (RadioGroup) findViewById(R.id.group_mode);

        mAppWidgetId = getIntent()
                .getIntExtra(AppWidgetManager.EXTRA_APPWIDGET_ID,
                        AppWidgetManager.INVALID_APPWIDGET_ID);

        setResult(RESULT_CANCELED);
    }

    public void onAddClick(View v) {
        SharedPreferences.Editor prefs =
                getSharedPreferences(String.valueOf(mAppWidgetId),
                MODE_PRIVATE).edit();
        RemoteViews views = new RemoteViews(getPackageName(),
                R.layout.list_widget_layout);
        switch (mModeGroup.getCheckedRadioButtonId()) {
        case R.id.mode_image:
            prefs.putString(ListWidgetService.KEY_MODE,
                    ListWidgetService.MODE_IMAGE).commit();
            views.setTextViewText(R.id.text_title, "Image Collection");
            break;
        case R.id.mode_video:
            prefs.putString(ListWidgetService.KEY_MODE,
                    ListWidgetService.MODE_VIDEO).commit();
            views.setTextViewText(R.id.text_title, "Video Collection");
            break;
        default:
            Toast.makeText(this, "Please Select a Media Type.",
                    Toast.LENGTH_SHORT).show();
            return;
        }

        Intent intent = new Intent(this, ListWidgetService.class);
        intent.putExtra(AppWidgetManager.EXTRA_APPWIDGET_ID, mAppWidgetId);
        intent.setData(Uri.parse(intent.toUri(Intent.URI_INTENT_SCHEME)));
```

```java
            //通过指向RemoteViewService的Intent为列表关联一个适配器,从而填充数据
            views.setRemoteAdapter(mAppWidgetId, R.id.list, intent);
            //设置列表为空白视图
            views.setEmptyView(R.id.list, R.id.list_empty);

            Intent viewIntent = new Intent(Intent.ACTION_VIEW);
            PendingIntent pendingIntent = PendingIntent.getActivity(this, 0,
                viewIntent, 0);
            views.setPendingIntentTemplate(R.id.list, pendingIntent);

            AppWidgetManager manager = AppWidgetManager.getInstance(this);
            manager.updateAppWidget(mAppWidgetId, views);

            Intent data = new Intent();
            data.putExtra(AppWidgetManager.EXTRA_APPWIDGET_ID, mAppWidgetId);
            setResult(RESULT_OK, data);
            finish();
        }
    }
```

这个 Activity 的布局中有一个用来选择图片和视频的 RadioGroup,之后会在 AppWidget 的列表和 Add 按钮上显示选择的媒体类型。按照惯例,在我们进入 Activity 后会立即将结果设为 RESULT_CANCELED。这是因为如果用户进入 Activity 后不点击 Add 就离开,这时我们并不希望 AppWidget 显示在屏幕上。Android 框架会检测 Activity 的结果,从而决定是否添加 AppWidget。这里我们还得到了框架传过来的 AppWidget 的 ID,先保存起来供后面使用。

在用户做了选择并点击了 Add 按钮之后,用户的选择就会被分别保存到名为 AppWidget 的 ID 的 SharedPreferences 实例中。这是因为我们希望应用程序可以处理多个小部件,而这些小部件的配置值是独立的,所以我们没有使用默认的 SharedPreferences 保存此数据。

注意:
在 Android 4.1 中,可以使用一个 Bundle 向 AppWidget 传递配置数据。但为了兼容之前的版本,可以选择使用 SharedPreferences 传递数据。

然后可以开始构建 AppWidget 的 RemoteViews,根据用户选择的媒体类型设置 RemoteView 的标题。对于基于集合的 AppWidget,必须构建一个 Intent 来启动 RemoteViewsService 的实例,从而作为集合中数据的适配器,这一点类似于 ListAdapter。然后通过 setRemoteAdapter()将该适配器关联到 RemoteViews,这里需要传入适配器关联的 ListView 的 ID。在列表为空时,我们会使用 setEmptyView()只显示一个 TextView。

每个列表条目都会关联一个 PendingIntent,当用户点击列表中的条目时会触发这个 PendingIntent。框架需要知道你为每个条目设置的特定信息,因此这里为每个条目都使用一个填充的 PendingIntent 模板。我们为每个条目都创建了一个简单 ACTION_VIEW 的 Intent,然后通过 setPendingIntentTemplate()把它与列表关联起来,数据和额外信息字段会在稍后填充。

准备工作做好后，我们调用了 AppWidgetManager 的 updateAppWidget()。本例中，由于我们只想更新这个特定的 AppWidget，因此调用了只有一个 ID 参数而不是以 ComponentName 作为参数的 updateAppWidget()版本。然后，设置结果的值为 RESULT_OK 并结束操作，这时框架会将 AppWidget 添加到主屏幕上。让我们简单地看一下代码清单 6-70 所示的 AppWidgetProvider。

代码清单 6-70　列表 AppWidgetProvider

```java
public class ListAppWidget extends AppWidgetProvider {

    /*
     * 更新这个提供程序创建的小部件时，这个方法会被调用
     * 通常情况下，以下情况会调用该方法：
     * 由于添加了配置Activity，因此在第一次添加小部件时并不会调用该方法。但在下面这
     *   种情况下还会被调用：
     * 达到AppWidgetProviderInfo中定义的updatePeriodMillis时间间隔
     */
    @Override
    public void onUpdate(Context context, AppWidgetManager appWidgetManager,
            int[] appWidgetIds) {
        //更新这个提供程序创建的所有小部件
        for(int i=0; i < appWidgetIds.length; i++) {
            Intent intent = new Intent(context, ListWidgetService.class);
            intent.putExtra(AppWidgetManager.EXTRA_APPWIDGET_ID, appWidgetIds[i]);
            intent.setData(Uri.parse(intent.toUri(Intent.URI_INTENT_SCHE-ME)));

            RemoteViews views = new RemoteViews(context.getPackageName(),
                    R.layout.list_widget_layout);
            //根据小部件的配置设置标题
            SharedPreferences prefs =
                    context.getSharedPreferences(String.valueOf
                        (appWidgetIds[i]), Context.MODE_PRIVATE);
            String mode = prefs.getString(ListWidgetService.KEY_MODE,
                    ListWidgetService.MODE_IMAGE);
            if(ListWidgetService.MODE_VIDEO.equals(mode)) {
                views.setTextViewText(R.id.text_title, "Video Collection");
            } else {
                views.setTextViewText(R.id.text_title, "Image Collection");
            }

            //通过指向RemoteViewService的Intent为列表关联一个适配器，从而填充数据
            views.setRemoteAdapter(appWidgetIds[i], R.id.list, intent);

            //为列表设置空白视图
            views.setEmptyView(R.id.list, R.id.list_empty);

            //为每个条目设置点击时的模板Intent
            Intent viewIntent = new Intent(Intent.ACTION_VIEW);
            PendingIntent pendingIntent = PendingIntent.getActivity(context, 0,
```

```java
                    viewIntent, 0);
            views.setPendingIntentTemplate(R.id.list, pendingIntent);
            appWidgetManager.updateAppWidget(appWidgetIds[i], views);
        }
    }

    /*
     * 当第一个小部件被添加到提供程序时会调用
     */
    @Override
    public void onEnabled(Context context) {
        //启动一个服务来监控MediaStore
        context.startService(new Intent(context, MediaService.class));
    }

    /*
     * 当所有的小部件从此提供程序中移除时会调用
     */
    @Override
    public void onDisabled(Context context) {
        //停止监控MediaStore的服务
        context.stopService(new Intent(context, MediaService.class));
    }

    /*
     * 当一个或多个关联的小部件从这个提供程序中移除时会调用
     */
    @Override
    public void onDeleted(Context context, int[] appWidgetIds) {
        //移除每个小部件时，同时移除为每个小部件创建的SharedPreferences
        for(int i=0; i < appWidgetIds.length; i++) {
            context.getSharedPreferences(String.valueOf(appWidgetIds[i]),
                    Context.MODE_PRIVATE)
                .edit()
                .clear()
                .commit();
        }

    }
}
```

除了将之前的更新动作换成现在读取用户配置当前值的动作，该提供程序中 onUpdate() 方法的代码和配置 Activity 中的代码几乎是一样的。这是因为我们希望后续更新会有相同的 AppWidget 结果。

这个提供程序还覆写了 onEnabled() 和 onDisabled() 方法。这两个方法分别会在第一个小部件被添加到提供程序以及最后一个小部件被移除时调用。这时提供程序会使用它们启动和停止一个需要长时间运行的服务，稍后会详细了解这个服务，但其作用是监控

MediaStore 的改动，从而可以相应更新 AppWidget。最后，每个 AppWidget 在移除时会回调 onDeleted()方法。在这个示例中，会在这个方法中清除在添加 AppWidget 时创建的 SharedPreferences。

现在查看一下代码清单 6-71，它定义了 RemoteViewsService 来为 AppWidget 列表提供数据。

代码清单 6-71　RemoteViews 适配器

```java
public class ListWidgetService extends RemoteViewsService {

    public static final String KEY_MODE = "mode";
    public static final String MODE_IMAGE = "image";
    public static final String MODE_VIDEO = "video";

    @Override
    public RemoteViewsFactory onGetViewFactory(Intent intent) {
        return new ListRemoteViewsFactory(this, intent);
    }

    private class ListRemoteViewsFactory implements
            RemoteViewsService.RemoteViewsFactory {
        private Context mContext;
        private int mAppWidgetId;

        private Cursor mDataCursor;

        public ListRemoteViewsFactory(Context context, Intent intent) {
            mContext = context.getApplicationContext();
            mAppWidgetId = intent.getIntExtra(
            AppWidgetManager.EXTRA_APPWIDGET_ID,
                    AppWidgetManager.INVALID_APPWIDGET_ID);
        }

        @Override
        public void onCreate() {
            //加载首选项以得到用户在添加小部件时设置的信息
            SharedPreferences prefs =
                    mContext.getSharedPreferences(String.valueOf(
                        mAppWidgetId), MODE_PRIVATE);
            //得到用户的配置信息，默认为图片模式
            String mode = prefs.getString(KEY_MODE, MODE_IMAGE);
            //根据用户的配置信息查询相应的媒体类型数据
            if(MODE_VIDEO.equals(mode)) {
                //在 MediaStore 中查询视频数据
                String[] projection = {MediaStore.Video.Media.TITLE,
                    MediaStore.Video.Media.DATE_TAKEN,
                    MediaStore.Video.Media.DATA};
                mDataCursor = MediaStore.Images.Media.query(getContentResolver(),
                    MediaStore.Video.Media.EXTERNAL_CONTENT_URI, projection);
```

```java
        } else {
            //在MediaStore中查询图片数据
            String[] projection = {MediaStore.Images.Media.TITLE,
                    MediaStore.Images.Media.DATE_TAKEN,
                    MediaStore.Images.Media.DATA};
            mDataCursor = MediaStore.Images.Media.query(getContentResolver(),
                    MediaStore.Images.Media.EXTERNAL_CONTENT_URI,
        projection);
        }
    }

    /*
     * 这个方法会在onCreate()后被调用,但在外部调用AppWidgetManager.notify-
       AppWidgetViewDataChanged()来刷新小部件数据时也会调用这个方法
     */
    @Override
    public void onDataSetChanged() {
        //刷新Cursor数据
        mDataCursor.requery();
    }

    @Override
    public void onDestroy() {
        //不再需要时关闭Cursor
        mDataCursor.close();
        mDataCursor = null;
    }

    @Override
    public int getCount() {
        return mDataCursor.getCount();
    }

    /*
     *如果数据来自于网络或其他地方,可能需要一些时间来下载,这里可以返回加载视图
     *加载视图在getViewAt()阻塞时会一直显示,直到getViewAt()返回时才消失
     */
    @Override
    public RemoteViews getLoadingView() {
        return null;
    }

    /*
     *返回集合中每个条目的视图。在这个方法里可以放心执行长时间的操作
     *该方法返回之前,会一直显示加载视图
     */
    @Override
    public RemoteViews getViewAt(int position) {
        mDataCursor.moveToPosition(position);
```

```java
        RemoteViews views = new RemoteViews(getPackageName(),
            R.layout.list_widget_item);
        views.setTextViewText(R.id.line1, mDataCursor.getString(0));
        views.setTextViewText(R.id.line2, DateFormat.format("MM/dd/yyyy",
            mDataCursor.getLong(1)));

        SharedPreferences prefs = mContext.getSharedPreferences(
            String.valueOf(mAppWidgetId), MODE_PRIVATE);
        String mode = prefs.getString(KEY_MODE, MODE_IMAGE);
        String type;
        if(MODE_VIDEO.equals(mode)) {
            type = "video/*";
        } else {
            type = "image/*";
        }

        Uri data = Uri.fromFile(new File(mDataCursor.getString(2)));

        Intent intent = new Intent();
        intent.setDataAndType(data, type);
        views.setOnClickFillInIntent(R.id.list_widget_item, intent);

        return views;
    }

    @Override
    public int getViewTypeCount() {
        return 1;
    }

    @Override
    public boolean hasStableIds() {
        return false;
    }

    @Override
    public long getItemId(int position) {
        return position;
    }
  }
}
```

RemoteViewsService 必须返回的 RemoteViewsFactory 实现和 ListAdapter 非常相似。很多方法，如 getCount()和 getViewTypeCount()，在本地列表中执行的功能是一样的。当 RemoteViewsFactory 开始创建时，我们会检查用户在配置阶段选择的设置值，然后会在系统的 MediaStore 内容提供程序中查询相应的 Cursor 来显示图片或视频数据。当不需要 RemoteViewsFactory 而要销毁时，会关闭刚才查询到的 Cursor。当由于外部原因通知 AppWidgetManager 需要更新数据时，就会调用 onDataSetChanged()方法。要刷新数据时，我们只需要重新查询 Cursor。

getViewAt()方法用来获得列表中每个条目的视图。这个方法可以放心地执行长时间的操作(例如网络 I/O)，这时 Android 框架在 getViewAt()方法返回之前会一直显示 getLoadingView()方法中指定的内容。这个示例中，我们需要更新 RemoteViews 中每行布局中的标题和文本表示(指定条目的创建日期)。然后，我们必须填充在最初的更新中设置的 PendingIntent 模板。我们设置了图片或视频数据的文件路径以及将 data 字段设置为相应的 MIME 类型。再加上 ACTION_VIEW，就可以在设备的 Gallery 应用程序(或当条目被点击时，打开其他可以处理媒体数据的应用程序)中打开相应的文件了。

你可能已经注意到，这个示例在检索 Cursor 数据时并没有使用明确的字段名。这主要是因为在两种媒体类型之间的投影中，相应字段的名称是不同的，所以直接使用索引访问会更加高效。最后，让我们看一下代码清单 6-72，它展示了由 AppWidgetProvider 进行启动和停止的后台服务代码。

代码清单 6-72　更新监控服务

```java
public class MediaService extends Service {

    private ContentObserver mMediaStoreObserver;

    @Override
    public void onCreate() {
        super.onCreate();
        //服务启动后，创建并注册MediaStore的一个观察者
        mMediaStoreObserver = new ContentObserver(new Handler()) {
            @Override
            public void onChange(boolean selfChange) {
                //更新AppWidgetProvider当前所关联的所有小部件
                AppWidgetManager manager =
                        AppWidgetManager.getInstance(MediaService.this);
                ComponentName provider = new ComponentName(MediaService.this,
                        ListAppWidget.class);
                int[] appWidgetIds = manager.getAppWidgetIds(provider);
                //这个方法会触发RemoteViewsService中的onDataSetChanged()方法
                manager.notifyAppWidgetViewDataChanged(appWidgetIds, R.id.list);
            }
        };
        //注册图片和视频的观察者
        getContentResolver().registerContentObserver(
                MediaStore.Images.Media.EXTERNAL_CONTENT_URI, true,
                mMediaStoreObserver);
        getContentResolver().registerContentObserver(
                MediaStore.Video.Media.EXTERNAL_CONTENT_URI, true,
                mMediaStoreObserver);
    }

    @Override
```

```
public void onDestroy() {
    super.onDestroy();
    //当服务停止后，解除观察者的注册
    getContentResolver().unregisterContentObserver(mMediaStoreObserver);
}

/*
 * 我们不需要绑定这个服务，方法返回 null 即可
 */
@Override
public IBinder onBind(Intent intent) {
    return null;
}
```

这个服务的作用就是在有 AppWidget 可用时为 MediaStore 注册 ContentObserver。这样，在添加或删除图片或视频时，就可以及时更新我们的小部件列表。每当触发 ContentObserver 时，我们都会对当前关联的每个小部件调用 AppWidgetManager 的 notifyAppWidgetViewDataChanged() 方法。这时就会触发 onDataSetChanged() 中的 onDataSetChanged() 回调方法来刷新列表。在图 6-12 和图 6-13 中可以看到运行结果。

可以看到，只需要在 AppWidgetProviderInfo 中添加简单的 resize 属性，用户就可以调整 AppWidget 的大小了。每个列表都是可滚动的，点击列表中的条目会打开默认的应用程序来浏览图片或播放视频。

图 6-12　添加 AppWidget 前出现的配置 Activity

图 6-13　两种媒体类型的 AppWidget(左图)和调整大小后的效果(右图)

6.18　支持受限制的配置文件

6.18.1　问题

应用程序的目标使用者是各种年龄层次的人和残疾人士，你需要为此提供控制权，修改应用程序的行为来满足每个特殊的用户。

6.18.2　解决方案

(API Level 18)

UserManager 提供了关于系统级功能的一些常规信息，如果通过 getUserRestrictions() 将用户配置文件设置为受限制，则这些系统级功能不会提供给该配置文件。此外，应用程序可以定义应该在受限制环境中可配置的自定义功能集，然后通过调用 getApplication-Restrictions()，从 UserManager 中获得设备的当前设置。

每个应用程序都可以定义一组 RestrictionEntry 元素，系统会在用户设置中将这些元素展示给设备所有者，以便其为"受限制的配置文件"配置应用程序。每个元素定义了设置的类型(布尔选项、单选或多选)以及应在设置中可见的数据。

支持多个用户账户的 Android 设备为设备所有者(其定义为在设备上建立的第一个账户)提供了创建额外用户或"受限制的配置文件"的能力。这些辅助用户有自己的应用程序、数据空间，并且能够像所有者一样管理设备，只不过管理的是其他用户账户。

"受限制的配置文件"在 Android 4.3 中引入，是对作为所有者账户一部分的应用程序和数据提供受限制访问的方式。这些配置文件没有自己的应用程序空间或关联的数据。相反，它们是一组控件，所有者可以将这些控件放在相应的位置，在此位置可以限制使用他们自己的哪些应用程序以及访问这些应用程序的哪些功能。例如，明显的用例是家长控制，但也可以使用"受限制的配置文件"将设备暂时置于信息亭模式(kiosk mode)。

提示：
模拟器映像中一般不会启用多用户账户，通常只在平板设备上支持多用户账户。这些类型的功能在手机上并不常见。

6.18.3 实现机制

为说明利用受限用户环境优点的应用程序，我们为儿童和年轻人构造了一个简单的绘图应用程序。我们将使用应用程序级的限制来移除和修改某些应用程序功能。代码清单6-73包含了用户界面的布局。

代码清单 6-73 res/layout/activity_main.xml

```xml
<?xml version="1.0" encoding="utf-8"?>
<FrameLayout xmlns:android="http://schemas.android.com/apk/res/android"
    android:layout_width="match_parent"
    android:layout_height="match_parent" >

    <com.androidrecipes.restrictedprofiles.DrawingView
        android:id="@+id/drawing_surface"
        android:layout_width="match_parent"
        android:layout_height="match_parent"/>

    <Button
        android:id="@+id/button_purchase"
        android:layout_width="wrap_content"
        android:layout_height="wrap_content"
        android:layout_gravity="right"
        android:text="$$$$"
        android:onClick="onPurchaseClick"/>
    <SeekBar
        android:id="@+id/full_slider"
        android:layout_width="match_parent"
        android:layout_height="wrap_content"
        android:layout_gravity="bottom"
        android:max="45"/>
    <RadioGroup
        android:id="@+id/simple_selector"
        android:layout_width="match_parent"
        android:layout_height="wrap_content"
        android:layout_gravity="bottom"
        android:orientation="horizontal">
        <RadioButton
            android:id="@+id/option_small"
            android:layout_width="0dp"
```

```xml
            android:layout_height="wrap_content"
            android:layout_weight="1"
            android:textColor="#555"
            android:text="Small" />
        <RadioButton
            android:id="@+id/option_medium"
            android:layout_width="0dp"
            android:layout_height="wrap_content"
            android:layout_weight="1"
            android:textColor="#555"
            android:text="Medium" />
        <RadioButton
            android:id="@+id/option_large"
            android:layout_width="0dp"
            android:layout_height="wrap_content"
            android:layout_weight="1"
            android:textColor="#555"
            android:text="Big" />
        <RadioButton
            android:id="@+id/option_xlarge"
            android:layout_width="0dp"
            android:layout_height="wrap_content"
            android:layout_weight="1"
            android:textColor="#555"
            android:text="Really Big" />
    </RadioGroup>
</FrameLayout>
```

在此例中，我们创建了一个绘图表面，用户可在该表面上使用手指绘画(我们将很快查看这个自定义视图)；创建了一个按钮，让用户从我们的虚构商店中购买升级的内容(在此是更好的颜色)；还在界面底部创建了一些 UI 元素，用于调整绘图的线宽(滑块和一组单选按钮)。我们将使用应用程序限制来控制后两种功能。参见代码清单 6-74 以了解 Activity 元素。

代码清单 6-74　受限制配置文件的 Activity

```java
public class MainActivity extends Activity implements
        OnSeekBarChangeListener, OnCheckedChangeListener {

    private Button mPurchaseButton;
    private DrawingView mDrawingView;
    private SeekBar mFullSlider;
    private RadioGroup mSimpleSelector;

    /* 配置文件限制值 */
    private boolean mHasPurchases;
    private int mMinAge;
    /* 内容购买标志 */
    private boolean mHasCanvasColors = false;
```

```java
    private boolean mHasPaintColors = false;

    @Override
    protected void onCreate(Bundle savedInstanceState) {
        super.onCreate(savedInstanceState);
        setContentView(R.layout.activity_main);

        mPurchaseButton = (Button) findViewById(R.id.button_purchase);
        mDrawingView = (DrawingView) findViewById(R.id.drawing_surface);
        mFullSlider = (SeekBar) findViewById(R.id.full_slider);
        mSimpleSelector = (RadioGroup) findViewById(R.id.simple_selector);

        mFullSlider.setOnSeekBarChangeListener(this);
        mSimpleSelector.setOnCheckedChangeListener(this);

        if(Build.VERSION.SDK_INT >= Build.VERSION_CODES.JELLY_BEAN_MR2) {
            UserManager manager = (UserManager) getSystemService(USER_SERVICE);
            // 检查系统级的限制
            Bundle restrictions = manager.getUserRestrictions();
            if(restrictions != null && !restrictions.isEmpty()) {
                showSystemRestrictionsDialog(restrictions);
            }
        }
    }

    @Override
    protected void onStart() {
        super.onStart();
        /*
         * 当应用程序在后台时，限制可能改变，因此需要在每次返回时检查此限制
         */
        updateRestrictions();
        // 基于限制改变更新 UI
        updateDisplay();
    }

    public void onPurchaseClick(View v) {
        AlertDialog.Builder builder =
                new AlertDialog.Builder(this);
        builder.setTitle("Content Upgrades")
                .setMessage(
                        "Tap any of the following items to add them.")
                .setPositiveButton("Canvas Colors $2.99",
                        mPurchaseListener)
                .setNeutralButton("Paint Colors $2.99",
                        mPurchaseListener)
                .setNegativeButton("Both Items $4.99",
                        mPurchaseListener).show();
    }
```

```java
        private DialogInterface.OnClickListener mPurchaseListener =
                new DialogInterface.OnClickListener() {
            @Override
            public void onClick(DialogInterface dialog, int which) {
                switch(which) {
                    case DialogInterface.BUTTON_POSITIVE:
                        mHasCanvasColors = true;
                        break;
                    case DialogInterface.BUTTON_NEUTRAL:
                        mHasPaintColors = true;
                        break;
                    case DialogInterface.BUTTON_NEGATIVE:
                        mHasCanvasColors = true;
                        mHasPaintColors = true;
                        break;
                }
                Toast.makeText(getApplicationContext(), "Thank You For Your
                   Purchase!", Toast.LENGTH_SHORT).show();
                updateDisplay();
            }
        };

        private void showSystemRestrictionsDialog(Bundle restrictions) {
            StringBuilder message = new StringBuilder();
            for(String key : restrictions.keySet()) {
                // 确保限制的值为 true
                if(restrictions.getBoolean(key)) {
                    message.append(RestrictionsReceiver.getNameForRestriction(key));
                    message.append("\n");
                }
            }

            AlertDialog.Builder builder = new AlertDialog.Builder(this);
            builder.setTitle("System Restrictions")
                .setMessage(message.toString())
                .setPositiveButton("OK", null)
                .show();
        }

        @Override
        public void onCheckedChanged(RadioGroup group, int checkedId) {
            float width;
            switch(checkedId) {
                default:
                case R.id.option_small:
                    width = 4f;
                    break;
                case R.id.option_medium:
                    width = 12f;
                    break;
```

```java
            case R.id.option_large:
                width = 25f;
                break;
            case R.id.option_xlarge:
                width = 45f;
                break;
        }
        mDrawingView.setStrokeWidth(width);
    }

    @Override
    public void onProgressChanged(SeekBar seekBar, int progress,
            boolean fromUser) {
        mDrawingView.setStrokeWidth(progress);
    }

    @Override
    public void onStartTrackingTouch(SeekBar seekBar) { }

    @Override
    public void onStopTrackingTouch(SeekBar seekBar) { }

    private void updateDisplay() {
        // 显示/隐藏购买按钮
        mPurchaseButton.setVisibility(
                mHasPurchases ? View.VISIBLE : View.GONE);

        // 更新年龄限制内容
        mFullSlider.setVisibility(View.GONE);
        mSimpleSelector.setVisibility(View.GONE);
        switch (mMinAge) {
            case 18:
                // 完整范围的滑块
                mFullSlider.setVisibility(View.VISIBLE);
                mFullSlider.setProgress(4);
                break;
            case 10:
                // 4个选项
                mSimpleSelector.setVisibility(View.VISIBLE);
                findViewById(R.id.option_medium).setVisibility(View.VISIBLE);
                findViewById(R.id.option_xlarge).setVisibility(View.VISIBLE);
                mSimpleSelector.check(R.id.option_medium);
                break;
            case 5:
                // 大/小选项
                mSimpleSelector.setVisibility(View.VISIBLE);
                findViewById(R.id.option_medium).setVisibility(View.GONE);
                findViewById(R.id.option_xlarge).setVisibility(View.GONE);
```

```java
                    mSimpleSelector.check(R.id.option_small);
                    break;
                case 3:
                default:
                    // 无选择项
                    break;
            }

            // 使用购买的内容更新显示
            mDrawingView.setPaintColor(mHasPaintColors ? Color.BLUE : Color.GRAY);
            mDrawingView.setCanvasColor(mHasCanvasColors ? Color.GREEN : Color.BLACK);
        }

        private void updateRestrictions() {
            // 检查限制
            if(Build.VERSION.SDK_INT >= Build.VERSION_CODES.JELLY_BEAN_MR2) {
                UserManager manager = (UserManager)
                        getSystemService(USER_SERVICE);
                Bundle restrictions = manager
                        .getApplicationRestrictions(getPackageName());
                if(restrictions != null) {
                    // 读取限制设置
                    mHasPurchases = restrictions.getBoolean(
                            RestrictionsReceiver.RESTRICTION_PURCHASE, true);
                    try {
                        mMinAge = Integer.parseInt(restrictions.getString(
                                RestrictionsReceiver.RESTRICTION_AGERANGE, "18"));
                    } catch(NumberFormatException e) {
                        mMinAge = 0;
                    }
                } else {
                    // 没有限制
                    mHasPurchases = true;
                    mMinAge = 18;
                }
            } else {
                // 我们未处于支持限制的系统上
                mHasPurchases = true;
                mMinAge = 18;
            }
        }
    }
```

1. 系统功能限制

创建 Activity 时，在确认运行在目标平台为 API Level 18 或以后版本的设备上之后，我们使用 UserManager.getUserRestrictions()确定是否存在系统级的限制。这会返回一个 Bundle，如果没有任何限制(即作为设备所有者或另一个完整权限用户运行此设备时)，则该 Bundle 为空。然而，如果确实存在限制，我们就将这些限制收集到一起，并在屏幕上显示一个对话框。Bundle 中的唯一键和布尔值描述了每个限制。对于每个可能的键，如果值为 true，则应用相应限制；如果限制未得到应用，则值可能为 false，或者键可能完全没有出现在 Bundle 中。下面是可能的系统限制的列表：

- DISALLOW_CONFIG_BLUETOOTH：该配置文件不能配置蓝牙。
- DISALLOW_CONFIG_CREDENTIALS：该配置文件不能配置系统用户凭证。
- DISALLOW_CONFIG_WIFI：该配置文件不能修改 Wi-Fi 访问点配置。
- DISALLOW_INSTALL_APPS：该配置文件不能安装新的应用程序。
- DISALLOW_INSTALL_UNKNOWN_SOURCES：该配置文件不能在设备设置中启用"未知来源"以安装应用程序。
- DISALLOW_MODIFY_ACCOUNTS：该配置文件不能添加或移除设备上的账户。
- DISALLOW_REMOVE_USER：该配置文件不能移除其他用户。
- DISALLOW_SHARE_LOCATION：该配置文件不能切换位置共享设置。
- DISALLOW_UNINSTALL_APPS：该配置文件不能卸载应用程序。
- DISALLOW_USB_FILE_TRANSFER：该配置文件不能通过 USB 传输文件。

我们显示的描述信息提取自 RestrictionsReceiver 内的辅助实用方法，这是一个 BroadcastReceiver，我们在代码清单 6-75 中进行了定义。

代码清单 6-75　RestrictionsReceiver

```java
public class RestrictionsReceiver extends BroadcastReceiver {

    public static final String RESTRICTION_PURCHASE = "purchases";
    public static final String RESTRICTION_AGERANGE = "age_range";

    private static final String[] AGES = {"3+", "5+", "10+", "18+"};
    private static final String[] AGE_VALUES = {"3", "5", "10", "18"};

    @Override
    public void onReceive(Context context, Intent intent) {
       ArrayList<RestrictionEntry> restrictions = new
           ArrayList<RestrictionEntry>();

       RestrictionEntry purchase = new
           RestrictionEntry(RESTRICTION_PURCHASE, false);
       purchase.setTitle("Content Purchases");
       purchase.setDescription("Allow purchasing of content in the
           application.");
       restrictions.add(purchase);

       RestrictionEntry ages =
               new RestrictionEntry(RESTRICTION_AGERANGE, AGE_VALUES[0]);
       ages.setTitle("Age Level");
       ages.setDescription("Difficulty level for application content.");
       ages.setChoiceEntries(AGES);
       ages.setChoiceValues(AGE_VALUES);
```

```java
        restrictions.add(ages);

        Bundle result = new Bundle();
        result.putParcelableArrayList(Intent.EXTRA_RESTRICTIONS_LIST,
            restrictions);

        setResultExtras(result);
    }

    /*
     * 用于从限制键获取可读字符串的实用方法
     */
    public static String getNameForRestriction(String key) {
        if(UserManager.DISALLOW_CONFIG_BLUETOOTH.equals(key)) {
            return "Unable to configure Bluetooth";
        }
        if(UserManager.DISALLOW_CONFIG_CREDENTIALS.equals(key)) {
            return "Unable to configure user credentials";
        }
        if(UserManager.DISALLOW_CONFIG_WIFI.equals(key)) {
            return "Unable to configure Wifi";
        }
        if(UserManager.DISALLOW_INSTALL_APPS.equals(key)) {
            return "Unable to install applications";
        }
        if(UserManager.DISALLOW_INSTALL_UNKNOWN_SOURCES.equals(key)) {
            return "Unable to enable unknown sources";
        }
        if(UserManager.DISALLOW_MODIFY_ACCOUNTS.equals(key)) {
            return "Unable to modify accounts";
        }
        if(UserManager.DISALLOW_REMOVE_USER.equals(key)) {
            return "Unable to remove users";
        }
        if(UserManager.DISALLOW_SHARE_LOCATION.equals(key)) {
            return "Unable to toggle location sharing";
        }
        if(UserManager.DISALLOW_UNINSTALL_APPS.equals(key)) {
            return "Unable to uninstall applications";
        }
        if(UserManager.DISALLOW_USB_FILE_TRANSFER.equals(key)) {
            return "Unable to transfer files";
        }

        return "Unknown Restriction: "+key;
    }
}
```

2. 应用程序特有限制

除了托管描述的实用方法之外，RestrictionsReceiver 的主要作用是为此应用程序定义一组自定义限制，我们需要向设备所有者明确揭示这些限制。在查找揭示出来的限制时，框架将使用 android.intent.action.GET_RESTRICTION_ENTRIES 动作字符串发送一个有序的广播 Intent。然后，接收器负责过滤此动作以构造 RestrictionEntry 元素的列表，并在结果 Bundle 中返回该列表。

提示：

如果限制字符串相当复杂，无法简化为一些可选择的条目，或者你更愿意更好地提升用户体验，可以使用 EXTRA_RESTRICTIONS_INTENT 作为键，在接收器的结果 Bundle 中返回 Intent。该 Intent 应引用一个 Activity，设备设置需要启动该 Activity 以便为应用程序建立限制。在此情况下，应通过 Activity 的结果返回限制的键/值数据。

我们定义了两个要揭示的限制设置：一个用于允许用户从应用程序中进行购买操作，另一个用于根据用户的年龄层次修改应用程序体验。第一个设置创建为布尔类型，其默认值为 false(即该限制的值默认为 false)；第二个设置是单项选择，其年龄选项从 3 岁以上到 18 岁以上。代码清单 6-76 显示了必须放在清单中的<receiver>代码片段，这样此接收器才可以正确发布。

代码清单 6-76　用于限制接收器的清单代码片段

```
<receiver android:name=".RestrictionsReceiver">
    <intent-filter>
        <action android:name="android.intent.action.GET_RESTRICTION_ENTRIES"/>
    </intent-filter>
</receiver>
```

安装此应用程序后，现在这两个设置将显示给设备所有者，以便其在建立受限制用户配置文件时进行配置。

在代码清单 6-74 中，可以看到当 Activity 启动时，通过调用 UserManager.getApplicationRestrictions()检查当前的应用程序限制集以获得另一个 Bundle。这个 Bundle 包含在接收器中定义的设置的键/值对列表。使用此 Bundle 中的值更新 Activity 的内部状态，该 Activity 控制用户界面如何显示。如果没有任何限制(假如是设备所有者)，该方法将返回 null。

因为该应用程序在设备所有者和受限制配置文件之间共享，我们必须假设这些设置值可以在 Activity 处于后台时更改。出于该原因，在 onStart()中完成这些检查，而不是在 onCreate()或其他一些一次性的初始化例程中进行检查。

Purchases 设置控制顶部角落的 Money 按钮是否可见。如果允许购买，用户就可以点击此按钮并从虚构的店面中进行选择，获得新的线条颜色、背景色或两者，从而为他们的绘图增添趣味。

Age Level 设置控制用户可以执行哪些操作来更新线宽。对于每名幼儿，该设置将会产生妨碍，因此保持固定线宽并隐藏所有相关控件。随着儿童年龄的增长，我们希望为他们提供一些选项，因此给出一组单选按钮，其中包含两个或 4 个宽度选择项。如果最小年龄始终设置为 18 岁以上，则使用完整滑块元素更换此 UI，以便用户准确选择所需的线宽(单个像素精度)。

为结束本例，代码清单 6-77 显示了自定义的手指绘图视图。

代码清单 6-77　手指绘图的视图

```java
public class DrawingView extends View {

    private Paint mFingerPaint;
    private Path mPath;

    public DrawingView(Context context) {
        super(context);
        init();
    }

    public DrawingView(Context context, AttributeSet attrs) {
        super(context, attrs);
        init();
    }

    public DrawingView(Context context, AttributeSet attrs,
            int defStyle) {
        super(context, attrs, defStyle);
        init();
    }

    private void init() {
        // 设置画笔
        mFingerPaint = new Paint(Paint.ANTI_ALIAS_FLAG);
        mFingerPaint.setStyle(Style.STROKE);
        mFingerPaint.setStrokeCap(Cap.ROUND);
        mFingerPaint.setStrokeJoin(Join.ROUND);
        // 默认的笔划宽度
        mFingerPaint.setStrokeWidth(8f);
    }

    public void setPaintColor(int color) {
        mFingerPaint.setColor(color);
    }

    public void setStrokeWidth(float width) {
        mFingerPaint.setStrokeWidth(width);
    }
```

```java
    public void setCanvasColor(int color) {
        setBackgroundColor(color);
    }

    @Override
    public boolean onTouchEvent(MotionEvent event) {
        switch(event.getActionMasked()) {
            case MotionEvent.ACTION_DOWN:
                mPath = new Path();
                // 在触摸位置开始绘图
                mPath.moveTo(event.getX(), event.getY());
                // 重新绘图
                invalidate();
                break;
            case MotionEvent.ACTION_MOVE:
                // 添加事件之间的所有触摸点
                for(int i=0; i < event.getHistorySize(); i++) {
                    mPath.lineTo(event.getHistoricalX(i),
                            event.getHistoricalY(i) );
                }
                // 重新绘图
                invalidate();
                break;
            default:
                break;
        }
        return true;
    }


    @Override
    protected void onDraw(Canvas canvas) {
        // 绘制背景
        super.onDraw(canvas);
        // 绘制画笔
        if(mPath != null) {
            canvas.drawPath(mPath, mFingerPaint);
        }
    }
}
```

这是基本的 View 实现，它跟踪所有的触摸事件并将它们转换为待绘制的 Path。在每个新的触摸手势中，抛弃旧的 Path 并将初始触摸点添加到新的 Path。在每个随后的移动事件中，当用户滑动手指时，使用沿着触摸事件轨迹的线条更新 Path，并且使该视图失效(这会再次触发 onDraw())。我们在每个新的手势中抛弃旧的内容，因此视图仅绘制当前笔划，并且当再次触摸视图时，现有内容将清除。

此外，我们添加了外部设置器，根据在 UI 中做出的选择更新笔划宽度和颜色参数。这些值仅是对用于绘制结果线条的 Paint 进行的修改。图 6-14 显示了运行在设备所有者账户上的应用程序，其中所有功能都不受限制地运行。

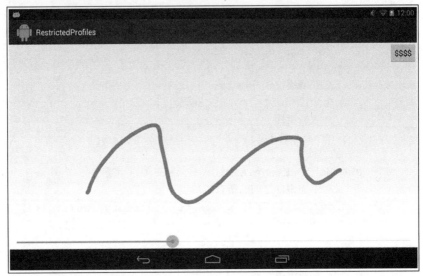

图 6-14　不受限制用户的绘制应用程序 UI

如果在此设备上创建了受限制的用户，则部分配置设置将用于对该配置文件启用应用程序，然后设定适合于目标用户的设置。参见图 6-15，查看 Nexus 7 设备上这些设置的示例。

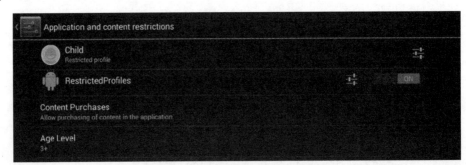

图 6-15　受限制配置文件的内容设置

最后，使用如图 6-15 所示的限制，图 6-16 显示了运行在受限制配置文件下的相同应用程序。

首先可以看到显示系统级限制的对话框，然后是主应用程序 UI。在此可以注意到，Purchase 按钮不再可见，并且笔划宽度控件已经更换为更简单的选择项。

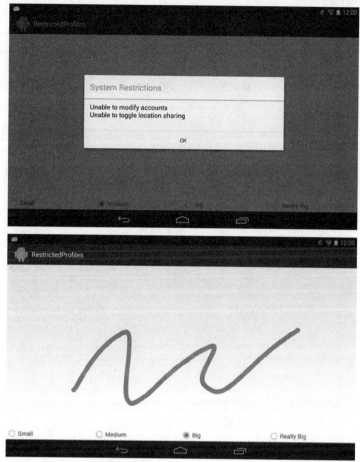

图 6-16 显示了系统限制的对话框(上图),处于受限制模式的应用程序 UI(下图)

6.19 小结

本章介绍了如何让应用程序直接与 Android 操作系统互动,讨论了几种将操作提交到后台运行的方法,这些操作所需的运行时间长短不一。接下来你了解了应用程序之间如何分工合作、相互调用,从而才能更好地完成任务。最后介绍了系统如何将其核心应用程序收集的数据提供给你的应用程序使用。在第 7 章中,你将学习如何利用各种公共可用的 Java 库来开发更加强大的应用程序。

第 7 章

图形和绘图

Android 的 UI 工具包明确地重视灵活性，能够创建在所有可能设备类型之间无缝缩放的内容。前面介绍了视图系统如何帮助开发人员适应这种需求。在本章中，我们将讨论使用 Android 的可扩展图形对象 Drawable 增强应用程序用户界面的其他技术，其中会介绍如何利用 Drawable 创建灵活的动画图形元素，这些元素可以与任何视图关联。本章主要就是研究如何利用应用程序中较少的资源完成较多的工作。

7.1 用 Drawable 做背景

7.1.1 问题

应用程序需要创建可缩放和适合任意视图空间的背景图片，而你并不想为此耗费时间来生成大量图片文件。

7.1.2 解决方案

(API Level 1)
用 Android 最强力的 XML 资源系统实现：创建形状 Drawable。如果能做到这点，用 XML 资源创建这些视图就很有意义了，因为用 XML 创建的 Drawable 本身就是可缩放的。如果将其设为背景，它会自动填充视图的边界。

在 XML 中用<shape>标签定义 Drawable 时，实际上是创建了一个 GradientDrawable 对象。对象的形状可以是矩形、椭圆、线条或圆圈，最常用的背景形状是矩形。具体来说，在使用矩形时，可以用下面这些参数定义形状：

- 角半径：定义所有 4 个圆角的半径，或是分别定义各个圆角的半径。
- 渐变：线性、放射或扫描线渐变，支持两个或三个颜色值。方向可以是 45°的任何倍数(0 就是从左到右，90 就是从下到上，依此类推)。

- 单色：用一种颜色填充形状。如果同时定义了渐变的话，效果会受影响。
- 边线：对象形状的边界。可以定义边界的宽度和颜色。
- 大小和填充。

7.1.3 实现机制

创建视图的静态背景图片是非常麻烦的事情，需要准备各种尺寸的图片才能确保背景在所有的设备上正常显示。如果视图的大小还需要根据其中的内容动态变化，事情会变得更复杂。

为了避免这个问题，在 res/drawable 中创建一个描述形状的 XML 文件，然后可以通过 android:background 属性将其应用为视图的背景。

1. 渐变的 ListView 行

第一个示例是创建一个渐变的矩形，可以将其设为 ListView 中各行的背景。代码清单 7-1 中是定义该形状的 XML。

代码清单 7-1　res/drawable/backgradient.xml

```xml
<?xml version="1.0" encoding="utf-8"?>
<shape xmlns:android="http://schemas.android.com/apk/res/android"
    android:shape="rectangle">
    <gradient
        android:startColor="#EFEFEF"
        android:endColor="#989898"
        android:type="linear"
        android:angle="270"
    />
</shape>
```

这里我们选择使用两种灰阶之间的线性渐变，从上到下变化。如果要在渐变中加入第三种颜色，需要在<gradient>标签中添加 android:middleColor 属性。

现在，可以在构建自定义 ListView 的视图或布局中引用这个 Drawable 资源。要将这个 Drawable 设为背景，可以在视图的 XML 文件中加入 android:background="@drawable/backgradient"，也可以在 Java 代码中调用 View.setBackgroundResource(R.drawable.backgradient)。

高级提示：

XML 中颜色的种类限制是 3 种，但在 GradientDrawable 的构造函数中，颜色参数是一个整型数组 int[]，传递多少颜色都可以。

将该 Drawable 设为 LisrView 中各行的背景，结果如图 7-1 所示。

图 7-1　行背景是渐变的 Drawable

2. 圆角视图组

另一种流行的技术是用 XML Drawable 创建布局的背景，将若干个小部件组织在一起。常用的样式是圆角加上细边框。在 XML 中定义该形状的代码如代码清单 7-2 所示。

代码清单 7-2　res/drawable/roundback.xml

```xml
<?xml version="1.0" encoding="utf-8"?>
<shape xmlns:android="http://schemas.android.com/apk/res/android"
    android:shape="rectangle">
    <solid
        android:color="#FFF"
    />
    <corners
        android:radius="10dip"
    />
    <stroke
        android:width="5dip"
        android:color="#555"
    />
</shape>
```

在这个示例中，填充色是白色，边框线条是灰色。如前一个示例所示，在任何视图和布局中都可以在 XML 中加入属性 android:background="@drawable/roundback"，或是在 Java 代码中调用 View.setBackgroundResource(R.drawable.roundback) 以引用这个 Drawable。

将该背景应用到视图，结果如图 7-2 所示。

图 7-2 带边框的圆角矩形视图背景

3. Drawable 模式

我们要查看的关于 Drawable 的下一个分类就是模式。通过 XML，我们可以定义一些规则来定义小图片的重复平铺模式。这对于处理全屏背景图片是非常有用的，因为这种方式不需要向内存中加载大的 Bitmap 对象。

应用程序可以通过设置<bitmap>元素的 tileMode 属性来创建一种模式，属性值如下：

- clamp：复制源位图的边缘像素。
- repeat：源位图会在横向和纵向两个方向重复地平铺。
- mirror：源位图会重复地进行平铺，但按照源位图和镜像位图交替的方式平铺。

图 7-3 显示了两张小的正方形图片，它们将成为 Drawable 模式的源图片。

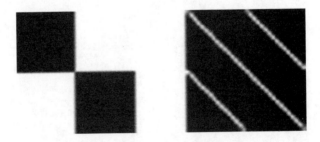

图 7-3 Drawable 模式的源位图

代码清单 7-3 和 7-4 显示了如何将 XML 模式定义为背景。

代码清单 7-3　res/drawable/pattern_checker.xml

```xml
<?xml version="1.0" encoding="utf-8"?>
<bitmap xmlns:android="http://schemas.android.com/apk/res/android"
    android:src="@drawable/checkers"
    android:tileMode="repeat" />
```

代码清单 7-4　res/drawable/pattern_stripes.xml

```xml
<?xml version="1.0" encoding="utf-8"?>
<bitmap xmlns:android="http://schemas.android.com/apk/res/android"
    android:src="@drawable/stripes"
    android:tileMode="mirror" />
```

提示：

Drawable 模式只可以使用有固定边界的位图，例如外部图片，而 XML 形状不能作为 Drawable 模式的源位图。

图 7-4 显示了视图在应用每种模式后的背景效果。

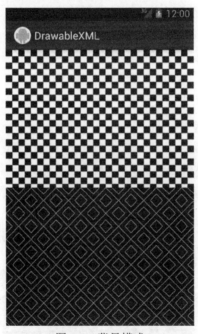

图 7-4　背景模式

你会看到棋盘图片会按原样重复，而条纹图片会在水平和垂直方向进行镜像，之后再平铺到整个屏幕上，从而创建图 7-4 所示的钻石效果。

4. 9-patch 图片

NinePatchDrawable 是 Android 最大的优势之一，使其可在所有设备上都灵活调整 UI。9-patch 是一种特殊的图片，通过指定可缩放和不可缩放的图片区域来实现某个特定区域的拉伸。事实上，这种图片类型得名于 9 个拉伸区域，这些区域会创建图片映射(在某一时刻)。

让我们看一个示例以更好地了解它的实现机制。图 7-5 显示了两张图片，左边的是原始图片，右边的是转换后生成的 9-patch 图片。

图 7-5　语音气泡源图片 speech_background.png(左图)和 9-patch 转换图片 speech_background.9.png(右图)

注意右边图片上各条边上的黑色标记。一个有效的 9-patch 图片文件就是一张简单的 PNG 图片，该图片最外边的 1 个像素只会是黑色像素或透明像素。每条边上的黑色像素定义了图片该如何拉伸并保护图片的一部分内容不被拉伸。

- 左侧：这里的黑色像素定义了图片垂直方向拉伸的区域。这个区域的内容会重复平铺以实现拉伸。图 7-5 的示例图片定义了该拉伸区域。
- 顶部：这里的黑色像素定义了图片水平方向拉伸的区域。这个区域的内容会重复平铺以实现拉伸。图 7-5 的示例图片定义了两个这样的拉伸区域。
- 右侧：这里的黑色像素定义了图片垂直方向的内容区域，也就是视图内容的显示区域。事实上，它决定了顶部和底部的填充值，但保留了背景图片区域。
- 底部：这里的黑色像素定义了图片水平方向的内容区域，也就是视图内容的显示区域。事实上，它决定了左侧和右侧的填充值，但保留了背景图片区域。这里必须包含一行，组成该行的实体像素定义了该区域。

这张图片是使用 Android SDK 提供的 draw9patch 工具创建的。为了更好地看清楚这些标记如何影响图片的显示效果，让我们看一下该图片加载到此工具中的效果。参见图 7-6。

图 7-6　draw9patch 工具中的语音气泡图片

现在你已知道 9-patch 名称的由来。图片的非高亮部分是不可以拉伸的。每张图片的高亮区域将会在某个方向(水平方向或垂直方向，根据它们的方向)进行拉伸，交叉高亮的区域则可以在每个方向都进行拉伸。在各个方向至少拥有一张拉伸区域的图片，这将会在图片中创建 9 个独立的映射区域：4 个角是不能修改的、4 个中间区域拉伸一次、1 个中间区域拉伸两次。

创建 NinePatchDrawable 和使用它作为背景并不需要任何特殊的代码；图片文件只需要以特殊的.9.png 扩展名命名，这样 Android 就可以正确地处理它。代码清单 7-5 显示了如何将这张图片设置为背景，而图 7-7 显示了将这张图片设置为 TextView 的背景后的效果。

图 7-7　将语音气泡作为 TextView 的背景

代码清单 7-5　res/layout/main.xml

```
<?xml version="1.0" encoding="utf-8"?>
<RelativeLayout
    xmlns:android="http://schemas.android.com/apk/res/android"
    android:layout_width="match_parent"
    android:layout_height="match_parent" >
    <TextView
        android:layout_width="match_parent"
        android:layout_height="wrap_content"
        android:layout_centerVertical="true"
        android:gravity="center"
        android:text="This is a text speech bubble"
        android:background="@drawable/speech_background"/>
</RelativeLayout>
```

注意两个 3 像素宽的横向拉伸区如何围绕语音气泡的原点中心均匀分配它们之间多余的空间。要在两个拉伸点之间创建偏移区域，可以通过改变它们到图片中心的距离或改变它们的大小来实现。如果有个区域是三个像素宽，而另一个区域只有一个像素宽，那么前者在拉伸时会占据三倍于后者的空间。

7.2 创建自定义状态的 Drawable

7.2.1 问题

想要自定义诸如 Button 或 CheckBox 之类的有多个状态(默认、按下、选中等)的元素。

7.2.2 解决方案

(API Level 1)

创建一个状态列表(state-list)的 Drawable，将其应用到要自定义的元素。无论是用 XML 定义 Drawable 图形还是使用图片，都可以通过 Android 提供的另一个 XML 元素<selector>创建一个引用，指向多张图片以及多张图片可见的条件。

(API Level 21)

使用 AnimatedStateListDrawable 和 StateListAnimator 为附加视图定义动画和状态过渡。还可以使用此 API Level 中的 RippleDrawable 在此视图上提供动画形式的触摸反馈作为逐渐增强的波纹效果。

7.2.3 实现机制

先看看下面这个状态列表的 Drawable 的示例，然后再深入讨论它的各个部分：

```xml
<?xml version="1.0" encoding="utf-8"?>
<selector
    xmlns:android="http://schemas.android.com/apk/res/android">
    <item android:state_enabled="false"
        android:drawable="@drawable/disabled" />
    <item android:state_pressed="true"
        android:drawable="@drawable/selected" />
    <item android:state_focused="true"
        android:drawable="@drawable/selected" />
    <!-- 默认状态 -->
    <item android:drawable="@drawable/default" />
</selector>
```

注意：
<selector>是有序的。Android 在遍历列表时，会返回第一个状态完全匹配的 Drawable。在将状态属性应用到各条目时要注意这一点。

列表的每个条目中所引用的 Drawable 都对应一种状态，如果某个条目需要匹配多个状态值的话，可以加入多个状态参数。Android 会遍历这个列表，选择其中第一个能匹配当前视图各项指标的状态，将其对应的可绘制资源应用到视图。从这个角度考虑，最好是将正常状态(即没有任何附加指标的默认状态)放在列表的最底部。

下面列出了最常用的状态属性，它们都有布尔值：
- state_enabled：isEnabled()返回的视图的值。
- state_pressed：用户在触摸屏上单击了该视图。
- state_focused：视图获得了焦点。
- state_selected：用户通过按键或 D-pad 选中了视图。
- state_checked：isChecked()返回的可选中视图的值。

现在来查看如何将这些状态列表的 Drawable 应用到各种视图。

1. Button 和其他可单击小部件

当视图在上述状态间切换时，诸如 Button 之类的小部件的背景 Drawable 会发生变化。所以，XML 中的 android:background 属性或 view.setBackgroundDrawable()方法都可用于添加状态列表。代码清单 7-6 中的示例是 res/drawable/中定义的 button_states.xml 文件。

代码清单 7-6　res/drawable/button_states.xml

```xml
<?xml version="1.0" encoding="utf-8"?>
<selector
    xmlns:android="http://schemas.android.com/apk/res/android">
    <item android:state_enabled="false"
        android:drawable="@drawable/disabled" />
    <item android:state_pressed="true"
        android:drawable="@drawable/selected" />
    <!-- 默认状态 -->
    <item android:drawable="@drawable/default" />
</selector>
```

这里列出的 3 个@drawable 资源是项目中的图片，选择器就在这些图片间切换。如前所述，如果所有的项都不能与当前视图的状态匹配，那么最后一项就会作为默认值返回，所以最后一项不需要标明其所匹配的状态。下面将这个状态列表添加到 XML 中定义的一个视图：

```xml
<Button
    android:layout_width="wrap_content"
    android:layout_height="wrap_content"
    android:text="My Button"
    android:background="@drawable/button_states" />
```

2. CheckBox 和其他可选择小部件

很多小部件都实现了可单击的界面，例如 CheckBox 和其他的 CompoundButton 子类，它们的状态切换机制都大同小异。在这些小部件中，背景跟状态没有关联，而是通过另一个名为 button 的属性实现用自定义的 Drawable 来表示"选中"状态。在 XML 中，这就是 android:button 属性，在代码中则是用 CompundButton.seBbuttonDrawable()方法来实现的。

代码清单 7-7 中的示例是 res/drawable/中定义的 check_stater.xml 文件。再提醒一下，其中罗列的@drawable 资源就是要切换的引用图片。

代码清单 7-7　res/drawable/check_states.xml

```xml
<?xml version="1.0" encoding="utf-8"?>
<selector
    xmlns:android="http://schemas.android.com/apk/res/android">
    <item android:state_enabled="false"
        android:drawable="@drawable/disabled" />
    <item android:state_checked="true"
        android:drawable="@drawable/checked" />
    <!-- 默认状态 -->
    <item android:drawable="@drawable/unchecked" />
</selector>
```

在 XML 中将其关联到 CheckBox 上：

```xml
<CheckBox
    android:layout_width="wrap_content"
    android:layout_height="wrap_content"
    android:button="@drawable/check_states" />
```

3. 制作状态过渡动画

(API Level 21)

接下来使用一些动画触摸反馈修饰之前的按钮示例。代码清单 7-8 使用<ripple>元素定义了更新的后台 XML(用于构造 RippleDrawable)：

代码清单 7-8　res/drawable_v21/background_button.xml

```xml
<?xml version="1.0" encoding="utf-8"?>
<ripple xmlns:android="http://schemas.android.com/apk/res/android"
    android:color="#0CC">
    <!-- 显示的默认 Drawable -->
    <item android:drawable="@drawable/button_default"/>

    <!-- 匹配默认值的波纹效果剪切遮罩 -->
    <item
        android:id="@android:id/mask"
        android:drawable="@drawable/button_default"/>
</ripple>
```

RippleDrawable 采用多个子 Drawable 作为层，并且按顺序绘制这些 Drawable。我们在此例中包括了两层，第一层是静态 Drawable，它引用与之前相同的默认图片。当按钮处于默认状态时，用户就会看到此图片。

默认情况下，RippleDrawable 绘制一个圆形波纹动画，在用户点击包含视图时，此波纹会从触摸点向外发散(此触摸点也称为波纹的热点)。除非提供遮罩，否则此圆形波纹不会被视图的边界(或其他边界)剪切。

通过向第二个<item>层添加 android:id/mask 属性，就可以告诉 Android 框架，此 Drawable 代表要用于剪切波纹效果的边界。此条目不会被绘制。

通过在根元素上使用 android:color 属性，就可以应用希望用于波纹效果的颜色。Android 框架会将给定的颜色划分为稍微透明的版本和不透明的覆盖层，前者用于直接突出显示视图，后者用于制作动画。动画的速率取决于触摸反馈：对于点击事件动画速率较快，而对于长按事件则较慢。

代码清单 7-9 使用 StateListAnimator 添加了额外的反馈层。在按下按钮时，StateListAnimator 将使此按钮的尺寸稍微收缩。

代码清单 7-9　res/animator/button_press.xml

```xml
<?xml version="1.0" encoding="utf-8"?>
<selector xmlns:android="http://schemas.android.com/apk/res/android">
    <item android:state_enabled="true" android:state_pressed="true">
        <set android:ordering="together">
            <objectAnimator
                android:duration="@android:integer/config_shortAnimTime"
                android:propertyName="scaleX"
                android:valueTo="0.8"
                android:valueType="floatType" />
            <objectAnimator
                android:duration="@android:integer/config_shortAnimTime"
                android:propertyName="scaleY"
                android:valueTo="0.8"
                android:valueType="floatType" />
        </set>
    </item>
    <!-- 默认状态 -->
    <item>
        <set android:ordering="together">
            <objectAnimator
                android:duration="@android:integer/config_shortAnimTime"
                android:propertyName="scaleX"
                android:valueTo="1.0"
                android:valueType="floatType" />
            <objectAnimator
                android:duration="@android:integer/config_shortAnimTime"
                android:propertyName="scaleY"
                android:valueTo="1.0"
                android:valueType="floatType" />
        </set>
    </item>
</selector>
```

此 XML 结构使用与前面相同的<selector>，但是在 res/animator 目录中，这代表动画器实例而非 Drawable 的状态列表集合。在此例中，每个状态是一对 ObjectAnimator 实例，用于同时在两个主轴中缩放视图。现在对我们的按钮视图应用这对状态过渡实例，如下面的代码所示：

```xml
<Button
    android:layout_width="wrap_content"
    android:layout_height="wrap_content"
    android:text="My Button"
    android:background="@drawable/background_button"
    android:stateListAnimator="@animator/button_press" />
```

我们还可以使用过渡动画修饰复选框。在代码清单 7-10 中，我们将可复选的 Drawable 更新为 AnimatedStateListDrawable(通过 XML 标记<animatedselector>)，该 Drawable 将一系列关键帧图片序列化为默认状态和选中状态之间的过渡动画。

代码清单 7-10　res/drawable-v21/background_checkable.xml

```xml
<?xml version="1.0" encoding="utf-8"?>
<animated-selector
    xmlns:android="http://schemas.android.com/apk/res/android">
    <item android:state_enabled="false"
        android:drawable="@drawable/check_disabled" />
    <item android:id="@+id/state_checked"
        android:state_checked="true"
        android:drawable="@drawable/check_checked" />
    <!-- 默认状态 -->
    <item android:id="@+id/state_default"
        android:drawable="@drawable/check_default" />

    <!--
        这些过渡仅支持 AnimationDrawable、AnimatedVectorDrawable 或作为子元素的
        另一个 Animatable
    -->
    <transition android:fromId="@id/state_default"
                android:toId="@id/state_checked">
        <animation-list>
            <item android:duration="15"
                android:drawable="@drawable/check_default" />
            <item android:duration="15"
                android:drawable="@drawable/check_to_checked_01" />
            <item android:duration="15"
                android:drawable="@drawable/check_to_checked_02" />
            <item android:duration="15"
                android:drawable="@drawable/check_to_checked_03" />
            <item android:duration="15"
                android:drawable="@drawable/check_to_checked_04" />
            <item android:duration="15"
                android:drawable="@drawable/check_to_checked_05" />
            <item android:duration="15"
                android:drawable="@drawable/check_to_checked_06" />
```

```xml
            <item android:duration="15"
                android:drawable="@drawable/check_to_checked_07" />
            <item android:duration="15"
                android:drawable="@drawable/check_to_checked_08" />
            <item android:duration="15"
                android:drawable="@drawable/check_checked" />
        </animation-list>
    </transition>
</animated-selector>
```

此文件最初的部分没有变化，其中包含的每个条目声明应用于关联每种视图状态的 Drawable。但是请注意，我们为其中两个状态提供了唯一的 ID，这就使我们可以单独声明应用作这两种状态之间可视过渡的动画。此动画必须封装在<transition>元素中，并且支持如下类型：

- AnimationDrawable(XML 中的<animation-list>)：这是在其包含的每个元素之间序列化的关键帧动画。
- AnimatedVectorDrawable(XML 中的<animated-vector>)：应用于向量路径集合的变形动画(本章后面将更详细地进行讨论)。

提示：

应用<transition>时，附加到<item>的 android:drawable 属性代表在动画结束之后显示的最终 Drawable。

我们使用提供的关键帧定义了从默认状态(通过 fromId)到选中状态(通过 toId)的一种过渡。在相同状态之间以相反方向移动时，Android 框架默认以倒序运行相同的过渡，因此可以获得另一种过渡！然而，如果你更愿意使用另一个动画控制不同的状态过渡(如倒序)，则可以在 XML 中定义任意数量的过渡。

7.3 将遮罩应用于图片

7.3.1 问题

需要将裁剪遮罩应用于图片或形状，定义应用中另一张图片的可见边界。

7.3.2 解决方案

(API Level 1)

利用 2D 图形和 PorterDuffXferMode，可以将各种遮罩(以另一张位图的形式)应用于某张位图。本范例的基本步骤如下所示：

(1) 创建一个可变的空白 Bitmap 实例，以及在其中绘图的 Canvas。
(2) 首先在 Canvas 上画好遮罩模式。

(3) 将 PorterDuffXferMode 应用到 Paint。

(4) 用传输模式将原图绘制到 Canvas 上。

其中的关键就是 PorterDuffXferMode，它在绘图操作中既会考虑到原图的当前状态，也会考虑到目标对象的当前状态。目标对象就是 Canvas 中已有的数据，原图就是应用到当前操作的图形数据。

有很多模式参数可以应用到 PorterDuffXferMode，得到各种不同的效果。在这里只需要使用 PorterDuff.Mode.SRC_IN 模式即可。这个模式只会在原图和目标图重叠的地方绘制图形，绘制的像素来自原图；换句话说，就是会根据目标边界对原图进行裁剪。

使用图片作为 BitmapShader 将内容绘制到另一个元素中，也可以实现相同的效果。通过这种方式，就可以将图片像素视为用于绘制形状或元素的"颜色"，这些形状或元素将组成遮罩图片。我们在本节中将介绍这两种方法。

7.3.3 实现机制

1. 圆角位图

图片遮罩的一个常见应用就是在 ImageView 中显示位图之前，为其添加圆角。为了演示这个示例，我们将给图 7-8 中的原图加上遮罩。

图 7-8　原图

为了说明这一点，我们创建了一个接收图片的自定义视图，使用 BitmapShader 将其作为圆角矩形绘制到提供的 Canvas 中。该视图也管理将图片在自定义视图内部居中显示所需的尺寸调整数学函数。

代码清单 7-11 和 7-12 显示了自定义视图及其在 Activity 内部的使用。

代码清单 7-11　向位图应用圆角矩形遮罩的视图

```
public class RoundedCornerImageView extends View {

    private Bitmap mImage;
    private Paint mBitmapPaint;
```

```java
    private RectF mBounds;
    private float mRadius = 25.0f;

    public RoundedCornerImageView(Context context) {
        super(context);
        init();
    }

    public RoundedCornerImageView(Context context,
            AttributeSet attrs) {
        super(context, attrs);
        init();
    }

    public RoundedCornerImageView(Context context,
            AttributeSet attrs, int defStyle) {
        super(context, attrs, defStyle);
        init();
    }

    private void init() {
        // 创建图片涂绘
        mBitmapPaint = new Paint(Paint.ANTI_ALIAS_FLAG);
        // 创建作为绘图边界的矩形
        mBounds = new RectF();
    }

    @Override
    protected void onMeasure(int widthMeasureSpec,
            int heightMeasureSpec) {
        int height, width;
        height = width = 0;

        // 所请求大小是图片内容的大小
        int imageHeight, imageWidth;
        if(mImage == null) {
            imageHeight = imageWidth = 0;
        } else {
            imageHeight = mImage.getHeight();
            imageWidth = mImage.getWidth();
        }
        // 获得最佳测量值并在视图上设置该值
        width = getMeasurement(widthMeasureSpec, imageWidth);
        height = getMeasurement(heightMeasureSpec, imageHeight);

        setMeasuredDimension(width, height);
    }

    /*
```

```
     * 用于测量宽度和高度的辅助方法
     */
    private int getMeasurement(int measureSpec, int contentSize) {
        int specSize = MeasureSpec.getSize(measureSpec);
        switch(MeasureSpec.getMode(measureSpec)) {
            case MeasureSpec.AT_MOST:
                return Math.min(specSize, contentSize);
            case MeasureSpec.UNSPECIFIED:
                return contentSize;
            case MeasureSpec.EXACTLY:
                return specSize;
            default:
                return 0;
        }
    }

    @Override
    protected void onSizeChanged(int w, int h,
            int oldw, int oldh) {
        if(w != oldw || h != oldh) {
            // 我们要使图片居中,因此在视图改变大小时偏移值
            int imageWidth, imageHeight;
            if(mImage == null) {
                imageWidth = imageHeight = 0;
            } else {
                imageWidth = mImage.getWidth();
                imageHeight = mImage.getHeight();
            }
            int left = (w - imageWidth) / 7;
            int top = (h - imageHeight) / 7;
            // 设置边界以偏移圆角矩形
            mBounds.set(left, top, left+imageWidth,
                    top+imageHeight);
            // 偏移着色器以在矩形内部绘制位图
            // 如果没有此步骤,位图将在视图中的(0,0)处
            if(mBitmapPaint.getShader() != null) {
                Matrix m = new Matrix();
                m.setTranslate(left, top);
                mBitmapPaint.getShader().setLocalMatrix(m);
            }
        }
    }

    public void setImage(Bitmap bitmap) {
        if(mImage != bitmap) {
            mImage = bitmap;
            if(mImage != null) {
                BitmapShader shader = new BitmapShader(mImage,
                        TileMode.CLAMP, TileMode.CLAMP);
                mBitmapPaint.setShader(shader);
```

```
            } else {
                mBitmapPaint.setShader(null);
            }
            requestLayout();
        }
    }

    @Override
    protected void onDraw(Canvas canvas) {
        // 让视图绘制背景等对象
        super.onDraw(canvas);
        // 使用计算得出的值绘制图片
        if(mBitmapPaint != null) {
            canvas.drawRoundRect(mBounds, mRadius, mRadius,
                    mBitmapPaint);
        }
    }
}
```

代码清单 7-12　显示 RoundedCornerImageView 的 Activity

```
public class ShaderActivity extends Activity {

    @Override
    protected void onCreate(Bundle savedInstanceState) {
        super.onCreate(savedInstanceState);
        RoundedCornerImageView iv =
                new RoundedCornerImageView(this);
        Bitmap source = BitmapFactory.decodeResource(
                getResources(), R.drawable.dog);

        iv.setImage(source);
        setContentView(iv);
    }
}
```

　　在自定义视图中，通过 setImage()传入新图片时，创建了一个 BitmapShader 来封装图片像素，并在用于绘图的画笔上进行相应的设置。测量和布局视图之后，我们将对具有大小的视图调用 onSizeChanged()；在此测量边界以满足图片在视图内居中显示的要求。我们还必须使用 Matrix 偏移着色器，否则圆角矩形遮罩会在中间位置绘制，但图片仍然位于视图的左上角。

　　现在，当视图为绘制做好准备时，我们就可以使用计算得出的边界和预先配置的画笔对 Canvas 调用 drawRoundRect()。这会在视图中绘制一个圆角矩形，但使用来自位图的颜色对形状着色。这些操作的结果如图 7-9 所示。

图 7-9　应用圆角矩形遮罩的图片

2. 任意遮罩图片

接下来看一个稍微更加有趣的示例。在此采用两张图片：原图和代表所要应用遮罩的图片(在此是上下颠倒的三角形)，如图 7-10 所示。

图 7-10　原始图片(左图)和应用的任意遮罩图片(右图)

所选择的遮罩图片并非必须是我们所选的样式，只要把遮罩的地方设为黑色，其他地方设为透明就可以。不过，最好选择确保能在系统上正确显示的图片。

首先在 Canvas 中绘制三角形图片，这就是我们的图片"遮罩"。然后，在同一个 Canvas 上绘制原图时应用 PorterDuff.Mode.SRC_IN 转换，得到的就是带圆角的原图。

这是因为 SRC_IN 转换模式就是告诉 Paint 对象，只在 Canvas 上原图和目标图(已经画好的三角形)重叠的地方绘制像素点，像素点则来自原图。代码清单 7-13 是遮罩图片并在视图中显示的简单 Activity 代码。

代码清单 7-13　将任意遮罩应用到一张位图的 Activity

```
public class MaskActivity extends Activity {
```

```
@Override
public void onCreate(Bundle savedInstanceState) {
    super.onCreate(savedInstanceState);
    ImageView iv = new ImageView(this);

    // 创建并加载图片(通常是不可修改的)
    Bitmap source = BitmapFactory.decodeResource(getResources(),
        R.drawable.dog);
    Bitmap mask = BitmapFactory.decodeResource(getResources(),
        R.drawable.triangle);

    // 创建一个可修改的位置以及一个在其中绘图的 Canvas
    Bitmap result = Bitmap.createBitmap(source.getWidth(),
        source.getHeight(), Config.ARGB_8888);
    Canvas canvas = new Canvas(result);
    Paint paint = new Paint(Paint.ANTI_ALIAS_FLAG);
    paint.setColor(Color.BLACK);

    canvas.drawBitmap(mask, 0, 0, paint);
    paint.setXfermode(new PorterDuffXfermode(Mode.SRC_IN));
    canvas.drawBitmap(source, 0, 0, paint);
    paint.setXfermode(null);

    iv.setImageBitmap(result);
    setContentView(iv);
}
```

得到的结果如图 7-11 所示。

图 7-11　使用了遮罩的图片

3. 视图的轮廓

(API Level 21)

在运行 Android 5.0 及其更高版本的设备上，Android 框架支持通过动态阴影表明视图的提高(通过 elevation 和 translationZ 属性)。为使该功能正常运作，框架必须了解视图的可视边界。在简单的示例中，可以在内部进行此处理，但如果应用任意遮罩，则还必须使用匹配的 ViewOutlineProvider 指示在何处产生阴影。代码清单 7-14 给出了图片遮罩的修改版，其中还为轮廓提供了适当的阴影。

代码清单 7-14　带有视图轮廓的遮罩 Activity

```java
public class MaskActivity extends Activity {

    @Override
    public void onCreate(Bundle savedInstanceState) {
        super.onCreate(savedInstanceState);
        ImageView iv = new ImageView(this);
        iv.setScaleType(ImageView.ScaleType.CENTER);
        //创建并加载图片(通常是不可变的)
        Bitmap source = BitmapFactory.decodeResource(getResources(),
            R.drawable.dog);
        Bitmap mask = BitmapFactory.decodeResource(getResources(),
            R.drawable.triangle);

        //创建"可变的"位置，并且在其中绘制 Canvas
        final Bitmap result =
            Bitmap.createBitmap(source.getWidth(), source.getHeight(),
                Config.ARGB_8888);
        Canvas canvas = new Canvas(result);
        Paint paint = new Paint(Paint.ANTI_ALIAS_FLAG);
        paint.setColor(Color.BLACK);

        canvas.drawBitmap(mask, 0, 0, paint);
        paint.setXfermode(new PorterDuffXfermode(Mode.SRC_IN));
        canvas.drawBitmap(source, 0, 0, paint);
        paint.setXfermode(null);

        iv.setImageBitmap(result);
        if(Build.VERSION.SDK_INT >= Build.VERSION_CODES.LOLLIPOP) {
            //提高视图以建立可见阴影
            iv.setElevation(32f);
            //绘制匹配遮罩的轮廓，从而提供适当的阴影
            iv.setOutlineProvider(new ViewOutlineProvider() {
                @Override
                public void getOutline(View view, Outline outline) {
                    int x = (view.getWidth() - result.getWidth()) / 2;
                    int y = (view.getHeight() - result.getHeight()) / 2;

                    Path path = new Path();
```

```
                path.moveTo(x, y);
                path.lineTo(x+result.getWidth(), y);
                path.lineTo(x+result.getWidth()/2, y+result.getHeight());
                path.lineTo(x, y);
                path.close();

                outline.setConvexPath(path);
            }
        });
    }
    setContentView(iv);
}
```

ViewOutlineProvider 有一个必需的方法 getOutline()，如果由于大小或配置发生变化而需要更新轮廓，就会调用该方法。为提高效率，我们提供了 Outline 实例，在此实例中填充适当的形状。我们的三角形是不规则的轮廓形状，因此必须构造一个代表三角形的 Path 并使用 setConvexPath()应用此路径。

请注意，我们还在新的示例中设置了静态高程值(static elevation value)。这样做就可以查看和验证阴影效果，如图 7-12 所示。

图 7-12　带有轮廓阴影的图片遮罩

如果轮廓足够简单，Android 还可以将其用作视图的剪切遮罩。只需要调用 setClipToOutline(true)，即可表明视图应使用其轮廓作为剪切遮罩。代码清单 7-15 使用此方式通过圆形遮罩剪切图片。

注意：

Android 目前仅支持通过矩形、圆形和圆角矩形轮廓进行剪切。例如，刚才建立的三角形轮廓不能用作剪切。

代码清单 7-15　对视图应用圆形轮廓剪切的 Activity

```java
public class OutlineActivity extends Activity {

    @Override
    protected void onCreate(Bundle savedInstanceState) {
        super.onCreate(savedInstanceState);
        ImageView iv = new ImageView(this);
        iv.setScaleType(ImageView.ScaleType.CENTER);

        //提高视图以建立可见阴影
        iv.setElevation(32f);

        iv.setImageResource(R.drawable.dog);

        //告诉视图使用其轮廓作为剪切遮罩
        iv.setClipToOutline(true);

        //为剪切和阴影提供圆形视图轮廓
        iv.setOutlineProvider(new ViewOutlineProvider() {
            @Override
            public void getOutline(View view, Outline outline) {
                ImageView iv = (ImageView) view;
                int radius = iv.getDrawable().getIntrinsicHeight() / 2;
                int centerX = (view.getRight() - view.getLeft()) / 2;
                int centerY = (view.getBottom() - view.getTop()) / 2;

                outline.setOval(centerX - radius,
                        centerY - radius,
                        centerX + radius,
                        centerY + radius);
            }
        });

        setContentView(iv);
    }
}
```

此轮廓更加直观，它围绕居中的图片产生居中的圆形遮罩，如图 7-13 所示。

图 7-13 带有剪切轮廓的遮罩

7.4 在视图内容上绘制

7.4.1 问题

需要在当前可见视图的顶部显示内容,但不会插入视图或以其他方式修改现有的视图层次结构。

7.4.2 解决方案

(API Level 1)

在新的临时窗口 PopupWindow 中放置内容,在该窗口中可以放置将显示在当前 Activitiy 窗口顶部的视图。PopupWindow 可以显示在屏幕上的任意位置,方法是提供明确的显示位置,或者提供 PopupWindow 应该停靠的现有视图。

(API Level 18)

还可以使用新的 ViewOverlay 在视图顶部绘制内容。ViewOverlay 用于向父视图管理的私有层添加任意数量的 Drawable 对象。这些对象将绘制在相应视图的顶部,前提是它们的边界在父视图的边界之内。

7.4.3 实现机制

为了在视图层次结构的顶部绘制内容,首先需要创建待显示的内容。代码清单 7-16 构造了一组简单的视图，它们将是我们的 PopupWindow 的内容。

代码清单 7-16 res/layout/popup.xml

```xml
<?xml version="1.0" encoding="utf-8"?>
<LinearLayout
    xmlns:android="http://schemas.android.com/apk/res/android"
    android:layout_width="wrap_content"
    android:layout_height="wrap_content"
    android:orientation="vertical">
    <TextView
        android:layout_width="wrap_content"
        android:layout_height="wrap_content"
        android:layout_gravity="center_horizontal"
        android:text="This is a PopupWindow" />
    <EditText
        android:layout_width="250dp"
        android:layout_height="wrap_content"
        android:layout_gravity="center_horizontal" />
    <Button
        android:layout_width="wrap_content"
        android:layout_height="wrap_content"
        android:layout_gravity="center_horizontal"
        android:text="Close" />
</LinearLayout>
```

在停靠到视图的弹出窗口中显示此内容时，PopupWindow 默认将左对齐显示在视图下方。然而，如果视图下方没有足够的空间显示 PopupWindow，它将改为显示在停靠视图的上方。为使弹出窗口在这两种情况下有不同的外观，可以提供一个自定义背景 Drawable 对象，该对象会开启 android:state_above_anchor 属性。代码清单 7-17 和图 7-14 说明了用于此例的 Drawable 对象。

代码清单 7-17 res/drawable/popup_background.xml

```xml
<?xml version="1.0" encoding="utf-8"?>
<selector
    xmlns:android="http://schemas.android.com/apk/res/android" >
    <item android:state_above_anchor="true"
        android:drawable="@drawable/speech_background_top" />
    <!-- 默认状态 -->
    <item
        android:drawable="@drawable/speech_background_bottom" />
</selector>
```

图 7-14 后台的 9-patch Drawable 对象

可以从 7.1 节的语音气泡可拉伸图片示例中认出这些背景图片。我们稍微修改了拉伸区域，以使扩展点始终在相同一侧。

代码清单 7-18 和 7-19 说明了一个示例 Activity 和布局，用于构造和显示响应按钮单击的 PopupWindow。在此例中，PopupWindow 显示为停靠到所单击按钮的旁边。

代码清单 7-18　res/layout/activitiy_main.xml

```xml
<?xml version="1.0" encoding="utf-8"?>
<FrameLayout
    xmlns:android="http://schemas.android.com/apk/res/android"
    android:layout_width="match_parent"
    android:layout_height="match_parent">
    <Button
        android:id="@+id/button"
        android:layout_width="match_parent"
        android:layout_height="wrap_content"
        android:text="Show PopupWindow"
        android:onClick="onShowWindowClick" />
    <Button
        android:layout_width="match_parent"
        android:layout_height="wrap_content"
        android:layout_gravity="bottom"
        android:text="Show PopupWindow"
        android:onClick="onShowWindowClick" />
</FrameLayout>
```

代码清单 7-19　显示 PopupWindow 的 Activity

```java
public class MainActivity extends Activity
        implements View.OnTouchListener {
    private PopupWindow mOverlay;

    @Override
    protected void onCreate(Bundle savedInstanceState) {
        super.onCreate(savedInstanceState);
```

```java
        setContentView(R.layout.activity_main);

        // 填充弹出窗口内容布局：我们尚无权访问父视图，
        // 因此传递 null 作为容器视图参数。
        View popupContent =
                getLayoutInflater().inflate(R.layout.popup, null);

        mOverlay = new PopupWindow();
        // 弹出窗口应该封装内容视图
        mOverlay.setWindowLayoutMode(
                WindowManager.LayoutParams.WRAP_CONTENT,
                WindowManager.LayoutParams.WRAP_CONTENT);
        // 设置内容和背景
        mOverlay.setContentView(popupContent);
        mOverlay.setBackgroundDrawable(getResources()
                .getDrawable(R.drawable.popup_background));

        // 默认行为是不允许 PopupWindow 中的任何元素具有交互性，
        // 但是允许将触摸事件直接传递给 PopupWindow。
        // 所有外部触摸将传递给主(Activity)窗口
        mOverlay.setTouchInterceptor(this);

        // 调用 setFocusable()以使 PopupWindow 中的元素获得焦点，
        // 这也允许在任何外部触摸上解除 PopupWindow
        mOverlay.setFocusable(true);

        // 如果要使外部触摸可以自动解除 PopupWindow,
        // 但又不想 PopupWindow 内部的元素获得焦点，
        // 则调用 setOutsideTouchable()
        mOverlay.setOutsideTouchable(true);
    }

    @Override
    protected void onPause() {
        super.onPause();
        // PopupWindow 类似于对话框，如果在 Activity 结束时希望其保持可见，
        // 将会产生泄露
        mOverlay.dismiss();
    }

    @Override
    public boolean onTouch(View v, MotionEvent event) {
        // 处理传递给 PopupWindow 的直接触摸事件
        return true;
    }
```

```
    public void onShowWindowClick(View v) {
        if(mOverlay.isShowing()) {
            // 解除弹出窗口
            mOverlay.dismiss();
        } else {
            // 显示停靠在所单击按钮旁边的 PopupWindow，如果有足够的空间，
            // 它将显示在按钮下方，否则就显示在上方
            mOverlay.showAsDropDown(v);
        }
    }
}
```

在此例中，我们创建带有两个按钮的简单布局，这两个按钮都设置为触发相同的动作。单击任意一个按钮时，使用showAsDropDown()方法显示停靠在视图旁边的PopupWindow。

提醒：

也可以使用showAtLocation()方法在指定位置显示PopupWindow。类似于showAsDrop-Down()，该方法采用一个View参数，但仅用于获得窗口信息。

在按下按钮时，该例的结果如图7-15所示。

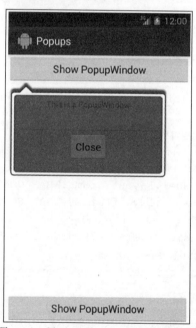

图7-15 显示PopupWindow的Activity

初次创建Activity时，初始化PopupWindow并将布局模式设置为WRAP_CONTENT。我们必须在代码中完成此工作，即使在布局XML中已经进行了相关定义，这是因为在使用空的父视图容器手动填充期间，XML中的布局参数会被删除。接下来，我们提供内容视图和创建的自定义背景。在后面还会讨论在覆盖层上设置的其他标识。

目前，如果尝试原样运行此应用程序，你可能注意到在按下底部的按钮时

PopupWindow 没有显示。原因在于我们设置了 WRAP_CONTENT 布局模式。请记住，如果停靠视图下方没有可用空间，弹出窗口应该显示在其上方。然而，这是通过比较弹出窗口的大小与主窗口的剩余空间来确定的。如果没有为该窗口定义大小，它将测量剩余的空间并尝试压缩包含的内容。为了修正此问题，我们将在项目中添加 dimens.xml 文件，并且修改 Activity 的 onCreate()，如代码清单 7-20 和 7-21 所示。

代码清单 7-20　res/values/dimens.xml

```xml
<?xml version="1.0" encoding="utf-8"?>
<resources>
    <dimen name="popupWidth">350dp</dimen>
    <dimen name="popupHeight">250dp</dimen>
</resources>
```

代码清单 7-21　针对固定大小修改的 onCreate()

```java
@Override
protected void onCreate(Bundle savedInstanceState) {
    super.onCreate(savedInstanceState);
    setContentView(R.layout.activity_main);

    // 填充弹出窗口内容布局，
    // 我们尚无权访问父视图，
    // 因此传递 null 作为容器视图参数
    View popupContent =
            getLayoutInflater().inflate(R.layout.popup, null);

    mOverlay = new PopupWindow();
    // 弹出窗口应该封装内容视图
    mOverlay.setWindowLayoutMode(
            WindowManager.LayoutParams.WRAP_CONTENT,
            WindowManager.LayoutParams.WRAP_CONTENT);
    mOverlay.setWidth(getResources()
            .getDimensionPixelSize(R.dimen.popupWidth));
    mOverlay.setHeight(getResources()
            .getDimensionPixelSize(R.dimen.popupHeight));
    // 设置内容和背景
    mOverlay.setContentView(popupContent);
    mOverlay.setBackgroundDrawable(getResources()
            .getDrawable(R.drawable.popup_background));

    // 默认行为是不允许 PopupWindow 中的任何元素具有交互性，
    // 但是允许将触摸事件直接传递给 PopupWindow。
    // 所有外部触摸将传递给主(Activity)窗口。
    mOverlay.setTouchInterceptor(this);

    // 调用 setFocusable() 以使 PopupWindow 中的元素获得焦点，
```

```
        // 这也允许在任何外部触摸上解除 PopupWindow。
        mOverlay.setFocusable(true);

        // 如果要使外部触摸可以自动解除 PopupWindow,
        // 但又不想 PopupWindow 内部的元素获得焦点,
        // 调用 setOutsideTouchable()。
        mOverlay.setOutsideTouchable(true);
    }
```

现在我们的覆盖层窗口具有已定义的大小，我们已从尺寸资源中取出该大小，以便在不同的设备之间保持像素密度无关性。运行该例并单击按钮，就可以看到 PopupWindow 显示在顶部按钮下方和底部按钮上方，如图 7-16 所示。

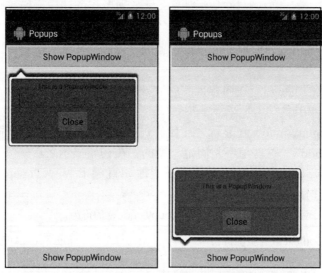

图 7-16　PopupWindow 停靠到每个按钮旁边的 Activity

1. 操作 PopupWindow 的行为

PopupWindow 有大量有用的标志，我们可以设置这些标志来掌控它的行为和交互点。默认情况下，弹出窗口是可触摸的，这意味着它可以接收直接触摸事件。为操作这些事件，我们调用 setTouchInterceptor()来提供作为这些触摸事件目标的 onTouchListener。

弹出窗口默认是不可聚焦的，这意味着该窗口内的视图不能接收焦点(而 EditText 或 Button 可以接收焦点)。我们在内容视图中放置了这些小部件，因此设置可聚焦标志为 true，以使用户可以与这些元素交互。最后一个标志是 setOutsideTouchable，我们也启用了该标志。该标志的值默认为 false，但可以设置为 true，从而将弹出窗口内容区域外部的触摸事件发送到 PopupWindow 而非其下的主窗口。这样做就可以使 PopupWindow 在任何外部触摸事件上取消自身。如果不想在弹出窗口上启用焦点，但又想要取消该窗口提供的行为，则使用该标志是最常见的做法。

可以使用少量的构造函数创建新的 PopupWindow，我们在此使用的是不带参数的基本版本，但还有一些版本带有 Context 参数。向构造函数传递 Context 参数时，框架会创建一个包含默认系统背景的 PopupWindow，而此处使用的版本不会这样创建。我们不需要系统背景,因为会提供自定义的 Drawable。值得注意的是,无论发生何种情况(框架或应用程序为弹出窗口提供背景),提供给 PopupWindow 的内容视图都会封装在另一个私有的 ViewGroup 中,以便管理背景实例。这一点非常重要,因为额外的容器也会稍微修改覆盖层的行为。

结合所选的标志和 PopupWindow 的创建选项,大量用户交互行为将会改变。我们在此研究如下行为:

1) 接收触摸事件:在通过 setTouchInterceptor()提供的 onTouchListener 中接收和处理事件。

2) 允许在视图内部交互:内容视图内部可聚焦的小部件(如 Button)将具备可交互性,并且能够接收焦点。

3) 在外部触摸上自动取消:在内容视图区域外部发生的任何触摸事件都会自动取消弹出窗口。

4) 在单击"返回"按钮时取消:单击设备的"返回"按钮将取消弹出窗口,而不是结束当前 Activity。

5) 允许外部触摸主窗口:在内容视图区域外部触摸时,触摸事件会传递给 Activity 主窗口而非直接产生作用。

表 7-1 根据 PopupWindow 在显示之前如何进行初始化,概述了这些操作中的哪一个将应用于此 PopupWindow。这些值不是固定不变的;在初始显示之后,可以修改这些值。如果在 PopupWindow 显示时修改标志,则此次改动直到下一次 PopupWindow 显示或其 update()方法被调用时才会生效。

表 7-1 PopupWindow 的行为

	使用 Context 或背景图片创建			标准 PopupWindow	
操作	默认	可聚焦	外部触摸	默认	可聚焦
1	×	×	×		
7		×			×
3		×	×		
4		×			
5	×			×	×

除了上述讨论的内容之外,从此信息中还可以了解到,如果内容覆盖层需要处理触摸事件,就要确保提供 Context 或背景图片。

2. 制作 PopupWindow 的动画

完成上述示例之后,你可能已注意到 PopupWindow 在显示或取消时具有相关的默认动画。可以自定义或移除该动画,方法是通过 setAnimationStyle()传递新的资源。该方法采用引用样式的资源 ID,该样式定义了一对儿动画,一个动画用于窗口进入,另一个动画用于窗口退出。代码清单 7-22 说明了我们需要创建以自定义 PopupWindow 动画的样式资源。

代码清单 7-22   res/values/styles.xml

```xml
<resources>
    <!-- 在现有主题下方定义此元素 -->
    <style name="PopupAnimation">
        <item name="android:windowEnterAnimation">
            @android:anim/slide_in_left</item>
        <item name="android:windowExitAnimation">
            @android:anim/slide_out_right</item>
    </style>
</resources>
```

提示：

不一定要定制自己的动画样式才能自定义过渡动画。Android.R.style 中定义了许多样式，框架使用这些样式建立其他标准窗口类型(如对话框或 Toast)的过渡动画。为使用这些动画，只需要传递相关的 ID，如 Android.R.style.Animation_Dialog 或 Android.R.style.Animation_Toast。

每个条目都可以是一个引用，指向使用 XML 定义的动画或框架中已有的动画。此处选择引用框架中已有的滑进和滑出动画。在代码清单 7-23 中，我们接下来修改示例 Activity 的 onCreate()以应用自定义动画。为简洁起见，我们还删除了配置标志。

代码清单 7-23   使用自定义动画显示 PopupWindow 的 Activity onCreate()方法

```java
@Override
protected void onCreate(Bundle savedInstanceState) {
    super.onCreate(savedInstanceState);
    setContentView(R.layout.activity_main);

    // 填充弹出窗口内容布局，
    // 我们尚无权访问父视图，
    // 因此传递 null 作为容器视图参数。
    View popupContent =
            getLayoutInflater().inflate(R.layout.popup, null);
    mOverlay = new PopupWindow();
    // 弹出窗口应该封装内容视图
    mOverlay.setWindowLayoutMode(
            WindowManager.LayoutParams.WRAP_CONTENT,
            WindowManager.LayoutParams.WRAP_CONTENT);
    mOverlay.setWidth(getResources()
            .getDimensionPixelSize(R.dimen.popupWidth));
    mOverlay.setHeight(getResources()
            .getDimensionPixelSize(R.dimen.popupHeight));
    // 设置内容和背景
    mOverlay.setContentView(popupContent);
    mOverlay.setBackgroundDrawable(getResources()
            .getDrawable(R.drawable.popup_background));

    // 设置自定义动画进入/退出对，或设为 0 以表示禁用动画。
    // 还可以使用在平台中定义的动画样式，
```

```
    // 比如 android.R.style.Animation_Toast
    mOverlay.setAnimationStyle(R.style.PopupAnimation);

    // 默认行为是不允许 PopupWindow 中的任何元素具有交互性，
    // 但是允许将触摸事件直接传递给 PopupWindow。
    // 所有外部触摸将传递给主(Activity)窗口。
    mOverlay.setTouchInterceptor(this);
}
```

提示：
还可以调用 setAnimationStyle(0)来彻底删除动画，或者使用 setAnimationStyle(-1)来重设默认动画。

现在再次运行应用程序，这时会使用自定义滑动动画来将 PopupWindow 过渡到屏幕中和移出屏幕。

3. 使用 ViewOverlay

(API Level 18)

在视图上绘制内容的另一种简单方法是使用较新的 ViewOverlay 实现。ViewOverlay 及其表亲 ViewGroupOverlay 用于添加可在视图顶部绘制的任意数量的 Drawable 对象。应用程序不能直接创建 ViewGroupOverlay，而是在层次结构中的任意视图上调用 getOverlay() 来获得 ViewOverlay。视图被约束为只能在其边界内绘制，因此如果覆盖层中的内容延伸出驻留覆盖层的视图的边界，则超出边界的部分会被裁剪。

为说明此功能，我们创建了一个简单的应用程序，该应用程序在 Activity 中主视图的顶部绘制标记内容。我们特意选择普通的视图进行绘制，以便指出任意 View 子类(无论是显示文本、HTML、图片，还是显示一些自定义的内容)都可以操作覆盖层。应用程序会在可交互视图上用户触摸的位置放置箭头标志或可调整大小的方框。只要用户按下手指并拖动，就可以移动标志或调整方框的大小。释放触摸之后，一个标记会持续显示在视图上，直到再次点击该标记，这会彻底将其删除。

首先，通过代码清单 7-24 和图 7-17 查看使用的资源。

图 7-17　res/drawable/flag_arraw.png

代码清单 7-24　res/drawable/box.xml

```xml
<?xml version="1.0" encoding="utf-8"?>
<shape xmlns:android="http://schemas.android.com/apk/res/android"
    android:shape="rectangle">
    <solid
        android:color="@android:color/transparent"/>
    <stroke
```

```
            android:width="3dp"
            android:color="#F00" />
</shape>
```

代码清单 7-25 显示了用于主 Activity 的布局。我们创建了一个主视图，其中包含将要绘制的一些文本(@+id/textview)，而底部的选择器用于确定放置何种类型的标记。

代码清单 7-25 res/layout/activity_main.xml

```
<LinearLayout
    xmlns:android="http://schemas.android.com/apk/res/android"
    android:layout_width="match_parent"
    android:layout_height="match_parent"
    android:orientation="vertical" >

    <TextView
        android:id="@+id/textview"
        android:layout_width="match_parent"
        android:layout_height="0dp"
        android:layout_weight="1"
        android:gravity="center"
        android:text="Android Recipes" />

    <RadioGroup
        android:id="@+id/container_options"
        android:layout_width="match_parent"
        android:layout_height="wrap_content"
        android:orientation="horizontal"
        android:background="#CCC">
        <RadioButton
            android:id="@+id/option_box"
            android:layout_width="0dp"
            android:layout_height="wrap_content"
            android:layout_weight="1"
            android:text="Box" />
        <RadioButton
            android:id="@+id/option_arrow"
            android:layout_width="0dp"
            android:layout_height="wrap_content"
            android:layout_weight="1"
            android:text="Arrow" />
    </RadioGroup>

</LinearLayout>
```

最后，我们通过代码清单 7-26 创建 Activity。此 Activity 查找主视图，设置一个 OnTouchListener 来监控通过该视图的触摸事件。这会造成对每个触摸事件调用 onTouch()方法，该方法用于确定用户是否在主视图上触摸、拖动或释放手指。

代码清单 7-26　具有交互式 ViewOverlay 的 Activity

```java
public class MainActivity extends Activity implements View.OnTouchListener {

    private RadioGroup mOptions;

    private ArrayList<Drawable> mMarkers;
    private Drawable mTrackingMarker;
    private Point mTrackingPoint;

    @Override
    protected void onCreate(Bundle savedInstanceState) {
        super.onCreate(savedInstanceState);
        setContentView(R.layout.activity_main);

        // 接收要在其上绘制的视图的触摸事件
        findViewById(R.id.textview).setOnTouchListener(this);
        mOptions =
            (RadioGroup)findViewById(R.id.container_options);

        mMarkers = new ArrayList<Drawable>();
    }

    /*
     * 所监控视图中的触摸事件将在此处传递
     */
    @Override
    public boolean onTouch(View v, MotionEvent event) {
        switch(mOptions.getCheckedRadioButtonId()) {
            case R.id.option_box:
                handleEvent(R.id.option_box, v, event);
                break;
            case R.id.option_arrow:
                handleEvent(R.id.option_arrow, v, event);
                break;
            default:
                return false;
        }
        return true;
    }

    /*
     * 当用户选择绘制方框时处理触摸事件
     */
    private void handleEvent(int optionId, View v,
            MotionEvent event) {
        int x = (int)event.getX();
        int y = (int)event.getY();
        switch(event.getAction()) {
            case MotionEvent.ACTION_DOWN:
```

```java
                    Drawable current = markerAt(x, y);
                    if(current == null) {
                        // 在新的触摸上添加新的标记
                        switch(optionId) {
                            case R.id.option_box:
                                mTrackingMarker = addBox(v, x, y);
                                mTrackingPoint = new Point(x, y);
                                break;
                            case R.id.option_arrow:
                                mTrackingMarker = addFlag(v, x, y);
                                break;
                        }
                    } else {
                        // 移除现有的标记
                        removeMarker(v, current);
                    }
                    break;
                case MotionEvent.ACTION_MOVE:
                    // 在移动时更新当前标记
                    if(mTrackingMarker != null) {
                        switch(optionId) {
                            case R.id.option_box:
                                resizeBox(v, mTrackingMarker,
                                        mTrackingPoint, x, y);
                                break;
                            case R.id.option_arrow:
                                offsetFlag(v, mTrackingMarker, x, y);
                                break;
                        }

                    }
                    break;
                case MotionEvent.ACTION_UP:
                case MotionEvent.ACTION_CANCEL:
                    // 当手势结束时清除状态
                    mTrackingMarker = null;
                    mTrackingPoint = null;
                    break;
            }
        }

        /*
         * 在给定坐标处添加新的可调整大小的方框
         */
        private Drawable addBox(View v, int x, int y) {
            Drawable box = getResources().getDrawable(R.drawable.box);

            // 在触摸点首先创建一个大小为 0 的方框
            Rect bounds = new Rect(x, y, x, y);
            box.setBounds(bounds);
```

```java
        // 添加到 ViewOverlay
        mMarkers.add(box);
        v.getOverlay().add(box);

        return box;
    }

    /*
     * 更新现有方框以基于给定坐标调整大小
     */
    private void resizeBox(View v, Drawable target,
            Point trackingPoint, int x, int y) {
        Rect bounds = new Rect(target.getBounds());
        // 如果新的触摸点位于跟踪点的左侧,则向左增大;
        // 否则,向右增大
        if(x < trackingPoint.x) {
            bounds.left = x;
        } else {
            bounds.right = x;
        }

        // 如果新的触摸点位于跟踪点的上方,则向上增大;
        // 否则,向下增大
        if(y < trackingPoint.y) {
            bounds.top = y;
        } else {
            bounds.bottom = y;
        }

        // 更新 Drawable 的边界并重绘
        target.setBounds(bounds);
        v.invalidate();
    }

    /*
     * 在给定坐标处添加新的标记
     */
    private Drawable addFlag(View v, int x, int y) {
        // 建立新的标记 Drawable
        Drawable marker =
            getResources().getDrawable(R.drawable.flag_arrow);

        // 创建符合图片大小的边界
        Rect bounds = new Rect(0, 0,
                marker.getIntrinsicWidth(),
                marker.getIntrinsicHeight());
        // 在坐标周围底部居中标记
        bounds.offset(x - (bounds.width() /7),
                y - bounds.height());
```

```
            marker.setBounds(bounds);
            // 添加到覆盖层
            mMarkers.add(marker);
            v.getOverlay().add(marker);

            return marker;
        }

        /*
         * 更新现有标记的位置
         */
        private void offsetFlag(View v, Drawable marker,
                int x, int y) {
            Rect bounds = new Rect(marker.getBounds());
            // 移动Drawable的边界以对齐新的坐标
            bounds.offset(x - bounds.left - (bounds.width() / 2),
                    y - bounds.top - bounds.height());
            // 更新并重绘
            marker.setBounds(bounds);
            v.invalidate();
        }

        /*
         * 移除请求的标记条目
         */
        private void removeMarker(View v, Drawable marker) {
            mMarkers.remove(marker);
            v.getOverlay().remove(marker);
        }

        /*
         * 查找包含所请求坐标的第一个标记(如果存在的话)
         */
        private Drawable markerAt(int x, int y) {
            // 返回找到的包含给定点的第一个标记
            for(Drawable marker : mMarkers) {
                if(marker.getBounds().contains(x, y)) {
                    return marker;
                }
            }

            return null;
        }
    }
```

在 onTouch()内部，检查从 RadioGroup 选择的条目以确定标记类型。对于初始的 ACTION_DOWN，我们调用 addBox()或 addFlag()以创建新的 Drawable，使用 setBounds() 设置它的大小和位置，将其应用于主视图的 ViewOverlay。只需要调用 add()，就可以将标记添加到覆盖层。ViewOverlay 没有提供合适的方法来跟踪所添加的条目，因为我们维护自己的条目列表，该列表稍后可用于基于触摸查找标记。

随着手指在 ACTION_DOWN 中四处移动，我们要么更新标志的位置，要么调整方框的大小以适应初始触摸点和当前触摸位置。在这两种情况下，通过再次更新 Drawable 的边界 Rect 来实现操作。一旦释放手指，就清除所有跟踪状态，标记现在位于其持久放置的位置。

注意：

Drawable 元素通过边界 Rect 获得其大小和位置。来自图片资源(如 PNG)的位图内容具有内在的高度和宽度，可以用来生成边界的大小部分；但是，取自 XML 的内容没有内在的大小，必须明确设置其大小。无论是何种情况，特别是在 ViewOverlay 中使用时，边界用于协助在正确的位置放置内容，因此一定要记住为每个添加的元素至少调用 setBounds() 一次。

如果新的触摸将现有标记的位置(使用 marketAt()辅助方法检查)下移，则通过将其从 ViewOverlay 中移除来删除该标记。图 7-18 中的左图显示了初始布局，右图则显示了添加到覆盖层的一些标记。

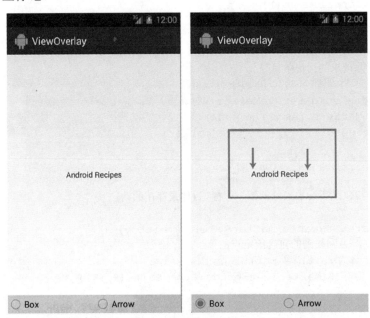

图 7-18　在 ViewOverlay 内具有 Drawable 内容的 Activity

警告：

在 ViewGroup 上调用 getOverlay()会返回 ViewGroupOverlay，ViewGroupOverlay 具有附加的 add()和 remove()方法，用于操作视图而非 Drawable。需要注意的是，其工作方式与本节描述的并不相同。它无法用于在现有层次结构的顶部添加新的视图，但可以用于仅将容器中已有的现存视图提升为覆盖层。它还可以在将视图添加到覆盖层时从容器中移除该视图，这会修改 ViewGroup 的布局。如果要在主窗口的顶部放置视图，应使用本节前面介绍的 PopupWindow 相关技巧。

7.5 高性能绘制

7.5.1 问题

应用程序需要在屏幕上渲染和绘制复杂的场景或动画，这些通常通过后台线程来完成。

7.5.2 解决方案

(API Level 1)

使用 SurfaceView 或 TextureView 将内容从后台线程渲染到屏幕上。在 Android UI 开发中一般遵循这样的规定：不要在主线程之外的线程中修改任何与 View 相关的属性。这两个类则不遵循这个规定，它们专门设计用来在后台线程中执行绘制命令，并将绘制内容展现到屏幕上。在后面的章节中，你将看到 Android 框架使用这两个类来渲染相机的预览数据和视频输出。但是，现在我们会关注如何实现我们的绘图。

SurfaceView 非常独特，与传统 View 的原理有很大差异。当实例化一个 SurfaceView 时，实际上会在 View 的位置创建另一个窗口，该窗口位于当前窗口的下方，然后 View 控件会在顶层窗口简单地"打一个洞"来透明地显示下方窗口的内容。这种方式的优势在于，可以在没有任何硬件加速支持的情况下实现高性能绘图。当然，这同时意味着 SurfaceView 是一种非常静态的视图，无法对动画或任何形式的变换做出很好的响应。

TextureView 适用于 Android 4.0 和之后的版本，在很多情况下可以顺便作为 SurfaceView 的替代品来使用。TextureView 的行为更像传统的 View，可以对绘制在它上面的内容实现动画和变换。但要求运行它的环境是硬件加速的，这可能会导致某些应用程序的兼容性问题。

7.5.3 实现机制

让我们看一个示例应用程序，其中一个后台线程将很多对象渲染到 SurfaceView 上。在这个示例中，我们会在屏幕上显示几个图标的连续动作动画。参见代码清单 7-27 和 7-28。

代码清单 7-27　res/layout/main.xml

```xml
<FrameLayout
    xmlns:android="http://schemas.android.com/apk/res/android"
    android:layout_width="match_parent"
    android:layout_height="match_parent"
    android:orientation="vertical" >
    <Button
        android:id="@+id/button_erase"
        android:layout_width="match_parent"
        android:layout_height="wrap_content"
        android:text="Erase" />
```

```xml
<SurfaceView
    android:id="@+id/surface"
    android:layout_width="300dp"
    android:layout_height="300dp"
    android:layout_gravity="center" />

</FrameLayout>
```

代码清单 7-28 Surface 绘制的 Activity

```java
public class SurfaceActivity extends Activity implements View.OnClickListener,
        View.OnTouchListener, SurfaceHolder.Callback {

    private SurfaceView mSurface;
    private DrawingThread mThread;

    @Override
    public void onCreate(Bundle savedInstanceState) {
        super.onCreate(savedInstanceState);
        setContentView(R.layout.main);
        //为按钮关联监听器
        findViewById(R.id.button_erase).setOnClickListener(this);

        //通过触摸监听器和回调来设置Surface
        mSurface = (SurfaceView) findViewById(R.id.surface);
        mSurface.setOnTouchListener(this);
        mSurface.getHolder().addCallback(this);
    }

    @Override
    public void onClick(View v) {
        mThread.clearItems();
    }

    public boolean onTouch(View v, MotionEvent event) {
        if(event.getAction() == MotionEvent.ACTION_DOWN) {
            mThread.addItem((int) event.getX(), (int) event.getY());
        }
        return true;
    }

    @Override
    public void surfaceCreated(SurfaceHolder holder) {
        mThread = new DrawingThread(holder,
                BitmapFactory.decodeResource(getResources(),
                        R.drawable.ic_launcher));
        mThread.start();
    }

    @Override
    public void surfaceChanged(SurfaceHolder holder, int format, int width,
```

```java
        int height) {
    mThread.updateSize(width, height);
}

@Override
public void surfaceDestroyed(SurfaceHolder holder) {
    mThread.quit();
    mThread = null;
}

private static class DrawingThread extends HandlerThread implements
        Handler.Callback {
    private static final int MSG_ADD = 100;
    private static final int MSG_MOVE = 101;
    private static final int MSG_CLEAR = 102;

    private int mDrawingWidth, mDrawingHeight;

    private SurfaceHolder mDrawingSurface;
    private Paint mPaint;
    private Handler mReceiver;
    private Bitmap mIcon;
    private ArrayList<DrawingItem> mLocations;

    private class DrawingItem {
        //当前位置的标识
        int x, y;
        //动作方向的标识
        boolean horizontal, vertical;

        public DrawingItem(int x, int y, boolean horizontal, boolean
                vertical) {
            this.x = x;
            this.y = y;
            this.horizontal = horizontal;
            this.vertical = vertical;
        }
    }

    public DrawingThread(SurfaceHolder holder, Bitmap icon) {
        super("DrawingThread");
        mDrawingSurface = holder;
        mLocations = new ArrayList<DrawingItem>();
        mPaint = new Paint(Paint.ANTI_ALIAS_FLAG);
        mIcon = icon;
    }

    @Override
    protected void onLooperPrepared() {
        mReceiver = new Handler(getLooper(), this);
```

```java
        //开始渲染
        mRunning = true;
        mReceiver.sendEmptyMessage(MSG_MOVE);
    }

    @Override
    public boolean quit() {
        //退出之前清除所有的消息
        mRunning = false;
        mReceiver.removeCallbacksAndMessages(null);

        return super.quit();
    }

    @Override
    public boolean handleMessage(Message msg) {
        switch(msg.what) {
        case MSG_ADD:
            //在触摸的位置创建一个新的条目,该条目的开始方向是随机的
            DrawingItem newItem = new DrawingItem(msg.arg1, msg.arg2,
                Math.round(Math.random()) == 0,
                Math.round(Math.random()) == 0);
            mLocations.add(newItem);
            break;
        case MSG_CLEAR:
            //删除所有对象
            mLocations.clear();
            break;
        case MSG_MOVE:
          if(!mRunning) return true;

            //渲染一帧
            Canvas c = mDrawingSurface.lockCanvas();
            if(c == null) {
                break;
            }
            //首先清空 Canvas
            c.drawColor(Color.BLACK);
            //绘制每个条目
            for(DrawingItem item : mLocations) {
                //更新位置
                item.x += (item.horizontal ? 5 : -5);
                if(item.x >= (mDrawingWidth - mIcon.getWidth()) ) {
                    item.horizontal = false;
                } if(item.x <= 0) {
                    item.horizontal = true;
                }
                item.y += (item.vertical ? 5 : -5);
                if(item.y >= (mDrawingHeight - mIcon.getHeight()) ) {
                    item.vertical = false;
```

```
                } if(item.y <= 0) {
                    item.vertical = true;
                }

                c.drawBitmap(mIcon, item.x, item.y, mPaint);
            }
            mDrawingSurface.unlockCanvasAndPost(c);
            break;
        }
        //发送下一帧
        if(mRunning) {
            mReceiver.sendEmptyMessage(MSG_MOVE);
        }
        return true;
    }

    public void updateSize(int width, int height) {
        mDrawingWidth = width;
        mDrawingHeight = height;
    }

    public void addItem(int x, int y) {
        //通过Message参数将位置传给处理程序
        Message msg = Message.obtain(mReceiver, MSG_ADD, x, y);
        mReceiver.sendMessage(msg);
    }

    public void clearItems() {
        mReceiver.sendEmptyMessage(MSG_CLEAR);
    }
}
```

这个示例构造了一个简单的后台 DrawingThread 以在 SurfaceView 上渲染和绘制内容。DrawingThread 是 HandlerThread 的子类，而 HandlerThread 是一个方便的框架辅助类，用来生成后台工作线程以处理收到的消息。我们在第 6 章详细探讨了这种模式，在此只要说明后台线程的响应操作是通过发送给处理程序的 handleMessage() 的消息决定的。SurfaceView 实际上由两部分组成：一个窗口下方的 Surface，一个布局结构中的空白 View。要实现绘图功能，实际上需要访问底层的 Surface，它被封装在一个 SurfaceHolder 中。

直到 View 关联到当前窗口后，Surface 才会开始实际构建。相反，当创建、销毁或改变 Surface 时，SurfaceHolder 会有一个回调接口，所以我们可以通过 SurfaceHolder 来管理依赖它的控件(本例中是 DrawingThread)的生命周期。本例中会等待 surfaceCreated() 创建一个新的 DrawingThread 并开始渲染，在 surfaceDestroyed() 中停止渲染(因为此时 Surface 已经无效)。只在最后的回调函数 surfaceChanged() 中提供 Surface 的尺寸值，所以会在这里根据尺寸值(可用时)修改绘制代码。

我们为线程定义了 3 个不同的响应指令：add、clear 和 move。用户点击 SurfaceView 时会触发 add 相关的方法，此时会在显示列表中添加一个绘制条目，它的初始位置即为触点的位置。当按下按钮时，clear 方法会将显示列表中的所有条目都删除。

当线程向 SurfaceView 渲染每帧画面时，move 方法十分高效。每一个绘图操作都需要以 lockCanvas()开始，lockCanvas()可以提供一个 Canvas 来应用绘图调用。之后线程会迭代显示列表中的每一个条目，并将每个条目更新到新的位置，然后在 Canvas 中每个更新后的位置绘制一个图标。线程同样会检查是否有条目超过了 Surface 边界，如果有，就将方向设置为相反的方向。在绘制每幅画面之前必须调用 drawColor()方法清除之前画面的内容。如果没有这样操作，当图标移动时会在图标之前的位置出现拖尾痕迹。某些应用程序可能需要这种拖尾效果(比如绘画应用程序，它会在一个事件的上面添加另一个事件)，但我们的示例并不需要这种效果。当所有的绘制调用完成之后，应用程序需要调用 unlockCanvasAndPost()方法来将数据真正地渲染到屏幕上。

通过不断地给自己发送 MSG_MOVE 消息，DrawingThread 在整个处理过程中会一直运行直到被应用程序退出。通过 HandlerThread 来完成这个处理过程的好处是，可以随时调用 quit()方法来取消操作，而且线程可以销毁得很干净，这种方式比尝试中断线程的执行要好很多。

可以在图 7-19 中看到应用程序的运行结果。用户可以单击黑色区域(不限制次数)，然后就可以看到许多飞行的图标叠加在一起。因为绘制代码只是用一张位图来绘制所有的图标，该视图能够支持很多的绘制条目，所以不用担心遇到内存问题。

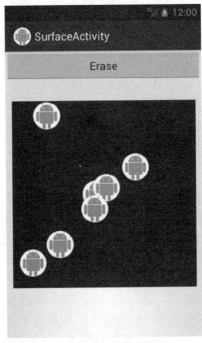

图 7-19　SurfaceView 绘图场景

TextureView

(API Level 14)

如果应用程序的目标版本为 Android 4.0 和之后的版本，也可以使用 TextureView，它有几个附加的属性可以使得你的应用更加完美。最有用的属性就是它可以进行变换。看一看代码清单 7-29 和 7-30，其中我们使用 TextureView 对前一个示例进行了修改。

代码清单 7-29 res/layout/texture.xml

```xml
<FrameLayout
    xmlns:android="http://schemas.android.com/apk/res/android"
    android:layout_width="match_parent"
    android:layout_height="match_parent">
    <Button
        android:id="@+id/button_transform"
        android:layout_width="match_parent"
        android:layout_height="wrap_content"
        android:text="Rotate" />
    <TextureView
        android:id="@+id/surface"
        android:layout_width="300dp"
        android:layout_height="300dp"
        android:layout_gravity="center" />

</FrameLayout>
```

代码清单 7-30 Texture 绘制的 Activity

```java
@TargetApi(Build.VERSION_CODES.ICE_CREAM_SANDWICH)
public class TextureActivity extends Activity implements View.OnClickListener,
        View.OnTouchListener, TextureView.SurfaceTextureListener {

    private TextureView mSurface;
    private DrawingThread mThread;

    @Override
    public void onCreate(Bundle savedInstanceState) {
        super.onCreate(savedInstanceState);
        setContentView(R.layout.texture);
        //为按钮关联监听器
        findViewById(R.id.button_transform).setOnClickListener(this);

        //通过触摸监听器和回调来设置 Surface
        mSurface = (TextureView) findViewById(R.id.surface);
        mSurface.setOnTouchListener(this);
        mSurface.setSurfaceTextureListener(this);
    }

    @Override
    public void onClick(View v) {
        //旋转整个绘制视图
        mSurface.animate()
            .rotation(mSurface.getRotation() < 180.f ? 180.f : 0.f);
```

```java
            }

            public boolean onTouch(View v, MotionEvent event) {
                if(event.getAction() == MotionEvent.ACTION_DOWN) {
                    mThread.addItem((int) event.getX(), (int) event.getY());
                }
                return true;
            }

            @Override
            public void onSurfaceTextureAvailable(SurfaceTexture surface, int width,
                    int height) {
                mThread = new DrawingThread(new Surface(surface),
                        BitmapFactory.decodeResource(getResources(),
                                R.drawable.ic_launcher));
                mThread.updateSize(width, height);
                mThread.start();
            }

            @Override
            public void onSurfaceTextureSizeChanged(SurfaceTexture surface, int width,
                    int height) {
                mThread.updateSize(width, height);
            }

            @Override
            public void onSurfaceTextureUpdated(SurfaceTexture surface) {}

            @Override
            public boolean onSurfaceTextureDestroyed(SurfaceTexture surface) {
                mThread.quit();
                mThread = null;

                //返回 true 并允许框架释放 Surface
                return true;
            }

            private static class DrawingThread extends HandlerThread implements
                    Handler.Callback {
                private static final int MSG_ADD = 100;
                private static final int MSG_MOVE = 101;
                private static final int MSG_CLEAR = 107;

                private int mDrawingWidth, mDrawingHeight;
                private boolean mRunning = false;

                private Surface mDrawingSurface;
                private Rect mSurfaceRect;
                private Paint mPaint;
```

```java
private Handler mReceiver;
private Bitmap mIcon;
private ArrayList<DrawingItem> mLocations;

private class DrawingItem {
    //当前位置标识
    int x, y;
    //运动方向的标识
    boolean horizontal, vertical;

    public DrawingItem(int x, int y, boolean horizontal,
            boolean vertical) {
        this.x = x;
        this.y = y;
        this.horizontal = horizontal;
        this.vertical = vertical;
    }
}

public DrawingThread(Surface surface, Bitmap icon) {
    super("DrawingThread");
    mDrawingSurface = surface;
    mSurfaceRect = new Rect();
    mLocations = new ArrayList<DrawingItem>();
    mPaint = new Paint(Paint.ANTI_ALIAS_FLAG);
    mIcon = icon;
}

@Override
protected void onLooperPrepared() {
    mReceiver = new Handler(getLooper(), this);
    //开始渲染
    mRunning = true;
    mReceiver.sendEmptyMessage(MSG_MOVE);
}

@Override
public boolean quit() {
    //退出之前清除所有消息
    mRunning = false;
    mReceiver.removeCallbacksAndMessages(null);

    return super.quit();
}

@Override
public boolean handleMessage(Message msg) {
    switch(msg.what) {
```

```java
case MSG_ADD:
    //在触摸的位置创建一个新的条目，该条目的开始方向是随机的
    DrawingItem newItem = new DrawingItem(msg.arg1, msg.arg7,
            Math.round(Math.random()) == 0,
            Math.round(Math.random()) == 0);
    mLocations.add(newItem);
    break;
case MSG_CLEAR:
    //删除所有的对象
    mLocations.clear();
    break;
case MSG_MOVE:
    //如果取消，则不做任何事情
    if (!mRunning) return true;

    //渲染一帧
    try {
        Canvas c = mDrawingSurface.lockCanvas(mSurfaceRect);
        //首先清空 Canvas
        c.drawColor(Color.BLACK);
        //绘制每个条目
        for(DrawingItem item : mLocations) {
            //更新位置
            item.x += (item.horizontal ? 3 : -3);
            if(item.x >= (mDrawingWidth - mIcon.getWidth()) ) {
                item.horizontal = false;
            } if(item.x <= 0) {
                item.horizontal = true;
            }
            item.y += (item.vertical ? 3 : -3);
            if(item.y >= (mDrawingHeight - mIcon.getHeight()) ) {
                item.vertical = false;
            } if(item.y <= 0) {
                item.vertical = true;
            }

            c.drawBitmap(mIcon, item.x, item.y, mPaint);
        }

        mDrawingSurface.unlockCanvasAndPost(c);
    } catch(Exception e) {
        e.printStackTrace();
    }
    break;
}
//发送下一帧
if (mRunning) {
mReceiver.sendEmptyMessage(MSG_MOVE);
}
```

```
            return true;
        }

        public void updateSize(int width, int height) {
            mDrawingWidth = width;
            mDrawingHeight = height;
            mSurfaceRect.set(0, 0, mDrawingWidth, mDrawingHeight);
        }

        public void addItem(int x, int y) {
            //通过Message参数将位置传给处理程序
            Message msg = Message.obtain(mReceiver, MSG_ADD, x, y);
            mReceiver.sendMessage(msg);
        }

        public void clearItems() {
            mReceiver.sendEmptyMessage(MSG_CLEAR);
        }
    }
}
```

在这个修改过的示例中，布局有一个 TextureView 实例。和 SurfaceView 类似，直到视图关联到当前的窗口，底层的 Surface 才会开始构建，所以我们必须依赖一个回调才能访问它。对于 TextureView 来说，这个回调就是 SurfaceTextureListener。它的大多数方法都和 SurfaceHolder.Callback 中的方法很像，例如 onSurfaceTextureAvailable()、onSurface-TextureChanged() 和 onSurfaceTextureDestroyed()。但是有一个额外的回调方法 onSurfaceTextureUpdated()在本例中并没有用到。当 SurfaceTexture 渲染一个新的帧时会调用该方法。

TextureView 提供的界面绘制 Surface 略有不同，它不用通过 SurfaceHolder 封装来访问。相反，我们可以访问一个 SurfaceTexture 实例，该实例会封装一个新的 Surface 来进行绘图。因此，需要对我们的 DrawingThread 做一点小的修改。SurfaceHolder 中有一个方便的无参版本的 lockCanvas()方法可以将整个 Surface 标记为脏对象。但当直接使用 Surface 时，该方法并不存在。所以，我们需要向 lockCanvas()传入一个 Rect，从而确定 Surface 的哪一部分在进行新的渲染时可以作为 Canvas 返回。因为我们仍然希望返回整个 Surface，所以在 updateSize()中维护 Rect 的大小，这个方法在 Surface 发生改变时会被监听事件调用。

为了生动地展示在 SurfaceTexture 渲染过程中对其进行变换，我们用 Rotate 按钮取代了 Erase 按钮。每次单击该按钮都会使 TextureView 做半周的旋转动画。当前动画运行时单击按钮会取消该动画并从当前的点开始新的旋转，所以如果快速单击按钮，视图会旋转出一些非常古怪的角度。整个过程中 SurfaceTexture 都会保持动态而且很流畅。在图 7-20 中可以看到 TextureView 倒着旋转的应用程序。

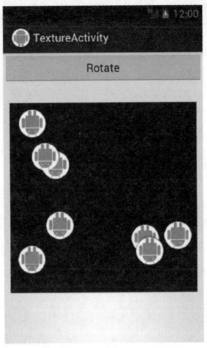

图 7-20　TextureView 绘制场景

7.6　提取图片调色板

7.6.1　问题

要使用从用户图片内容提取的颜色方案动态设计应用程序界面的主题。

7.6.2　解决方案

(API Level 7)

使用 Palette 支持包分析图片的像素数据，并为旨在补充此图片的背景和文本产生颜色样本。Palette 会尝试为每张图片生成如下样本颜色方案：

- Vibrant
- Vibrant Dark
- Vibrant Light
- Muted
- Muted Dark
- Muted Light

取决于图片内容，并不是所有 6 种颜色方案都可用。每种颜色方案都返回为 Palette.Swatch，如果图片示例包含的颜色不足以产生兼容的方案，则对应的颜色方案可能为 null。

注意：

调色板仅作为 Android 支持库中的模块提供，它不是核心框架的一部分。然而，目标平台为 API Level 7 或之后版本的任意应用程序都可以通过包含支持库使用调色板。有关在项目中包含支持库组件的更多信息，请参考如下网址：http://developer.android.com/tools/support-library/index.html。

7.6.3 实现机制

为展示此功能，我们将创建一个图片图块的列表，其中每个图块的背景和标题将通过 Palette 主题化。结果如图 7-21 所示。

图 7-21 使用调色板颜色的列表

代码清单 7-31 和 7-32 定义了该列表的条目布局和适配器。

代码清单 7-31　res/layout/item_list.xml

```
<?xml version="1.0" encoding="utf-8"?>
<LinearLayout xmlns:android="http://schemas.android.com/apk/res/android"
    android:id="@+id/root"
    android:layout_width="match_parent"
    android:layout_height="match_parent"
    android:orientation="vertical"
    android:padding="16dp">
```

```xml
<ImageView
    android:id="@+id/image"
    android:layout_width="match_parent"
    android:layout_height="102dp"
    android:scaleType="centerCrop"/>
<TextView
    android:id="@+id/text"
    android:layout_width="match_parent"
    android:layout_height="72dp"
    android:gravity="center"
    android:textAppearance="?android:textAppearanceLarge"/>
</LinearLayout>
```

代码清单 7-32 适配器中的调色板颜色

```java
public class ColorfulAdapter extends ArrayAdapter<String> {
    private static final int[] IMAGES = {
            R.drawable.bricks, R.drawable.flower,
            R.drawable.grass, R.drawable.stones,
            R.drawable.wood, R.drawable.dog
    };

    private static final String[] NAMES = {
            "Bricks", "Flower",
            "Grass", "Stones",
            "Wood", "Dog"
    };

    private SparseArray<Bitmap> mImages;
    private SparseArray<Palette.Swatch> mBackgroundColors;

    public ColorfulAdapter(Context context) {
        super(context, R.layout.item_list, NAMES);
        mImages = new SparseArray<Bitmap>(IMAGES.length);
        mBackgroundColors = new SparseArray<Palette.Swatch>(IMAGES.length);
    }

    @Override
    public View getView(int position, View convertView, ViewGroup parent) {
        if(convertView == null) {
            convertView = LayoutInflater.from(getContext())
                    .inflate(R.layout.item_list, parent, false);
        }

        View root = convertView.findViewById(R.id.root);
        ImageView image = (ImageView) convertView.findViewById(R.id.image);
        TextView text = (TextView) convertView.findViewById(R.id.text);

        int imageId = IMAGES[position];
        if(mImages.get(imageId) == null) {
            new ImageTask().execute(imageId);
```

```java
            text.setTextColor(Color.BLACK);
        } else {
            image.setImageBitmap(mImages.get(imageId));

            Palette.Swatch colors = mBackgroundColors.get(imageId);
            if(colors != null) {
                root.setBackgroundColor(colors.getRgb());
                text.setTextColor(colors.getTitleTextColor());
            }
        }

        text.setText(NAMES[position]);

        return convertView;
    }

    private class ImageResult {
        public int imageId;
        public Bitmap image;
        public Palette.Swatch colors;

        public ImageResult(int imageId, Bitmap image, Palette.Swatch colors) {
            this.imageId = imageId;
            this.image = image;
            this.colors = colors;
        }
    }

    private void updateImageItem(ImageResult result) {
        mImages.put(result.imageId, result.image);
        mBackgroundColors.put(result.imageId, result.colors);
    }

    private class ImageTask extends AsyncTask<Integer, Void, ImageResult> {

        @Override
        protected ImageResult doInBackground(Integer… params) {
            int imageId = params[0];
            //确保图片缩略图不会太大
            BitmapFactory.Options options = new BitmapFactory.Options();
            options.inSampleSize = 4;
            Bitmap image = BitmapFactory.decodeResource(getContext().
                    getResources(),imageId, options);

            Palette colors = Palette.generate(image);
            Palette.Swatch selected = colors.getVibrantSwatch();
            if(selected == null) {
                selected = colors.getMutedSwatch();
```

```
            }
            return new ImageResult(imageId, image, selected);
        }

        @Override
        protected void onPostExecute(ImageResult result) {
            updateImageItem(result);
            notifyDataSetChanged();
        }
    }
}
```

适配器处理此例中的所有繁重工作。我们有 6 张本地图片，并且在网格内显示的每个元素都有一个字符串标题。在适配器的 getView() 回调中，可以看到文本颜色和背景容器颜色都从 Palette.Swatch 中读取，Palette.Swatch 可能已存在，也可能尚未存在。

因为从磁盘加载图片和使用 Palette 分析这些图片的过程会花费一些时间，所以我们要在后台执行此工作以避免阻塞主线程太长时间；为此，我们将此工作封装在 AsyncTask 中，从而确保在后台执行。

后台加载的结果存储在一对 SparseArray 实例中。当适配器在未产生结果的情况下遇到一个图片条目时，它会触发新的后台任务以加载内容。通过 BitmapFactory 解码图片之后，可以使用 Palette.Generate() 收集颜色方案数据，该方法将在分析图片时阻塞。

提示：

如果在生成颜色方案时未准备好后台线程，Palette 的 generateAsync() 方法会接受回调。

完成上述操作之后，可以使用各种 getter 方法访问颜色样本。首先尝试使用 getVibrantSwatch() 获得 Vibrant 样本。然而，如果图片颜色不兼容 Vibrant 颜色方案(即方法返回 null)，我们就降低要求，获取一种颜色减弱的样本。每次后台任务完成时，系统就使用 notifyDataSetChanged() 通知适配器重新填充视图。

注意：

解码的图片和选择的样本封装在持有者对象(ImageResult)中，这样有助于从 AsyncTask 返回多个条目，AsyncTask 仅支持返回单个类型。

回到 getView() 中，当条目成功返回现有的样本之后，可以分别使用 getTitleTextColor() 和 getRgb() 获得文本颜色和背景颜色。参见代码清单 7-33，其中将适配器附加到 Activity 中的 GridView 以完成示例。

代码清单 7-33 彩色调色板 Activity

```
public class ColorfulListActivity extends ActionBarActivity {

    private GridView mGridView;
```

```
    @Override
    protected void onCreate(Bundle savedInstanceState) {
        super.onCreate(savedInstanceState);
        mGridView = new GridView(this);
        mGridView.setNumColumns(2);
        mGridView.setAdapter(new ColorfulAdapter(this));

        setContentView(mGridView);
    }
}
```

代码清单 7-34 显示了为编译此例而必须存在于 build.gradle 文件中的所需部分。

代码清单 7-34　build.gradle 的部分代码

```
apply plugin: 'com.android.application'

android {
    compileSdkVersion 21
    ...

    defaultConfig {
        applicationId "com.androidrecipes.palette"
        ...
    }
    ...
}

dependencies {
    compile 'com.android.support:appcompat-v7:21.0.+'
    compile 'com.android.support:palette-v7:21.0.+'
}
```

为使用 Palette 库，必须将其作为依赖项添加进来。此外，我们使用 ActionBarActivity 来全面支持较早的 API Level 平台，这也需要包括 AppCompat 作为依赖项。

7.7　平铺 Drawable 元素

7.7.1　问题

你希望通过在运行时对基础资源动态改变颜色，避免重复仅在颜色方面有所变化的常见资源。

7.7.2　解决方案

(API Level 1)
使用颜色滤镜，对任意 Drawable 实例应用颜色遮罩。Drawable 颜色滤镜通常是完全不透明的，但 Android 框架还支持通过 PorterDuff.XferMode 进行部分混合。只可以从 Java 代码中执行此方法。

(API Level 21)

通过在 XML 中调用 android:tint 或在 Java 代码中调用 setTint()，可以使用任意 Drawable 实例上的原生色调功能。在此例中，可以通过 android:tintMode 或 setTintMode()应用混合；两者都采用 Porter-Duff 常量表示变换模式。

7.7.3 实现机制

在此例中，我们将采用图 7-22 中的 4 张图片，并且对这些图片动态应用颜色以产生如图 7-23 所示的输出。

图 7-22　着色前的基础图标(放在 res/drawable 中)

图 7-23　带有已着色 Drawable 的 Activity

注意：

为正确应用 Porter-Duff 颜色混合，不应修改的图标区域应完全透明而非纯白色。其他像素不需要为黑色，但它们必须完全不透明。

代码清单 7-35 显示了用于在 Activity 中放置这些图标的简单布局。

代码清单 7-35　res/layout/activity_filter.xml

```xml
<?xml version="1.0" encoding="utf-8"?>
<LinearLayout xmlns:android="http://schemas.android.com/apk/res/android"
    android:layout_width="match_parent"
    android:layout_height="match_parent"
    android:orientation="horizontal"
    android:padding="8dp">

    <ImageView
        android:id="@+id/icon_marker"
        android:layout_width="0dp"
        android:layout_height="wrap_content"
        android:layout_weight="1"
        android:src="@drawable/ic_marker"/>

    <ImageView
        android:id="@+id/icon_gear"
        android:layout_width="0dp"
        android:layout_height="wrap_content"
        android:layout_weight="1"
        android:src="@drawable/ic_gear"/>

    <ImageView
        android:id="@+id/icon_check"
        android:layout_width="0dp"
        android:layout_height="wrap_content"
        android:layout_weight="1"
        android:src="@drawable/ic_check"/>

    <ImageView
        android:id="@+id/icon_heart"
        android:layout_width="0dp"
        android:layout_height="wrap_content"
        android:layout_weight="1"
        android:src="@drawable/ic_heart"/>
</LinearLayout>
```

此布局中需要注意的一点是，在此没有应用任何特殊的滤镜。对于这种初级技巧，不能通过 XML 应用颜色。代码清单 7-36 展示了一个 Activity，其中的代码用于对图标着色。

代码清单 7-36　颜色滤镜 Activity

```java
public class ColorFilterActivity extends ActionBarActivity {

    @Override
    protected void onCreate(Bundle savedInstanceState) {
        super.onCreate(savedInstanceState);
```

```java
        setContentView(R.layout.activity_filter);

        applyIconFilters();
    }

    private void applyIconFilters() {

        ImageView iconView = (ImageView) findViewById(R.id.icon_marker);
        //绘制纯紫色图标
        iconView.getDrawable().setColorFilter(0xFFAA00AA,
            PorterDuff.Mode.SRC_ATOP);

        iconView = (ImageView) findViewById(R.id.icon_gear);
        //绘制纯绿色图标
        iconView.getDrawable().setColorFilter(0xFF00AA00,
            PorterDuff.Mode.SRC_ATOP);

        iconView = (ImageView) findViewById(R.id.icon_check);
        //绘制纯蓝色图标
        iconView.getDrawable().setColorFilter(0xFF0000AA,
            PorterDuff.Mode.SRC_ATOP);

        iconView = (ImageView) findViewById(R.id.icon_heart);
        //绘制纯红色图标
        iconView.getDrawable().setColorFilter(0xFFAA0000,
            PorterDuff.Mode.SRC_ATOP);
    }
}
```

使用 setColorFilter()方法，可以通过色调绘制任何 Drawable。此方法最简单的版本(在此使用的就是该版本)接受 ARGB 颜色值和 PorterDuff.Mode 以进行像素变换和混合。我们选择的 SRC_ATOP 确保完全绘制所选的颜色并忽略原始的图片像素。

提示：
如果基础图片存在要显露出来的变化情况(如渐变)，可选取部分透明的滤镜颜色和/或尝试使用不同的 PorterDuff.Mode 值，如 MULTIPLY。

setColorFilter()方法的另一个版本接受 ColorFilter 实例以实现更加复杂的覆盖层。例如，LightingColorFilter 旨在模拟光源效果。

关于恒定状态的说明

如果在同一个 Activity 中(例如在列表中)多次重复使用相同的图片资源，并且尝试设置多个色调值，就会发现所有 Drawable 将仅显示最近设置的颜色。这是因为存在所谓的共享恒定状态。此状态假设通过同一个资源创建的每个 Drawable 都共享一个公共的状态对象，因为这些 Drawable 在大多数情况下都被假设为不可变的对象，这就节省了内存资源。

然而，这种假设的负面效果就是更改某个 Drawable 状态的属性也会影响所有其他的

Drawable。为避免出现这种问题，在进行影响单个 Drawable 状态的修改时，都应首先调用其 mutate()方法。这会在进行修改之前创建状态对象的副本。可以随所选的颜色修改程序一起内联调用此方法，例如下面的代码：

```
iconView.getDrawable().mutate().setColorFilter(...)
```

这将确保所执行的颜色修改不会影响通过同一个资源创建的其他 Drawable。

最后要说明的是，Drawable 通常只可以变化一次，因此最好不要将此方法用作复制或克隆机制。仍然应该通过资源获得独特的实例，并且在对这些实例进行更改之前调用 mutate()方法。

原生着色

(API Level 21)

从 Android 5.0 开始，可以使用 android:tint 属性或在代码中通过 setTint()方法，从 XML 中将相同的效果应用于多个 Drawable。隐藏在代码下的操作是：Android 框架使用前面介绍的相同技巧，但是稍微更加高效，因为框架现在可以共享具有公共色调的状态。代码清单 7-37~7-41 重新将图标资源定义为着色的 Drawable。

代码清单 7-37 res/drawable/tinted_marker.xml

```xml
<?xml version="1.0" encoding="utf-8"?>
<bitmap xmlns:android="http://schemas.android.com/apk/res/android"
    android:src="@drawable/ic_marker"
    android:tint="#FFAA00AA" />
```

代码清单 7-38 res/drawable/tinted_gear.xml

```xml
<?xml version="1.0" encoding="utf-8"?>
<bitmap xmlns:android="http://schemas.android.com/apk/res/android"
    android:src="@drawable/ic_gear"
    android:tint="#FF00AA00" />
```

代码清单 7-39 res/drawable/tinted_check.xml

```xml
<?xml version="1.0" encoding="utf-8"?>
<bitmap xmlns:android="http://schemas.android.com/apk/res/android"
    android:src="@drawable/ic_check"
    android:tint="#FF0000AA" />
```

代码清单 7-40 res/drawable/tinted_heart.xml

```xml
<?xml version="1.0" encoding="utf-8"?>
<bitmap xmlns:android="http://schemas.android.com/apk/res/android"
    android:src="@drawable/ic_heart"
    android:tint="#FFAA0000" />
```

现在当我们将这些 Drawable 插入 Activity 布局时，不需要执行进一步的着色工作。参见代码清单 7-41 和 7-42。

代码清单 7-41　res/layout/activity_tinted.xml

```xml
<?xml version="1.0" encoding="utf-8"?>
<LinearLayout xmlns:android="http://schemas.android.com/apk/res/android"
    android:orientation="horizontal"
    android:layout_width="match_parent"
    android:layout_height="match_parent"
    android:padding="8dp">
    <ImageView
        android:id="@+id/icon_marker"
        android:layout_width="0dp"
        android:layout_height="wrap_content"
        android:layout_weight="1"
        android:src="@drawable/tinted_marker" />
    <ImageView
        android:id="@+id/icon_gear"
        android:layout_width="0dp"
        android:layout_height="wrap_content"
        android:layout_weight="1"
        android:src="@drawable/tinted_gear"/>
    <ImageView
        android:id="@+id/icon_check"
        android:layout_width="0dp"
        android:layout_height="wrap_content"
        android:layout_weight="1"
        android:src="@drawable/tinted_check"/>
    <ImageView
        android:id="@+id/icon_heart"
        android:layout_width="0dp"
        android:layout_height="wrap_content"
        android:layout_weight="1"
        android:src="@drawable/tinted_heart"/>
</LinearLayout>
```

代码清单 7-42　已着色 Drawable 的简单 Activity

```java
public class TintActivity extends ActionBarActivity {

    @Override
    protected void onCreate(Bundle savedInstanceState) {
        super.onCreate(savedInstanceState);
        setContentView(R.layout.activity_tinted);
    }
}
```

着色的图片会替换布局 XML 内的原始图片，并且此处不再有之前用于应用滤镜的 Java 代码。

提示：
仍然可以在原生色调 API 中采用变换模式。使用 android:tintMode 属性，可以将 Porter-Duff 常量与着色配对。省略此属性的效果等同于应用 android:tintMode="src_in"。

7.8 使用可缩放的向量资源

7.8.1 问题

希望通过使用可缩放的向量资源而非静态图片来减少应用程序包括的资源文件数。

7.8.2 解决方案

(API Level 21)

通过图片资源的路径数据构造一个 VectorDrawable。Android 并不原生支持读取常见的向量文件格式，如 SVG(主要是由于缺少 CSS 支持)，但其支持相同的路径数据语法。这意味着通过一些次要的转换，就可以创建在所有屏幕密度之间完全缩放的单个 XML 向量资源。

注意：
有关 SVG 路径数据语法的更多信息，请参阅 W3C 参考手册：http://www.w3.org/TR/SVG11/paths.html。

Android 还支持使用 AnimatedVectorDrawable 制作路径操作动画。这个类提供了用于将 ObjectAnimator 实例映射到 VectorDrawable 内目标路径的支持程序。

7.8.3 实现机制

在此例中，我们将采用如图 7-24 所示的 SVG 图片，并且将其转换为可以作为 VectorDrawable 显示的对象。

图 7-24　从 Adobe Illustrator 导出的 SVG 图片

SVG 文件实际上只是 XML 内容，代码清单 7-43 显示了原始文件数据。

代码清单 7-43   assets/examples.svg

```xml
<?xml version="1.0" encoding="utf-8"?>
<!DOCTYPE svg PUBLIC "-//W3C//DTD SVG 1.1//EN" "http://www.w3.org/Graphics/SVG/1.1/DTD/svg11.dtd">
<svg version="1.1" id="Layer_2" xmlns="http://www.w3.org/2000/svg"
    xmlns:xlink="http://www.w3.org/1999/xlink" x="0px" y="0px"
         width="163px" height="154px" viewBox="0 0 163 154"
         style="enable-background:new 0 0 163 154;"
         xml:space="preserve">

<style type="text/css">
        .st0{fill:#3355CC;stroke:#000000;stroke-width:0.25;stroke-
            miterlimit:10;}
</style>
<g>
        <circle class="st0" cx="81.5" cy="77" r="14.7"/>
        <path class="st0" d="M81.5,3c-11.2,0-19.2,9.1-19.2,20.3
            c0,8.3,6.7,15.4,13,18.6c-0.9-1.3-1.4-2.8-1.4-4.4c0-4.2,
            3.4-7.6,7.6-7.6c4.2,0,7.6,3.4,7.6,7.6c0,1.8-0.6,3.5-1.7,
            4.8c6.8-2.9,13.3-10.3,13.3-19C100.7,12.1,92.7,3,81.5,3z"/>
        <path class="st0" d="M29.2,24.7c-7.9,7.9-7.1,20,0.8,28
            c5.9,5.9,15.7,6.2,22.3,4c-1.5-0.3-3-1-4.2-2.1c-3-3-3-7.8,
            0-10.8c3-3,7.8-3,10.8,0c1.3,1.3,2,2.9,2.2,4.6c2.7-6.9,
            2.1-16.7-4-22.8C49.2,17.5,37.1,16.7,29.2,24.7z"/>
        <path class="st0" d="M7.5,77c0,11.2,9.1,19.2,20.3,19.2
            c8.3,0,15.4-6.7,18.6-13c-1.3,0.9-2.8,1.4-4.4,1.4c-4.2,
            0-7.6-3.4-7.6-7.6c0-4.2,3.4-7.6,7.6-7.6c1.8,0,3.5,0.6,
            4.8,1.7c-2.9-6.8-10.3-13.3-19-13.3C16.6,57.8,7.5,65.8,
            7.5,77z"/>
        <path class="st0" d="M29.2,129.3c7.9,7.9,20,7.1,28-0.8
            c5.9-5.9,6.2-15.7,4-22.3c-0.3,1.5-1,3-2.1,4.2c-3,3-7.8,
            3-10.8,0c-3-3-3-7.8,0-10.8c1.3-1.3,2.9-2,4.6-2.2c-6.9-
            2.7-16.7-2.1-22.8,4C22,109.3,21.2,121.4,29.2,129.3z"/>
        <path class="st0" d="M81.5,151c11.2,0,19.2-9.1,19.2-20.3
            c0-8.3-6.7-15.4-13-18.6c0.9,1.3,1.4,2.8,1.4,4.4c0,4.2-3.4,
            7.6-7.6,7.6c-4.2,0-7.6-3.4-7.6-7.6c0-1.8,0.6-3.5,1.7-4.8
            c-6.8,2.9-13.3,10.3-13.3,19C62.3,141.9,70.3,151,81.5,151
            z"/>
        <path class="st0" d="M133.8,129.3c7.9-7.9,7.1-20-0.8-28
            c-5.9-5.9-15.7-6.2-22.3-4c1.5,0.3,3,1,4.2,2.1c3,3,3,7.8,
            0,10.8s-7.8,3-10.8,0c-1.3-1.3-2-2.9-2.2-4.6c-2.7,6.9-2.1,
            16.7,4,22.8C113.8,136.5,125.9,137.3,133.8,129.3z"/>
        <path class="st0" d="M155.5,77c0-11.2-9.1-19.2-20.3-19.2
            c-8.3,0-15.4,6.7-18.6,13c1.3-0.9,2.8-1.4,4.4-1.4c4.2,0,
```

```
            7.6,3.4,7.6,7.6c0,4.2-3.4,7.6-7.6,7.6c-1.8,0-3.5-0.6-4.8
            -1.7c2.9,6.8,10.3,13.3,19,13.3C146.4,96.2,155.5,88.2,
            155.5,77z"/>
        <path class="st0" d="M133.8,24.7c-7.9-7.9-20-7.1-28,0.8
            c-5.9,5.9-6.2,15.7-4,22.3c0.3-1.5,1-3,2.1-4.2c3-3,7.8-3,
            10.8,0c3,3,3,7.8,0,10.8c-1.3,1.3-2.9,2-4.6,2.2c6.9,2.7,
            16.7,2.1,22.8-4C141,44.7,141.8,32.6,133.8,24.7z"/>
    </g>
</svg>
```

这段代码看起来有点乱，特别是在不熟悉 SVG 路径语法的情况下。好消息是这段代码的几乎全部内容都可以原封不动地复制到 XML 资源中。SVG 中的每个<path>元素都在 VectorDrawable 中有对应的元素。Android 向量不支持 SVG 的特殊类型，如<circle>，因此我们必须手动将其构造为新的路径。Android 同样不支持 CSS 样式化，因此用于设置每条路径上笔画/填充颜色的 class 属性必须转换为 Android 属性。查看代码清单 7-44 以了解转换后的向量图片。

代码清单 7-44 res/drawable/svg_converted.xml

```
<?xml version="1.0" encoding="utf-8"?>
<vector xmlns:android="http://schemas.android.com/apk/res/android"
    android:width="163dp"
    android:height="154dp"
    android:viewportWidth="163"
    android:viewportHeight="154">
    <!-- 这是来自 SVG 文件的 circle 元素 -->
    <path
        android:fillColor="#3355CC"
        android:pathData="M66.8,77 a14.7,14.7 0 0,1 29.4,0 a14.7,14.7 0 0,1
                         -29.4,0z"/>

    <!-- 剩余的路径逐字复制 -->
    <path
        android:fillColor="#3355CC"
        android:pathData="M81.5,3c-11.2,0-19.2,9.1-19.2,20.3
            c0,8.3,6.7,15.4,13,18.6c-0.9-1.3-1.4-2.8-1.4-4.4c0-4.2,
            3.4-7.6,7.6-7.6c4.2,0,7.6,3.4,7.6,7.6c0,1.8-0.6,3.5-1.7,
            4.8c6.8-2.9,13.3-10.3,13.3-19C100.7,12.1,92.7,3,81.5,3z"/>
    <path
        android:fillColor="#3355CC"
        android:pathData="M29.2,24.7c-7.9,7.9-7.1,20,0.8,28
            c5.9,5.9,15.7,6.2,22.3,4c-1.5-0.3-3-1-4.2-2.1c-3-3-3-7.8,
            0-10.8c3-3,7.8-3,10.8,0c1.3,1.3,2,2.9,2.2,4.6c2.7-6.9,
            2.1-16.7-4-22.8C49.2,17.5,37.1,16.7,29.2,24.7z"/>
    <path
        android:fillColor="#3355CC"
```

```xml
        android:pathData="M7.5,77c0,11.2,9.1,19.2,20.3,19.2
            c8.3,0,15.4-6.7,18.6-13c-1.3,0.9-2.8,1.4-4.4,1.4c-4.2,
            0-7.6-3.4-7.6-7.6c0-4.2,3.4-7.6,7.6-7.6c1.8,0,3.5,0.6,
            4.8,1.7c-2.9-6.8-10.3-13.3-19-13.3C16.6,57.8,7.5,65.8,
            7.5,77z"/>
    <path
        android:fillColor="#3355CC"
        android:pathData="M29.2,129.3c7.9,7.9,20,7.1,28-0.8
            c5.9-5.9,6.2-15.7,4-22.3c-0.3,1.5-1,3-2.1,4.2c-3,3-7.8,
            3-10.8,0c-3-3-3-7.8,0-10.8c1.3-1.3,2.9-2,4.6-2.2c-6.9-
            2.7-16.7-2.1-22.8,4C22,109.3,21.2,121.4,29.2,129.3z"/>
    <path
        android:fillColor="#3355CC"
        android:pathData="M81.5,151c11.2,0,19.2-9.1,19.2-20.3
            c0-8.3-6.7-15.4-13-18.6c0.9,1.3,1.4,2.8,1.4,4.4c0,4.2-
            3.4,7.6-7.6,7.6c-4.2,0-7.6-3.4-7.6-7.6c0-1.8,0.6-3.5,
            1.7-4.8c-6.8,2.9-13.3,10.3-13.3,19C62.3,141.9,70.3,151,
            81.5,151z"/>
    <path
        android:fillColor="#3355CC"
        android:pathData="M133.8,129.3c7.9-7.9,7.1-20-0.8-28
            c-5.9-5.9-15.7-6.2-22.3-4c1.5,0.3,3,1,4.2,2.1c3,3,3,7.8,
            0,10.8s-7.8,3-10.8,0c-1.3-1.3-2-2.9-2.2-4.6c-2.7,6.9-2.1,
            16.7,4,22.8C113.8,136.5,125.9,137.3,133.8,129.3z"/>
    <path
        android:fillColor="#3355CC"
        android:pathData="M155.5,77c0-11.2-9.1-19.2-20.3-19.2
            c-8.3,0-15.4,6.7-18.6,13c1.3-0.9,2.8-1.4,4.4-1.4c4.2,0,
            7.6,3.4,7.6,7.6c0,4.2-3.4,7.6-7.6,7.6c-1.8,0-3.5-0.6-4.8
            -1.7c2.9,6.8,10.3,13.3,19,13.3C146.4,96.2,155.5,88.2,
            155.5,77z"/>
    <path
        android:fillColor="#3355CC"
        android:pathData="M133.8,24.7c-7.9-7.9-20-7.1-28,0.8
            c-5.9,5.9-6.2,15.7-4,22.3c0.3-1.5,1-3,2.1-4.2c3-3,7.8-3,
            10.8,0c3,3,3,7.8,0,10.8c-1.3,1.3-2.9,2-4.6,2.2c6.9,2.7,
            16.7,2.1,22.8-4C141,44.7,141.8,32.6,133.8,24.7z"/>
</vector>
```

VectorDrawable XML 定义的根元素是<vector>。原始<svg>元素中的宽度、高度和视图框属性已作为 android:width、android:height、android:viewportWidth 和 android:viewportHeight 移动到此处，用于定义图片的实际尺寸。

此处唯一涉及的 CSS 样式参数是填充颜色，将其作为每条路径上的 android:fillColor 属性添加进来。最后，所有的路径数据都从 SVG 的 d 属性移动到 android:pathData 中。原始的<circle>元素已经被等价的<path>属性取代，因为这是 Android 支持的唯一向量类型。

SVG 路径语法入门

在<circle>的 SVG 示例中，我们必须手动执行一次转换，因此接下来详细分析此次转换以更好地了解 SVG 路径语法。在原始 SVG 文件中，我们有如下元素：

`<circle class="st0" cx="81.5" cy="77" r="14.7"/>`

Android 向量中的如下路径取代了<circle>元素：

`android:pathData="M66.8,77 a14.7,14.7 0 0,1 29.4,0 a14.7,14.7 0 0,1 -29.4,0z"`

该路径是串联在一起的一系列绘制命令。每个字母字符表示新命令的开始，因此可以将路径视为：

```
M66.8,77
a14.7,14.7 0 0,1 29.4,0
a14.7,14.7 0 0,1 -29.4,0
z
```

该路径包含 3 条不同的命令：

- M：表示 Moveto 命令，将画笔移到特定的点，通过以 x,y 形式分隔的后续数字表明这些点。
- a：表示 Arc 命令，使用指定的半径，从当前画笔位置到指定的点绘制一个圆弧。
- z：闭合路径。不一定要闭合路径，但如果当前位置与开始位置不同，则会在两个位置之间绘制一条线。

初始的<circle>元素定义了位于(81.5,77)的中心点(cx,cy)，同时定义半径为 14.7。新的路径开始于相同垂直点处圆形的左侧(减去一个半径)，因此首先移动到(66.8,77)。然后使用相同的半径绘制一个圆弧，圆弧的终点在相同垂直位置处右侧 29.4 个单位的新点；这就绘制了圆形的上半部分。再次重复绘制相同的圆弧，移回到左侧 29.4 个单位处，以绘制圆形的下半部分。两个圆弧中间的数值(0 0,1)用作标志，表明圆弧不应预先旋转，并且圆弧应沿顺时针扫掠两次。最终，该路径闭合以形成实线圆。

向量动画

在此例中，我们使用路径动画将 X 形状变形为选中标记。两个元素都通过 SVG 语法定义为向量路径。代码清单 7-45 将路径定义为字符串资源。

代码清单 7-45 　res/values/strings.xml

```
<resources>
    <string name="path_cross">M24,0 l 0,48 M0,24 l 48,0</string>
    <string name="path_check">M9,36 l 20,0 M27,36 l 0,-36</string>
</resources>
```

这些路径引入了一条新的命令，即 lineto，其简写形式为小写的"L"。此命令从当前位置绘制一条线，线的长度为字母l后面定义的相对距离。例如，"交叉"路径绘制两条直线：一条是从(24,0)到(24,48)的垂直线，另一条是从(0,24)到(48,24)的水平线。同样，"选中"路径绘制两条对角线以形成相应形状。这些路径已抽象化为资源，因为在此例的多个位置都需要引用它们。

代码清单 7-46 定义了用于形成初始 X 形状的 VectorDrawable。

代码清单 7-46　res/drawable/vector_cross.xml

```xml
<?xml version="1.0" encoding="utf-8"?>
<vector xmlns:android="http://schemas.android.com/apk/res/android"
    android:height="64dp"
    android:width="64dp"
    android:viewportHeight="48"
    android:viewportWidth="48">
    <group
        android:name="rotateContainer"
        android:pivotX="24.0"
        android:pivotY="24.0"
        android:rotation="45.0">
        <path
            android:name="cross"
            android:strokeColor="#A00"
            android:strokeWidth="4"
            android:pathData="@string/path_cross" />
    </group>
</vector>
```

此图片使用"交叉"路径，然后将其旋转 45°，使其看起来像 X 而非交叉线。路径元素自身不能通过转换、旋转或缩放进行变换。相反，它们必须封装在<group>中，该元素支持对其包含的所有路径进行这些变换。每个元素都有一个名称，我们将很快使用此名称来引用个别组件。

按照实际情况来说，可以将此图片放入视图中，并且绘制为静态图片。然而，为了对其制作动画，需要将此向量封装到 AnimatedVectorDrawable 中，与动画对象配对。参见代码清单 7-47。

代码清单 7-47　res/drawable/animated_check.xml

```xml
<?xml version="1.0" encoding="utf-8"?>
<animated-vector
    xmlns:android="http://schemas.android.com/apk/res/android"
    android:drawable="@drawable/vector_cross">
```

```xml
    <target
        android:name="cross"
        android:animation="@animator/check_animation"/>

    <target
        android:name="rotateContainer"
        android:animation="@animator/rotate_check_animation" />
</animated-vector>
```

AnimatedVectorDrawable 将动画对象与所提供向量(作为 android:drawable 传递)内的目标关联起来。XML 代码按其名称引用每个目标元素(无论是路径还是组),并且将元素与要应用的动画进行匹配。代码清单 7-48 和 7-49 显示了动画的定义。

代码清单 7-48   res/drawable/check_animator.xml

```xml
<?xml version="1.0" encoding="utf-8"?>
<set xmlns:android="http://schemas.android.com/apk/res/android"
    android:ordering="sequentially">

    <!-- 推进路径动画 -->
    <objectAnimator
        android:duration="@android:integer/config_longAnimTime"
        android:propertyName="pathData"
        android:valueFrom="@string/path_cross"
        android:valueTo="@string/path_check"
        android:valueType="pathType" />

    <!-- 反转路径动画 -->
    <objectAnimator
        android:duration="1000"
        android:propertyName="pathData"
        android:valueFrom="@string/path_check"
        android:valueTo="@string/path_cross"
        android:valueType="pathType"
        android:startOffset="@android:integer/config_longAnimTime"/>
</set>
```

代码清单 7-49   res/animator/rotate_check_animation.xml

```xml
<?xml version="1.0" encoding="utf-8"?>
<set xmlns:android="http://schemas.android.com/apk/res/android"
    android:ordering="sequentially">

    <objectAnimator
        android:duration="@android:integer/config_longAnimTime"
```

```xml
            android:propertyName="rotation"
            android:valueFrom="45"
            android:valueTo="405"/>
    <objectAnimator
            android:duration="1000"
            android:propertyName="rotation"
            android:valueFrom="405"
            android:valueTo="45"
            android:startOffset="@android:integer/config_longAnimTime"/>
</set>
```

应用于目标路径的动画集包含两个 pathData 动画,这就使我们可将路径从一个命令集变形为另一个;在此例中是从交叉变为选中,然后又变回交叉。只要两个路径有相同数量的数据点,就可以完成这种变形;换句话说,就是按相同顺序执行具有相同数量参数的相同命令。

应用于容器组的动画器简单地将整个图片旋转一个完整的圆,在路径变形继续执行的过程中为此动画提供更强的视觉吸引力。两个动画都会在暂停后以更长的持续时间反转执行,以便我们轻松地确切了解 Android 框架如何转换每个路径。

注意:

AnimatedVectorDrawable 不支持以代码方式访问其动画器。这意味着不能附加侦听器,也不能在内部独立于其附加的 Drawable 重新启动动画。

为完成本例,我们给出了代码清单 7-50 和 7-51,提供可以放置这些新 Drawable 的 Activity 和布局。

代码清单 7-50　res/layout/activity_vector.xml

```xml
<?xml version="1.0" encoding="utf-8"?>
<LinearLayout xmlns:android="http://schemas.android.com/apk/res/android"
    xmlns:tools="http://schemas.android.com/tools"
    android:orientation="vertical"
    android:layout_width="match_parent"
    android:layout_height="match_parent"
    tools:context=".VectorActivity">
    <ImageView
        android:id="@+id/image_static"
        android:layout_width="match_parent"
        android:layout_height="0dp"
        android:layout_weight="2" />
    <ImageView
        android:id="@+id/image_animated"
        android:layout_width="match_parent"
```

```xml
        android:layout_height="0dp"
        android:layout_weight="1"
        android:scaleType="center"/>
    <Button
        android:layout_width="match_parent"
        android:layout_height="wrap_content"
        android:text="Morph Drawable"
        android:onClick="onMorphClick"/>
</LinearLayout>
```

代码清单 7-51　显示 VectorDrawable 的 Activity

```java
public class VectorActivity extends Activity {

    private AnimatedVectorDrawable mAnimatedDrawable;

    @Override
    protected void onCreate(Bundle savedInstanceState) {
        super.onCreate(savedInstanceState);
        setContentView(R.layout.activity_vector);

        //将转换的 SVG 向量设置为静态图片
        ImageView imageView = (ImageView) findViewById(R.id.image_static);
        imageView.setImageResource(R.drawable.svg_converted);

        //创建向量路径变形动画
        imageView = (ImageView) findViewById(R.id.image_animated);

        mAnimatedDrawable = (AnimatedVectorDrawable) getResources()
                .getDrawable(R.drawable.animated_check);
        imageView.setImageDrawable(mAnimatedDrawable);
    }

    public void onMorphClick(View v) {
        mAnimatedDrawable.start();
    }
}
```

该布局包含两个 ImageView 实例和一个 Button。转换后的 SVG 图片静态放置于顶部视图中,而其动画条目放置于底部视图中。当用户按下按钮时,路径变形动画将会运行,结果如图 7-25 所示。

图 7-25　包含向量图片的 Activity

7.9　小结

在本章，我们探讨了使用 Drawable 作为灵活背景、传达状态更改以及作为可缩放向量路径的方式。我们查看了各种图片处理技术，如图片遮罩和调色板提取。最后，我们介绍了一些更加高级的直接绘制方法，即使用覆盖层、表面和纹理。在最后一章，我们将研究一些高性能的技巧，即使用原生的(C/C++)代码和 RenderScript 大量处理应用程序中的代码。

第 8 章

使用 Android NDK 和 RenderScript

开发人员通常完全使用 Java 来编写 Android 应用程序。但是，也会出现部分代码需要(或必须)使用其他语言(尤其是 C 或 C++)来编写的情况。这可能是需要访问仅在原生代码中提供的资源(例如内核驱动程序)，或是需要通过代码的关键部分提升性能。Google 提供了 Android 本地开发包(Native Development Kit，NDK)和 RenderScript 来处理这种情况。

8.1 Android NDK

Android NDK 是 Android SDK 的补充，它提供了一个工具集让用户可以使用 C 和 C++ 这样的原生代码语言实现部分应用程序。对于构建原生 Activity、处理用户输入、使用硬件传感器等操作，NDK 都提供了相应的头文件和库。NDK 通常应用在以下场景：
- 提升用于执行大量处理或计算的 CPU 绑定代码的性能。RenderScript(本章后面会讨论)也为此提供了解决方案。
- 访问未通过运行时 API 直接提供的系统资源。这可能包括特定的设备驱动程序或内核资源。
- 提高随 C/C++ 开发环境提供的跨平台兼容性。许多游戏开发引擎/框架都出于此原因而利用 NDK。

原生代码与平台相关，因此 Android NDK 必须为部署应用程序的每个设备架构交叉编译代码。最新版本的 Android NDK(在编写本书时为 r10b)支持如下设备架构：armv5、armv7、x86、arm64-v8a、x86_64 和 mips64。这些架构也称为设备的应用二进制接口或 ABI。

构建 NDK 代码

在项目中构建原生代码的默认方法是通过命令行编写。ndk-build 命令将所有原生源代码编译为适当数量的平台相关二进制文件。从命令行调用 build 命令时，生成工具需要两个 makefile：Android.mk 和 Application.mk。

- Android.mk：该文件控制 NDK 需要编译和链接的模块，它将定义输出文件、源文件、包含文件以及构建工具可能需要的其他链接器或编译器标志。
- Application.mk：该文件主要用于描述顶层的构建属性，它最常见的用途是定义构建输出应支持哪些 ABI(一个、多个或全部)。

运行 ndk-build 会将输出文件放入项目的 libs/目录中，按照 ABI 进行分隔：

```
libs/
    armeabi/
        libSomethingAwesome.so
    armeabi-v7a/
        libSomethingAwesome.so
    mips/
        libSomethingAwesome.so
    x86/
        libSomethingAwesome.so
```

ADT/Eclipse 环境支持 NDK 代码的构建。在此情况下，IDE 的内部 make 脚本会自动调用 ndk-build，因此项目中必须有适当的 Android.mk 和 Application.mk 文件。

NDK 的 Gradle 集成目前仍在开发中。现在可以向应用程序 build.gradle 文件的 DSL 添加非常基本的 NDK 组件，但本书中的示例在 Gradle 下还没有得到完全支持。因此，如果正在使用 Android Studio 或构建服务器上的 Gradle 进行开发，那么目前来说，直接调用 ndk-build 可能是更好的选择。

Android NDK 将为每个支持的 ABI 生成共享的库文件，最终构建工具可以将这些 ABI 打包放入最终的 APK 中。通常情况下，应用程序在同一个 APK 中包括所有这些二进制文件以简化分发。然而，如果应用程序中的原生代码太大，可能会使 APK 文件超出所选分配器的大小限制。例如，Google Play 的 APK 大小限制为 50MB。

为降低 APK 的大小，许多开发人员将他们的构建版本拆分为多个 APK 文件，每个 APK 对应一个支持的 ABI。许多著名的应用商店，如 Google Play，还支持上传多个 APK，并且负责处理将适当二进制文件分发到相应设备的工作。

使用 Gradle 构建系统，可以通过为每个 ABI 提供独特的产品风格来管理这种拆分：

```
productFlavors {
    x86 {
        ndk {
            abiFilter "x86"
        }
    }
    arm {
        ndk {
            abiFilter "armeabi-v7a"
        }
    }
    mips {
        ndk {
```

```
            abiFilter "mips"
        }
    }
}
```

接下来，导出每个 ABI 的独特 APK 只需要构建并导出另一种产品风格即可。

注意：

Android NDK 并没有通过 SDK 管理器分发，必须单独下载。可以从如下网址获得在环境中安装 NDK 的最新说明：https://developer.android.com/tools/sdk/ndk/index.html。

8.2 使用 JNI 添加原生位

8.2.1 问题

希望在 Java 应用程序项目的较大环境中执行少量原生代码。

8.2.2 解决方案

Android 支持使用 Java 原生接口(Java Native Interface，JNI)将 Java 代码与原生执行连接起来。使用 JNI 时，可以将原生共享库动态载入运行时，从而实现在应用程序的 Java 类中声明的特定方法。

为将 JNI 添加到项目中，必须执行如下步骤：

(1) 使用 native 关键字，在 Java 类中声明将以 C/C++代码实现的方法。

(2) 在共享库(.so 文件)中编写和编译原生代码，可以使用 Android NDK 动态加载此文件。

(3) 在调用任何原生方法之前使用 System.loadLibrary()通知运行时加载原生代码。

下面是一个示例 Java 代码片段，其中说明了此实现：

```
package com.androidrecipes.app;
public class NativeWrapper {

    //将带有 C/C++实现的方法声明为原生的
    public static native void nativeMethod();

    static {
        //告诉运行时加载共享库，在此例中是"libnative_wrapper.so"
        System.loadLibrary("native_wrapper");
    }
}
```

为使此代码正确运作，应用程序运行时必须通过某种方式将原生方法与来自 Java 代码的调用绑定在一起。JNI API 提供了两种不同的机制来将原生方法映射到 Java：

- 使用默认的类-名识别编码 JNI 规范定义原生方法名。在前面的示例中，原生方法名是 Java_com_androidrecipes_app_NativeWrapper_nativeMethod()。
    ➢ JDK 命令行工具 javah 将自动生成这些代码。
- 使用显式方法表将 Java 方法映射到原生方法签名。在本例中，原生代码中的方法名可以是任意所选的名称，但必须手动编写全部代码(即不使用工具生成样板文件代码)。

在接下来的示例中将探讨如何实现这两种机制。

注意：
Android 的 libc 实现(称为 bionic)不具备与其他常见实现(如 glibc 或 uClibc)的二元兼容性。如果从其他平台移植原生代码，必须使用 NDK 重新编译代码，产生适合于 Android 运行时的二进制文件。

8.2.3 实现机制

在本例中，我们使用特定于 NDK 的 API，其名为 cpufeatures。该库用于检查设备的 CPU 架构信息，以及确定是否支持某些指令集(如 NEON 或 SSE3)。我们希望通过以 Java 编写的 Activity 代码揭示此信息，因此需要实现一些 JNI 代码以绑定两者。

注意：
《开发人员指南》中归档了可用的 NDK API，该指南位于 NDK 安装目录的 docs/目录中。该指南中的"稳定的 API"一节列出了每种功能以及最低平台需求。

首先，我们必须从一个 Java 类开始，将其用作 JNI 绑定的起点(参见代码清单 8-1)。在需要从 Java 调用原生代码时，就通过此类完成操作。

代码清单 8-1 定义了原生方法的 Java 类

```java
package com.androidrecipes.ndkjni;

public class NativeLib {
    /**
     * 返回设备上可用的核心数
     */
    public static native int getCpuCount();

    public static native String getCpuFamily();

    static {
        System.loadLibrary("features");
    }
}
```

请注意，使用 native 关键字声明计划以 C/C++代码实现的每个方法。该关键字告诉运行时应在原生共享库中的某处查找此方法。其次，我们必须通知运行时哪个共享库包含我们的代码。在本例中,我们告诉运行时在应用程序的目录路径中查找名为 libfeatures.so 的文件(稍后将构建)。

现在必须创建这两个已声明方法的原生实现。代码清单 8-2 和 8-3 显示了原生头文件和实现文件,这些文件使用对应的 JNI 签名描述这些方法。

代码清单 8-2　src/main/jni/NativeLib.h

```
#include <jni.h>
/* 类 com_androidrecipes_ndkjni_NativeLib 的头文件 */

#ifndef _Included_com_androidrecipes_ndkjni_NativeLib
#define _Included_com_androidrecipes_ndkjni_NativeLib
#ifdef __cplusplus
extern "C" {
#endif
/*
 * 类: com_androidrecipes_ndkjni_NativeLib
 * 方法: getCpuCount
 * 签名: ()I
 */
JNIEXPORT jint JNICALL
  Java_com_androidrecipes_ndkjni_NativeLib_getCpuCount(JNIEnv *, jclass);

/*
 * 类: com_androidrecipes_ndkjni_NativeLib
 * 方法: getCpuFamily
 * 签名: ()Ljava/lang/String;
 */
JNIEXPORT jstring JNICALL
  Java_com_androidrecipes_ndkjni_NativeLib_getCpuFamily(JNIEnv *, jclass);

#ifdef __cplusplus
}
#endif
#endif
```

代码清单 8-3　src/main/jni/NativeLib.c

```
#include "NativeLib.h"

#include <android/log.h>
#include <cpu-features.h>

JNIEXPORT jint JNICALL
  Java_com_androidrecipes_ndkjni_NativeLib_getCpuCount
  (JNIEnv *env, jclass clazz)
{
    return android_getCpuCount();
}
```

```c
JNIEXPORT jstring JNICALL
  Java_com_androidrecipes_ndkjni_NativeLib_getCpuFamily
  (JNIEnv *env, jclass clazz)
{
    AndroidCpuFamily family = android_getCpuFamily();
    switch(family)
    {
        case ANDROID_CPU_FAMILY_ARM:
            return (*env)->NewStringUTF(env, "ARM (32-bit)");
        case ANDROID_CPU_FAMILY_X86:
            return (*env)->NewStringUTF(env, "x86 (32-bit)");
        case ANDROID_CPU_FAMILY_MIPS:
            return (*env)->NewStringUTF(env, "MIPS (32-bit)");
        case ANDROID_CPU_FAMILY_ARM64:
            return (*env)->NewStringUTF(env, "ARM (64-bit)");
        case ANDROID_CPU_FAMILY_X86_64:
            return (*env)->NewStringUTF(env, "x86 (64-bit)");
        case ANDROID_CPU_FAMILY_MIPS64:
            return (*env)->NewStringUTF(env, "MIPS (64-bit)");
        case ANDROID_CPU_FAMILY_UNKNOWN:
        default:
            return (*env)->NewStringUTF(env, "Vaporware");
    }
}
```

使用 javah

如果倾向于不亲自编写原生签名,一种选择是使用 javah 完成此工作。javah 是 JDK 提供的命令行工具,它将检查给定 Java 类中的原生方法,并且产生包含每个方法签名的头文件。对于我们的示例,为了让 javah 生成头文件,可执行如下命令:

```
$ cd <project_directory>/src/main
$ javah -jni -classpath <project_bin_directory> -d jni
    com.androidrecipes.ndkjni.NativeLib
```

我们将此命令分解为多个部分,以便更好地了解其执行的操作:

- –jni: 告诉工具生成 Java 类的头文件。
- –classpath <project_bin_directory>: 在项目中引用此目录,其中存放编译的 Java 类输出。ADT/Eclipse 使用目录 bin/classes,而 Android Studio 使用 build/intermediates/classes/<build_type>。
- –d jni: 将输出置于 jni/目录(即 src/main/jni)中。
- com.androidrecipes.ndkjni.NativeLib: 待检查 Java 类的完全限定类名。

这将产生 com_androidrecipes_ndkjni_NativeLib.h 文件,其中的内容与代码清单 8-2 中的相同。文件名本身并不重要,可以将其重命名为更易于管理的名称。

注意用于唯一确认共享库中每个方法的长方法名。这些方法名中的印刷错误会在运行时产生 UnsatisfiedLinkError，因为 JNI API 将不能够找到适当的原生实现。每个方法名都带有 Java 前缀，后跟包名、Java 类名，最后是 Java 方法名。

对于 Java 方法声明的每个参数，每个原生方法也有对应的参数，但是额外增加了两项。JNIEnv 指针是 JNI API 函数的引用，可以从原生代码调用这些函数。jclass 是附加此方法的 Java 类，该类存在是因为我们的 Java 方法定义为 static。如果它们是 Java 中的实例方法，则第二个参数会是 jobject，并且引用方法调用所属的对象实例。最后，方法的返回值已转换为 JNI 友好的类型(int -> jint 和 String -> jstring)。

涉及 JNI API 时，还必须在实现中包括 jni.h 头文件。在我们的示例中以及对于使用 javah 的任何人，将在生成的头文件中自动完成此工作。这样，原生代码就可以访问提供的功能以连接 Java 代码和原生代码。

注意：

JNI 编程是一个庞大的主题，本书中无法全面介绍。如果想要更加熟悉使用 JNI 进行编程，请参考 Oracle JNI 文档：http://docs.oracle.com/javase/7/docs/technotes/guides/jni/。

android_getCpuCount()和 android_getCpuFamily()函数由 cpufeatures 定义，我们已通过在代码中包括 cpu-features.h 来访问 cpufeatures。我们的 JNI 方法仅是这些函数的封装器，用于将结果传回到 Java 中。CPU 计数将以整数方式返回设备具有的核心数。可以直接返回该值，因为 jint 是简单整数类型。

在返回 CPU 系列时，该值是一个枚举，必须将其转换为可读的字符串。C 语言中的字符串值与 Java 存在很大区别；前者是空值终止字符数组，而后者是完整的对象，除了字符之外还包含额外的元数据。因此，无法从 C 返回一个字符串字面值到 Java。必须调用 JNIEnv 提供的 NewStringUTF()转换函数，根据字符数据分配新的 Java 字符串对象(jstring)。

为构建原生代码，必须构造两个 makefile，即 Android.mk 和 Application.mk；代码清单 8-4 和 8-5 给出了这两个 makefile。

代码清单 8-4　src/main/jni/Android.mk

```
LOCAL_PATH := $(call my-dir)
include $(CLEAR_VARS)

LOCAL_MODULE := features
LOCAL_SRC_FILES := NativeLib.c
LOCAL_LDLIBS := -llog
LOCAL_STATIC_LIBRARIES := cpufeatures

include $(BUILD_SHARED_LIBRARY)

$(call import-module,android/cpufeatures)
```

代码清单 8-5　src/main/jni/Application.mk

```
APP_ABI := all
```

Android.mk 告诉 NDK 将单个的 NativeLib.c 文件编译到名为 features 的模块(即名为 libfeatures.so 的文件)中。请注意此名称如何匹配之前提供给 System.loadLibrary()的名称。该 makefile 还定义了访问 NDK 提供的 cpufeatures API 所需的导入语句。

注意：
NDK 文档将 cpufeatures 库格式称为导入模块，它要求使用 import-module 宏以及在 LOCAL_STATIC_LIBRARIES 下声明的库名称导入该模块。后面的声明通知构建系统必须首先构建库，因为它与我们的 NDK 模块存在依赖关系。

Application.mk 文件只是通知构建系统，它应该为每个支持的 ABI 架构生成一个 output.so 文件。如果没有此文件(或者在命令行上传递 APP_ABI)，默认的构建代码将仅为 ARMv5 输出一个.so 文件。

构建原生代码

必须从命令行使用 ndk-build 构建原生代码。然后，IDE 构建工具会在设备上运行时将构建的二进制文件复制到最终的 APK 中。在命令行中执行如下步骤：

```
$ cd <project_directory>/src/main
$ ndk-build
```

可以看到类似于如下的输出：

```
[armeabi-v7a] Compile thumb : features <= NativeLib.c
[armeabi-v7a] Compile thumb : cpufeatures <= cpu-features.c
[armeabi-v7a] StaticLibrary : libcpufeatures.a
[armeabi-v7a] SharedLibrary : libfeatures.so
[armeabi-v7a] Install : libfeatures.so => libs/armeabi-v7a/libfeatures.so
[armeabi] Compile thumb : features <= NativeLib.c
[armeabi] Compile thumb : cpufeatures <= cpu-features.c
[armeabi] StaticLibrary : libcpufeatures.a
[armeabi] SharedLibrary : libfeatures.so
[armeabi] Install : libfeatures.so => libs/armeabi/libfeatures.so
[x86] Compile : features <= NativeLib.c
[x86] Compile : cpufeatures <= cpu-features.c
[x86] StaticLibrary : libcpufeatures.a
[x86] SharedLibrary : libfeatures.so
[x86] Install : libfeatures.so => libs/x86/libfeatures.so
[mips] Compile : features <= NativeLib.c
[mips] Compile : cpufeatures <= cpu-features.c
[mips] StaticLibrary : libcpufeatures.a
[mips] SharedLibrary : libfeatures.so
[mips] Install : libfeatures.so => libs/mips/libfeatures.so
```

所有原生共享库都应位于项目的 src/main/libs 目录中，并且准备好供用户使用。

成功包括原生代码的最后一个要求是修改项目的 build.gradle 文件，以告诉 Gradle 目前在何处定位预先构建的共享库。参见代码清单 8-6。

代码清单 8-6 build.gradle 的部分代码

```
android {
    ...

    //禁止构建 JNI 代码，仅在 libs 中复制
    sourceSets.main {
        jni.srcDirs = []
        jniLibs.srcDir 'src/main/libs'
    }

    ...
}
```

准备好起作用的共享库之后，我们转向代码清单 8-7 和 8-8，其中给出了一个简单的 Activity，用于在视图中显示当前设备的此信息。

代码清单 8-7 res/layout/activity_main.xml

```xml
<?xml version="1.0" encoding="utf-8"?>
<FrameLayout xmlns:android="http://schemas.android.com/apk/res/android"
    android:layout_width="match_parent"
    android:layout_height="match_parent">

    <TextView
        android:id="@+id/text_info"
        android:layout_width="wrap_content"
        android:layout_height="wrap_content"
        android:layout_gravity="center"/>
</FrameLayout>
```

代码清单 8-8 调用 JNI 库的 Activity

```java
public class MainActivity extends Activity {

    private TextView mInfoText;

    @Override
    protected void onCreate(Bundle savedInstanceState) {
        super.onCreate(savedInstanceState);
        setContentView(R.layout.activity_main);

        mInfoText = (TextView) findViewById(R.id.text_info);

        getInfo();
    }

    private void getInfo() {
        String text = String.format("%s CPU with %d core(s)",
                NativeLib.getCpuFamily(),
                NativeLib.getCpuCount() );
```

```
        mInfoText.setText(text);
    }
}
```

在应用程序代码中，我们只需要像调用其他 Java 方法一样调用原生方法。Android 运行时将完成调用原生代码并在执行完成后返回 Java 的工作。在你的设备上运行此例，可以看到定义处理器架构和可用核心数量的文本字符串！

原生方法表

作为类-名识别编码的备选实现方式，我们可以使用显式的方法表创建原生方法的引用，该引用更易于读取和维护。在本例中，如代码清单 8-9 所示，我们将利用 JNI_OnLoad 回调。此方法在加载库时会被调用，为我们提供方法映射的良好初步观点。

代码清单 8-9　src/main/jni/NativeLibAlternate.c

```c
//通过之前的自定义头文件包括的 JNI API
#include <jni.h>

#include <android/log.h>
#include <cpu-features.h>

static jint native_getCpuCount(JNIEnv *env, jclass clazz)
{
    return android_getCpuCount();
}

static jstring native_getCpuFamily(JNIEnv *env, jclass clazz)
{
    AndroidCpuFamily family = android_getCpuFamily();
    switch(family)
    {
        case ANDROID_CPU_FAMILY_ARM:
            return (*env)->NewStringUTF(env, "ARM (32-bit)");
        case ANDROID_CPU_FAMILY_X86:
            return (*env)->NewStringUTF(env, "x86 (32-bit)");
        case ANDROID_CPU_FAMILY_MIPS:
            return (*env)->NewStringUTF(env, "MIPS (32-bit)");
        case ANDROID_CPU_FAMILY_ARM64:
            return (*env)->NewStringUTF(env, "ARM (64-bit)");
        case ANDROID_CPU_FAMILY_X86_64:
            return (*env)->NewStringUTF(env, "x86 (64-bit)");
        case ANDROID_CPU_FAMILY_MIPS64:
            return (*env)->NewStringUTF(env, "MIPS (64-bit)");
        case ANDROID_CPU_FAMILY_UNKNOWN:
        default:
            return (*env)->NewStringUTF(env, "Vaporware");
```

```
        }
    }

    //构造一个表,将Java方法签名映射到原生函数指针
    static JNINativeMethod method_table[] = {
        { "getCpuCount", "()I", (void *) native_getCpuCount },
        { "getCpuFamily", "()Ljava/lang/String;", (void *) native_getCpuFamily }
    };

    //使用OnLoad初始化程序向运行时注册方法表
    jint JNI_OnLoad(JavaVM* vm, void* reserved) {
        JNIEnv* env;
        if((*vm)->GetEnv(vm, (void**)&env, JNI_VERSION_1_6) != JNI_OK) {
            return JNI_ERR;
        } else {
            jclass clazz = (*env)->FindClass(env,
                    "com/androidrecipes/ndkjni/NativeLib");
            if(clazz) {
                jint ret = (*env)->RegisterNatives(env, clazz, method_table,
                        sizeof(method_table)/sizeof(method_table[0]));
                if(ret == 0) {
                    return JNI_VERSION_1_6;
                }
            }
            return JNI_ERR;
        }
    }
```

JNI_OnLoad 中样板文件代码的主旨是查找指向我们创建的 Java 类的引用(使用反射),并且使用来自 JNI 环境的 **RegisterNatives** 函数向其附加方法表。该方法采用 JNINatives 结构的数组作为参数,每个结构包含如下数据(按顺序):

- Java 类上方法的名称(以字符串形式)。
- Java 类上方法的签名。该字符串定义 Java 方法采用的参数和返回类型,从而唯一标识重载的方法版本。
- 指向创建用于处理调用的原生函数的函数指针。

注意:
关于如何构造 JNI 方法签名字符串的更多信息,请参考如下 Oracle JNI 文档:
http://docs.oracle.com/javase/7/docs/technotes/guides/jni/spec/types.html。

请注意,原生方法参数必须仍然保持相同(它们仍然需要有适当的参数列表,包括环境和类指针)。然而,现在我们可以任意命名方法,因为方法表负责执行映射。与往常一样,方法名的输入错误将造成 UnsatisfiedLinkError。然而,与前面不同的是,当运行时类加载器将 NativeLib Java 类载入内存中时,这将造成代码执行失败。相比于等到实际调用方法,这样可以更快地产生失败。

8.3 构建纯原生 Activity

8.3.1 问题

应用程序 UI 需要与原生代码更加紧密地集成，而使用 NDK 可以更加简单地构建整个 Activity。

8.3.2 解决方案

在应用程序中使用 NativeActivity 实现。NDK 提供了名为 android_native_app_glue 的导入库，用于将 Activity API 与原生实现关联。该库为处理输入事件(如触摸和传感器事件)提供生命周期内的回调和处理程序。

应用程序在技术上仍然有一个基于 Java 的 Activity 在运行，而 NativeActivity 负责所有 JNI 关联和生命周期内的转发工作，因此可以对 Activity 的行为完全采用 C/C++代码，从而采用更加深入的方式整合 NDK API。

除了生命周期内的回调之外，android_native_app_glue 还为处理输入事件提供了事件处理程序。可以直接在原生代码中处理排队的触摸事件、关键事件和传感器(如加速度计)事件。该过程比 Java 稍微复杂一些，因为原生代码也负责轮询事件并在其得到处理时取消排队；这就产生了比使用 Java 时更多的样板文件代码。

8.3.3 实现机制

在本例中，我们将在原生代码中构造一个 Activity，该 Activity 通过在屏幕上渲染不同的颜色(以 OpenGL 代码的形式)来响应触摸事件。当用户的手指围绕视图滑动时，颜色将基于最新的触摸位置发生变化。即使是如此简单的目标，我们也会看到，在放弃 Java 框架之后，需要更多的代码来实现此目标。

注意：
OpenGL 编程的主题超出了本书的范围，并且不是此例的关注重点。此例中使用的 OpenGL 直接来自适用于 OpenGL 的 Android NDK 示例。

此项目不包含任何用于 Activity 或其他方面的 Java 源代码。相反，NativeActivity 支持在应用程序清单文件中使用元数据定义共享库，共享库应由 Android 框架加载，并且紧密关联到 Activity 的生命周期中。代码清单 8-10 突出强调了 AndroidMainfest.xml 文件中我们需要具有的部分。

代码清单 8-10 AndroidManifest.xml

```xml
<manifest xmlns:android="http://schemas.android.com/apk/res/android"
    package="com.androidrecipes.nativeactivity"
    android:versionCode="1"
    android:versionName="1.0" >
```

```xml
<application
    android:hasCode="false"
    ... >
    <activity
        android:name="android.app.NativeActivity"
        android:configChanges="orientation"
        android:label="@string/app_name" >
        <!-- Where to find NativeActivity implementation -->
        <meta-data
            android:name="android.app.lib_name"
            android:value="native-activity" />

        <intent-filter>
            <action android:name="android.intent.action.MAIN" />
            <category android:name="android.intent.category.LAUNCHER" />
        </intent-filter>
    </activity>
</application>
```

```xml
</manifest>
```

此清单中的关键元素如下：

- Activity 元素必须是 android.app.NativeActivity。
  - 还可以子类化此元素并使用自定义的实现，但根元素必须是 NativeActivity。
- <meta-data>元素应存在，其包含 android.app.lib_name 名称和代表共享库名的值。
  - 库名与传递给前一个示例中 System.loadLibrary()的名称相同。它应该匹配 APK 中名为 lib<name>.so 的文件。
  - 在本例中，具体的约定是代码将位于 libnative-activity.so 中。
- 可选：如果在应用程序中未提供其他 Java 代码，可以在<application>元素中将 android:hasCode 设置为 false。
  - 如果应用未提供任何 Java 代码，那么这只是启动优化。如果程序包含有 Java 类，则不能设置此属性！

接下来，我们感兴趣的代码块将位于最终内置于共享库的原生实现中。下面看一下代码清单 8-11 中的代码。

代码清单 8-11　src/main/jni/activity.c

```c
#include <EGL/egl.h>
#include <GLES/gl.h>

#include <android/log.h>
#include <android/window.h>
#include <android_native_app_glue.h>

#define LOGI(...) ((void)__android_log_print(ANDROID_LOG_INFO, \
    "AndroidRecipes", __VA_ARGS__))
```

```c
#define LOGD(...) ((void)__android_log_print(ANDROID_LOG_DEBUG, \
    "AndroidRecipes", __VA_ARGS__))
#define LOGW(...) ((void)__android_log_print(ANDROID_LOG_WARN, \
    "AndroidRecipes", __VA_ARGS__))

//存放最近已知触摸位置的数据结构
struct touch_state
{
    int32_t x;
    int32_t y;
};

//存储Activity全局状态的数据结构
struct driver
{
    struct android_app* app;
    struct touch_state state;
    EGLDisplay display;
    EGLSurface surface;
    EGLContext context;
    int32_t width;
    int32_t height;
};

/**
 * 以OpenGL代码方式渲染下一个颜色帧的辅助函数
 */
static void render_frame(struct driver* driver)
{
    if(driver->display == NULL) {
        // 无显示
        return;
    }

    float red = (float)driver->state.x / driver->width;
    float green = (float)driver->state.y / driver->height;
    float blue = 1 - (float)driver->state.x / driver->width;
    //基于触摸位置渲染新颜色
    glClearColor(red, green, blue, 1.0f);
    //告诉OpenGL刷新颜色缓冲区
    glClear(GL_COLOR_BUFFER_BIT);
    //将新的帧置于颜色缓冲区之上
    eglSwapBuffers(driver->display, driver->surface);
}

/**
 * 为当前显示初始化EGL环境
 */
static int engine_init_display(struct driver* driver) {
    //初始化ES和EGL
```

```
/*
 * 在此指定所需配置的属性。接下来,我们选择一个 EGLConfig,
 * 其中对每个兼容屏幕窗口的颜色组成部分至少包含 8 位
 */
const EGLint attribs[] = {
    EGL_SURFACE_TYPE, EGL_WINDOW_BIT,
    EGL_BLUE_SIZE, 8,
    EGL_GREEN_SIZE, 8,
    EGL_RED_SIZE, 8,
    EGL_NONE
};
EGLint w, h, dummy, format;
EGLint numConfigs;
EGLConfig config;
EGLSurface surface;
EGLContext context;
EGLDisplay display = eglGetDisplay(EGL_DEFAULT_DISPLAY);

eglInitialize(display, 0, 0);

/* 在此应用程序选择它所需的配置。在本例中,
 * 我们采用一个非常简化的选择过程,即选择符合标准的第一个 EGLConfig
 */
eglChooseConfig(display, attribs, &config, 1, &numConfigs);

/* EGL_NATIVE_VISUAL_ID 是 EGLConfig 的一个属性,
 * 用于保证其被 ANativeWindow_setBuffersGeometry()接受。
 * 只要选择了 EGLConfig,就可以使用 EGL_NATIVE_VISUAL_ID 安全地重新
 * 配置 ANativeWindow 缓冲区以进行匹配 */
eglGetConfigAttrib(display, config, EGL_NATIVE_VISUAL_ID, &format);

ANativeWindow_setBuffersGeometry(driver->app->window, 0, 0, format);

surface = eglCreateWindowSurface(display, config, driver->app->window,
    NULL);
context = eglCreateContext(display, config, NULL, NULL);

if(eglMakeCurrent(display, surface, surface, context) == EGL_FALSE) {
    LOGW("Unable to eglMakeCurrent");
    return -1;
}

eglQuerySurface(display, surface, EGL_WIDTH, &w);
eglQuerySurface(display, surface, EGL_HEIGHT, &h);

driver->display = display;
driver->context = context;
driver->surface = surface;
driver->width = w;
```

```c
        driver->height = h;

        // 初始化 GL 状态
        glHint(GL_PERSPECTIVE_CORRECTION_HINT, GL_FASTEST);
        glEnable(GL_CULL_FACE);
        glShadeModel(GL_SMOOTH);
        glDisable(GL_DEPTH_TEST);

        return 0;
}

/**
 * 解除当前与显示关联的 EGL 环境
 */
static void engine_term_display(struct driver* driver) {
    if(driver->display != EGL_NO_DISPLAY) {
        eglMakeCurrent(driver->display, EGL_NO_SURFACE, EGL_NO_SURFACE,
            EGL_NO_CONTEXT);
        if(driver->context != EGL_NO_CONTEXT) {
            eglDestroyContext(driver->display, driver->context);
        }

        if(driver->surface != EGL_NO_SURFACE) {
            eglDestroySurface(driver->display, driver->surface);
        }
        eglTerminate(driver->display);
    }

    driver->display = EGL_NO_DISPLAY;
    driver->context = EGL_NO_CONTEXT;
    driver->surface = EGL_NO_SURFACE;
}

/*
 * 此事件处理程序将接收封闭 Activity 实例的生命周期事件
 */
static void handle_cmd(struct android_app* app, int32_t cmd)
{
    struct driver* driver = (struct driver*)app->userData;
    switch(cmd)
    {
        case APP_CMD_SAVE_STATE:
            LOGI("Save state");
            // 系统已要求我们保存当前状态，保存此状态。
            driver->app->savedState = malloc(sizeof(struct touch_state));
            *((struct touch_state*)driver->app->savedState) =
                driver->state;
            driver->app->savedStateSize = sizeof(struct touch_state);
            break;
```

```c
        case APP_CMD_INIT_WINDOW:
            LOGI("Init window");
            // 窗口正在显示，准备好该窗口
            if(driver->app->window != NULL) {
                engine_init_display(driver);
                render_frame(driver);
            }
            break;

        case APP_CMD_TERM_WINDOW:
            LOGI("Terminate window");
            // 窗口正在隐藏或关闭，清除该窗口
            engine_term_display(driver);
            break;

        case APP_CMD_PAUSE:
            LOGI("Pausing");
            break;

        case APP_CMD_RESUME:
            LOGI("Resuming");
            break;

        case APP_CMD_STOP:
            LOGI("Stopping");
            break;

        case APP_CMD_DESTROY:
            LOGI("Destroying");
            break;

        case APP_CMD_LOST_FOCUS:
            LOGI("Lost focus");
            break;

        case APP_CMD_GAINED_FOCUS:
            LOGI("Gained focus");
            break;
    }
}

/*
 * 此事件处理程序将被触发以处理由 main 函数中的轮询循环接收的输入事件
 */
static int32_t handle_input(struct android_app* app, AInputEvent* event)
{
    struct driver* driver = (struct driver*)app->userData;
    //保存最近的触摸事件以用于渲染
    if(AInputEvent_getType(event) == AINPUT_EVENT_TYPE_MOTION)
    {
```

```c
            driver->state.x = AMotionEvent_getX(event, 0);
            driver->state.y = AMotionEvent_getY(event, 0);
            return 1;
        }
    else if (AInputEvent_getType(event) == AINPUT_EVENT_TYPE_KEY)
    {
        LOGI("Received key event: %d", AKeyEvent_getKeyCode(event));
        if(AKeyEvent_getKeyCode(event) == AKEYCODE_BACK)
        {
            //结束此Activity
            if(AKeyEvent_getAction(event) == AKEY_EVENT_ACTION_UP)
            {
                ANativeActivity_finish(app->activity);
            }
        }
        return 1;
    }
    return 0;
}

/*
 * 这是原生代码的主要入口点。此代码在单独的线程上被调用,
 * 该线程由native_app_glue API 创建
 */
void android_main(struct android_app* state)
{
    struct driver driver;

    app_dummy(); // 防止取消关联


    memset(&driver, 0, sizeof(driver));
    //在app 结构中保存指向状态驱动程序的引用
    state->userData = &driver;
    //定义app 事件处理程序
    state->onAppCmd = &handle_cmd;
    state->onInputEvent = &handle_input;

    driver.app = state;

    if(state->savedState != NULL) {
        // 我们从前一个保存的状态开始,从该状态还原
        driver.state = *(struct touch_state*)state->savedState;
    }

    while(1)
    {
        int ident;
        int fdesc;
        int events;
```

```
            struct android_poll_source* source;

            //用于在消息队列中查询传入事件的无限循环
            while((ident = ALooper_pollAll(0, &fdesc, &events,
        (void**)&source)) >= 0)
            {
                //每个事件将在附加的处理程序中得到处理
                if(source)
                    source->process(state, source);

                //在销毁 Activity 时将对此进行设置
                if(state->destroyRequested)
                    return;
            }

            // 在每个循环中，渲染下一帧...
            // OpenGL 对此进行限制,
            // 从而主循环将以帧缓冲区更新率(在大多数情况下是 16.7 毫秒)高效运行
            render_frame(&driver);
        }
    }
```

使用原生 Activitiy API 时，android_main()函数是代码中的入口点。该函数在 Activity 的 onCreate()阶段得到调用，它在单独的线程上被创建。这意味着此方法内部发生的任何操作都不会与 Activity 中的其他生命周期事件同步。

除非应销毁 Activity，否则代码应使此方法保持激活状态。在本例中，作为典型方式，我们使用此方法无限循环，处理排队的输入事件并渲染显示帧。我们将在后面详细查看每个输入事件。

在初始化时，我们传递 android_app 数据结构。如果此 Activity 的创建是配置变更(即循环)的一部分，则该结构将包含任何保存的状态，并且提供挂钩(hook)以关联额外事件的回调函数。

onAppCmd 函数指针应引用可以在其中触发生命周期事件的回调，例如暂停、恢复、保存实例状态等。我们已附加 handle_cmd()以接收这些回调事件。此函数的回调将发生在**应用程序的主线程上**。

此例的 OpenGL 设置和销毁工作在 engine_init_display()和 engine_term_display()函数中完成。当生命周期告诉我们已创建(APP_CMD_INIT_WINDOW)或终止(APP_CMD_TERM_WINDOW) Activity 窗口时，这些方法就会被调用。

我们还使用了 APP_CMD_SAVE_STATE，在配置变更事件期间调用此方法(如同在 Java 中使用 onSaveInstanceState)。在本例中，我们使用此事件持久保存当前状态，从而可以在循环后重新构造相同的颜色。该值通过 android_app 数据结构提供给 android_main()中的新 Activity。

onInputEvent 回调将负责操作主循环从队列中推出的任何已处理事件。每次主循环使用 ALoop_pollAll()从输入源中解除事件排队时，该调用就会激活，并对输入源调用 process()。我们定义了 handle_input()来管理这些事件。框架提供的所有事件类型将在此处终止,因此必须首先区分哪些是触摸事件(即 AINPUT_EVENT_TYPE_MOTION)。对于接收的每个事件,我们仅在全局状态中保存 x/y 值。

在通过主循环的每次迭代中,render_frame()将被调用,该方法读取最新的 x/y 触摸位置并计算颜色值。通过 OpenCL 命令 glClearColor()将此颜色渲染到显示屏上;当系统调用 eglSwapBuffers()时,该帧就会实际地显示。

提示:

Android 框架以 GPU 的更新率(通常是 60fps)限制缓冲区交换方法的调用,eglSwapBuffers()将阻塞,直到完全渲染一帧后提供新的缓冲区时才被调用。这就意味着 eglSwapBuffers()以 60fps 对循环定时,而不采用其他计时方式。

使用 NDK 构建原生代码需要通过 Android.mk 和 Application.mk 文件来定义构建模块。代码清单 8-12 和 8-13 给出了此项目的上述文件。

代码清单 8-12   src/main/jni/Android.mk

```
LOCAL_PATH := $(call my-dir)
include $(CLEAR_VARS)

LOCAL_MODULE := native-activity
LOCAL_SRC_FILES := activity.c
LOCAL_LDLIBS := -llog -landroid -lEGL -lGLESv1_CM
LOCAL_STATIC_LIBRARIES := android_native_app_glue

include $(BUILD_SHARED_LIBRARY)

$(call import-module,android/native_app_glue)
```

代码清单 8-13   src/main/jni/Application.mk

```
APP_ABI := all
APP_PLATFORM := android-19
```

Android.mk 文件告诉 NDK 将单个的 activity.c 文件编译到名为 native-activity 的模块(即 libnative-activity.so 文件)中。请注意此名称如何匹配前面提供给清单文件的名称。此 makefile 也定义了访问 NDK 提供的 API native_app_glue 所需的导入语句。

注意:

NDK 文档将 native_app_glue 库格式称为导入模块,它要求使用 import-module 宏以及在 LOCAL_STATIC_LIBRARIES 下声明的库名称导入该模块。后面的声明通知构建系统必须首先构建库,因为它与我们的 NDK 模块存在依赖关系。

Application.mk 文件只是通知构建系统,它应该为每个支持的 ABI 架构生成一个 output.so 文件。如果没有此文件(或者在命令行上传递 APP_ABI),默认的构建代码将仅为 ARMv5 输出一个.so 文件。Android 2.3 及以上版本的 NDK 中引入了 NDK API native_app_glue。我们添加了 APP_PLATFORM 属性来确保针对具有这些 API 的平台版本进行编译。大于 NDK 版本所支持的 android-10 的任何值都是可接受的。

构建原生代码

必须从命令行使用 ndk-build 构建原生代码。然后，IDE 构建工具会在设备上运行时将构建的二进制文件复制到最终的 APK 中。在命令行中执行如下步骤：

```
$ cd <project_directory>/src/main
$ ndk-build
```

可以看到类似于如下的输出：

```
   [armeabi-v7a] Compile thumb  : native-activity <= activity.c
   [armeabi-v7a] Compile thumb  : android_native_app_glue <=
android_native_app_glue.c
   [armeabi-v7a] StaticLibrary  : libandroid_native_app_glue.a
   [armeabi-v7a] SharedLibrary  : libnative-activity.so
   [armeabi-v7a] Install        : libnative-activity.so =>
libs/armeabi-v7a/libnative-activity.so
   [armeabi] Compile thumb  : native-activity <= activity.c
   [armeabi] Compile thumb  : android_native_app_glue <=
android_native_app_glue.c
   [armeabi] StaticLibrary  : libandroid_native_app_glue.a
   [armeabi] SharedLibrary  : libnative-activity.so
   [armeabi] Install        : libnative-activity.so =>
libs/armeabi/libnative-activity.so
   [x86] Compile  : native-activity <= activity.c
   [x86] Compile  : android_native_app_glue <= android_native_app_glue.c
   [x86] StaticLibrary  : libandroid_native_app_glue.a
   [x86] SharedLibrary  : libnative-activity.so
   [x86] Install  : libnative-activity.so => libs/x86/libnative-activity.so
   [mips] Compile  : native-activity <= activity.c
   [mips] Compile  : android_native_app_glue <= android_native_app_glue.c
   [mips] StaticLibrary  : libandroid_native_app_glue.a
   [mips] SharedLibrary  : libnative-activity.so
   [mips] Install  : libnative-activity.so => libs/mips/libnative-activity.so
```

所有原生共享库都应位于项目的 src/main/libs 目录中，并且准备好供用户使用。

成功包括原生代码的最后一个要求是修改项目的 build.gradle 文件，以告诉 Gradle 目前在何处定位预先构建的共享库。参见代码清单 8-14。

代码清单 8-14　build.gradle 的部分代码

```
android {
    ...

    //禁止 JNI 代码的构建，仅在 libs 中复制
    sourceSets.main {
        jni.srcDirs = []
```

```
            jniLibs.srcDir 'src/main/libs'
    }
    ...
}
```

在项目中构建有效的原生库之后，现在就可以运行示例并观察在屏幕上拖动手指时的颜色变化，所有操作无须编写一行 Java 代码！

8.4 RenderScript

Google 发布的 RenderScript 对 Android 的高性能计算算法进行了升级。随着设备硬件功能的进步，CPU 和 GPU(甚至是专用的处理器)上提升的处理能力增加了我们在设备上执行的处理类型(而不是将繁重的工作交给服务器来完成)。多核编程也提升了代码线程模型和架构的复杂性。

RenderScript 包含 3 个主要组成部分：
- 基于 C99 的脚本语言，该语言专注于将算法分解为可以轻松并行化的多个"内核"。
- 编译器引擎，用于将脚本内核映射为 Java 类，开发人员可以使用这些 Java 类直接从 Java 中调用他们的脚本，而无须使用 NDK。
- 带有制造商提供驱动程序的运行时引擎，OEM 可以使用该引擎报告其设备具有的数据处理硬件功能(主要是报告可用的核心数)。

RenderScript 内核是在大型输入数据集之间调用的函数，称为 allocation。RenderScript 引擎对输入 allocation 中的每一项调用给定的内核，在第二个输出 allocation 中产生结果。作为示例，如果内核被设计为处理输入数据，输入 allocation 就是待处理的原始位图，而输出 allocation 将包含已处理的图片数据。下面给出内核函数的一个示例：

```
#pragma version(1)
#pragma rs java_package_name(com.example.renderscript)

float multiplier;

void root(const uchar4 *in_element, uchar4 *out_element, uint32_t x, uint32_t y)
{
    //在输入中处理，在输出中设置结果
}
```

每次调用此函数时，in_element 参数指向输入 allocation 中的当前元素。x 和 y 参数指示当前元素在 allocation 数据集内的位置。内核函数执行的工作是设置 out_element 的适当值，基于希望在返回前应用的算法，该参数代表输出 allocation 中的对应元素。这些参数的类型取决于预期的 allocation 类型；uchar4 是经常用于 ARGB 像素数据 allocation 的向量类型。

脚本文件可以只有一个内核函数，但可以添加需要在调用内核之前设置的其他设置函数或全局字段。RenderScript 编译器将添加到脚本的函数或全局字段映射为 Java 方法(字段将映射为 setter 方法)。例如，此脚本中的 multiplier 字段将成为所映射 Java 类中的

set_multiplier()。在调用内核之前,可以使用这些方法为脚本提供参数或者执行其他设置。

使用 RenderScript 支持包

从 Android 3.0 (API Level 11)开始,RenderScript 才成为公共 API。但你可能注意到,Google 的 Android 团队喜欢反向移植他们的框架,以便开发人员可以在旧设备上使用这些框架。RenderScript 支持包让应用程序可以在 Android 2.2 (FroYo)设备上使用其功能。

为此,构建工具在应用程序 APK 中包括了一组预编译的 NDK 库,以便安装在并不原生支持 RenderScript 支持包中所有可用 RenderScript 功能的设备上。

> 注意:
> 目前 RenderScript 支持包仅包括用于 ARMv7、x86 和 MIPS 的 NDK 库,而不支持 ARMv5 设备。

使用 RenderScript 支持包稍微不同于在编译时复制额外的 Java 库或资源。构建工具包括挂钩程序,用于在构建到 APK 之后复制适当的库,而不需要将这些库置于应用程序源树中。为通知构建工具需要执行此步骤,必须将如下代码添加到 build.gradle 文件的 defaultConfig 中:

```
defaultConfig {

    targetSdkVersion 18
    renderscriptTargetApi 18
    renderscriptSupportMode true
}
```

最低目标版本是 API 18,而 API 18.1.0 是最低支持的构建工具版本。请注意,这并不意味着应用程序必须将其最低 SDK 设置为 Android 4.3;但其表示应用程序应将其目标 SDK 至少设置为此级别。

设置这些参数之后,构建工具将处理所有剩余工作。唯一额外的必需步骤是使用应用程序的 android.support.v8.renderscript 包中的类,而不是使用原生的类。

> 要点:
> 必须在代码中导入 android.support.v8.renderscript.*包而非 android.renderscript 包。

本章接下来介绍使用 RenderScript 包。然而,在大多数情况下,只需要更改包括的导入语句,即可改为使用原生版本。

8.5 使用 RenderScript 过滤图片

8.5.1 问题

应用程序需要通过简单方式对图片应用常见滤镜。

8.5.2 解决方案

RenderScript 具有一个大型的脚本内联函数集合，或是预制的和封装的 RenderScript 内核，这些内核旨在执行常见的任务。可以使用这些内联函数对 RenderScript 进行计算，甚至不需要编写脚本代码！随着每个新的 Android 版本发布，系统中也会添加新的内联函数，并且会建立有用函数的库，以便从中提取。本节将研究三个最常见的内联函数：

- ScriptInstrinsicBlur：对输入 allocation 中的每个元素应用高斯模糊。在脚本中可配置模糊的半径。
- ScriptInstrinsicColorMatrix：对输入 allocation 中的每个元素应用颜色矩阵滤镜。该函数与应用于 Drawable 的 ColorFilter 类似，它具有用于设置灰度的额外便利方法。
- ScriptInstrinsicConvolve3x3：对输入 allocation 中的每个元素应用 3×3 卷积矩阵。该矩阵经常用于创建照片滤镜效果，例如锐化、浮雕和边缘检测。

8.5.3 实现机制

接下来研究一个示例应用程序，该应用程序使用 RenderScript 向图片资源应用滤镜。如图 8-1 所示，该应用程序具有一个包含 6 张图片的网格，对于网格中的每张图片，都会应用不同的滤镜。

图 8-1　RenderScript 图片滤镜(从上到下，从左到右)：无、灰度、边缘检测、模糊、锐化、浮雕

代码清单 8-15 中给出了 Activity 的布局，该 Activity 将构造此简单网格。

代码清单 8-15　res/layout/activity_main.xml

```xml
<?xml version="1.0" encoding="utf-8"?>
<TableLayout xmlns:android="http://schemas.android.com/apk/res/android"
    android:layout_width="match_parent"
    android:layout_height="match_parent"
    android:stretchColumns="*">
    <TableRow>
        <ImageView
            android:id="@+id/image_normal"
            android:layout_weight="1"
            android:layout_margin="5dp"
            android:scaleType="fitCenter" />
        <ImageView
            android:id="@+id/image_blurred"
            android:layout_weight="1"
            android:layout_margin="5dp"
            android:scaleType="fitCenter" />
    </TableRow>

    <TableRow>
        <ImageView
            android:id="@+id/image_greyscale"
            android:layout_weight="1"
            android:layout_margin="5dp"
            android:scaleType="fitCenter" />
        <ImageView
            android:id="@+id/image_sharpen"
            android:layout_weight="1"
            android:layout_margin="5dp"
            android:scaleType="fitCenter" />
    </TableRow>

    <TableRow>
        <ImageView
            android:id="@+id/image_edge"
            android:layout_weight="1"
            android:layout_margin="5dp"
            android:scaleType="fitCenter" />
        <ImageView
            android:id="@+id/image_emboss"
            android:layout_weight="1"
            android:layout_margin="5dp"
            android:scaleType="fitCenter" />
    </TableRow>

</TableLayout>
```

该网格中的每个单元格将填入相同的图片,但对图片应用不同的滤镜。在第一个单元格中,我们将显示没有进行任何过滤的基础图片。接下来的两个单元格将分别使用 ScriptInstrinsicBlur 和 ScriptInstrinsicColorMatrix 进行过滤。剩下的单元格将使用各种矩阵和 ScriptInstrinsicConvolve3x3 进行过滤。代码清单 8-16 中给出了 Activity 代码。

代码清单 8-16　图片滤镜 Activity

```java
import android.support.v8.renderscript.*;

public class MainActivity extends Activity {

    private enum ConvolutionFilter {
        SHARPEN, LIGHTEN, DARKEN, EDGE_DETECT, EMBOSS
    };

    @Override
    protected void onCreate(Bundle savedInstanceState) {
        super.onCreate(savedInstanceState);
        setContentView(R.layout.activity_main);
        // 创建源数据以及用于存放过滤结果的目标
        Bitmap inBitmap = BitmapFactory.decodeResource(getResources(),
        R.drawable.dog);
        Bitmap outBitmap = inBitmap.copy(inBitmap.getConfig(), true);
        // 显示普通图片
        setImageInView(outBitmap.copy(outBitmap.getConfig(), false),
        R.id.image_normal);

        // 建立 RenderScript 上下文
        final RenderScript rs = RenderScript.create(this);
        // 为输入和输出数据创建 allocation
        final Allocation input = Allocation.createFromBitmap(rs, inBitmap,
                Allocation.MipmapControl.MIPMAP_NONE,
                Allocation.USAGE_SCRIPT);
        final Allocation output = Allocation.createTyped(rs, input.getType());

        // 运行模糊脚本
        final ScriptIntrinsicBlur script = ScriptIntrinsicBlur
                .create(rs, Element.U8_4(rs));
        script.setRadius(4f);
        script.setInput(input);
        script.forEach(output);
        output.copyTo(outBitmap);
        setImageInView(outBitmap.copy(outBitmap.getConfig(), false),
        R.id.image_blurred);

        // 运行灰度脚本
        final ScriptIntrinsicColorMatrix scriptColor =
                ScriptIntrinsicColorMatrix.create(rs, Element.U8_4(rs));
        scriptColor.setGreyscale();
```

```java
        scriptColor.forEach(input, output);
        output.copyTo(outBitmap);
        setImageInView(outBitmap.copy(outBitmap.getConfig(), false),
    R.id.image_greyscale);

        // 运行锐化脚本
        ScriptIntrinsicConvolve3x3 scriptC = ScriptIntrinsicConvolve3x3
                .create(rs, Element.U8_4(rs));
        scriptC.setCoefficients(getCoefficients(ConvolutionFilter.SHARPEN));
        scriptC.setInput(input);
        scriptC.forEach(output);
        output.copyTo(outBitmap);
        setImageInView(outBitmap.copy(outBitmap.getConfig(), false),
    R.id.image_sharpen);

        // 运行边缘检测脚本
        scriptC = ScriptIntrinsicConvolve3x3.create(rs, Element.U8_4(rs));
        scriptC.setCoefficients(getCoefficients(ConvolutionFilter.EDGE_DETECT));
        scriptC.setInput(input);
        scriptC.forEach(output);
        output.copyTo(outBitmap);
        setImageInView(outBitmap.copy(outBitmap.getConfig(), false),
    R.id.image_edge);

        // 运行浮雕脚本
        scriptC = ScriptIntrinsicConvolve3x3.create(rs, Element.U8_4(rs));
        scriptC.setCoefficients(getCoefficients(ConvolutionFilter.EMBOSS));
        scriptC.setInput(input);
        scriptC.forEach(output);
        output.copyTo(outBitmap);
        setImageInView(outBitmap.copy(outBitmap.getConfig(), false),
    R.id.image_emboss);

        // 撤消RenderScript上下文
        rs.destroy();
}

private void setImageInView(Bitmap bm, int viewId) {
    ImageView normalImage = (ImageView) findViewById(viewId);
    normalImage.setImageBitmap(bm);
}

/*
 * 用于获得每种卷积图片滤镜的矩阵系数的辅助函数
 */
private float[] getCoefficients(ConvolutionFilter filter) {
    switch(filter) {
```

```
            case SHARPEN:
                return new float[] {
                        0f, -1f, 0f,
                        -1f, 5f, -1f,
                        0f, -1f, 0f
                };
            case LIGHTEN:
                return new float[] {
                        0f, 0f, 0f,
                        0f, 1.5f, 0f,
                        0f, 0f, 0f
                };
            case DARKEN:
                return new float[] {
                        0f, 0f, 0f,
                        0f, 0.5f, 0f,
                        0f, 0f, 0f
                };
            case EDGE_DETECT:
                return new float[] {
                        0f, 1f, 0f,
                        1f, -4f, 1f,
                        0f, 1f, 0f
                };
            case EMBOSS:
                return new float[] {
                        -2f, -1f, 0f,
                        -1f, 1f, 1f,
                        0f, 1f, 2f
                };
            default:
                return null;
        }
    }
}
```

在过滤图片之前，必须使用 RenderScript.create()初始化 RenderScript 上下文，还必须创建两个 Allocation 实例——一个实例用于输入数据，另一个用于输出结果。这些都是每个 RenderScript 内核将操作的缓冲区。可以采用便利的函数，根据框架中许多常见的数据结构创建 Allocation。在此例中，我们选择直接根据输入图片 Bitmap 建立一个 Allocation 实例。

可以看到每个脚本都遵循类似的模式。必须首先创建想要运行的脚本，使用将要用于 Allocation 的数据大小对该脚本进行初始化。在此选择 Element.U8_4()，这是因为我们的位图具有 ARGB 像素数据，因此每个元素(即像素)的大小为 4 个无符号字节。然后必须设置脚本所需的任何参数，必须通过调用 forEach()执行该脚本。脚本执行完毕后，将来自输出 Allocation 的结果复制到新的 Bitmap 中，从而显示在 ImageView 中。

对于模糊滤镜,半径是唯一可配置的参数。相应的内联函数接受 0 到 25 之间的半径值。通过在设置期间调用 setGreyscale(),我们使用颜色矩阵滤镜建立图片的灰度滤镜。如果我们要提供不同的矩阵,则改为使用 setColorMatrix()传递矩阵。

提示:
ScriptInstrinsicColorMatrix 也具备在 YUV 和 RGB 颜色空间之间执行转换的功能。

最后,应用剩余的滤镜,方法是获得给定滤镜的系数矩阵并通过 setCoefficients()将这些矩阵传递给脚本。这些滤镜的 3×3 矩阵广为人知,很容易在 Web 上获得。当脚本在 Allocation 中的每个像素上移动时,矩阵中的值定义了应如何基于输入像素及其相邻像素的当前值对输出像素的值进行乘法操作。矩阵中心点的值代表当前像素,周围的值则代表相邻像素。

例如,暗化滤镜将当前像素的值减半(乘数为 0.5),但周围的像素不影响结果。锐化滤镜将初始值放大 5 倍,然后从上、下、左、右像素中减去该值,从而实现相应的效果。

提示:
在操作卷积矩阵时,所有矩阵值的和应该为 1,以此保留图片的原始亮度。如果和大于 1,图片将变得更亮;如果和小于 1,则图片会变暗。示例中的边缘检测滤镜具有净总和值 0,这就是该图片非常暗的原因。

8.6 使用 RenderScript 操作图片

8.6.1 问题

应用程序需要利用 RenderScript 的强大功能与性能,但没有编写脚本 Intrinsic 来完成适当的任务。

8.6.2 解决方案

可以构造自己的脚本内核实现,构建工具将此实现编译和映射为 Java 类,可以采用与前一个示例中的 Intrinsic 相同的方式调用这些 Java 类。

8.6.3 实现机制

在本例中,我们将构建一个 RenderScript 内核来对图片应用水的波纹效果,如图 8-2 所示。

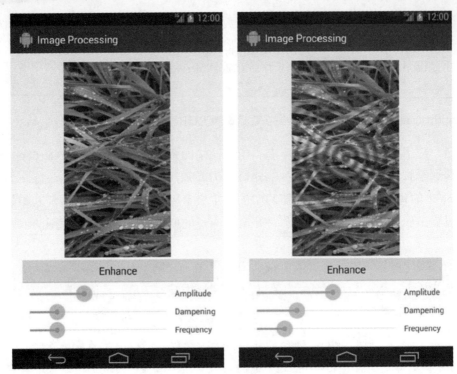

图 8-2 图片处理示例，左边是未应用过滤器的图片，右边是应用了水波纹的图片

用户将能够使用三个滑块(分别用于调整波纹振幅、阻尼和频率)控制波纹效果。点击 Enhance 按钮将触发 RenderScript，使用选择的参数应用过滤器。查看代码清单 8-17 以了解内核实现。

代码清单 8-17   src/main/ripple.rs

```
#pragma version(1)
#pragma rs java_package_name(com.androidrecipes.imageprocessing)

float centerX;
float centerY;
float minRadius;

//波峰的振幅控制
float scalar;
//波纹阻尼器，较大的值会较快减弱波纹振动效果
float damper;
//正弦频率，较大的值会显示更多的波纹
float frequency;

void root(const uchar4* v_in, uchar4* v_out, const void* usrData,
        uint32_t x, uint32_t y)
{
    //计算离中心的距离
    float dx = x - centerX;
```

```
    float dy = y - centerY;
    float radius = sqrt(dx*dx + dy*dy);

    if(radius < minRadius)
    {
        //使用原点像素
        *v_out = *v_in;
    }
    else
    {
        float4 f4 = rsUnpackColor8888(*v_in);
        float shiftedRadius = radius - minRadius;

        //基于距离确定正弦函数乘数
        float multiplier = (scalar * exp(-shiftedRadius * damper)
                * -sin(shiftedRadius * frequency)) + 1;

        //在定义的最小/最大范围内亮化或暗化像素
        float3 transformed = f4.rgb * multiplier;
        *v_out = rsPackColorTo8888(transformed);
    }
}
```

root()函数是主内核,它将在图片(输入 allocation)的每个像素上执行。通过如下方式应用波纹效果:确定输入像素与中心点的距离,并且基于衰减的正弦波函数计算是否应亮化或暗化此像素。该函数使波纹在远离中心点时消失。我们还定义了 damper、scalar 和 frequency 系数,可以在外部设置这些系数以控制波纹输出的效果。该脚本也支持对波纹的中心点进行控制,以及波纹应从距离中心的什么位置开始。

注意:
衰减正弦波函数是 $y=e^{-kt}sin(wt)$,其中 t 代表相对于中心点的距离。

我们使用 RenderScript 提供的函数 rsUnpackColor8888()将输入像素转换为 ARGB 浮点向量,从而可以对其他浮点数轻松执行数学运算。float4 和 float3 等类型是 RenderScript 定义的向量类型,这意味着它们代表一组浮点数。这些类型通常也包括特殊的存取函数,用于对数据的特定部分进行访问。例如,在此例中使用 float4.rgb 返回 RGB 像素数据的较小向量,对 alpha 通道进行遮罩。

此正弦波函数提供了 0f~2f 的乘数,可以将此乘数应用于像素,将其亮度修改为稍微接近黑色或白色。使用向量类型时,该乘法应用为点积,因此每个颜色值乘以相同的系数。在将修改后的像素放入输出 allocation 之前,必须使用 rsPackColorTo8888(),以适当的格式将其重新打包。

提示:
如果算法要求分析 allocation 中的相邻数据,也可以将 allocation 设置为脚本全局对象。这样就可以通过 rs_GetElementAt_type()读取 allocation 所需的任何项以执行计算。

构建工具将编译此脚本，并且一个新的 Java 类 ScriptC_ripple 在项目中生成。该类用于访问需要设置的所有全局对象，其提供的新方法 forEach_root()用于调用内核处理。代码清单 8-18 和 8-19 显示了如何将此类内置于 Activity 中。

代码清单 8-18　src/main/res/layout/activity_main.xml

```xml
<?xml version="1.0" encoding="utf-8"?>
<LinearLayout xmlns:android="http://schemas.android.com/apk/res/android"
    android:orientation="vertical"
    android:layout_width="match_parent"
    android:layout_height="match_parent"
    android:padding="16dp">
    <ImageView
        android:id="@+id/image"
        android:layout_width="match_parent"
        android:layout_height="0dp"
        android:layout_weight="1"
        android:scaleType="centerInside"/>
    <GridLayout
        android:layout_width="match_parent"
        android:layout_height="wrap_content">
        <Button
            android:id="@+id/button_enhance"
            android:layout_width="match_parent"
            android:layout_height="wrap_content"
            android:layout_columnSpan="2"
            android:text="Enhance"/>
        <TextView
            android:layout_width="wrap_content"
            android:layout_height="wrap_content"
            android:layout_row="1"
            android:layout_column="1"
            android:layout_gravity="center_vertical"
            android:text="Amplitude"/>
        <SeekBar
            android:id="@+id/control_amplitude"
            android:layout_width="wrap_content"
            android:layout_height="wrap_content"
            android:layout_row="1"
            android:layout_column="0"
            android:layout_gravity="fill_horizontal"/>
        <TextView
            android:layout_width="wrap_content"
            android:layout_height="wrap_content"
            android:layout_row="2"
            android:layout_column="1"
            android:layout_gravity="center_vertical"
            android:text="Dampening"/>
        <SeekBar
            android:id="@+id/control_dampening"
```

```xml
            android:layout_width="wrap_content"
            android:layout_height="wrap_content"
            android:layout_row="2"
            android:layout_column="0"
            android:layout_gravity="fill_horizontal"/>
        <TextView
            android:layout_width="wrap_content"
            android:layout_height="wrap_content"
            android:layout_row="3"
            android:layout_column="1"
            android:layout_gravity="center_vertical"
            android:text="Frequency"/>
        <SeekBar
            android:id="@+id/control_frequency"
            android:layout_width="wrap_content"
            android:layout_height="wrap_content"
            android:layout_row="3"
            android:layout_column="0"
            android:layout_gravity="fill_horizontal"/>
    </GridLayout>
</LinearLayout>
```

代码清单 8-19 图片处理 Activity

```java
public class MainActivity extends Activity implements
        View.OnClickListener {

    private ImageView mImage;
    private SeekBar mAmplitude, mDampening, mFrequency;

    @Override
    public void onCreate(Bundle savedInstanceState) {
        super.onCreate(savedInstanceState);
        setContentView(R.layout.activity_main);
        mImage = (ImageView) findViewById(R.id.image);
        mAmplitude = (SeekBar) findViewById(R.id.control_amplitude);
        mDampening = (SeekBar) findViewById(R.id.control_dampening);
        mFrequency = (SeekBar) findViewById(R.id.control_frequency);

        /*
         * 设置范围
         * A = 0.01 - 1.0
         * D = 0.0001 - 0.01
         * F = 0.01 - 0.5
         */

        mAmplitude.setProgress(40);
        mDampening.setProgress(20);

        mFrequency.setProgress(10);
        mFrequency.setMax(50);
```

```java
        mImage.setImageResource(R.drawable.background);

        findViewById(R.id.button_enhance).setOnClickListener(this);
    }

    @Override
    public void onClick(View v) {
        drawRipples(mImage, R.drawable.background);
    }

    private void drawRipples(ImageView iv, int imID) {
        Bitmap bmIn = BitmapFactory.decodeResource(
                getResources(), imID);
        Bitmap bmOut = Bitmap.createBitmap(bmIn.getWidth(),
                bmIn.getHeight(), bmIn.getConfig());
        //初始化 RenderScript 环境
        RenderScript rs = RenderScript.create(this);
        //创建数据 allocation
        Allocation allocIn = Allocation.createFromBitmap(rs, bmIn,
                Allocation.MipmapControl.MIPMAP_NONE,
                Allocation.USAGE_SCRIPT);
        Allocation allocOut = Allocation.createTyped(rs,
                allocIn.getType());
        //获得脚本实例和初始参数
        ScriptC_ripple script = new ScriptC_ripple(rs,
                getResources(), R.raw.ripple);

        //设置波纹控制值
        script.set_centerX(bmIn.getWidth() / 2);
        script.set_centerY(bmIn.getHeight() / 2);
        script.set_minRadius(0f);
        //抓取 UI 中的用户控件
        float amplitude = Math.max(0.01f, mAmplitude.getProgress() / 100f);
        script.set_scalar(amplitude);
        float dampening = Math.max(0.0001f, mDampening.getProgress() /
10000f);
        script.set_damper(dampening);
        float frequency = Math.max(0.01f, mFrequency.getProgress() / 100f);
        script.set_frequency(frequency);

        //运行脚本
        script.forEach_root(allocIn, allocOut);

        allocOut.copyTo(bmOut);
        iv.setImageBitmap(bmOut);
        //销毁 RenderScript 环境
        rs.destroy();
    }

}
```

在此构造了一个基本的 Activity 界面，该界面包含的视图显示了当前图片和三个滑块，用户通过这些滑块控制波纹过滤器的振幅、阻尼和频率。在 onClick() 处理程序内部，初始化 RenderScript 并为脚本创建两个 allocation。根据初始位图构造输入 allocation，而输出 allocation 则是空集合，后面会将其转换回图片。我们还创建了空的(和可变的)位图，输出数据将被复制到其中。

最后，分配了新的 ScriptC_ripple 实例以开始设置过滤器。定义为脚本全局对象的所有参数都适当地具有针对 ScriptC_ripple 的 setter 方法。将中心点配置为图片的中间，并且从用户界面中的滑块取出正弦波控制参数。完成所有设置之后，调用 forEach_root() 遍历图片 allocation，并且应用波纹过滤器。

提示：
RenderScript 编译器自动根据脚本文件生成 ScriptC Java 代码。如果在 Java 中无法看到这些类，就可能需要快速重新构建或重新同步项目，以使 RenderScript 编译器再次运行。

现在必须从输出 allocation 中获取数据，并放入可以使用的某个对象中。使用 Allocation.copyTo()将内容复制到可变的空位图中，并且通过将此位图放回 ImageView 中来显示结果。脚本的执行完成之后，最好销毁 RenderScript 环境，以便回收资源。

尝试进行不同的设置以了解它们如何改变结果，还可以考虑添加对中心点和内径的用户控制！

8.7 使用模糊滤镜仿造透明覆盖层

8.7.1 问题

想要使用部分透明的露珠或模糊效果提供一种幻象，即一个视图覆盖在另一个视图上。

8.7.2 解决方案

我们可以再次调用 ScriptIntrinsicBlur，同时通过一些定制的 View 和 Drawable 代码建立背景图片的模糊副本，并且应用该副本以提供部分透明覆盖层的形象外观。实时渲染生动的模糊覆盖层需要花费大量计算资源，应用程序的性能因此会受到影响。所以，我们改为预先计算背景内容的模糊图片，并且采用绘图技巧实现相同的效果，同时保持应用程序的响应灵敏性。

8.7.3 实现机制

在此例中，我们在全彩背景图片的顶部显示一个 ListView。此 ListView 带有用于偏移列表内容的定制表头视图，从而在滚动到顶部时，第一项位于屏幕的底部。随着列表向上滚动，我们会演示用于创建模糊覆盖层的两项技术：第一项技术会将图片逐渐从清晰变得模糊，第二项技术将模糊覆盖层与列表一起向上滑动，直到其完全覆盖。

为形象演示我们如何实现此效果，请查看图 8-3 和 8-4。

图 8-3　褪色模糊示例：初始为清晰(左图)以及在滚动时部分模糊(右图)

图 8-4　滑动模糊示例：模糊覆盖层跟随列表

可以看到，在褪色模糊示例中，背景最初完全清晰。随着列表内容向上滚动，模糊变得更加明显，并在整个视图中保持统一。在滑动模糊示例中，模糊覆盖层始终跟随列表项内容；随着更多的列表项显示出来，模糊占据更多的视图区域。接下来首先查看此应用程序中的资源。请参见代码清单 8-20 和 8-21。

代码清单 8-20   res/layout/activity_blur.xml

```xml
<?xml version="1.0" encoding="utf-8"?>
<FrameLayout xmlns:android="http://schemas.android.com/apk/res/android"
    android:layout_width="match_parent"
    android:layout_height="match_parent">
    <!-- 每种模糊类型的背景视图 -->
    <com.androidrecipes.backgroundblur.BackgroundOverlayView
        android:id="@+id/background_slide"
        android:layout_width="match_parent"
        android:layout_height="match_parent"
        android:scaleType="centerCrop" />
    <ImageView
        android:id="@+id/background_fade"
        android:layout_width="match_parent"
        android:layout_height="match_parent"
        android:scaleType="centerCrop"
        android:visibility="gone" />

    <ListView
        android:id="@+id/list"
        android:layout_width="match_parent"
        android:layout_height="match_parent"
        android:cacheColorHint="@android:color/transparent"
        android:scrollbars="none"/>
</FrameLayout>
```

代码清单 8-21   res/menu/blur.xml

```xml
<?xml version="1.0" encoding="utf-8"?>
<menu xmlns:android="http://schemas.android.com/apk/res/android" >
    <item android:id="@+id/menu_slide"
        android:title="Sliding Blur" />
    <item android:id="@+id/menu_fade"
        android:title="Fading Blur" />
</menu>
```

应用程序的布局仅仅是一些图片内容的顶部有一个 ListView。我们的列表后面有两个背景视图，用于在两个模糊类型示例之间轻松切换，因此在任意时间点只有一个视图可见。在本例中，使用选项菜单在两个模式之间切换，因此我们也创建了包含两个选项的简单 <menu> 元素。代码清单 8-22 显示了存放 RenderScript 代码的 Activity。

代码清单 8-22   模糊覆盖层 Activity

```java
public class BlurActivity extends Activity implements
        AbsListView.OnScrollListener,
        AdapterView.OnItemClickListener {

    private static final String[] ITEMS = {
            "Item One", "Item Two", "Item Three", "Item Four", "Item Five",
            "Item Six", "Item Seven", "Item Eight", "Item Nine", "Item Ten",
            "Item Eleven", "Item Twelve", "Item Thirteen", "Item Fourteen",
```

```java
            "Item Fifteen"};

    private BackgroundOverlayView mSlideBackground;
    private ImageView mFadeBackground;
    private ListView mListView;
    private View mHeader;

    @Override
    protected void onCreate(Bundle savedInstanceState) {
        super.onCreate(savedInstanceState);
        setContentView(R.layout.activity_blur);

        mSlideBackground = (BackgroundOverlayView)
            findViewById(R.id.background_slide);
        mFadeBackground = (ImageView) findViewById(R.id.background_fade);
        mListView = (ListView) findViewById(R.id.list);

        // 应用清晰的表头视图,以将列表元素的开始位置向下移动
        mHeader = new HeaderView(this);
        mListView.addHeaderView(mHeader, null, false);
        mListView.setAdapter(new ArrayAdapter<String>(this,
                android.R.layout.simple_list_item_1, ITEMS));

        mListView.setOnScrollListener(this);
        mListView.setOnItemClickListener(this);

        initializeImage();
    }

    @Override
    public boolean onCreateOptionsMenu(Menu menu) {
        getMenuInflater().inflate(R.menu.blur, menu);
        return true;
    }

    @Override
    public boolean onOptionsItemSelected(MenuItem item) {
        // 基于选择项,显示适当的背景视图
        switch(item.getItemId()) {
            case R.id.menu_slide:
                mSlideBackground.setVisibility(View.VISIBLE);
                mFadeBackground.setVisibility(View.GONE);
                return true;
            case R.id.menu_fade:
                mSlideBackground.setVisibility(View.GONE);
                mFadeBackground.setVisibility(View.VISIBLE);
                return true;
            default:
                return super.onOptionsItemSelected(item);
        }
```

```java
    }

    /*
     * 透明度技巧的核心思想。我们获得背景图片的普通副本和预模糊副本
     */
    private void initializeImage() {
        Bitmap inBitmap = BitmapFactory.decodeResource(getResources(),
            R.drawable.background);
        Bitmap outBitmap = inBitmap.copy(inBitmap.getConfig(), true);

        // 创建 RenderScript 上下文
        final RenderScript rs = RenderScript.create(this);
        // 创建用于输入和输出数据的 Allocation
        final Allocation input = Allocation.createFromBitmap(rs, inBitmap,
            Allocation.MipmapControl.MIPMAP_NONE,
            Allocation.USAGE_SCRIPT);
        final Allocation output = Allocation.createTyped(rs, input.getType());
        // 以最大支持半径(25f)运行模糊效果
        final ScriptIntrinsicBlur script = ScriptIntrinsicBlur.create(rs,
            Element.U8_4(rs));
        script.setRadius(25f);
        script.setInput(input);
        script.forEach(output);
        output.copyTo(outBitmap);

        // 撤消 RenderScript 上下文
        rs.destroy();

        // 对自定义 Drawable 对象应用两个副本以实现褪色
        OverlayFadeDrawable drawable = new OverlayFadeDrawable(
                new BitmapDrawable(getResources(), inBitmap),
                new BitmapDrawable(getResources(), outBitmap));
        mFadeBackground.setImageDrawable(drawable);

        // 对自定义 ImageView 应用两个副本以实现滑动
        mSlideBackground.setImagePair(inBitmap, outBitmap);
    }

    @Override
    public void onItemClick(AdapterView<?> parent, View v, int position,
long id) {
        // 在点击事件中，动态滚动的列表返回顶部
        mListView.smoothScrollToPosition(0);
    }

    @Override
    public void onScroll(AbsListView view, int firstVisibleItem,
        int visibleItemCount, int totalItemCount) {
        // 确保首先测量视图
        if(mHeader.getHeight() <= 0) return;
```

```
            // 基于滚动位置调整滑动效果裁剪点
            int topOffset;
            if(firstVisibleItem == 0) {
                // 表头仍然可见
                topOffset = mHeader.getTop() + mHeader.getHeight();
            } else {
                // 表头已分离，此时应向上滑动
                topOffset = 0;
            }
            mSlideBackground.setOverlayOffset(topOffset);

            // 基于滚动位置调整褪色效果
            // 一旦85%的表头滚动消失，模糊就完成了
            float percent = Math.abs(mHeader.getTop()) / (mHeader.getHeight() * 0.85f);
            int level = Math.min((int)(percent * 10000), 10000);
            mFadeBackground.setImageLevel(level);
        }

        @Override
        public void onScrollStateChanged(AbsListView view,
                int scrollState) { }
}
```

创建 Activity 之后，我们应用一个非常简单的列表适配器，其中包含一些静态数据元素。我们还对列表应用一个作为表头的自定义 HeaderView；代码清单 8-23 给出了该 HeaderView 的实现，该实现用于在图 8-3 和 8-4 的初始视图中将列表项向下移动。

代码清单 8-23　清晰的列表 HeaderView

```
public class HeaderView extends View {

    public HeaderView(Context context) {
        super(context);
    }

    public HeaderView(Context context, AttributeSet attrs) {
        super(context, attrs);
    }

    public HeaderView(Context context, AttributeSet attrs, int defStyle) {
        super(context, attrs, defStyle);
    }

    /*
     * 测量此视图的高度，使其始终为父视图(ListView)的测量高度的 85%
     */
    @Override
    protected void onMeasure(int widthMeasureSpec, int heightMeasureSpec) {
```

```
        View parent = (View) getParent();
        int parentHeight = parent.getMeasuredHeight();

        int height = Math.round(parentHeight * 0.85f);
        int width = MeasureSpec.getSize(widthMeasureSpec);

        setMeasuredDimension(width, height);
    }
}
```

在此没有多少需要讨论的代码；这仅是一个视图，目的是将其高度测量为父视图(在本例中始终是 ListView)高度的 85%。这样我们就可以将该视图用作经过测量的间隔空间，即使它不包含任何真正的内容。相比于硬编码固定的视图高度，该方法可以更灵活地使设备屏幕产生差异。

返回到代码清单 8-22，在 initializeImage()内部使用 ScriptIntrinsicBlur 函数创建背景图片的模糊副本。在前面关于图片滤镜的主题中讨论过，模糊半径确定扭曲程度，该半径可以取大于 0 的值，最大为 25。

RenderScript 完成模糊效果之后，获取图片对(初始图片和模糊图片)并将它们发送到两个位置。第一个位置是自定义 OverlayFadeDrawable 实例，我们将该实例用于褪色示例。第二个位置是 BackgroundOverlayView，我们将其用于滑动示例。接下来将简单查看这些项。

Activity 负责监控列表滚动并向背景视图报告这些改动。此 Activity 注册为 ListView 的 OnScrollListener，因此在视图滚动时，会定期调用 onScroll()方法。在此方法内部，基于表头视图计算偏移位置，并将相关数据填入两个背景视图。最后，Activity 也会被设置为接收每个列表项的点击事件。当发生点击事件时，列表以动画方式滚动回顶部。

为了解如何绘制模糊过渡，首先查看代码清单 8-24 中的 Drawable。

代码清单 8-24　覆盖层褪色 Drawable

```
public class OverlayFadeDrawable extends LayerDrawable {
    /*
     * Drawable 容器的实现，用于将普通图片和模糊图片作为层进行保存
     */
    public OverlayFadeDrawable(Drawable base, Drawable overlay) {
        super(new Drawable[] {base, overlay});
    }

    /*
     * 当从外部改变层值时，强制进行重绘
     */
    @Override
    protected boolean onLevelChange(int level) {
        invalidateSelf();
        return true;
    }

    @Override
```

```java
public void draw(Canvas canvas) {
    final Drawable base = getDrawable(0);
    final Drawable overlay = getDrawable(1);
    // 获得作为最大值百分比的层值
    final float percent = getLevel() / 10000f;
    int setAlpha = Math.round(percent * 0xFF);

    // 优化最终案例以避免过多绘制
    if(setAlpha == 255) {
        overlay.draw(canvas);
        return;
    }
    if(setAlpha == 0) {
        base.draw(canvas);
        return;
    }

    // 如果值在两者之间，则绘制组合图片
    base.draw(canvas);

    overlay.setAlpha(setAlpha);
    overlay.draw(canvas);
    overlay.setAlpha(0xFF);
}
```

可以回顾一下第 2 章，Drawable 仅是待显示内容的抽象化。我们选择扩展框架中的 LayerDrawable，这是 N 个元素的容器，这些元素默认按顺序作为层绘制。我们没有利用默认的绘制行为，而是使用 LayerDrawable 作为基础，这样框架可以处理验证每一层的其他一些复杂逻辑。

为更新状态，我们使用该项的 level 参数。回顾一下，此 Drawable 在 ImageView 上设置，当滚动位置改变时，调用 setImageLevel()更新背景。层值直接传入此实例，在每次调用 draw()时，会检测层值以确定如何混合两幅图片。我们明确地优化两个案例，其中 alpha 值为 0 或 100%，从而尽量减少像素的过度绘制(如果元素完全不透明，则绘制另一个元素就是浪费)。然而，如果 alpha 取中间值，我们将首先绘制初始图片，在其顶部绘制部分可见的模糊副本。现在让我们查看代码清单 8-25 中滑动模糊的绘制技巧。

代码清单 8-25　背景覆盖层视图

```java
public class BackgroundOverlayView extends ImageView {

    private Paint mPaint;
    private Bitmap mOverlayImage;
    private int mClipOffset;

    /*
     * ImageView 的定制版本使我们可以绘制两幅图片的组合版本，
     * 但仍然利用框架的所有图片调整功能
```

```java
         */
        public BackgroundOverlayView(Context context) {
            super(context);
            init();
        }

        public BackgroundOverlayView(Context context, AttributeSet attrs) {
            super(context, attrs);
            init();
        }

        public BackgroundOverlayView(Context context, AttributeSet attrs, int
defStyle) {
            super(context, attrs, defStyle);
            init();
        }

        private void init() {
            mPaint = new Paint(Paint.ANTI_ALIAS_FLAG);
        }

        /*
         * 在我们的视图中设置普通图片和模糊图片
         */
        public void setImagePair(Bitmap base, Bitmap overlay) {
            mOverlayImage = overlay;

            /* 将普通图片应用于基础 ImageView,
             * 这将使其可以应用 ScaleType 并提供矩阵,
             * 我们在后面可以使用该矩阵相应地绘制调整过的两幅图片,
             * 这也会验证视图以触发新的绘制
             */
            setImageBitmap(base);
        }

        /*
         * 调整普通副本和模糊副本应进行切换的垂直点
         */
        public void setOverlayOffset(int overlayOffset) {
            mClipOffset = overlayOffset;
            invalidate();
        }

        @Override
        protected void onDraw(Canvas canvas) {
            //首先绘制基础图片,并裁剪顶部
            //我们裁剪基础图片是为了避免在视图的底部进行不必要的过度绘制
            canvas.save();
            canvas.clipRect(getLeft(), getTop(), getRight(), mClipOffset);
            super.onDraw(canvas);
```

```
            canvas.restore();

            //获取用于调整基础图片的矩阵，并将其应用于模糊覆盖层图片
            //这样两幅图片才会匹配
            final Matrix matrix = getImageMatrix();
            canvas.save();
            canvas.clipRect(getLeft(), mClipOffset, getRight(), getBottom());
            canvas.drawBitmap(mOverlayImage, matrix, mPaint);
            canvas.restore();
        }
    }
```

这是执行一些自定义绘制的 ImageView 扩展版本。在此使用 ImageView 是因为它包含许多我们希望利用的图片调整逻辑。在我们的布局中，将 scaleType 参数设置为 centerCrop，这样视图就可以负责精细地调整和放置背景。请注意，Activity 在此视图上设置图片时，基础图片作为主图片直接传递给基础实现。这样做有两点原因：主要是因为该框架可以执行图片调整数学运算，另一个原因则是我们可以通过在后面调用 super.onDraw() 来轻松绘制基础图片。

每次滚动位置发生改变时，通过 setOverlayOffset() 将偏移传入此视图；该方法也会验证视图，强制执行新的绘制轮次。在视图的绘制部分，利用偏移标记为 Canvas 创建两个裁剪遮罩。这样做的主要目的是仅绘制模糊覆盖层的一部分，我们想要显示该部分以匹配列表位置。然而，我们也可以使用相同的偏移裁剪基础图片绘图。这样就再次避免了视图中的像素过度绘制，从而使应用程序更加顺畅地执行。即使显示的效果是正在绘制半透明的覆盖层，我们也不会在此视图中多次绘制任意像素。

关于我们已创建的视图，最后要提醒一点：在图 8-5 显示的应用程序中，列表向上滚动完成，而模糊覆盖层涵盖了完整的视图。

图 8-5　完整显示的模糊覆盖层

8.8 小结

使用 Android NDK 时,开发人员可以向应用程序添加适当数量的原生 C/C++代码,对 Java SDK 形成有益的补充。无论是使用 JNI 的大量函数还是整个 Activity 类,NDK 都提供了必需的功能来使优秀的应用程序变得更好。使用 RenderScript,开发人员可以充分利用移动设备的真实计算能力,而无须浪费时间处理多核和多线程同步问题。这样就可以在设备上完成更多的图片处理和算法计算,而不需要将这些工作转移到远程服务器上完成。